PERSPECTIVES IN CONDENSED MATTER PHYSICS
A Critical Reprint Series

Condensed Matter Physics is certainly one of the scientific disciplines presently characterized by a high rate of growth, both qualitatively and quantitatively. As a matter of fact, being updated on several topics is getting harder and harder, especially for junior scientists. Thus, the requirement of providing the readers with a reliable guide into the forest of printed matter, while recovering in the original form some fundamental papers suggested us to edit critical selections on appealing subjects.

In particular, the present Series is conceived to fill a cultural and professional gap between University graduate studies and current research frontiers. To this end each volume provides the reader with a critical selection of reprinted papers on a specific topic, preceded by an introduction setting the historical view and the state of art. The choice of reprints and the perspective given in the introduction is left to the expert who edits the volume, under the full responsibility of the Editorial Board of the Series. Thus, even though an organic approach to each subject is pursued, some important papers may be omitted just because they lie outside the editor's goal.

The Editorial Board

PERSPECTIVES IN CONDENSED MATTER PHYSICS
A Critical Reprint Series: Volume 1

Editorial Board

Executive Board

F. Bassani, Scuola Normale di Pisa (*Chairman*)
L. Miglio, Università di Milano (*Executive Editor*)
E. Rimini, Università di Catania
A. Stella, Università di Pavia
M.P. Tosi, Università di Trieste

Advisory Board

P.N. Butcher, University of Warwick
F. Capasso, AT & T Bell Laboratories, Murray Hill
M.L. Cohen, University of California, Berkeley
F. Flores, Universidad Autonoma de Madrid
J. Friedel, Université de Paris Sud, Orsay
G. Harbeke, RCA Laboratories, Zurich
N. Kroo, Central Research Institut for Physics, Budapest
F. Levy, Ecole Polytechnique Fédérale, Lausanne
M.J. Mayer, Cornell University, Ithaca
T.M. Rice, Eidgenossische Technische Hochschule, Zurich
R.M. Thomson, National Bureau of Standards, Washington
J.P. Toennies, Max-Planck Institut für Stromongsforschung, Göttingen

ELECTRONIC STRUCTURE OF SEMICONDUCTOR HETEROJUNCTIONS

edited by

Giorgio Margaritondo

University of Wisconsin-Madison

KLUWER ACADEMIC PUBLISHERS

Jaca Book

PHYSICS

03386922

Electronic structure of semiconductor heterojunctions.

(Perspectives in condensed matter phyisics; 1)
1. Semiconductors-Junctions. 2. Electronic structure.
I. Margaritondo, Giorgio, 1946- II. Series.
QC611.6.J85E38 1988 537.6'22 88-8222

ISBN 90-277-2823-2 (Kluwer)
ISBN 90-277-2824-0 (Kluwer: pbk.)

ISBN 88-16-96001-9 (Jaca Book)

per informazioni sulle opere pubblicate e in programma
ci si può rivolgere a Editoriale Jaca Book spa
via A. Saffi 19, 20123 Milano, telefono 4982341

To Laura and Francesca

Preface

E se non che di ciò son vere prove
Per più e più autori, che saranno
Per i miei versi nominati altrove,

Non presterei alla penna la mano
Per notar ciò ch'io vidi, con temenza
Che non fosse da altri casso e vano;

Ma la lor chiara e vera esperienza
Mi assicura nel dir, come persone
Degne di fede ad ogni gran sentenza.

And were it not for the true evidence
Of many authors who will be
Mentioned elsewhere in my rhyme

I would not lend my hand to the pen
And describe my observations, for fear
That they would be rejected and in vane;

But these authors' clear and true experience
Encourages me to report, since they
Should always be trusted for their word.

[From "*Dittamondo*", by Fazio degli Uberti]

Heterojunction interfaces, the interfaces between different semiconducting materials, have been extensively explored for over a quarter of a century. The justification for this effort is clear — these interfaces could become the building blocks of many novel solid-state devices. Other interfaces involving semiconductors are already widely used in technology. These are, for example, metal-semiconductor and insulator-semiconductor junctions and homojunctions. In comparison, the present applications of heterojunction interfaces are limited, but they could potentially become much more extensive in the near future.

The path towards the widespread use of heterojunctions is obstructed by several obstacles. Heterojunction interfaces appear deceptively simple whereas they are intrinsically complicated. After years of research, the simple problem of understanding the energy lineup of the two band structures at a semiconductor-semiconductor heterojunction is still a challenge for solid-state theory. The complex character of the interface properties has been a stimulating factor in heterojunction research, since it adds fundamental interest to an already interesting technological problem.

In the past five years, the extensive work of experimentalists and theorists has produced unprecedented progress in heterojunction research. This progress has mostly occurred in five areas:

- New heterojunction growth techniques based on Molecular Beam Epitaxy (MBE) and Metallo-Organic Chemical Vapor Deposition (MOCVD) have been developed or refined.

- New kinds of devices have been developed. The advances have been spectacular in superlattices, graded-composition structures and bond-stretched overlayers.

- New kinds of experimental techniques, and in particular synchrotron-radiation photoemission, have produced an extensive data base for the band lineup problem.

- New and sophisticated theories have replaced the simplified schemes that had been used for years to treat the band lineup problem.

- Several research groups have been successful in modifying the band lineup between two materials with extrinsic factors, opening the way for a possible "tuning" of the heterojunction parameters in future devices.

This fast progress makes it difficult to provide a comprehensive description of the *status* of heterojunction research. In the early 1970's, such a description was provided by a classic treatise of Milnes and Feucht.[1] Fifteen years afterwards, Federico Capasso and I were the editors of a comprehensive series of articles dedicated to different aspects of heterojunction research.[2] The present book provides a general description of the *status* of this area of research, using reprints of several classic articles in this field. Therefore, the present book can be used as a complement to Ref. 2, as well as a stand-alone, elementary introduction to this fascinating field.

The reprinted articles included in the book have had a major impact on heterojunction research. The primary selection criterion was a clear and effective presentation of the recent evolution of this field. Regrettably, I could not include many fundamental articles which were too specialized, too long, or did not fit the presentation scheme. This is *not*, therefore, a complete collection of the "best" articles published in heterojunction research.

The rationale of the book is the following. First, for historical prospective, I present the fundamental 1962 work of R. L. Anderson which, in my opinion, marks the birth of heterojunction research.[R1] Second, I explain why heterojunction research is so important in technology. This is done with the help of several review and research articles on devices and device fabrication. Then I initiate the treatment of band lineups.

The first part of such a treatment deals with the experimental methods to measure band lineups. Then I review some of the simplest ideas concerning this problem, and recent experimental results which almost invariantly lead to their breakdown. Special attention is dedicated to one of these ideas, the *common-anion rule*, and to the experimental and theoretical work which reveals its limits. The next group of reprinted articles is dedicated to more sophisticated theoretical treatments of band lineups. These can be divided in three main areas: tight-binding theories, theories based on induced gap states, and self-consistent calculations of the electronic structure.

These fundamental theories are followed by semi-empirical and empirical approaches, with particular emphasis on the tests of the predictions of "linear" band lineup models, and on the links between heterojunctions and metal-semiconductor junctions. The book is then completed by reports of the recent successes in controlling band lineups, using doping profiles or ultrathin intralayers.

The reprinted articles are preceeded by a discussion, which interprets their message and explains why they are included in the books. I found interpreting the messages of other authors the most difficult task in developing this book. I suspect that some colleagues will not agree with *my* interpretation of *their* work. Disagreements, after all, are quite common in a very active field of research. Such controversies notwithstanding, my hope is that this book will provide a simple and reasonably complete introduction to heterojunction research. In particular, I hope that it will stimulate the imagination and creativity of many young scientists. The more we advance in the understanding of heterojunctions, the more we realize how complicated they are — and future progress in this field requires the fresh contribution of young investigators.

I cannot attribute part of the blame for my choices and interpretations to somebody else, since I was directly responsible for both. I did, however, profit from many discussions with outstanding colleagues and with my own collaborators. Among the latter, I wish to thank Ahmad Katnani, Ned Stoffel, Bob Daniels, Mike Kelly, Te-Xiu Zhao, Dave Niles, Doug Kilday, Yeh Chang, Elio Colavita, Paolo Perfetti, Mario Capozi and Claudio Quaresima. I also thank the National Science Foundation, the Office of Naval Research and the Wisconsin Alumni Research Foundation for having provided the necessary support of my own work, as well as all the funding agencies which support this crucial research area throughout the world. Finally, I am grateful to the copyright holders who released the reprinted articles, making it possible to include them in this presentation.

Table of Contents

G. Margaritondo, Electronic Structure of Semiconductor Heterojunctions

Giorgio Margaritondo
Electronic Structure of Semiconductor Heterojunctions

Introduction

The central problem in heterojunction research can be summarized by the following question: "How do the band structures of two semiconductors line up in energy with respect to each other, when the materials are joined together to form a heterojunction?". This deceivingly simple question has profound fundamental and technological implications. For over twenty-five years, a definite answer has been sought by experimentalists and theorists. We have recently seen substantial progress towards a solution of this problem. The main purpose of this book is to illustrate the recent progress, after establishing a general background necessary for the understanding of the problem and its implications.

The technological implications of the band lineup problem can be easily appreciated with the help of Fig. 1. This figure shows the energy band diagram of a p-n heterojunction, composed of two different semiconductors with gaps E_g^1 and E_g^2. The difference $E_g^1 - E_g^2$ must be accommodated by discontinuities in the edges of the valence and conduction bands, ΔE_v and ΔE_c. If you imagine carriers crossing the interfaces, it should be clear that such discontinuities play the leading role in determining the transport properties of the heterojunction. They influence other properties, such as the optical response, and in general determine the behavior and the performances of the corresponding heterojunction devices.

Some features of the heterojunction energy diagram of Fig. 1 are quite similar to the corresponding features of other classes of semiconductor interfaces. For example, band bending at the two sides of the interfaces is also present for p-n and metal-semiconductor interfaces. The band discontinuity are, on the contrary, peculiar to heterojunction interfaces. On one hand, they add to the flexibility in designing devices tailored to particular tasks. On the other hand, they also add to the complexity of the interfaces and of the devices.

These facts explain the entire evolution of heterojunction research. The design flexibility, due to the presence of two different semiconductors with two different sets of parameters, is a powerful incentive for the development of heterojunction technology. Potentially, heterojunction devices could revolutionize solid-state electronics, and introduce an unprecedented degree of freedom in tailoring devices to their applications. However, the complexity of the heterojunction interfaces has made it impossible to use the same empirical approach that has been so successful for other kinds of interfaces. We make very extensive use of systems such as the Schottky barrier without completely understanding their physical properties (although progress has been recently made in this field too). Heterojunctions are not so forgiving. As we will see, virtually all of the simplistic ideas fail in their case, not only to explain details of the observed phenomena, but even to provide the minimum background for the development of devices.

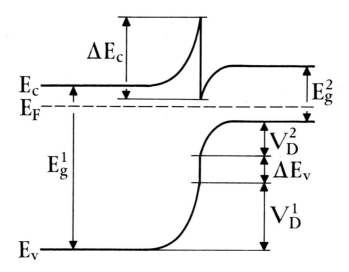

Fig. 1 – Schematic energy diagram of the interface between two different semiconductors with forbidden gaps E_g^1, E_g^2. E_c and E_v are the conduction and valence band edges, and E_F is the Fermi level. V_D^1 and V_D^2 measure the band bendings of the two sides of the junction (see next section). The difference between the two gaps is accommodated by the valence and conduction band discontinuities, ΔE_v and ΔE_c. The exact values of such discontinuities depends on the lineup of the two band structures.

The central question concerning the heterojunction band discontinuities can be formulated using Fig. 1. The figure shows that:

$$\Delta E_v + \Delta E_c = E_g^1 - E_g^2, \tag{1}$$

but we do not know *how large* are the two discontinuities *relative to each other*. Note that Eq. 1 is written using the following conventions: (i) ΔE_v is positive if the valence-band edge of the small-gap semiconductor is above that of the large-gap semiconductor, and negative *vice-versa*; (ii) the opposite convention is valid for ΔE_c; (iii) E_g^1 is larger than E_g^2. These conventions are not universally adopted, and the reader should take *cum grano salis* the signs in heterojunction literature.

Before initiating the discussion of the band-lineup problem, we must address two basic questions. First: does the problem have physical meaning, in the sense that abrupt interfaces like the picture of Fig. 1 really exist? This question has been definitely answered by the modern transmission electron micrographs. In principle, "abrupt" for a heterojunction interface means that the interface region must be thin with respect to the carrier diffusion lengths. The micrographs show that, for high-quality interfaces, the interface region is extremely sharp. In fact, it consists in some cases of two atomic planes. Thus, "atomically sharp" interfaces are not just a product of the experimentalists' imagination, but a solid reality.

The second question is: is the claim, made above, that band lineups are technologically important substantiated by facts? From the reader's point of view, the same question can be re-formulated as: why should I invest my time in reading this book? In either form, the question is answered by a series of articles, discussed in the next section and included in the reprints.

The point is that the actual and potential heterojunction devices are truly exceptional. They could perform a variety of tasks much beyond the present limitations of solid-state electronics. If we consider the enormous economic and social impact of today's microelectronics, we can easily understand the potential importance of heterojunction technology and heterojunction research in general. This, by itself, would provide ample justification for the effort to explain the band lineup problem. In addition, the band lineup is a problem of great fundamental interest. We will see that it touches our very understanding of the chemical bonding process in condensed systems, and of the corresponding electronic states.

After discussing the reasons for studying heterojunctions, the book will then directly address the band lineup problem. Several aspects will be considered: experimental measurements, theoretical solutions and empirical approaches. The final part of the book is dedicated to the most important objective of this research, the *control* of band lineups. For many years, scientists wondered if the band lineup is an intrinsic, *i.e.*, unchangeable, property of each pair of semiconductors. We now know that this is not true. We have been able to modify band lineups. This potentially increases the flexibility in designing heterojunction devices, and their corresponding technological applications. This concept had been postulated by Capasso at AT&T Bell Labs, at a time when it appeared to be little more than a dream. The last reprints in this book demonstrate that Capasso's *bandgap engineering* is not necessarily a dream — and that it could become a guideline for the development of solid-state technology.

Heterojunctions in Technology

The possible device applications have, since its very first steps, provided the motivation for heterojunction research. Proposals for devices based on two different semiconductors were made in the 1950's by Gubanov, Schokley and Herbert Kroemer (see literature quoted in Ref. R1). In 1962, R. R. Anderson published his landmark article (the first reprint in this book), which proposed a coherent model for heterojunctions.[R1] Such a model was essentially an extension of the Schottky model for metal-semiconductor diodes and, like the Schottky model, explained the basic parameters of the junction in terms of the parameters of the two component materials. The Schottky model predicts,[1] for example, that

$$\phi_n = \Phi_m - \chi, \tag{2}$$

where ϕ_n is the Schottky barrier for the interface between a given metal and an n-type semiconductor, Φ_m is the metal work function and χ is the electron affinity of the semiconductor.

Anderson identified the band discontinuities and the "built-in potential" as the fundamental parameters of a heterojunction. The latter is given by:

$$V_D = V_D^1 + V_D^2, \tag{3}$$

where V_D^1 and V_D^2 are, as shown in Fig. 1, the band-bending potentials of the two sides of the junction. The band bending is required to keep the Fermi energy constant everywhere in the system, while far from the junction its distance from the valence (or conduction) band edge is entirely determined by doping. In the specific case illustrated in Fig. 1, the band bending corresponds to an n-p heterojunction. Figure 1 of Ref. R1 illustrates the basic feature of Anderson's model, now known as the *electron affinity rule*:

$$\Delta E_c = \chi_2 - \chi_1, \tag{4}$$

where χ_1, χ_2 are the electron affinities of the two semiconductors (θ_1 and θ_2 in Ref. R1).

The electron affinity rule has been very widely used in heterojunction research, although it is now criticized and rejected by most authors. A detailed discussion of the rule and of the controversy which it has generated will be presented in the next sections. I emphasize, however, that the problems affecting the electron affinity rule do not diminish the fundamental importance of Anderson's article. Written at a time when solid-state electronics was in its infancy, this work correctly identified the essential issues in heterojunction research and even established a conventional nomenclature for the corresponding variables, which is still universally used. The article directly or indirectly stimulated much of the research on heterojunctions in the 1960's and 1970's.

Anderson's article stimulated a tremendous amount of research on heterojunction devices. The practical implementation of such devices, however, was negatively affected by the absence of the technology necessary to grow one semiconductor on top of another. The development of new deposition techniques, first liquid-phase epitaxy (LPE) and then MBE and MOCVD, removed this obstacle. The reprint R2 is an overview of MBE, the growth technique that plays a major role in the fabrication of novel heterojunction devices. The article also presents advanced characterization methods of the growth products, such as transmission electron microscopy of the interfaces, performed with atomic-level resolution.

Advanced techniques such as MBE are necessary for the production of the most advanced heterojunction structures. The reprint R3, written by the Nobel Laureate Leo Esaki, discusses two of the most important among such structures, superlattices and quantum wells. Heterojunction superlattices consist of alternating ultrathin layers of two different semiconductors. Pioneered by Esaki himself in the late 1960's, superlattices can now be fabricated with layers so thin that they consist of single atomic planes.

The reduction of the layer thickness produces interesting quantum phenomena which affect the properties of the heterojunction structure. Consider, for example, Fig. 1 (bottom) in reprint R3. The valence and conduction band discontinuities result in potential wells, both for the electrons and for the holes. The quantum particle-in-a-box problem predicts discrete energy levels for these one-dimensional wells, whose energy separations increase when the "size" of the well decreases. These separations become of the order of one-tenth of an electronvolt for well thicknesses in the range of tens of angstroms, reachable with today's deposition technologies. The presence of discrete levels causes several interesting effects such as *resonant tunneling*. This is an enhancement of the tunneling cross section through the well for an incident particle, when its energy matches one of the energy levels of the well. This phenomenon, as we will see, has been recently exploited in novel designs for logic devices.

The advances in the fabrication techniques are gradually enhancing our capability to control the microscopic interface parameters. Capasso proposed *"bandgap engineering"* as a general term for heterojunction device technology based on this capability. The reprint R4 discusses a series of novel devices that rely on the control of the forbidden gap, achieved through the modification of the composition of semiconducting thin layers. The most widely used material for these applications is the ternary semiconductor $Al_{1-x}Ga_xAs$. The forbidden gap of this material

5

changes with x, and modern deposition techniques such as MBE enable us to produce layers with graded composition and gap. Among the graded-composition devices discussed by Capasso,[R4] particularly important is the "staircase" solid-state photomultiplier.

The applications of resonant tunneling are discussed in detail by reprint R5, with particular emphasis on structures consisting of double barriers. Among these, the development of the first resonant tunneling bipolar transistor operating at room temperature has generated a great deal of interest. The reasons for the interest resonant tunneling transistors can be understood with the help of Fig. 2 in Ref. R5. For increasing values of the bias between base and emitter, one obtains a series of resonant-tunneling situations, each one corresponding to one of the levels in the quantum well. Each resonant tunneling situation produces a regime of negative differential resistivity, i.e., a drop in the resistivity as the bias voltage increases. The multiple resonant characteristics could be used for the implementation of logic elements with multiple values, as opposed to the binary logic element used in today's computers. In principle, this could revolutionize the design philosophy of future logic circuitry. The recent, successful test of a quantum well resonant bipolar transistor at room temperature has been a fundamental step towards the practical use of this kind of device.[3]

The overview provided by reprints R3-R5 explains the extensive research effort dedicated to heterojunctions — the current devices, and the new devices that could be produced by this effort, hold promise of a revolution of the microelectronics industry. Another message is also clear from these presentations. The band lineup between the two semiconductors, and the resulting band discontinuities, are the most crucial elements of a heterojunction interface. The following sections will be entirely dedicated to the experimental and theoretical aspects of the band lineup problem.

Band Lineup Measurements

The development of reliable measurement techniques has been one the major obstacles in heterojunction research. The first investigations were affected by a problem common to all areas of semiconductor interface research. Such investigations were performed with transport techniques, e.g., the study of current-voltage (I-V) or capacitance-voltage (C-V) characteristics. The interface parameters were deduced from the data using theoretical models for the transport properties of the system. The problem in this approach is that transport measurements intrinsically perform averages in space, while the interface properties are highly localized. Thus, this measurement technique is indirect, and it can easily produce errors.

The limitations of transport studies of the band lineups are discussed in detail by reprint R6. In this article, Herbert Kroemer – a pioneer in heterojunction research – uproots the common mistakes and assumptions which affect these measurements. As one can see from his analysis, transport measurements are highly reliable only when applied to certain sophisticated heterojunction structures. Kroemer's article[R6] is also an excellent general review of heterojunction research, and in particular of the role of band discontinuities in heterojunction devices.

The problems affecting transport techniques have stimulated the search for other methods to measure the band lineups. The resulting approaches can be divided in two general classes: *photoemission* techniques and *optical* techniques. Somewhat intermediate between transport and optical technique is a recent approach to measure ΔE_v, based on deep-level transient spectroscopy (DLTS). This approach is discussed in reprint R7.

The highly localized character of the band discontinuities explains the success of photoemission techniques in studying them. In a photoemission experiment, electrons are emitted from a surface bombarded with ultraviolet or soft-x-ray photons. The emitted photoelectrons are analyzed in vacuum, and in particular one measures their distribution in energy. In first approximation, this distribution reflects the distribution in energy of the electrons in the specimen, shifted upwards in energy by an amount equal to the energy of the photon, $h\nu$.

The electrons excited upon absorption of a photon have a very short mean-free-path in the sample, of the order of a few angstroms or tens of angstroms. Therefore, the photoelectron energy distribution curves or EDC's reflect the distribution in energy of electrons in a thin slab of the specimen close to its surface. Assume that the specimen consists of a semiconductor substrate with a thin overlayer of a second semiconductor. The EDC's contain contributions from substrate and overlayer. In particular, the region close to its upper leading edge reflects the presence of the *two* different valence-band edges, which give rise to the valence band discontinuity between the two materials.

This phenomenon is clearly visible in the EDC's of a cleaved CdS substrate covered by a Si overlayer, shown in Fig. 2. From the double edge structure of these curves, ΔE_v can be directly observed and measured. In first approximation, the measurements are performed by using linear extrapolation to derive the positions in energy of the band edges (see Fig. 2). More sophisticated methods, based on theoretical fitting of the double-edge lineshape, show that the linear extrapolation reaches an accuracy of the order of ± 0.1 eV. The studies of ΔE_v based on the observation of double edges are all the more effective if one enhances the surface sensitivity of the photoemission experiments, by shortening the mean-free-path of the excited electrons. This can be done by exploiting the dependence on the electron energy of the mean-free-path. In turn, the excited-electron energy can be controlled by tuning the photon energy, since it equals the initial energy of the electrons in the specimen plus $h\nu$. This approach, of course, requires an energy-tunable source of ultraviolet and soft-x-ray photons. Since the late 1960's, such sources are available — they are the *synchrotron radiation sources*, widely used in modern photoemission spectroscopy.

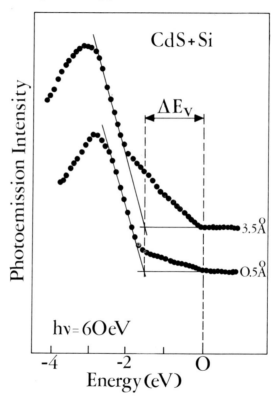

Fig. 2 – The double-edge structure of photoelectron energy distribution curves taken on cleaved CdS covered by a thin Si overlayer reflects the CdS-Si valence band discontinuity. The horizontal scale is referred to the Si valence band edge. The thickness of the overlayer is shown on the right-hand side of each curve.

8

The practical use of synchrotron-radiation photoemission to measure band discontinuities is not always as simple as in the case of Fig. 2. For the majority of the heterojunctions, ΔE_v is small, and the two edges cannot be separated from each other in the spectra. In these cases, the valence-band edge positions can be indirectly derived by measuring the position in energy of core-level peaks in the EDC's. In fact, the core-level peaks often track the valence-band edges. This is true, specifically, when the core-level binding energy is not affected by changes in the chemical status of the corresponding element during the interface formation process. Such an approach is not immune from complications and possible errors, as discussed in detail in Ref. 2. However, after several years of development, photoemission techniques are capable of measuring ΔE_v for virtually all heterojunctions, reliably and with an accuracy of ± 0.1 eV or better.

The use of photoemission to measure heterojunction band discontinuity was pioneered in 1978 by Perfetti *et al* at Berkeley, by Bauer and McMenamin at Xerox-Palo Alto, and by Grant and coworkers at Rockwell. The corresponding articles, which have historical as well as scientific interest, are included among the reprints.[R8-R10] The subsequent developments of this method involved not only EDC measurements of the valence band discontinuities, but also measurements of ΔE_c with partial-yield spectroscopy, a non-conventional photoemission mode made possible by the tunability of synchrotron radiation sources.[4]

Somewhat related to the photoemission methods to measure ΔE_v and ΔE_c is the approach described in reprint R11 by Abstreiter *et al.* This approach is based on *internal photoemission*, and derives the band discontinuities from the analysis of the plots of the photoconductivity of heterojunction structures, as a function of photon energy. As discussed in Ref. R11, the excitation ("internal photoemission") of carriers from one semiconductor to another gives rise to photocurrent thresholds, at energies related to the forbidden gaps and to the discontinuities.

The use of optical techniques to study heterojunction band lineups can be traced back to the article by Dingle *et al*, reprinted as Ref. R12. In essence, this method consists of studying the optical absorption and emission phenomena caused by the discrete level in heterojunction quantum wells. It is clear that the corresponding features in the optical spectra depend on the effective masses of the involved carriers and on well parameters. In turn, the latter depend in particular on the band discontinuities. From the analysis of quantum-well optical absorption spectra, Dingle *et al.*[R12] measured the ratio $\Delta E_c / (E_g^1 - E_g^2)$ for GaAs-Al$_{1-x}$Ga$_x$As heterojunctions. Subsequent experiments used not only absorption spectra, but also photoluminescence spectra.

In principle, optical measurements of this kind should be able to reach high accuracy. The numerical accuracy of measurements of photon energy is much higher than, for example, the accuracy in measuring photoelectron energies. The actual history of optical measurements suggests some caution. In the next section, we will discuss the controversial common-anion rule, which was supported by the results of Ref. R12, but is now rejected by most authors. This indicates that the *numerical* accuracy of the photon energy measurements does not necessarily coincide with the accuracy of the band discontinuities derived from them.

Over the past ten years, extensive measurements of band lineups have been performed with the techniques outlined in this section as well as with other approaches.[2] The results have produced an excellent data base, that can be used to test the theoretical models of band lineups and stimulate the development of new models. Margaritondo and Perfetti have recently analyzed the existing data on band discontinuities (see Ref. 2). Table I reports average values for a number of heterojunctions, deduced from their compilation of data. The signs in this table follow the same conventions adopted for Eq. 1.

Limitations of Simplified Models

The problem of understanding band lineups is complicated because it is related to the *local* electronic structure of the interface. Only in recent years has solid-state theory treated local electronic structures with calculations which approach realism. Previously, the pressing needs of technological research stimulated the use of oversimplified band lineup models. After years of studies of the intricacies of local electronic structures, these early models appear quite naive. Several of them, however, have been used in heterojunction research for such a long time that many scientists find it difficult to remove them from their minds, or at least to use them with prudence.

The electron affinity rule is a typical example of simplified approach. This author remembers the dogmatic statement made by the referee of an article he wrote in the mid-1970's about band lineups. In perfectly good faith, the referee questioned the wisdom of dedicating one's time to the study of heterojunction band discontinuities, since they were "explained by the electron affinity rule". At present, we certainly know better than that — although we are much less certain that we know what causes the band lineups.

The apparent naivety of the electron affinity rule could lead to a different mistake, *i.e.*, its outright rejection without serious consideration. In 1986, Mailhiot and Duke finally provided a sound theoretical background for the electron affinity rule as well as for the Schottky model, the corresponding theory of metal-semiconductor interfaces.[5] Under the assumption of no changes in the interface atomic position with respect to the bulk, Mailhiot and Duke derived, for a p-n heterojunction, the equation:

$$\Delta E_c = \chi_2 - \chi_1 + V_{dipole}, \qquad (5)$$

which coincides with the electron affinity rule except for the term V_{dipole}. This term is the net electrostatic potential drop across the interface, corrected for the conversion of the bulk chemical potentials to work functions measured with respect to vacuum. In essence, the electron affinity rule (as well as the Schottky model) was recovered from this approach because of the small magnitude of V_{dipole}, typically much less than 100 meV.

The treatment of Ref. 5 does not apply to "realistic" interfaces affected, for example, by chemical reactions between the two components. A good background for these cases is provided by the "effective work function model" of Freeouf and Woodall. The foundations of the model, in the case of metal-semiconductor interfaces, are presented in reprint R13. Freeouf and Woodall use a standard Schottky picture, but they assume that the relevant interface is not that between the pure metal and the semiconductor. Due to local chemical interactions, they argue that the interface consists of a mixture of microclusters of different phases, each one with its own work function. Thus, the work function of the pure metal should be replaced by an *effective* work function, Φ_{eff}, determined by the work functions of the interface phases. Freeouf and Woodall also argue that Φ_{eff} is dominated by the work function of the anion component of the semiconductor, which in turn dominates the interface species.

The reprint R14 includes the extension of the Freeouf-Woodall model to the case of heterojunctions, which essentially extends a modified version of the electron affinity rule to the case of chemically reacted interfaces. As we see, then, the electron affinity rule is less naive than it may seem, it survives sophisticated theoretical treatments, and therefore it cannot be lightly dismissed.

The main problems for the electron affinity rule arise from experiments. Many authors have reported discontinuity measurements which disagree with the rule's prediction. These tests, however, may be affected by a basic problem. The electron affinities used in Eq. 4 are measured on interfaces between semiconductors and vacuum. Most of the "old" data are heavily affected by insufficient surface characterization and therefore highly unreliable. This prompted Niles and Margaritondo to perform a complete test of Eq. 1 in a single experiment, using well characterized surfaces and interfaces, all prepared in the same system. The test, performed on the ZnSe-Ge interface, is described in reprint R15. Its results are in net disagreement with the electron affinity rule. This shows, at least, that the rule cannot be applied to all heterojunction interfaces.

The *common-anion rule* is another widely used, simplified approach to the problem of band lineups. This rule is based on the fact that the most important contributions to the valence band of a binary semiconductor arise from the s and p states of its anion. Thus, it may seem safe to assume that an interface between two semiconductors with the same anion, the same crystal structure and similar interatomic distances has a small valence band discontinuity. This rule applies to some of the most important "technological" interfaces, such as $Al_{1-x}Ga_xAs$-GaAs. The results decribed in Ref. R12 seemed to provide a solid experimental confirmation of this hypothesis. They indicated that the gap difference between GaAs and $Al_{1-x}Ga_xAs$ was mostly accommodated by ΔE_c, which accounted for 85% of $E_g^1 - E_g^2$. For many years, this "15-85%" rule stood unchallenged.

In 1984, however, photoluminescence measurements by Miller *et al.* (see reprint R16) produced results in disagreement with the "15-85%" rule, and in general with the common-anion rule. They indicated that the GaAs-$Al_{1-x}Ga_xAs$ gap difference is more evenly distributed between ΔE_v and ΔE_c. Many subsequent experiments[2] confirmed this fact. As shown in Table I, the average of recent measurements for GaAs-$Al_{1-x}Ga_xAs$ corresponds to $\Delta E_c/(E_g^1 - E_g^2) = 0.59$. The common-anion rule has been found to fail for other interfaces besides GaAs-$Al_{1-x}Ga_xAs$, such as GaSb-AlSb.[6]

The breakdown of the common-anion rule raises fundamental questions about our understanding of the electronic structure of compound semiconductors. After all, the rule was based on the simple assumption that the valence band is dominated by anion states! This puzzle stimulated several theorists to re-examine the role of all electronic states in the construction of the valence band of compound semiconductors. A breakthrough was obtained by Wei and Zunger, who explaned the failure of the common-anion rule in terms of the previously neglected role of the cation d orbitals. This explanation is reported in reprint R17. After including the cation d-state contributions in their first-principle electronic structure calculations, Wei and Zunger predicted valence band discontinuities in agreement with the recent experimental values, and in disagreement with the common-anion rule.

This approach was also tested by extending its applications to the case of ternary and quaternary semiconductors, as described in reprint R18. In essence, it was found that, for ternary semiconductors, the d-contributions of different cations often compensate each other — and this tends to restore the common-anion rule. Experimental measurements at $CuIn_xGa_{1-x}Se_2$-Ge and $CuAg_{1-x}InSe_2$-Ge interfaces indicated that ΔE_v is almost independent of x, in agreement with this prediction.

Thus, except in special cases involving semiconductors with more than two components, the common-anion rule fails to predict the correct band lineup. The problems affecting this rule and the electron affinity rule have had a sobering impact on heterojunction research. For technological purposes, the band lineup should be measured and theoretically modeled with an accuracy better than the thermal energy at room temperature, $k_BT \approx 0.025$ eV. Far from reaching these accuracies, the approaches treated in this section appear inaccurate by several tenths of an electronvolt. The experiments have taken years to detect such an inaccuracy in the case of $Al_{1-x}Ga_xAs$-GaAs. This demonstrates the urgent need for better theories and for more refined experimental approaches. Photoemission techniques do provide high reliability, since they can directly probe the interface electronic structure. However, their accuracy is limited at present to 0.1 eV, except in special cases.

General Theories of the Band Lineups Mechanism

In 1977, fifteen years after the formulation of Anderson's model, two new general theories of fundamental importance were published by Frensley and Kroemer[7] and by Walter Harrison (see reprint R19). In both cases, the authors used a philosophy similar to Anderson's model, in the sense that they tried to calculate the valence band discontinuity by first identifying the position of the band structure of each semiconductor on a well-defined energy scale — and then taking differences of the band edge positions of the two semiconductors.

Frensley and Kroemer used as a reference the "mean interstitial potential" of each semiconductors. This is not an absolute reference for the energy scale, and therefore the band-edge differences had to be corrected for the difference of the mean intestitial potentials for the two semiconductors. This correction term was related to a "charge transfer dipole" at the heterojunction interface. Frensley and Kroemer calculated the valence-band edge positions using a pseudopotential scheme. The calculated valence-band edges have reasonable accuracy, as suggested by a comparison between the corresponding differences and measured ΔE_v's. However, the subsequent attempt to estimate the dipole correction term made the accuracy worse.

Harrison[R19] used an absolute energy scale, and calculated the position of the valence band edges of each semiconductor in that scale with a simple tight-binding technique. ΔE_v for a given heterojunction was then simply estimated by taking the difference of the corresponding valence-band edge positions. The tight-binding position for the valence band edge of a binary (or elemental, as a special case) semiconductor is:

$$E_v = \frac{1}{2}(\varepsilon_p^c + \varepsilon_p^a) - \left[\left(\frac{\varepsilon_p^c - \varepsilon_p^a}{2}\right)^2 + V_{xx}^2\right]^{1/2}, \tag{6}$$

where ε_p^c and ε_p^a are the atomic energies of the cation and anion p states, and V_{xx} is an interatomic matrix element between atomic p states on adjacent atoms. In turn, V_{xx} can be empirically written as $V_{xx} = Cd^{-2}$, where d is the nearest cation-anion distance, and C is a constant which is determined, for example, by fitting the bands for Si and Ge.

Tests of Harrison's predicted ΔE_v's with measured values reveal that the model reaches a reasonable accuracy, certainly better than the original electron affinity rule. This accuracy has been improved by refinements, due to Harrison as well as to other authors.[8] The model, however, is affected by a limitation similar to the electron affinity rule. It calculates ΔE_v entirely in terms of *bulk* parameters of the component semiconductors, without taking into account the specific electronic structure of the interface.

More exactly, the electron affinity rule does not entirely neglect the interface electronic structure, since it uses electron affinities which are measured at interfaces between each semiconductor and vacuum, and are affected by surface effects. In a sense, therefore, the electron affinity rule tries to simulate the microscopic interface effects with a linear combination of similar effects affecting semiconductor-vacuum interfaces. As we have seen, the experimental evidence suggests that this attempt is not successful.

The magnitude of the microscopic interface effects was not known *a priori*, and some lines of reasoning suggested that they are small. Therefore, Harrison's approach of simply neglecting them was quite plausible. The debate about the importance of the microscopic interface charge distribution in the band lineups has been quite lively for several years. At present, however, most authors seem to converge towards the conclusion that such effects *cannot* be neglected. The experimental evidence in favor of this conclusion is discussed in the next sections and, in particular, in reprint R31.

A second, fundamental class of heterojunction band lineup models includes theories based on the metal-induced (or semiconductor-induced) gap states

(MIGS) and on the charge neutrality conditions. The origin of these theories can be traced back to Heine's landmark work on metal-semiconductor interfaces.[9] Heine suggested that the metal wave functions, tailing into the semiconductor gap, produce effects similar to those of localized states. In particular, they can affect and in fact determine the interface position of the Fermi level, E_F, in the gap of the semiconductor. In reprint R20, Jerry Tersoff applies Heine's hypothesis, and derives the concept of "midgap energy point". This is the energy in the gap of each semiconductor for which the character of the MIGS changes from valence-like to conduction-like. The midgap energy determines also, at least in first approximation, the interface position of E_F — and the Schottky barrier height.

The next reprint, R21, is Tersoff's extension of this approach to heterojunction interfaces. Tersoff simply argues that the two midgap energy points of the component semiconductor cannot be displaced with respect to each other without creating an interface dipole, which would cost much energy. Thus, the band lineup is a byproduct of the alignment of the two midgap energy points. If each midgap energy point is referred to the valence band edge of the corresponding semiconductor, then ΔE_v is given by the difference of the two midgap energies.

Tersoff's approach does not neglect the microscopic interface charge distribution like Harrison's model.[R21] On the contrary, it is considered the most important factor in the band lineup. In essence, this approach identifies the microscopic interface charge distribution effects with the effects of the MIGS. In this way, the microscopic effects have general chacteristics, that make it possible to formulate a "universal" theory like Tersoff's model — rather than calculating the specific charge distribution of each interface.

Tersoff's approach has generated a great deal of controversy. The comparison between its predictions and the measured ΔE_v's shows that the model reaches better accuracy than other kinds of theories. Tersoff actually underestimated the accuracy of his results, by limiting the comparison between theory and experiment to a small number of interfaces.[R21] A more extensive comparison, reported in reprint R22, fully reveals the accuracy of the predicted values.

As we have seen, one of the most controversial points in Tersoff's approach was the magnitude and role of the interface dipoles. After a long controversy, an article by Harrison and Tersoff (reprint R23) presented a clarification of this issue. At the same time, it explained the links and the differences between tight-binding theories and MIGS theories. This article, therefore, has fundamental importance in the development of heterojunction theory.

Tersoff's midgap-energy model is not the only theory in the general class generated by Heine's work.[9] In particular, Flores and Tejedor published in 1979 an article (reprint R24), that contained the elements of a MIGS theory of heterojunctions. The reader is cautioned about a possible wrong interpretation of this fundamental work as a mere correction of the electron affinity rule. A careful examination of the article reveals that its foundations are similar to those of the Heines-Tersoff approach. Furthermore, it clarifies the nature of Tersoff's midgap-energy point in terms of the charge neutrality condition.

The reprint R25 presents a very interesting new point of view in this class of models. In this article, Cardona and Christensen consider the problem of calculating screening effects on the hydrostatic deformation potentials. They argue that the screening response can be calculated by using the average of the conduction and valence energies at the Penn gap, also called the *dielectric midgap energy* or DME. The DME is related to the Tersoff-Heine-Flores midgap energy (charge-neutrality) point. Thus, DME's can be used, with an approach similar to that of Ref. R21, to calculate heterojunction band discontinuities. The work of Cardona and Christensen provides a deep insight into the physics of this class of theories, and therefore its importance goes well beyond that of a simple refinement of previous MIGS approaches.

The third, general class of heterojunction band lineups eliminates, at least in principle, all problems, by calculating directly the electronic structure of the interface. Readers not familiar with solid-state theory might ask why approximations such as the tight-binding and MIGS theories have been developed, rather than using the straightforward approach of this third class of theories. The answer is that realistic calculations of local electronic structures are very complicated, and are not – or perhaps not yet – able to solve the problem by brute force. The complications notwithstanding, realistic calculations of heterojunction interface electronic structures have produced fundamental advances.

The reprint R26 is one of the pioneering works in this area. It presents a self-consistent pseudopotential calculation of the ZnSe-Ge system by Pickett, Louie and Cohen. The model predicts, in particular, the formation of interface electronic states (see Fig. 6 in Ref. R26), for which experimental evidence was provided by subsequent photoemission studies.[10]

The next two reprints, R27 and R28, demonstrate the advances made by interface electronic structure calculations in the past ten years. These works are a sample of a series of recent, sophisticated papers in this area. The first reprint presents a general band lineup theory developed by Van de Walle and Martin. These authors first develop self-consistent density-functional calculations, using *ab initio* non-local pseudopotentials, to estimate the parameters of several heterojunction interfaces, including ΔE_v. Based on these results, the authors argue that their results are not inconsistent, within their accuracy, with an approach like that proposed by Frensley and Kroemer. Then they proceed with the formulation of an advanced version of such an approach. This work, therefore, is an interesting hybrid between a realistic calculation of the interface electronic structure, and a general-purpose model similar to the tight-binding or MIGS theories.

Reprint R28 presents some of the most sophisticated calculations the band lineups at GaAs-AlAs interfaces using a local-density scheme. The authors, Massidda, Min and Freeman, use core-level binding energies to calculate ΔE_v. This approach is somewhat similar to the approach used for many photoemission band lineup measurements. The results are closely related to the charge density interface distribution, and provide not only numerical estimates of the discontinuities, but also information on their nature. Quite interestingly, the results disagree with the common-anion rule, and therefore are in agreement with its experimental breakdown.

The reprints and the other articles discussed in this section by no means exhaust the list of important theoretical works on the band lineup problem. In particular, I would like to call the reader's attention to the work of Ruan and Ching[11] and to the dielectric electronegativity approach of J. A. Van Vechten.[12] For a different, more device-oriented point of view, one should also consider the work of Nussbaum and co-workers.[13] In general, we must conclude that no current theory appears able to reach the accuracy required for technological applications. A specific discussion of this point, based on the available experimental data, will be presented in the next section.

Empirical and Semi-Empirical Considerations

The two previous sections should have made clear that simplistic hypotheses fail to explain the heterojunction properties, and that full theories are quite complex. This has stimulated several authors to formulate empirical or semi-empirical solutions for the band lineup problem, or to find empirical ways to clarify some of its aspects.

Among the semi-empirical approaches, one of the most important is the deep-level model, formulated by Alex Zunger and his co-workers,[14] and, independently, by Langer and Heinrich.[15] The theoretical foundation of this model is provided by reprint R29. This article reveals that the deep energy levels produced by a given impurity in different semiconductors of the same family (*e.g.*, the III-V family) are independent of the materials, as long as they are measured *from the vacuum level*.

This interesting observation has several interesting implications.[R29] In particular, the deep impurity levels provide an empirical substitute for the vacuum level.

This implies that the absolute position of the band edges of each semiconductor correspond, at least in first approximation, to their distances in energy from each impurity level. If we consider a given impurity, and take the distance in energy between its impurity levels and the valence band edge in two different semiconductors, the difference of these two distances should provide a first-approximation estimate of ΔE_v for the corresponding heterojunction. In that regard, the deep impurity levels measured with respect to the valence band edge are empirical replacements for the tight-binding or pseudopotential edge positions discussed in the previous section. The accuracy reached by this approach is remarkably good.[14,15] However, the approach only considers the "natural" lineup, related to the relative positions of the band edges on an absolute scale — and it neglects the specific interface phenomena.

As we have seen, the relative weights of these two factors, absolute edge positions and microscopic interface contributions, is a fundamental problem for virtually all band lineup theories. Some theories simply neglect the interface contributions, assuming that their magnitude is small. This is a rather strong assumption, if one considers the small magnitude of $k_B T$ at room temperature. Several authors addressed the fundamental question of the magnitude of the interface contributions in a purely empirical way, and the results indicate that the interface contributions are not, or not always, negligible on a scale of 10 meV.

The key for this empirical approach is the *linearity* of the theories which neglect interface contributions. Although these theories differ substantially from each other, they are all based on a hypothesis of linearity, *i.e.*, that ΔE_v (or ΔE_c) can be expressed as the difference of parameters determined by the two semiconductors. Consider, for example, Harrison's tight-binding theory — ΔE_v is expressed as the difference of the tight-binding edge positions of the two semiconductors.[R19] Similarly, the electron affinity rule expresses ΔE_c as the difference of the electron affinities. Note that some of the theories which do not neglect interface contributions are also linear theories, *e.g.*, the MIGS models.

The hypothesis of linearity can be easily tested. For example, one of its implications is the *transitivity* of ΔE_v. For three semiconductors A, B and C, the transitivity requires that:

$$\Delta E_v^{AB} + \Delta E_v^{BC} + \Delta E_v^{CA} = 0, \qquad (7)$$

where ΔE_v^{XY} is the discontinuity for the interface between the semiconductors X and Y, and the signs are determined using the conventions discussed before.

In the simpler case of two materials, a similar rule says that the discontinuity is independent of the deposition sequence.

These implications of the hypothesis of linearity have been tested using experimental results. Reprint R30 discusses one of the first results of this empirical approach. The authors compared Eq. 7 to the measured discontinuities of the interfaces involving Ge, GaAs and CuBr. They found that the sum of the discontinuity deviated from zero by more than six tenths of an electronvolt.

In 1983, Katnani and Margaritondo (see reprint R31) used a more extensive data base, obtained with synchrotron-radiation photoemission, to test the hypothesis of linearity, based on Eq. 7 and on some of its other implications. The results of these tests were that the hypothesis of linearity fails on the scale of 0.1-0.2 eV per interface. Specifically, Ref. R31 estimated that such a hypothesis implies a built-in accuracy limit for the ΔE_v's predicted by all theories which adopt it, and this limit is, *on the average*, 0.15 eV per interface. This average limit does not prevent a given theory to reach better accuracy in predicting ΔE_v for a given interface — but it makes it impossible to reach better accuracy for all interfaces. This is a sobering conclusion, since 0.15 eV is much larger than $k_B T$ at room temperature. Of course, each linear theory does not necessarily reach this accuracy limit, since its own accuracy is affected by the specific assumptions and approximations. Another important point is that the average accuracy limit of 0.15 eV is much worse than the average experimental accuracy of the data used for the tests.

The conclusions of Ref. R31 apply to all linear theories. In particular, they apply to all theories which neglect interface contributions, or treat them with strong approximations (*e.g.*, the electron affinity rule). Therefore, 0.15 eV is also a reasonable estimate for the average magnitude of such contributions. It should be emphasized that this magnitude, although large on the "technological" scale set by $k_B T$, is not very large a more "fundamental" scale, whose magnitude is set by that of the semiconductor gaps. On this scale, the best linear theories do reasonably well in estimating the band discontinuities, reaching an accuracy not much worse than one tenth of an electronvolt. Actually, most linear theories have accuracies worse than the 0.15 eV limit, but this limit is reached by the theories of Ref. 11 and R21 (see Ref. R22).

The reprint R31 also proposes an empirical approach to the solution of the band lineup problem. This approach is, again, based on the hypothesis of linearity. Specifically, it assumes that ΔE_v^{XY} can be written:

$$\Delta E_v^{XY} = E_v^X - E_v^Y, \tag{8}$$

where E_v^X and E_v^Y are the positions in energy of the valence band edges of the two semiconductors. The same equation was used, for example, in Harrison's model.[R19] In this case, however, the valence band edge positions are *not* calculated using theory, but empirically derived from the experimental data on the band discontinuities.

Table II is a recent version of the list of empirical valence band edge positions, which uses the band edge of germanium as the reference point. With these empirical values, Eq. 8 provides values of ΔE_v with an average accuracy at the limit for linear models. Of course, this approach is not a theory, since it derives the band-edge terms from the data rather than from a theoretical model. As such, it does not provide any insight into the nature of band lineups, except a general confirmation of the built-in accuracy limits caused by the hypothesis of linearity.

Table II
Valence Band Edge Positions
(referred to the Ge edge)

Semiconductor	E_v (eV)	Semiconductor	E_v (eV)
Ge	0.00	CdS	-1.74
Si	-0.16	CdSe	-1.33
α-Sn	0.22	CdTe	-0.88
		ZnSe	-1.40
AlAs	-0.78	ZnTe	-1.00
AlSb	-0.61		
GaAs	-0.35	PbTe	-0.35
GaP	-0.89	HgTe	-0.75
GaSb	-0.21	CuBr	-0.87
InAs	-0.28	GaSe	-0.95
InP	-0.69	$CuInSe_2$	-0.33
InSb	-0.09	$CuGaSe_2$	-0.62
		$ZnSnP_2$	-0.48

The relation between Schottky barriers and heterojunctions is perhaps the most important issue explored with empirical approaches. The implications of this issue are far-reaching — a link between the two problems could clarify both of them, and pave the way for a generalized theory of semiconductor interfaces. It could also provide a clear-cut test for theories, either in favor of those that predict such a link or in favor of those which rule it out.

Consider, for example, the basic predictions of the electron affinity rule, Eq. 4, and of the Schottky model of metal-silicon interfaces, Eq. 2. If one takes two n-type semiconductors and the Schottky barriers between them and a given metal, ϕ_n^1 and ϕ_n^2, then these equations imply that, for an interface between these two semiconductors:

$$\Delta E_c = \phi_n^2 - \phi_n^1. \tag{9}$$

A similar equation links ΔE_v and p-type Schottky barriers:

$$\Delta E_v = \phi_p^2 - \phi_p^1. \tag{10}$$

These relations are also predicted by the MIGS models for heterojunctions and Schottky barriers. The question is — are such relations experimentally observed?

This question is still controversial. For example, evidence against such a correlation has been presented for the GaAs-Ge system,[16] whereas data supporting it have been published by Heiblum *et al.* for the $Al_{1-x}Ga_xAs$-GaAs system.[17] In reprint R32, a test of Eq. 10 is described, based on an extensive data base. As it can be seen from Fig. 1 in Ref. R32, the data suggest a correlation qualitatively similar to that predicted by Eqs. 9 and 10. However, there are also significant *and systematic* deviations with respect to the line of "perfect agreement" with Eq. 10.

This discrepancy has been analyzed by Tersoff,[18] and it could provide an interesting step towards a generalized theory of semiconductor interfaces. The element of a possible generalized theory are the following. Consider the Schottky model, Eq. 2. For many semiconductors, this equation disagrees with the experimental data, and the Schottky barrier, *i.e.*, the interface position of E_F, is determined by other factors. One of these possible factors is the Tersoff-Flores midgap-energy, E_B. Another possible factor is the pinning of E_F by defect states, as predicted by the *unified defect model*.[19] Furthermore, even when a linear dependence of the Schottky barrier on the metal work function, Φ_m, is observed, the proportionality factor, S, deviates from unity, the value predicted by Eq. 2. This factor is related to the reciprocal of the optical dielectric constant, and therefore to the midgap-energy.

In a pure MIGS model, the Schottky barrier would only be determined by MIGS. By adding to this term a "Schottky" term like that of Eq. 2, corrected for the factor S, Tersoff was able[18] to explain the deviations from Eq. 10, reported in Ref. R32. This approach is a first step towards a unified theory of semiconductor interfaces. A further step was made by Winfried Mönch, with the work described in the reprint R33. We invite the reader to examine in detail this article and its conclusions. The main message is that metal-semiconductor interfaces can be roughly divided in two groups, those dominated by MIGS and those dominated by defects. The discriminating element is the density of interface defects. For the first group, the main factor determining the Schottky barrier is the midgap energy. For the second group is the pinning position of E_F by defects. For both groups, the main factor is corrected by a Schottky-like dependence on the metal work function, with slope S.

The results of Ref. R32, and their explanation by Tersoff,[18] indicate that a similar, unified approach can possibly be extended to heterojunctions. In such a unified picture, defects, MIGS and Schottky terms all can play a role in the heterojunction band lineups and in the Schottky barrier heights — and the relative weights are determined by the parameter S (*i.e.*, by the optical dielectric constant), and by the density of interface defects. In particular, defects can play a fundamental role in the heterojunction band lineups, when their density is sufficiently large.

The reader must be cautioned that the experimental evidence in favor of this unified picture is still limited. The picture is quite appealing and provides a coherent explanation of many existing data. However, its final acceptance – or rejection – must be delayed until sufficient experimental tests have been performed.

As we have seen, empirical and semi-empirical approaches are making strong contributions to the understanding of the physics of heterojunction band lineups, and to the development of practical methods for estimating band discontinuities, and for testing heterojunction theories. Particularly important, among the latter, is the work described in reprint R34. This article describes an experimental and theoretical study of the *pressure* dependence of heterojunction band lineups. The study was performed on InAs-GaSb.

The authors argue that the observed pressure dependence is a powerful test of the band lineup theories. As we have seen, many different kinds of band lineup theories are able to predict band discontinuities with reasonable accuracy. The reason for this is probably that different theoretical approaches are based on electronic states which, although different in nature, are all directly or indirectly related to the band structures of the two components. The estimates of band discontinuities are obtained by taking energy *differences* between the two materials, and these differences probably tend to be more similar than the absolute terms. Therefore, even physically wrong models can accidentally produce reasonable estimates. Thus, the accuracy of a model in predicting band discontinuities is not a very sensitive test. On the contrary, the success or failure in predicting the dependence of the band discontinuities on external perturbations *is* a sensitive test.

These considerations explain the importance of the work of Ref. R34, which was performed using magneto-optical techniques. These results reveal that the offset between the valence-band edge of GaSb and the conduction-band edge of InAs decreases at a rate of 5.8 meV/KBar. This decrease cannot be entirely explained by the pressure dependence of the gaps of the two materials, and indicates that ΔE_c and ΔE_v are pressure dependent. Although the authors are careful not to overinterpret their data, it is quite clear that the results rule out some of the models, and in particular theories entirely based on tight-binding calculations of the band edge positions. Tersoff's version of the MIGS model[R31] appears able to justify the observed pressure dependence.

Control of Band Lineups

The final aim of heterojunction research is the production of new devices with performances which cannot be obtained from other devices. The band discontinuities are an important factor in the design of novel heterojunction systems, as we have seen from reprints R2-R6. This is a powerful motivation for the research on the band lineup mechanism. Of course, the flexibility in designing new heterojunction devices would be tremendously increased by the capability of *controlling* the band discontinuities.

The band discontinuity control is, in fact, a fundamental objective for heterojunction research — and for condensed matter research in general. This is, in particular, the underlying objective of the efforts devoted to the understanding of the nature of band lineups. For many years, however, it was not even clear if the band discontinuities between two given semiconducting materials could be modified at all. Many theoretical models calculated the discontinuities based only on the bulk parameters of the two semiconductors. The magnitude of the microscopic interface contributions to the band lineup, which could be potentially used to control the discontinuities, has been controversial. The recent indications that such contributions are *not* small on the scale of $k_B T$ at room temperature also implied that band lineup control is feasible.

The uncertainty about such a feasibility was definitely removed by two series of experiments, which established that band lineups can, indeed, be modified and potentially controlled. The first approach, due to Capasso *et al.*, is described in the reprint R35. This approach takes advantage of the increasing sophistication of the overlayer deposition techniques, specifically MBE. The authors fabricated heterojunction interfaces with a controlled doping profile. This introduces a doping interface dipole (DID) which, in turn, effectively modifies the band discontinuities and their influence on the interface behavior. The article discusses the practical implementation of this approach, the properties of the corresponding structures, and some applications.

The second approach is based on the deposition of ultrathin intralayers between the two sides of the interface. Ultrathin intralayers had already been used to modify the properties of metal-semiconductor interfaces.[20] The empirical attempts were motivated by the fact that intralayers can modify the charge distribution at the interface, and therefore affect the interface dipoles and the band lineup. The reprint R36 reports the success of this approach in modifying the valence band discontinuity of ZnSe-Ge and ZnSe-Si interfaces. Aluminum intralayers, of thickness ranging from 0.5 to 1 Å, caused increases in ΔE_v by up to 0.3 eV, measured with photoemission techniques.

The empirical success reported in Ref. R36 did not, of course, remove all problems related to band lineup control. First, it was not clear if the approach could be used in practical devices. This point must still be clarified. Second, the *mechanism* of the intralayer-induced band lineup modifications was not identified — and still is not, to some extent. In principle, intralayers can modify the band lineups with different mechanisms. For example, they can act as barriers or activators for microdiffusion processes of charged impurities, which in turn can produce interface dipoles. The chemical bonds involving the intralayer atoms can be another cause of interface dipoles.

The work described in the reprint R37 provides some clarification of the latter problem. This article describes very large modifications in the Si-SiO$_2$ valence band discontinuity, caused by cesium or hydrogen intralayers. The modifications occur in opposite directions for the two kinds of intralayers. Similarly large – although qualitatively different – modifications have been reported by Grunthaner et al.[21] The magnitude of these effects definitely established the use of intralayers as a feasible technique to modify band lineups.

Reference R37 also outlines a simple explanation of the phenomena, based on the charge-transfer dipoles associated to the chemical bonds at the interface. Basically, a non-diffusive intralayer replaces chemical bonds between the two sides of the interfaces with chemical bonds involving the intralayer atoms. The corresponding changes in the interface dipoles are estimated by using a simple approach to calculate the charge transfers due to the formation of chemical bonds. This model is remarkably successful in predicting the sign and magnitude of the observed changes in ΔE_v, for SiO$_2$-Si as well as for other interfaces.

Recent photoemission results[22] lend further support to this simple-minded approach. The experiments included successful tests of the non-diffusive character of Al intralayers. Then they found that the Al-Se chemical bonds formed by an Al intralayer between ZnSe and Ge are consistent with an Al_2Se_3-like configuration. This implies that the changes of ΔE_v increase with the intralayer thickness, d, up to d's corresponding to approximately one-third of a monolayer — "one monolayer" being defined as one intralayer atom per ZnSe substrate atom. The experimental plot of the ΔE_v changes as a function of d indicates, indeed, saturation at d's well below one monolayer, and consistent with the predicted 1/3 monolayer. This result can be observed in Fig. 3.

These facts indicate that simple charge transfers due to local chemical bonds can explain the intralayer-induced band lineup changes. A more sophisticated theoretical approach[23] indicates that the local charge re-distribution effects can be interpreted in terms of a change of the midgap-energy point, which in turn affects the band lineups within the framework of MIGS theories. Of course, these results cannot be automatically extended to all kinds of interfaces. In particular, diffusive interfaces can be affected by entirely different mechanisms. Nonetheless, these results are steps ahead towards the understanding of the intralayer-induced phenomena and, perhaps, towards their eventual practical use.

Some Considerations on Future Heterojunction Research

The reprints presented in the previous sections clearly show that heterojunction research, initiated because of practical considerations, now deals with questions at the foundations of condensed matter science. For example, the band lineup problems touch our very understanding of the nature of the electronic states in solids. Therefore, the present research on heterojunction is justified by fundamental as well as by practical motivations. Such motivations are likely to be present, and probably to be enhanced, in future years.

The previous sections have illustrated a series of recent breakthroughs, which are likely to affect future heterojunction research. The implementation of new, sophisticated growth techniques makes it possible to fabricate heterojunction structures with very advanced design characteristics. Current examples are doping-profile structures and quantum-well structures involving very thin, high-quality layers.

The fabrication of bond-stretched overlayers is another exciting development in growth technology.[24] These are thin films of a given material grown on top of a substrate with substantial lattice mismatch. In general, the lattice mismatch between substrate and overlayer is compensated by misfit dislocations. In several cases, however, it has been demonstrated that the overlayer grows free of misfit dislocations through strain accomodation up to a certain critical thickness. The overlayer atoms are in the positions corresponding to the substrate lattice, with their chemical bonds stretched with respect to a normal crystal. Thus, it becomes possible, in the plane parallel to the interface, to grow crystals of a given compound in the structure of another compound. When repeated layers of this type are grown, one obtains a strained-layer superlattice.

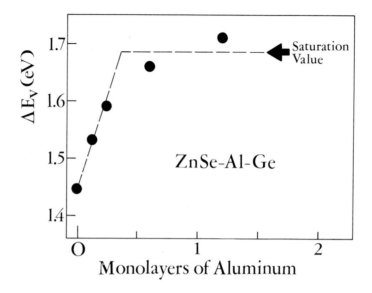

Fig. 3 – The measured changes in ΔE_v at a ZnSe-Ge interface, caused by an Al intralayer, as a function of the intralayer thickness. The horizontal arrow shows the saturation value for these changes. The saturation occurs for an intralayer thickness well below one monolayer, and consistent with the theoretically predicted 1/3 monolayer (dashed line).

These advanced growth techniques are the tools to implement novel ideas in heterojunction technology. The breakthroughs in the control of band lineups suggest that such techniques could be used to fabricate devices with controlled interface properties. Of course, much more work in research and development is necessary before achieving this exciting objective — or even determining if it is achievable. For example, the recent progress notwithstanding, our understanding of the band lineup control mechanisms is still very limited. The future research on heterojunctions must expand our knowledge of the systems that have already been explored in part. Furthermore, we must extend these studies to other systems, and in particular to diffusive intralayers. In essence, the first experiments in this area were "shots in the dark", which produced an excellent return. It is quite probable that further empirical explorations will discover new phenomena, potentially useful for the control of interface parameters.

Paradoxically, the interface control is being achieved even if we do not really understand the band lineup problem. The complete clarification of this problem remains, however, a central objective of heterojunction research. In this author's opinion, the "unified" approach for metal-semicondutor and heterojunction systems, discussed above, is the most promising direction for a complete understanding of heterojunction band lineups, as well as of other semiconductor interface properties. This approach is based on common sense and on basic physical intuitions. I repeat, however, that its final acceptance is subject to extensive experimental verification. This verification is a major objective for future heterojunction research. The perturbation-induced changes in band discontinuities are likely to play an important role in these tests, following the precedent recently established with pressure dependence studies.[R34]

These considerations make it easy to predict, to a certain extent, the future developments in heterojunction research. The predictive capabilities are limited, however, in a field that continuously produces novel and – to some extent – unexpected results. Who, for example, would have predicted five years ago the breakdown of the common-anion rule for binary semiconductors? Or the successful modifications of band discontinuities by up to half an electronvolt? Or the highly sophisticated level reached by the bond-stretched systems? It is prudent, therefore, to expect that a large part of heterojunction research will take place in areas not predictable at the present time. This uncertainty enhances the interest of this already exciting field, which has been for more than twenty-five years at the forefront of condensed matter science.

References

Reprinted Articles:

R1.   R. L. Anderson, Solid-State Electron. **5**, 341 (1962).

R2.   A. C. Gossard, IEEE J. Quantum Electron. **QE-22**, 1649 (1986).

R3.   L. Esaki, IEEE J. Quantum Electron. **QE-22**, 1611 (1986).

R4.   F. Capasso, Ann. Rev. Mat. Sci. **16**, 263 (1986).

R5.   F. Capasso, K. Mohammed and A. Y. Cho, IEEE J. Quantum Electron. **QE-22**, 1853 (1986).

R6.   H. Kroemer, Surface Sci. **132**, 543 (1983).

R7.   D. V. Lang, A. M. Sergent, M. B. Panish and H. Temkin, Appl. Phys. Lett. **49**, 812 (1986).

R8.   P. Perfetti, D. Denley, K. A. Mills and D. A. Shirley, Appl. Phys. Lett. **33**, 667 (1978).

R9.   R. S. Bauer and J. C. McMenamin, J. Vac. Sci. Technol. **15**, 1444 (1978).

R10.   R. W. Grant, J. R. Waldrop and E. A. Kraut, Phys. Rev. Lett. **40**, 656 (1978).

R11.   G. Abstreiter, U. Prechtel, G. Weimann and W. Schlapp, Physica **134B**, 433 (1985).

R12.   R. Dingle, W. Wiegman and C. H. Henry, Phys. Rev. Lett. **33**, 827 (1974).

R13.   J. L. Freeouf and J. M. Woodall, Appl. Phys. Lett. **39**, 727 (1981).

R14.   J. L. Freeouf and J. M. Woodall, Surface Sci. **168**, 518 (1986).

R15.   D. W. Niles and G. Margaritondo, Phys. Rev. **B34**, 2923 (1986).

R16.   R. C. Miller, A. C. Gossard, D. A. Kleinman and O. Munteanu, Phys. Rev. **B29**, 3740 (1984).

R17.   S.-H. Wei and A. Zunger, Phys. Rev. Lett. **59**, 144 (1987).

R18. D. G. Kilday, G. Margaritondo, T. F. Ciszek, S. K. Deb, S.-H. Wei and A. Zunger, Phys. Rev. **B36**, 9388 (1987).

R19. W. Harrison, J. Vac. Sci. Technol. **14**, 1016 (1977).

R20. J. Tersoff, Phys. Rev. Lett. **52**, 465 (1984).

R21. J. Tersoff, Phys. Rev. **B30**, 4874 (1984).

R22. G. Margaritondo, Phys. Rev. **B31**, 2526 (1985).

R23. W. A. Harrison and J. Tersoff, J. Vac. Sci. Technol. **B4**, 1068 (1986).

R24. F. Flores and C. Tejedor, J. Phys. **C12**, 731 (1979).

R25. M. Cardona and N. E. Christensen, Phys. Rev. **B35**, 6182 (1987).

R26. W. Pickett and M. L. Cohen, Phys. Rev. **B18**, 939 (1978).

R27. C. G. Van de Walle and R. M. Martin, Phys. Rev. **B35**, 8154 (1987).

R28. S. Massidda, B. I. Min and A. J. Freeman, Phys. Rev. **B 35**, 9871 (1987).

R29. M. J. Caldas, A. Fazzio and A. Zunger, Appl. Phys. Lett. **45**, 671 (1984).

R30. J. R. Waldrop and R. W. Grant, Phys. Rev. Lett. **43**, 1686 (1979).

R31. A. D. Katnani and G. Margaritondo, Phys. Rev. **B28**, 1944 (1983).

R32. D. W. Niles, E. Colavita, G. Margaritondo, P. Perfetti, C. Quaresima and M. Capozi, J. Vac. Sci. Technol. **A4**, 962 (1986).

R33. W. Mönch, Phys. Rev. Lett. **58**, 1260 (1987).

R34. L. M. Claessen, J. C. Maan, M. Altarelli, P. Wyder, L. L. Chang and L. Esaki, Phys. Rev. Lett. **57**, 2556 (1986).

R35. F. Capasso, K. Mohammed and A. Y. Cho, J. Vac. Sci. Technol. **B3**, 1245 (1985).

R36. D. W. Niles, G. Margaritondo, P. Perfetti, C. Quaresima and M. Capozi, Appl. Phys. Lett. **47**, 1092 (1985).

R37. P. Perfetti, C. Quaresima, C. Coluzza, C. Fortunato and G. Margaritondo, Phys. Rev. Lett. **57**, 2065 (1986).

Other References:

1. A. G. Milnes and D. L. Feucht, *Heterojunctions and Metal-Semiconductor Junctions* (Academic Press, New York 1972).

2. F. Capasso and G. Margaritondo: *Heterojunction Band Discontinuities: Physics and Device Applications* (North-Holland, Amsterdam 1987).

3. F. Capasso, S. Sen, A. C. Gossard, A. L. Hutchinson and J. H. English, IEEE Electron Device Lett. **EDL-7**, 573 (1986).

4. P. Perfetti, F. Patella, F. Sette, C. Quaresima, C. Capasso, A. Savoia and G. Margaritondo, Phys. Rev. **B 29**, 5941 (1984).

5. C. Mailhiot and C. B. Duke, Phys. Rev. **B 33**, 1118 (1986).

6. D. W. Niles, B. Lai, J. T. McKinley, G. Margaritondo, G. Wells, F. Cerrina, G. J. Gualtieri and G. P. Schwartz, J. Vac. Sci. Technol. **B5**, 1286 (1987), and the references therein.

7. W. R. Frensley and H. Kroemer, Phys. Rev. **B 16**, 6242 (1977).

8. W. A. Harrison, J. Vac. Sci. Technol. **B 3**, 1231 (1985); P. Vogl, H. P. Hjalmarson and J. D. Dow, J. Phys. Chem. Solids **44**, 365 (1983); Z. H. Chen, S. Margalit and A. Yariv, J. Appl. Phys. **57**, 2970 (1985); F. Bechstedt, R. Enderlein and O. Heinrich, Phys. Status Solidi **b 126**, 575 (1984).

9. V. Heine, Phys. Rev. **A 138**, 1689 (1965).

10. G. Margaritondo, F. Cerrina, C. Capasso, F. Patella, P. Perfetti, C. Quaresima and F. J. Grunthaner, Sol. State Commun. **52**, 495 (1984).

11. Y.-C. Ruan and W. Y. Ching, J. Appl. Phys. **60**, 4035 (1986).

12. J. A. Van Vechten, J. Vac. Sci. Technol. **B 3**, 1240 (1985), and Phys. Rev. **182**, 891 (1969).

13. A. Nussbaum, in *Semiconductors and Semimetals*, R. K. Willardson and A. C. Beer eds. (Academic Press, New York, 1981), Vol. 15, Chapter 2.

14. Alex Zunger, Ann. Rev. Mater. Sci. **15**, 411 (1985); Phys. Rev. Lett. **54**, 849 (1985); *Solid State Physics*, D. Ehrenreich and D. Turnbull Eds. (Academic Press, New York 1986), Vol. 39, Sect. VI.29, p. 275.

15. J. M. Langer and H. Heinrich, Phys. Rev. Lett. **55**, 1414 (1985), and Physica B+C **134**, 444 (1985).

16. A. D. Katnani, P. Chiaradia, H. W. Sang, Jr. and R. S. Bauer, J. Electronic Materials **14**, 25 (1985).

17. M. Heiblum, M. I. Nathan and M. Eizenberg, Appl. Phys. Lett. **47**, 503 (1985).

18. J. Tersoff, private communication.

19. W. E. Spicer, P. W. Chye, P. R. Skeath and I. Lindau, J. Vac. Sci. Technol. **16**, 1422 (1979).

20. L. J. Brillson, G. Margaritondo and N. G. Stoffel, Phys. Rev. Letters **44**, 667 (1980).

21. P. J. Grunthaner, F. J. Grunthaner, M. H. Hecht and N. M. Johnson, private communication.

22. D. W. Niles, G. Margaritondo, C. Quaresima, M. Capozi and P. Perfetti, unpublished.

23. J. C. Duràn, A. Muñoz, and F. Flores, Phys. Rev. **B35**, 7721 (1987).

24. J. C. Bean, T. T. Sheng, L. C. Feldman, A. T. Fiory and R. T. Lynch, Appl. Phys. Lett. **44**, 102 (1984); G. Osbourne, Phys. Rev. **B 27**, 5126 (1983).

REPRINTED ARTICLES

Solid-State Electronics Pergamon Press 1962. Vol. 5, pp. 341–351. Printed in Great Britain

EXPERIMENTS ON Ge–GaAs HETEROJUNCTIONS

R. L. ANDERSON*

International Business Machines Corporation, Yorktown Heights, New York

(Received 29 *January*; *in revised form* 23 *April* 1962)

Abstract—The electrical characteristics of Ge–GaAs heterojunctions, made by depositing Ge epitaxially on GaAs substrates, are described. *I–V* and electro-optical characteristics are consistent with a model in which the conduction- and valence-band edges at the interface are discontinuous. The forbidden band in heavily doped (*n*-type) germanium appears to shift to lower energy values.

Résumé—Les caractéristiques électriques des hétérojunctions de Ge–GaAs produites en déposant le Ge épitaxiallement sur les couches inférieures de l'AsGa sont décrites. Les caractéristiques *I–V* et électrooptiques sont consistantes avec un modèle dans lequel les bords des bandes de conduction et de valence à l'interface sont discontinus. La bande défendue dans le germanium (type-*n*) fortement dopé semble se déplacer vers des valeurs à énergie plus basse.

Zusammenfassung—Die elektrischen Kenngrössen von Ge–GaAs Hetero-Übergängen, die man durch epitaxiale Ablagerung von Ge auf GaAs-Substraten herstellt, werden beschrieben. *I–V* und elektro-optische Kenndaten entsprechen einem Modell, in dem die Ränder des Leitungs- und Valenzbandes an der Grenzfläche diskontinuierlich sind. Das verbotene Band in stark dotiertem (*n*-Typ) Germanium scheint sich nach niedrigeren Energiewerten zu verlagern.

1. INTRODUCTION

JUNCTIONS between two semiconductors of the same element but with different impurities present have been studied extensively. These junctions are reasonably well understood. The periodicity of the lattice is not disturbed at the junction and so the properties of the semiconductors at the junction can be expected to be the bulk properties.

Metal–semiconductor contacts, on the other hand, are not well understood. The chief difficulty is usually attributed to interface effects. Even though the semiconductor and the metal may each be monocrystalline, the crystal structures and lattice constants in general are different and so an expitaxial contact is not formed. Because of the abrupt change in the structure and periodicity of the lattice and the resultant disorder in the region near the interface, material properties are not the same here as they are in the bulk.

The theoretical voltage–current characteristic of a *p–n* junction or a metal semiconductor contact

is derived for most models to be of the form[1]

$$I = I_0[\exp(qV/kT) - 1] \qquad (1)$$

where I is the current due to an applied voltage V, I_0 is the saturation current or the current for large negative voltage, q is the electronic charge, k is Boltzmann's constant and T is the absolute temperature. The value of I_0 is reasonably independent of voltage in most derivations. The diode formula is often written in the form

$$I = I_0[\exp(qV/\eta kT) - 1] \qquad (2)$$

where η is an empirical factor which describes the disagreement between simple theory and experiment for forward bias ($V > 0$). The value of η is commonly about 2·3 for gallium arsenide *p–n* junctions, is between 2 and 4 in Ge point-contact diodes and approaches the theoretical value of unity only in Ge *p–n* junctions (and in silicon *p–n* junctions at elevated temperatures). The variation of current with reverse voltage is usually accounted for by permitting the term I_0 to vary slowly with voltage. These deviations from the theory have not been adequately explained.

---

* Now at the Department of Electrical Engineering, Syracuse University, Syracuse, New York.

Little work has been done on junctions between two semiconductors. GUBANOV has suggested that the *I–V* characteristics of copper oxide rectifiers might be indicative of semiconductor–semiconductor contacts.* SHOCKLEY[3] and KRÖMER[4] suggested using a semiconductor with a wide forbidden region as an emitter for a transistor which has base and collector of a narrower-gap semiconductor. The purpose of this is to obtain a high injection efficiency. JENNY[5] has described attempts to fabricate a GaP–GaAs wide-gap emitter by diffusing phosphorus into gallium arsenide. Little success has been reported.

This paper discusses the electrical characteristics of junctions formed between Ge and GaAs. These junctions are contained within a monocrystal. Ge was deposited epitaxially onto GaAs seeds by the Iodide Process.[6–8] These two materials have similar crystal structure, and virtually equal lattice constants (5·62 Å). As a result, it is expected that strain at the interface is negligible.

Junctions between two dissimilar materials will be referred to as "heterojunctions" in contrast to "homojunctions" where only one semiconductor is involved.

2. ENERGY-BAND PROFILE OF HETERO-JUNCTIONS

Consider the energy-band profile of two isolated pieces of semiconductor shown in Fig. 1. The two semiconductors are assumed to have different band gaps (E_g), different dielectric constants (ϵ), different work functions (ϕ) and different electron affinities (θ). Work function and electron affinity are defined, respectively, as that energy required to remove an electron from the Fermi level (E_f) and from the bottom of the conduction band (E_c) to a position just outside of the material (vacuum level). The top of the valence band is represented by E_v. The subscripts 1 and 2 refer to the narrow-gap and wide-gap semiconductors, respectively.

In Fig. 1, the band-edge profiles (E_{c1}, E_{c2}, E_{v1}, E_{v2}) are shown to be "horizontal". This is equivalent to assuming that space-charge neutrality exists in every region. The difference in energy of the conduction-band edges in the two materials is represented by ΔE_c and that in valence-band edges by ΔE_v.

---

* For a review of Gubanov's work see Ref. 2.

A junction formed between an *n*-type narrow-gap semiconductor and a *p*-type wide-gap semiconductor is considered first. This is referred to

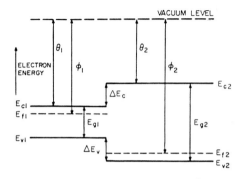

FIG. 1. Energy-band diagram for two isolated semiconductors in which space-charge neutrality is assumed to exist in every region.

as an *n–p* heterojunction. The energy-band profile of such a junction at equilibrium is shown in Fig. 2.

Within any single semiconductor the electrostatic potential difference between any two points

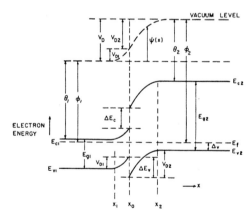

FIG. 2. Energy-band diagram of *n–p* heterojunction at equilibrium.

can be represented by the vertical displacement of the band edges between these two points, and the electrostatic field can be represented by the slope of the band edges on a diagram such as Fig. 2. Then the difference in the work functions of the two materials is the total built-in voltage (V_D). V_D is equal to the sum of the partial built-in voltages $(V_{D1} + V_{D2})$ where V_{D1} and V_{D2} are the electrostatic potentials supported at equilibrium by semiconductors 1 and 2, respectively. Since voltage is continuous in the absence of dipole layers, and since the vacuum level is parallel to the band edges, the electrostatic potential difference (ψ) between any two points is represented by the vertical displacement of the vacuum level between these two points. Because of the difference in dielectric constants in the two materials, the electrostatic field is discontinuous at the interface.

Since the vacuum level is everywhere parallel to the band edges and is continuous, the discontinuity in conduction-band edges (ΔE_c) and valence-band edges (ΔE_v) is invariant with doping in those cases where the electron affinity and band gap (E_g) are not functions of doping (i.e. non-degenerate material).

Solutions to Poisson's equation, with the usual assumptions of a Schottky barrier,* give, for the transition widths on either side of the interface for a step junction,

$$(X_0 - X_1) = \left[\frac{2}{q} \frac{N_{A2}\epsilon_1\epsilon_2(V_D - V)}{N_{D1}(\epsilon_1 N_{D1} + \epsilon_2 N_{A2})} \right]^{1/2} \quad (3a)$$

$$(X_2 - X_0) = \left[\frac{2}{q} \frac{N_{D1}\epsilon_1\epsilon_2(V_D - V)}{N_{A2}(\epsilon_1 N_{D1} + \epsilon_2 N_{A2})} \right]^{1/2} \quad (3b)$$

and the total width W of the transition region is

$$W = (X_2 - X_0) + (X_0 - X_1)$$

$$= \left[\frac{2\epsilon_1\epsilon_2(V_D - V)(N_{A2} + N_{D1})^2}{(q\epsilon_1 N_{D1} + \epsilon_2 N_{A2})N_{D1}N_{A2}} \right]^{1/2} \quad (4)$$

The relative voltages supported in each of the semiconductors are

$$\frac{V_{D1} - V_1}{V_{D2} - V_2} = \frac{N_{A2}\epsilon_2}{N_{D1}\epsilon_1} \quad (5)$$

where V_1 and V_2 are the portions of the applied voltage V supported by materials 1 and 2 respectively. Of course $V_1 + V_2 = V$. Then $(V_{D1} - V_1)$ and $(V_{D2} - V_2)$ are the total voltages (built in plus applied) for material 1 and material 2, respectively. We can see that most of the potential difference

---

* See Ref. 9 for details for calculations for homojunctions.

occurs in the most lightly doped region for nearly equal dielectric constants.

The transition capacitance is given by a generalization of the result for homojunctions:

$$C = \left[\frac{qN_{D1}N_{A2}\epsilon_1\epsilon_2}{2(\epsilon_1 N_{D1} + \epsilon_2 N_{A2})} \frac{1}{(V_D - V)} \right]^{1/2} \quad (6)$$

It can be seen from Fig. 2 that the barrier to electrons is considerably greater than that to holes, and so hole current will predominate.

The case of an n–n junction of the above two materials is somewhat different. Since the work function of the wide-gap semiconductor is the smaller, the energy bands will be bent oppositely to the n–p case (See Fig. 3). However, there are

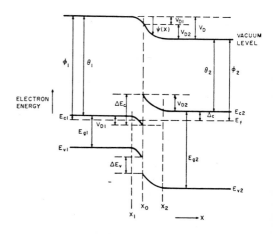

FIG. 3. Energy-band diagram of n–n heterojunction at equilibrium.

a negligible number of states available in the valence band and so the excess electrons in the material of greater work function will occupy states in the conduction band. Since there are a large number of states available in the conduction band, the transition region extends only a small distance into the narrow-band material and the voltage is supported mainly by the material with the smaller work function.

The voltage profile in the interface region can be determined by solving for the electric field strength (F) on either side of the interface and using the condition that the electric displacement ($D = \epsilon F$) is continuous at the interface. Assuming

Boltzman statistics in region 1,

$$\epsilon_1 F_1(X_0) = \left\{ 2\epsilon_1 q N_{D1} \left[\frac{kT}{q} \left(\exp \frac{q(V_{D1}-V_1)}{kT} -1 \right) \right. \right.$$

$$\left. \left. -(V_{D1}-V_1) \right] \right\}^{1/2} \qquad (7)$$

In region 2 the electric displacement at the interface is

$$\epsilon_2 F_2(X_0) = [2\epsilon_2 q N_{D2}(V_{D2}-V_2)]^{1/2} \qquad (8)$$

Equating equations (7) and (8) gives a relation between $(V_{D1}-V_1)$ and $(V_{D2}-V_2)$ which is quite complicated. However, it is reasonably easy to get an upper limit of $(V_{D1}-V_1)$. If the exponential in equation (1) is expanded in a Taylor series, the following inequality is obtained:

$$(V_{D1}-V_1) < \left[\frac{2kT}{q} \frac{\epsilon_2 N_{D2}}{\epsilon_1 N_{D1}}(V_{D2}-V_2) \right]^{1/2} \qquad (9)$$

From equation (9) we can see that the electrostatic potential will be supported mainly by semiconductor 2 unless $N_{D2} \gg N_{D1}$, or for high forward bias.

For n–n heterojunctions the transition capacitance is difficult to calculate. However, except for the cases mentioned above, the capacitance of a metal–semiconductor contact is a good approximation.

In the heterojunctions discussed here, the energy gap of the wide-gap material (Ga–As) "overlaps" that of the narrow-gap material, and the polarity of the built-in field (and of rectification) is dependent on the conductivity type of the wide-gap semiconductor. Fig. 4 shows the equilibrium energy-band diagrams for p–n and p–p heterojunctions.

3. PREDICTED I–V CHARACTERISTICS

Because of the discontinuities in the band edges at the interface, the barriers to the two types of carriers have different magnitudes, and so current in a heterojunction will in most cases consist almost entirely of electrons or of holes.

The variation of current with applied voltage for these heterojunctions (neglecting generation–recombination current) is

$$I = A \exp(-qV_{B2}/kT) - B \exp(-qV_{B1}/kT) \qquad (10)$$

where V_{B1} is the barrier that carriers in semiconductor 1 must overcome to reach semiconductor 2, and V_{B2} is the barrier to the carriers moving the opposite direction. The coefficients A and B depend on doping levels, on carrier effective mass and on the mechanism of current flow.

In the junctions depicted in Fig. 2–4, V_{B1} exists for the predominant current carrier and so

$$I = A \exp[-q(V_{D2})/kT]$$

$$[\exp(qV_2/kT) - \exp(-qV_1/kT)] \qquad (11)$$

where V_2 and V_1 are those portions of applied voltage appearing in materials 2 and 1, respectively. The first term in the brackets is important for forward bias and the second for reverse bias. If $V_2 = V/\eta$ then $V_1 = (1-1/\eta)V$ and the current varies approximately exponentially with voltage in both forward and reverse directions. It should be noticed however that at increased reverse voltage V_{B1} disappears, i.e. $(V_{D1}-V) > \Delta E_c$ (for the case of the p–n junction), and the current is expected to saturate. If $V_{D1} > \Delta E_c$ (again for a p–n heterojunction—see Fig. 5), $V_{B1} = 0$ and the

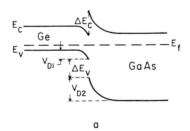

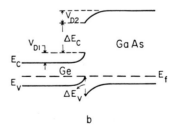

a   b

FIG. 4. Energy-band diagrams in the interface region for p–n and p–p heterojunctions. Electron energy is plotted vertically.

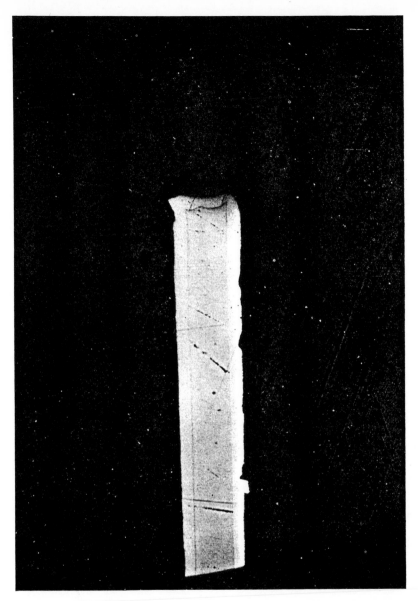

FIG. 6. Cross-sectional view of a wafer of GaAs on which Ge has been deposited. The thickness of the deposit is about 0·03 cm.

[*facing p.* 344

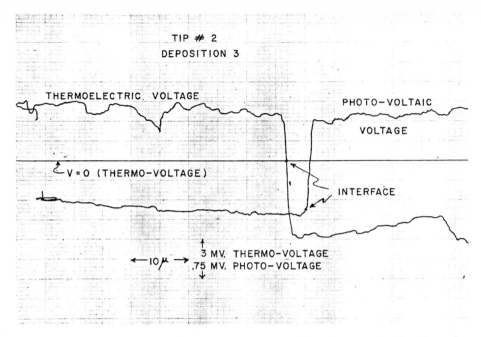

FIG. 7. Plot of photovoltage and thermoelectric voltage against distance normal to the surface for an *n–p* Ge–GaAs heterojunction. The surface is indicated by the extreme left of each trace. The junction position is indicated by a zero thermoelectric voltage.

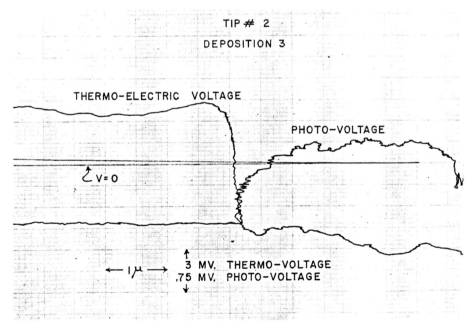

FIG. 8. Region near interface of Fig. 4 on expanded scale. The shape of the photovoltage plot indicates the transition region to be predominantly in the gallium arsenide.

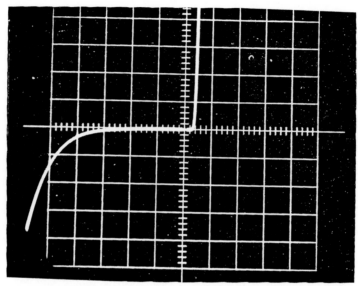

FIG. 9. *V–I* characteristics of a Ge–GaAs *n–n* heterojunction. The ordinate scale is 0·1 mA/div while the abscissa scale is 1·0 V/div.

entire applied voltage is effective in varying barrier height:

$$I = A \exp[-q(V_D - \Delta E_c)/kT][\exp(qV/kT) - 1] \quad (12)$$

Above a critical forward voltage in such a diode, V_{B1} will become finite $[(V_{D1} - V_1) < \Delta E_c]$ and the current will vary exponentially with $V_2 = V/\eta$ (see Section 4.2).

Since in the n–p heterojunction, the current is limited by the rate at which holes can diffuse in the narrow-gap material,[10]

$$A = XaqN_{A2}(D_p/\tau_p)^{1/2} \quad (13a)$$

where the transmission coefficient X represents the fraction of those carriers having sufficient energy to cross the barrier which actually do so.

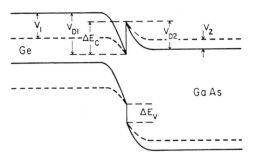

FIG. 5. Band diagram of p–n heterojunction in which no barrier exists for electrons going from Ge to GaAs (solid line) and for applied forward bias where now the barrier does exist (dashed line). The expected I–V characteristics are considerably different in the two regions.

D_p and τ_p are diffusion constant and lifetime, respectively, for holes in the narrow-gap material, and a represents junction area.

The case of the p–n heterojunction is analogous.

In the case of n–n and p–p heterojunctions, since V_{D1} and V_1 are small with respect to V_{D2} and V_2, respectively, and because the current is carried by majority carriers, we have, in analogy with the emission theory for metal–semiconductor diodes,[11]

$$A = XaqN_2\left(\frac{kT}{2\pi m^*}\right)^{1/2} \quad (13b)$$

where N_2 and m^* are, respectively, net impurity density and carrier effective mass in semiconductor 2.

The above formulae would be modified somewhat by generation–recombination[12] and "leakage" currents, by image and tunnel effects, and by interface states.

4. EXPERIMENTAL RESULTS

In this section the electrical characteristics of n–p, n–n, p–n and p–p heterojunctions are reported and interpreted with respect to the theory of Sections 2 and 3. It must be emphasized that the junctions reported here were made in two depositions. The n-type germanium in the n–p and n–n junctions is expected to be similar since this Ge (phosphorus doped) was deposited simultaneously on n- and p-type GaAs. Likewise the p-type Ge (gallium doped) in the p–n and p–p junction is expected to be similar. The p-type GaAs seeds in the n–p and p–p heterojunctions were cut from adjacent slices of a monocrystal. The same is true for the n-type GaAs seeds in the n–n and p–n junctions. Fig. 6 shows a cross-sectional view of a GaAs substrate surrounded by deposited Ge.

To fabricate a diode from such a wafer, the deposited Ge was removed from one side, and the wafer was then broken into chips. Ohmic contacts were made to both sides of the chip, and the chip was then mounted in a transistor header and etched to remove surface damage.

All heterojunctions tested showed rectification. For forward bias, the GaAs was biased negative (with respect to the Ge) for the n–n and p–n junctions and positive for n–p and p–p junctions. This is in agreement with the proposed model.

The junctions studied can be classified as being "good" diodes or "bad" diodes. The good units of each junction type all behave very nearly identically. The built-in voltages are equal and the electrical characteristics vary only slightly among units. The bad units, however, all appear to have somewhat lower built-in voltages which vary from unit to unit. Although the bad units have not been studied as intensively as the good units, it appears that if the reduced built-in voltage is taken into account, the electrical characteristics are similar to those of the good units. It is thought that the bad units contain defects at the interface which lower the barrier height. Many such bad units were transformed into good units by reducing the junction area and presumably eliminating defects. Only the good units will be discussed further.

That rectification actually occurs at the interface was determined by probing the material.[13] Fig. 7 shows a plot of thermoelectric and of photovoltaic potential versus distance from germanium surface of an *n–p* heterojunction. The germanium surface position is represented by the extreme left of either trace. The thermoelectric voltage null is indicated to be about 46·8 μ below the germanium surface. In the photovoltage plot, the position 46·8 μ from the Ge surface is as indicated. That this position indeed corresponds to the interface can be seen from Fig. 8 where the transition region is expanded to show that the junction (position of maximum slope in the photovoltage plot) is as indicated in Fig. 7. For the case of the *n–n* or *p–p* junctions, a similar method was used. Instead of the change in polarity for the thermoelectric voltage, an abrupt change in magnitude was observed at the junction.

The deposited *n*-type Ge (in the *n–p* and *n–n* junctions) was much more heavily doped than was the GaAs. The net donor concentration in the Ge was determined by resistivity measurements and was found to be about $10^{19}/cm^3$. Capacitance measurements on *n–p* and *p–p* heterojunctions indicate that the net acceptor concentration in the GaAs is constant for distance from the junction greater than $0·25\,\mu$ and is equal to $1·5 \times 10^{16}$ atoms/cm³. This is in agreement with Hall–effect data. In the *n–n* heterojunctions, capacity measurements indicate a net donor concentration in the GaAs varying as $x^{(4,7)}$ where x is the distance from the interface. At the edge of the transition region at equilibrium, the net donor concentration is about 4×10^{16} atoms/cm³. The resistivity of the *p*-type Ge (in *p–n* and *p–p* heterojunctions) was not measured. However, electrical characteristics indicate a net acceptor concentration in the neighborhood of 10^{16} atoms/cm³.

In the *n–n* and *p–p* junctions, the space charge in the Ge is composed of mobile carriers and so the voltage supported at the junction is expected to be almost entirely in the GaAs in these cases. However, since the *n*-type Ge is more heavily doped than the GaAs, and the *p*-type Ge is more lightly doped, the built-in voltage and transition region occur predominantly in the GaAs for *n–p* junctions and in the Ge for *p–n* junctions. This

can be seen for an *n–p* junction in a plot of photo voltage *vs.* position (see Fig. 8) where the position of maximum slope indicates an undetectable voltage is supported by the Ge.

4.1 *Alignment of bands at interface*

The built-in voltages at room temperature as determined from *I–V* and from *C–V* characteristics are presented (Table 1) for representative *n–n*, *n–p*, *p–p* and *p–n* heterojunctions. The

Table 1

Heterojunction	V_D	
	I–V	*C–V*
n–n	0·47 ± 0·02	0·48 ± 0·05
n–p	0·62 ± 0·02	0·85 ± 0·05
p–p	0·56 ± 0·03	0·70 ± 0·05
p–n	0·53 ± 0·03	0·55 ± 0·05

agreement between methods is good for *n–n* and *p–n* junctions but not for *n–p* or *p–p* heterojunctions.

Since similar germanium was used for *n–n* and *n–p* junctions, the model proposed predicts that the sum of the built-in voltages (V_D) for the two types of junctions plus the energy between the appropriate band edge and the Fermi level (Δ_c, Δ_v) adds up to the band gap of the GaAs. The same is true for *p–n* and *p–p* junctions. The values of Δ_v and Δ_c are calculated to be 0·19 and 0·07 eV assuming the magnitude of the hole effective mass is equal to that of a free electron (m_0) and using the published value of $0·078m_0$ for the electron effective mass.[14]

Then, with V_D obtained from *I–V* data,

$$0·62 + 0·47 + 0·19 + 0·07 = 1·35 \text{ eV}$$

for *n–p* and *n–n* junctions and

$$0·53 + 0·56 + 0·19 + 0·07 = 1·35 \text{ eV}$$

for *p–p* and *p–n* junctions, which is in good agreement with the published value of 1·36 eV for the band gap of GaAs.

The magnitude of ΔE_c and ΔE_v can be obtained only approximately from the data, because the position of the Fermi level with respect to the conduction band edge and the band gap of this degenerate germanium can only be estimated. If the density of states in the conduction band for degenerate Ge is that for non-degenerate Ge, and if the band gap of this Ge is assumed to be 0·48 eV as suggested by PANKOVE,[15] values of 0·56 and 0·32 eV are obtained for ΔE_c and ΔE_v, respectively, for degenerate n-type Ge and non-degenerate GaAs.

A calculation of the band edge discontinuities for non-degenerate p-type Ge and non-degenerate GaAs gives values of 0·15 and 0·55 eV, respectively, for ΔE_c and ΔE_v. These measurements indicate that with increased doping of germanium with phosphorus, the entire forbidden band is depressed to lower energies.

4.2 I–V characteristics

The heterojunctions studied have static I–V characteristics reasonably typical of those reported for homojunctions. The forward current varies approximately exponentially with applied voltage, and the reverse characteristics show a soft breakdown. The n–p junctions show an additional abrupt breakdown which is believed to be due to the avalanche effect. Fig. 9 shows the I–V characteristics of an n–n heterojunction at room temperature.

The I–V characteristics can generally be written as in equation (2) where the value of η indicates the deviation from ideal forward rectifier characteristics. For n–n and p–p junctions, the applied voltage is supported almost entirely by the GaAs, and as a result the factor η is expected to approach unity. The value of η is also expected to approach unity for the n–p junctions, because the Ge is so much more heavily doped than is the GaAs so that again the applied voltage is almost entirely supported by the GaAs.

However, in the p–n heterojunctions studied, the relative dopings indicate that η should equal unity for $V_{D1} - V_1 > \Delta E_c$ and approach V/V_2 for $V_{D1} - V_1 < \Delta E_c$. The plot of $\ln I$ vs. V is shown for respective junctions of these four classes in Fig. 10. The data was taken at elevated temperatures to reduce the influence of surface leakage and generation–recombination currents.

The plots of the n–n, n–p and p–p junctions can all be expressed by equation (2) where η is just slightly greater than unity as is expected. The I–V characteristic for the p–n heterojunction,

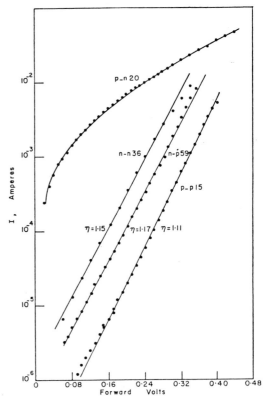

FIG. 10. Forward I–V characteristics for p–n, n–n, n–p and p–p heterojunctions. The indicated value of η is found by empirically fitting the expression $I = I_0 \exp(qV/\eta kT)$.

however, has a "sloppy" characteristic and a value of η of approximately 3·5, although it is not a constant. At 78°K the I–V characteristics of this diode are as shown in Fig. 11. There are three straight-line regions of this plot corresponding to three distinct values of η (equation 2). In region a, for applied voltage $V < 0·16$ V, $\eta = 2·1$. In the range $0·16 < V < 0·7$ V (region b), equation (2) is satisfied with $\eta = 16·7$. For $V > 0·7$ V (region c), the value of η doubles and becomes $\eta = 8·3$. These characteristics are interpreted as follows:

The barrier is decreased by the amount of the applied voltage in region a. A value of $\eta = 2·1$ results from recombination of carriers in the transition region. In regions b and c, the conduction-band edge in the Ge is lower than its peak in the GaAs (see Fig. 5). As a result, only that portion of applied voltage (V_2) appearing in the GaAs lowers the barrier. Then effectively η is increased. The halving of η at about $V = 0.7$ V results from the predominance of injected current above this voltage.

for the n–p and p–p homojunctions, and about 10^{-6} for the n–n junction. The value of X for the p–n junction was more difficult to determine. However, a value less than 10^{-3} with a value of ΔE_c in the neighborhood of 0.5 eV seems necessary to explain the experimental results.

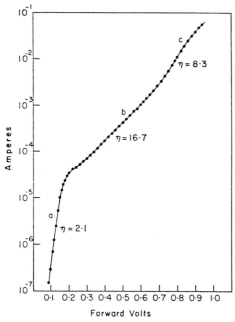

FIG. 11. Forward I–V characteristics at 78°K for p–n heterojunction having p-Ge less heavily doped than n-GaAs.

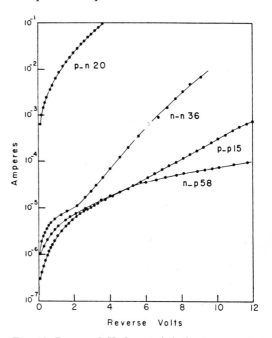

FIG. 12. Reverse I–V characteristic for p–n, n–n, p–p and n–p heterojunction at elevated temperatures. Soft breakdown is observed.

It must be pointed out that the value of ΔE_c as determined earlier would be expected to result in a change from region a to region b of the I–V characteristic at about 0.66 V, or, conversely, a value of ΔE_c of about 0.5 eV would be required to interpret the data as we have done. The reverse electrical I–V characteristics of representative n–n, n–p, p–p and p–n heterojunctions are shown in Fig. 12. Although the data was taken at elevated temperatures to minimize generation–recombination current, the reverse current does not saturate. The origin of this excess current is not known. From the magnitude of the current at a given voltage, the values of A in equations (11) and (12) can be experimentally determined. Comparing these values with equations (13a, b) give values of transmission coefficients (X) of about 10^{-3}

If the interpretation is correct, the small value of transmission coefficient X is a result of the radically different Bloch waves on either side of the interface. For the particular case of the n–n junctions discussed here, if it is assumed that all electrons in the Ge are reflected at the interface except those centered around the $K = (0, 0, 0)$ minimum, and that all these are transmitted, a value of $X = 3 \times 10^{-3}$ results. The actual transmission factor would be expected to be smaller than this because of additional reflection due to the discontinuities in band edges and in the periodicity of the potential-energy function at the interface.

4.3 Response to monochromatic radiation

A p–n heterojunction, which the electrical

characteristics suggest has a band profile as in Fig. 5, was illuminated with monochromatic radiation normally incident to the GaAs surface. The resultant photocurrent response is shown in Fig. 13 where the short-circuit photocurrent per

sufficiently low energy, the photons cannot excite electrons in the Ge and the current is again zero.

It is noticed that both the direct and the indirect absorption edges of the Ge are visible, although the data is not sufficiently accurate to

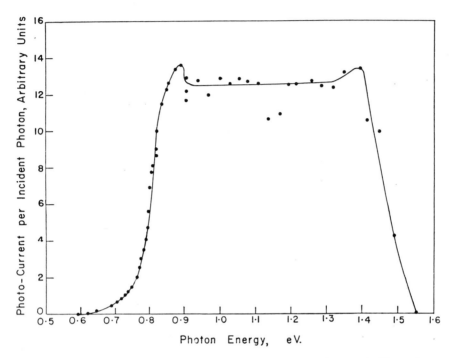

FIG. 13. Short-circuit photocurrent of a *p–n* heterojunction per incident photon vs. photon energy (see text).

incident photon is plotted against the photon energy.

The response shows a broad maximum between about 0·83 and 1·4 eV. This response may be explained as follows: the higher-energy photons are absorbed near the surface of the GaAs and do not contribute to the photocurrent. However, the GaAs is transparent to photons having energy less than that of the forbidden gap and these photons are transmitted to the interface. These photons having sufficient energy will excite carriers in the Ge and those which excite carriers in the transition region or within a carrier diffusion length of the transition region will contribute to the photocurrent. It is these photons which produce the photocurrent (see Fig. 5). At

see much "fine structure". The "flat-top" of this figure probably indicates that the incident radiation is entirely utilized in producing photo-current or else that the absorption coefficient is reasonably constant in this energy range. The decrease to zero in photocurrent in the high-energy region occurs at a value of about 1·55 eV instead of the expected value of the GaAs band gap (1·36 eV). This result is not understood.

5. HETEROJUNCTIONS AS DEVICES

The static *I–V* characteristics of the diodes studied are in general poorer than obtainable in homojunctions—principally because of the soft reverse breakdown. It is expected that this characteristic may be improved with more work.

6

An interesting effect in the pulse response is expected from certain heterojunctions. When a diode is abruptly switched from a state of forward bias to a state of reverse bias, no effects on the current due to minority carrier storage are expected. For the n–n and p–p junctions, this is because current is by majority carriers. For p–n and n–p junctions, however, minority-carrier storage exists as in homojunctions. Here, however, the discontinuity at the interface prevents the injected minority carriers from re-entering the GaAs when the diode is abruptly reverse biased.

Preliminary measurements on heterojunctions have detected no effects on the pulse response attributable to storage effects.

A heterojunction can be used as a photocell with a built-in filter, as indicated in Section 4.3. The cell is sensitive for only a narrow band of photon wavelengths.

With the GaAs as emitter, and Ge as base and collector, a wide-gap-emitter transistor seems possible. Such a transistor would be expected to have a high injection efficiency independent of impurity concentration ratio in base and collector. Attempts to construct such a transistor have not been successful.

Although the measurements reported above are on units made from two single depositions of Ge on GaAs, other depositions have been made.

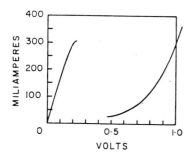

FIG. 14. I–V characteristics of a Ge–GaAs tunnel n–p heterodiode. The ordinate scale is 50 mA/div and the abscissa scale 0·1 V/div.

Degenerate n-type Ge was deposited on degenerate p-type GaAs and diodes were fabricated from this material.* The I–V plots of these units

* This work was carried out by J. C. MARINACE and F. H. DILL.

show the typical tunnel-diode characteristic (see Fig. 14). The value of V_D lies between the values obtained in Ge and in GaAs tunnel diodes and is approximately what would be expected from a consideration of the proposed band picture. Fig. 15 depicts the band picture suggested for a tunnel "heterodiode". Tunneling takes place between the Ge conduction band and the GaAs valence band as in tunnel "homodiodes".

The peak-to-valley current ratios for the tunnel heterodiodes at room temperature have been observed to be in excess of 20. Because of the magnitude of the built-in voltage V_D, the valleys are "wider" than the Ge units.

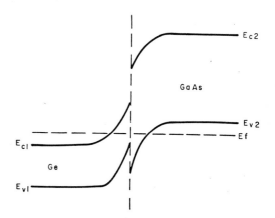

FIG. 15. Energy-band diagram of an n–p tunnel heterodiode at equilibrium.

SUMMARY

Germanium has been deposited on gallium arsenide by a process involving germanium–iodine compounds. The resultant structure is a monocrystal in which the junction between the Ge and GaAs is abrupt.

These junctions rectify. Probing of the junction region shows that the rectification occurs at the interface.

The electrical characteristics of these heterojunctions are roughly what is expected, assuming the conduction and valence band edges are discontinuous at the interface. For the case of degenerate Ge and non-degenerate GaAs, these discontinuities are approximately 0·56 and 0·32 eV respectively. The forbidden band in Ge appears to move to a region of higher energy as the doping decreases. The discontinuities for non-degenerate Ge and non-degenerate GaAs appear to be 0·15 and 0·55 eV, respectively. There is some evidence, however, which suggests that the discontinuity in conduction band is somewhat larger than this.

All the diodes tested had lower rectification ratios than have available homodiodes. However, unlike the case of homodiodes, no minority-carrier storage effects were observed for these heterojunctions upon switching from a state of forward to reverse bias.

The short-circuit current as a function of input photon energy shows the Ge absorption spectrum.

Tunnel heterodiodes have been fabricated which have *I–V* characteristics between those of Ge and GaAs tunnel homodiodes.

Acknowledgements—The author wishes to thank Miss ANNE BENORIC of the IBM Corporation, Yorktown Heights, New York, who fabricated most of the heterojunctions for his study. Gratitude is also expressed to J. C. MARINACE, M. J. O'ROURKE, J. A. SWANSON, P. J. PRICE, M. I. NATHAN and W. P. DUMKE for many helpful discussions. The author wishes also to thank J. BALTA-ELÍAS and J. F. GARCÍA DE LA BANDA who made their laboratories at the Consejo Superior de Investigaciones Cientificas (Madrid) available to him, and Professor BARCELÓ for aid in the electro-optical measurements.

REFERENCES

1. See, for example, A. VAN DER ZIEL, *Solid State Physical Electronics* Chaps. 12 and 13. Prentice-Hall, Englewood Cliffs (1957).
2. S. POGANSKI, *Halbleiterprobleme I* (Ed. by W. SCHOTTKY) p. 275. Friedr. Vieweg, Braunschweig (1954).
3. W. SHOCKLEY, *U.S. Pat.* 2,569, 347.
4. H. KRÖMER, *Proc. I.R.E.* **45**, 1535 (1957).
5. D. A. JENNY, *Proc. I.R.E.* **45**, 959 (1957).
6. R. P. RUTH, J. C. MARINACE and W. C. DUNLAP, JR., *J. Appl. Phys.* **31**, 995 (1960).
7. J. C. MARINACE, *IBM J. Res. Dev.* **4**, 248 (1960).
8. R. L. ANDERSON, *IBM J. Res. Dev.* **4**, 283 (1960).
9. A. VAN DER ZIEL, *Solid State Physical Electronics* p. 284. Prentice-Hall, Englewood Cliffs (1957).
10. W. SHOCKLEY, *Bell Syst. Tech. J.* **28**, 235 (1949).
11. H. C. TORREY and C. A. WHITMER, *Crystal Rectifiers.* McGraw-Hill, New York (1948).
12. C. SAH, R. W. NOYCE and W. SHOCKLEY, *Proc. I.R.E.* **45**, 1228 (1957).
13. R. L. ANDERSON and M. J. O'ROURKE, *An. fis. Quim.* **57**, 3 (1961).
14. W. G. SPITZER and J. M. WHELAN, *Phys. Rev.* **114**, 59 (1959).
15. J. I. PANKOVE, *Phys. Rev. Letters* **4**, 454 (1960).

Growth of Microstructures by Molecular Beam Epitaxy

A. C. GOSSARD

(*Invited Paper*)

Abstract—Molecular beam epitaxy is the most widely currently used technique for the growth of semiconductor microstructures. Multilayers with thicknesses and smoothness controlled near the monolayer level are being produced, including, recently, quantum wells with special shapes, quantum wells to which electric fields may be applied, new structures with enhanced carrier mobilities, structures for tunneling injection of carriers, and possible structures for achievement of quantum wires and dots. New crystal systems and new growth techniques are extending the range of accessible microstructures.

I. INTRODUCTION

SEMICONDUCTOR heterostructures are composed of materials which may have different bandgaps, electron affinities, and indexes of refraction, and thus can separately confine light and electrons in different regions of a crystal. The first principal application of heterostructures was in semiconductor heterostructure lasers, comprised of layers with thicknesses of the order of optical wavelengths. They were made possible by the development of liquid phase epitaxial growth techniques which could grow layers with 1000 Å thicknesses of controlled composition and doping and with good purity and crystal perfection. Subsequently, molecular beam epitaxy and metal-organic chemical vapor deposition crystal growth techniques were developed which can grow layers with finer dimensions, down to and including single monolayer thicknesses. They can also produce controlled doping profiles and good crystalline quality and are being widely used in the preparation of materials for quantized confinement of electrons and the quantum well studies described in this issue. It is the purpose of this paper to present recent developments in microstructure growth by molecular beam epitaxy, which is currently the most widely used means of preparation of quantum wells. Several books covering the subject of molecular beam epitaxy and heterostructures have recently been published [1], [2], and this author has previously reviewed the MBE growth of superlattices in thin films [3]. The field is developing rapidly, and many new structures have been produced and several new growth techniques have appeared, which are reviewed here.

II. MBE PROCESSES

Molecular beam epitaxy is an evaporation process which is carried out in an ultrahigh vacuum environment.

Manuscript received April 22, 1986.
The author is with AT&T Bell Laboratories, Murray Hill, NJ 07974.
IEEE Log Number 8609374.

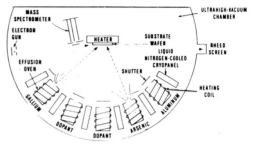

Fig. 1. Schematic illustration of molecular beam epitaxy growth apparatus (courtesy of M. B. Panish).

The evaporated constituents which control the crystal growth rate in most semiconductor MBE stick to whatever surfaces they strike. Thus, the beam flux which emanates from a molecular beam oven can usually be interrupted abruptly by insertion of a baffle or shutter into the path of the molecular beam (Fig. 1). Furthermore, the background pressure in the ultra high vacuum of an evaporating species which adheres to the surfaces of the growth apparatus and has a low vapor pressure at the temperature of the surfaces can be very low. These features make it possible to abruptly start and stop such molecular beams and their deposition on a crystal surface in molecular beam epitaxy. If the beams can be interrupted in times less than the deposition time for one monolayer, which is often of order one second, and the beam transit times, which are much less, then deposition can be controlled at monolayer thicknesses. The process is intrinsically repeatable and is suitable for computerized control.

The MBE crystal growth process is a two-step phenomenon in which the first step involves the incident atom sticking to the crystal surface and the second step involves motion on the surface to the point of incorporation into the crystal. These steps are species-dependent, temperature-dependent and crystal surface-dependent. At low temperature, motion on the surface is slow, and rough or noncrystalline growth may occur. At ideal temperatures, lateral motion occurs to atomic step edges where the atom is bound. The crystal surface may be strongly smoothed at the atomic level because growth proceeds more rapidly at the lowest points on the surface. At high temperatures, reevaporation from the surface becomes more important,

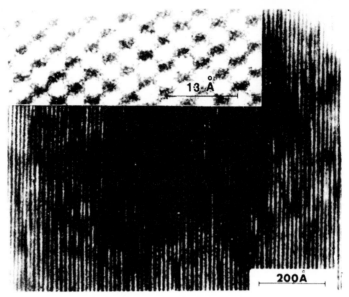

Fig. 2. Transmission electron microscope image of a cross section of a
$(GaAs)_2(AlAs)_2$ alternate bilayer deposition on a (111) face of GaAs
(courtesy of P. M. Petroff).

and surface roughening and interlayer interdiffusion may
occur. Since MBE growth is produced in an ultrahigh vacuum, the surface is available for surface analysis and electron diffraction which can shed light on these processes
and assist in the development of microstructures with good
quality interfaces.

These features of the MBE process are the most critical
elements for the growth of microstructures. Growth of a
number of these microstructures is discussed in the following section.

III. Microstructure Growth

A. Alternate Monolayer Structures

The ultimately fine microstructure is one in which the
individual layers have monolayer thickness. This is within
the capability of molecular beam epitaxy and multilayered
crystals containing thousands of such layers have been
grown. Growth of $(GaAs)_1(AlAs)_1$ alternate monolayer
depositions in which alternate monolayer composition
modulation was detected by X-ray and electron diffraction
[3] has been extended to other crystals. $(InAs)_1(GaAs)_1$
structures have been grown by MBE [4] and by MOCVD
[5]. $(AlSb)_2(GaSb)_{4.8}$ multilayers have been grown by
MBE and their structures studied by X-ray diffraction [6].
A transmission electron microscope image of a cross section of a $(GaAs)_2(AlAs)_2$ alternate bilayer deposition on a
(111) face of GaAs is shown in Fig. 2. Although the diffraction patterns show missing intensity at the alternate
single monolayer scale, these structures demonstrate the
possibility of growth of layers even thinner than needed

for electron quantum well confinement or tunneling. Evidence that some of the superlattice structures may have
energies lower than the energies of random alloys of
equivalent average composition has come from observations of spontaneous superlattice ordering under some
growth conditions [7]. The tendency to spontaneous order
may thus actually tend to stabilize abrupt interfaces.

B. Surface and Interface Smoothness

The smoothness of a crystal surface during epitaxial
growth can be qualitatively measured by high-energy reflection electron diffraction. Such measurements have recently shown that the surface smoothness after commencing or stopping growth is time dependent, and reflection
electron diffraction intensity oscillations are an active field
of study [8], [9]. Under many growth conditions, surface
roughness varies periodically upon commencing growth
with the roughness period equal to the deposition time for
one monolayer. Maximum roughness occurs at half-com-
pleted monolayers. Upon stopping growth, the surface
becomes smoother. This behavior is a result of the processes in which atoms move laterally on the growth surface to incorporate at the edges of islands or terraces [10].

The smoothness of quantum well interfaces and the uniformity of quantum well thicknesses may be gauged by
the quantum confinement energies of electrons in quantum wells. Interband quantum well exciton absorption
spectra are broadened inhomogeneously by interface
roughness and layer thickness nonuniformity. The reflection electron diffraction oscillations mentioned above have
suggested that smoother interfaces could be produced by

50

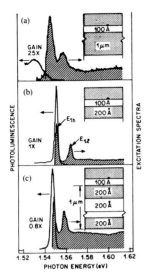

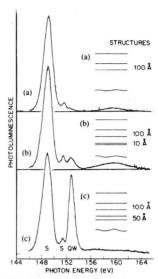

Fig. 3. Photoluminescence emission and excitation spectra of single 100 Å GaAs quantum wells grown on (a) 1 μm $Al_{0.3}Ga_{0.7}As$ layer. (b) 200 Å $Al_{0.3}Ga_{0.7}As$ layer. (c) 1 μm superlattice of alternate 200 Å GaAs and $Al_{0.3}Ga_{0.7}As$ layers. Emission is higher and more intrinsic, and excitation peaks are sharper in (b) and (c) [14].

Fig. 4. Effect of thin GaAs prelayers on photoluminescence of single GaAs 100 Å quantum wells. (a) No prelayer. (b) 10 Å GaAs prelayer separated from 100 Å GaAs quantum well by 100 Å $Al_{0.3}Ga_{0.7}As$. (c) 50 Å GaAs prelayer. Peak marked QW comes from the 100 Å quantum wells, peaks marked S come from substrate [14].

growing quantum well layers with integral number of monolayers thickness and by stopping crystal growth at each quantum well interface. Narrow emission peaks with energy splittings corresponding to monolayer fluctuations in quantum well widths have recently been seen in quantum wells grown with growth interruption [11]. But in no case have *single* lines with narrow widths been seen, so the goal of quantum well interfaces without monolayer fluctuations is not yet achieved. The narrowest widths for either continuous [12] or interrupted [11] growth can correspond to less than a monolayer width fluctuation by virtue of the fact that the exciton wavefunction averages over steps and islands smaller than the exciton diameter. Growth interruption may have deleterious effects, however, and has been reported to lead to increased incorporation of acceptors at interfaces and decreased quantum well luminescence efficiency [13].

C. Prelayers and Superlattices for Interface Improvement

Single quantum wells of GaAs grown on relatively thick (>1000 Å) (Al,Ga)As barrier layers frequently show higher impurity concentrations and rougher interfaces than wells grown on thinner barrier layers or in multiple quantum well sequences. Similarly, modulation-doped GaAs channels in which a doped, thick (Al,Ga)As barrier layer is grown before the GaAs channel often show stronger impurity effects than structures grown in the reverse order. These impurity effects and roughness are reduced by growth of a superlattice or of thin GaAs prelayers in the

(Al,Ga)As barrier layer just before the active GaAs quantum well [14] or channel [15] (Figs. 3 and 4).

The improvement is apparently produced by gettering of impurities at interfaces or cleaning and smoothing of the surface during GaAs prelayer growth. It demonstrates that multiple-layer structures may be grown with optical and electrical quality as high or higher than thicker layer structures.

D. Specially-Shaped Quantum Wells

Since thin-layer structures have strongly structure-dependent bandgaps and electron energy levels, they form a useful medium for tailoring of complex potential profiles with high purities. This has been illustrated recently by growth of pseudoparabolic-shape quantum wells in the (Al,Ga)As system. The pseudoparabolic wells were formed by alternate GaAs and $Al_{0.3}Ga_{0.7}As$ depositions in which the relative thicknesses of GaAs and $Al_{0.3}Ga_{0.7}As$ layers in each layer pair were adjusted to yield an average composition profile which is parabolic in shape [16]. The structures show sharp spectral peaks in emission and excitation with high luminescence efficiency. Their energy level spacings have the characteristic linear spacing of a parabolic potential well. The structure, energy levels and excitation spectrum are shown in Fig. 5. Related structures which consist of half parabolic potential wells have also been grown and studied and showed characteristic spectra [17]. The pulsed beam growth techniques offers essentially complete generality in the shape of quantum well potential shapes which can by synthesized.

51

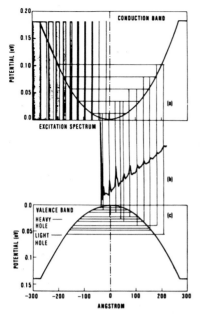

Fig. 5. Pseudoparabolic potential well bandedges, energy levels and photoluminescence excitation spectrum. The well consists of a GaAs-Al$_{0.3}$Ga$_{0.7}$As superlattice with a period of 25 Å and a quadratically varied duty cycle. Energy levels are for electrons in conduction band and heavy and light holes in valence band. The excitation spectrum was taken at 5 K for emission at the lowest intrinsic quantum well exciton transition from a sample with 10 wells separated by Al$_{0.3}$Ga$_{0.7}$As barriers of 240 Å width.

E. Structures for Application of Electric Fields

Other special quantum well structures produced recently with molecular beam epitaxy allow application of electric fields perpendicular to quantum wells [18]–[20]. They are produced by forming quantum wells in an undoped portion of the epitaxial structure, surrounded by n type and p type layers between which a reverse voltage may be applied. The n and p layers may contain wider bandgap, lower index materials which serve as window layers to transmit light or as layers to guide light. As discussed elsewhere in this issue, they are of importance in electrooptic studies and devices wherein the change in quantum well energy levels with electric field produce marked spectral changes. The structures are operable as optical modulators, lasers, and detectors and have the inherent potential for optical integration [21]. They require smooth, pure layers and low background doping in the updoped region containing the quantum wells in order to produce sharp optical transitions unscreened by background charge. Use of short-period superlattices for cleaning and smoothing the structures has proven to be useful [19], [20]. The structure of a sample used for electric-field experiments on quantum well excitons is shown in Fig. 6.

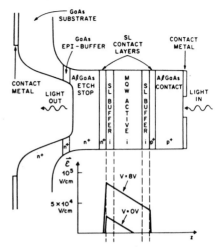

Fig. 6. Structure of multiple quantum well (MQW) sample used for optical studies under perpendicular electric field. MQW active region contains 100 GaAs quantum wells of thickness 95 Å with 98 Å Al$_{0.32}$Ga$_{0.68}$As barriers. Superlattice (SL) buffer and contact regions comprise thirty and twenty periods respectively of 29 Å GaAs layers alternating with 69 Å AlGaAs layers. The electric field distribution upon application of $V = 8$ V and 0 V for a background doping of 2×10^{15} cm^{-3} is shown at the bottom of the figure.

F. High-Carrier-Mobility Structures

Modulation doping of semiconductor microstructures is an extensively used means of obtaining high-mobility two-dimensional charge systems, and has been reviewed in [22]. A principal recent development in modulation doping has been the enormous increase in hole mobilities produced with modulated acceptor doping of Be in AlGaAs barrier layers in the GaAs quantum well system [23] (Fig. 7). Mobilities in excess of 200 000 cm^2/V · s have been reported at GaAs/AlGaAs selectively doped interfaces [24]. In addition to allowing high-speed p type and complementary transistor circuits, the high-mobility holes show the integral and fractional quantum Hall effects [25] and create the possibility for optical observation of hole heating, long minority-carrier lifetimes and high luminescence efficiencies from hole-containing layers as well as observation of electron-hole interaction phenomena not previously accessible.

G. Structures For Transport Across Layers

Molecular beam epitaxy can grow several species of structure in which the transport of charge between layers can be varied. These vertical transport structures are receiving increasing emphasis. In one type, electrons pass between layers by tunneling through thin barrier layers, often or order ten atom layers thick. When two such barriers are placed close together they form a sort of electron Fabry–Perot resonator, and resonant electron tunneling has been observed in such structures [26]. When such a penetrable array is used as a base in a transistor structure,

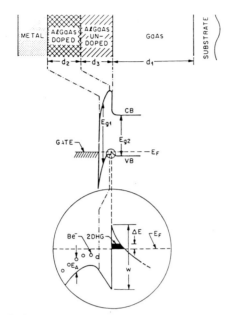

Fig. 7. Modulated acceptor doping structure for high hole mobilities. Two-dimensional hole gas is formed at GaAs-(Al,Ga)As interface by ionization of Be acceptors in doped (Al,Ga)As region [23].

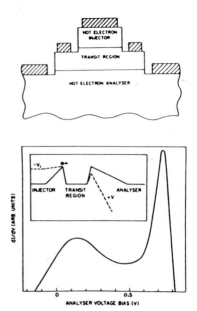

Fig. 8. Structure for hot electron spectroscopy. Injector, transit region, and base are formed by Si-doped n type GaAs layers. Barriers are formed by 100 Å p⁺ Be doped layers within undoped GaAs layers. Hot electron spectrum for a sample with a 650 Å transit region is shown at bottom. Peak near 0.1 V analyzer voltage results from nearly ballistic electron transit [31].

electrons are injected and cross the base at high temperature [27]. Superlattice structures with many penetrable barriers have been fabricated and permit study of electron systems with progressively more three-dimensional behavior, as in the case of study of the quantized Hall effect in the presence of three-dimensional coupling [28].

In a second type of vertical transport structure, the energy bandedge profiles are tailored so they may be modified by external electric fields. This is accomplished with graded potential steps produced either by use of planar impurity doping [29] or by use of compositional grading of the bandgaps [30]. It has, for example, recently been applied in structures for study of hot electron transport across thin GaAs layers [31] where one barrier serves as an injector of monoenergetic hot electrons and a second barrier is used as an energy analyzer for the electrons traversing the layers (Fig. 8).

H. Production of Fine Lateral Structures

The formation of microstructures with fine lateral resolution is only beginning to be developed. These structures will be especially interesting when dimensions small enough to produce lateral quantum confinement are achievable, providing quantum wires and, for confinement in three dimensions, quantum dots. Approaches which have been used to date are electron beam lithography, followed by reactive ion etching of epitaxial multilayer structures [32], photolithographic production of narrow ridges in multilayer structures with subsequent regrowth of an epitaxial confining cover layer [33] (Fig. 9),

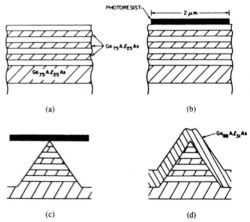

Fig. 9. Layer configuration, processing steps, and overgrowth employed in formation of quantum well wire structure [33].

and enhancement of interdiffusion in selected areas of multilayer structures by ion bombardment [34]. Approaches which have been proposed but for which experimental realization is not yet available are: 1) the induction of charge in a one-dimensional channel by epitaxial overgrowth of a doped barrier layer onto the edge of an undoped quantum well layer [35] and 2) the growth of sequential submonolayer coverages on an off-axis stepped

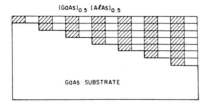

Fig. 10. Structure formed by alternate half-monolayer GaAs and AlAs depositions in terrace growth mode on an off-axis substrate [36].

substrate surface in a manner to produce monolayer-thick ribbons stacked in registry above each other to form quantum wires [36] (Fig. 10).

IV. NEW CRYSTAL SYSTEMS

Although a preponderance of microstructure growth and fabrication has used the closely matched GaAs-AlAs system on GaAs [100] oriented substrates, microstructures are also being developed in a number of other systems with less closely matched components and on other crystal faces. Modulation-doped GaAs-AlGaAs structures have recently been demonstrated on (110), (111), and a number of (N11) surface orientations with enhanced mobilities and high optical quality [37]. In these structures, the sign of the doping type for Si doping is found to reverse for some growth face crystal orientations, allowing a possible new degree of freedom in microstructure design [38]. Growth of nonlattice-matched heterostructures and strained-layer superlattices have been achieved with interesting results in III–V systems, such as GaP/InAs/GaAs [39], in group IV systems, such as Si/Ge [40], in group II–VI systems such as HgTe/CdTe [41], and with amorphous semiconductor layers [42]. Epitaxy also has been achieved for layers of compound semiconductors grown on elemental semiconductor substrates [43]. This offers interesting possibilities for the integration of GaAs and Si technologies on a single chip of material. These accomplishments with various new crystal systems are resulting in a considerable expansion of the scope of semiconductor microstructures.

V. NEW GROWTH TECHNIQUES

Growth techniques for molecular beam microstructure fabrication have also expanded recently. Of special interest are the development of gas-source MBE [44] and metalorganic MBE [45] in which gaseous compound sources replace the evaporation beam sources used in more conventional MBE. This is of particular importance in simultaneously generating beams of arsenic and phosphorus for use in precision growth of mixed arsenide and phosphide compounds such as $GaAs_{1-x}P_x$ [44] and in reducing defects generated by liquid gallium sources in GaAs MBE growth [45]. With the development of these techniques and of low pressure MOCVD, the gap between molecular beam epitaxy and MOCVD (Fig. 11) has effectively been converted to a continuum of growth methods. A further interesting development is the means of shut-

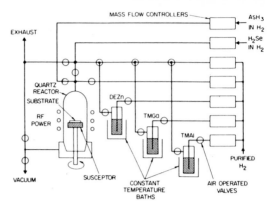

Fig. 11. Schematic illustration of metal-organic chemical vapor deposition apparatus (courtesy of R. D. Dupuis).

tling substrates between different vapor streams in the vapor deposition techniques, which also will provide enhanced capabilities for microstructure growth by those technologies.

We conclude that the capabilities of molecular beam epitaxy for growth of smooth and thin crystal layers is leading to a wide variety of new microstructures in the GaAs system. Growth is being extended to other crystal systems as ways are developed to circumvent limitations of lattice-matching and crystal-group matching. Modifications to the MBE growth technique are being made which will increase the range of accessible materials and structures.

REFERENCES

[1] "Molecular beam epitaxy and heterostructures," in NATO ASI Series, L. L. Chang and K. Ploog, Ed. Dordrecht, The Netherlands: Nijhoff, 1985.
[2] Technology and Physics of Molecular Beam Epitaxy. E. H. C. Parker, Ed. New York: Plenum, 1985.
[3] A. C. Gossard, "Molecular beam epitaxy of superlattices in thin films," Treatise on Materials Science and Technology, Vol. 24, Preparation and Properties of Thin Films, K. N. Tu and R. Rosenberg, Eds. New York: Academic, 1982, pp. 13–65.
[4] H. Ohno, R. Katsumi, T. Takama, and H. Hasegawa, "Growth of a $(GaAs)_n/(InAs)_m$ superlattice semiconductor by molecular beam epitaxy," J. Appl. Phys. (Japan), pp. L682–L684, Sept. 1985.
[5] T. Fukui and H. Saito, "$(InAs)_1(GaAs)_1$ layered crystal growth by MOCVD," J. Appl. Phys. (Japan), vol. 23, pp. L521–L523, Aug. 1984.
[6] Y. Hirayama, Y. Ohmori, and H. Okamoto, "AlSb-GaSb and AlAs-GaAs monolayer superlattices grown by molecular beam epitaxy," J. Appl. Phys., vol. 23, pp. L488–L489, July 1984.
[7] T. S. Kuan, T. F. Kuech, W. I. Wang, and E. L. Wilkie, "Long range order in Al$_x$Ga$_{1-x}$As," Phys. Rev. Lett., vol. 54, pp. 201–204, 1985.
[8] J. H. Neave, B. A. Joyce, P. J. Dobson, and N. Norton, "Dynamics of film growth of GaAs by MBE from RHEED oscillations," Appl. Phys., vol. A31, pp. 1–8, May 1983.
[9] J. M. van Hove, P. R. Pukite, and P. Cohen, "The dependence of RHEED oscillations on MBE growth parameters," J. Vac. Sci. Technol. B, vol. 3, pp. 563–567, Mar./Apr. 1985.
[10] A. Madhukar and S. V. Ghaisas, "Implications of the configuration-dependent reactive incorporation growth process for the group V pressure and substrate temperature dependence of III-V MBE growth and the dynamics of RHEED intensity," Appl. Phys. Lett., vol. 47, pp. 247–249, Aug. 1985.

[11] T. Fukunaga, K. L. I. Kobayashi, and H. Nakashima, "Reduction of well width fluctuation in AlGaAs-GaAs single quantum well by growth interruption during molecular beam epitaxy," *Surface Sci.*, in press.

[12] P. J. Pearah, J. Klem, C. K. Peng, T. Henderson, W. T. Masselink, H. Morkoc, and D. C. Reynolds, "Optical transitions and acceptor binding energies in GaAs/Al$_x$Ga$_{1-x}$As single quantum well heterostructures grown by molecular beam epitaxy," *Appl. Phys. Lett.*, vol. 47, pp. 166-168, July 1985.

[13] D. Bimberg, D. Mars, J. N. Miller, R. Bauer, and D. Oertel, "Enhanced interface incorporation of acceptors and luminescence efficiency degradation in GaAs quantum wells grown by molecular beam epitaxy using a growth interruption procedure," *J. Vac. Sci. Technol.*, in press.

[14] P. M. Petroff, R. C. Miller, A. C. Gossard, and W. Wiegmann, "Impurity trapping, interface structure, and luminescence of GaAs quantum wells grown by molecular beam epitaxy," *Appl. Phys. Lett.*, vol. 44, pp. 217-219, Jan. 1984; A. C. Gossard, W. Wiegmann, R. C. Miller, P. M. Petroff, and W. T. Tsang, "Growth of single-quantum-well structures by molecular beam epitaxy," in *Proc. MBE-CST-2*, Tokyo, Japan, 1982, pp. 39-42.

[15] W. T. Masselink, Y. L. Sun, R. Fischer, T. J. Drummond, Y. C. Chang, M. V. Klein, and H. Morkoc, "Improvement of the GaAs/AlGaAs heterointerface by the use of a graded superlattice," *J. Vac. Sci. Technol. B*, vol. B2, pp. 117-120, 1984.

[16] R. C. Miller, A. C. Gossard, D. A. Kleinman, and O. Munteanu, "Parabolic quantum wells with the GaAs-Al$_x$Ga$_{1-x}$As system," *Phys. Rev. B*, vol. 29, pp. 3740-3742, Mar. 15, 1984.

[17] R. C. Miller, A. C. Gossard, and D. A. Kleinman, "Band offsets from two special GaAs-Al$_x$Ga$_{1-x}$As quantum-well structures," *Phys. Rev. B*, vol. 32, pp. 5443-5446, Oct. 1985.

[18] F. Capasso, W. T. Tsang, A. L. Hutchinson, and G. F. Williams, "Enhancement of electron impact ionization in a superlattice: a new avalanche photodiode with a large ionization rate ratio," *Appl. Phys. Lett.*, vol. 40, pp. 38-40, 1982.

[19] D. A. B. Miller, D. S. Chemla, T. C. Damen, A. C. Gossard, W. Wiegmann, T. H. Wood, and C. A. Burrus, "Electric field dependence of optical absorption near the bandgap of quantum well structures," *Phys. Rev. B*, vol. 32, pp. 1043-1060, July 15, 1985.

[20] J. S. Weiner, D. A. B. Miller, D. S. Chemla, T. C. Damen, C. A. Burrus, T. H. Wood, A. C. Gossard, and W. Wiegmann, "Strong polarization-sensitive electroabsorption in GaAs/AlGaAs quantum well waveguides," *Appl. Phys. Lett.*, vol. 47, pp. 1148-1150, Dec. 1985.

[21] S. Tarucha and H. Okamoto, "Monolithic integration of a laser diode and an optical waveguide modulator having a GaAs/AlGaAs quantum well double heterostructure," *Appl. Phys. Lett.*, vol. 48, pp. 1-3, Jan. 1986.

[22] A. C. Gossard and A. Pinczuk, "Modulation-doped semiconductors," in *Synthetic Modulated Structures*, L. Chang and B. Giessen, Eds. New York: Academic, 1985, pp. 215-255.

[23] H. L. Stormer, A. C. Gossard, W. Wiegmann, R. Blondel, and K. Baldwin, "Temperature dependence of the mobility of two-dimensional hole systems in modulation-doped GaAs-(AlGa)As," *Appl. Phys. Lett.*, vol. 44, pp. 139-414, Jan. 1984.

[24] E. E. Méndez and W. I. Wang, "Temperature dependence of hole mobility in GaAs-Ga$_{1-x}$Al$_x$As heterojunctions," *Appl. Phys. Lett.*, vol. 46, pp. 1159-1161, June 1985.

[25] H. L. Stormer, A. Chang, D. C. Tsui, J. C. M. Hwang, A. C. Gossard, and W. Wiegmann, "Fractional quantization of the Hall effect," *Phys. Rev. Lett.*, vol. 50, pp. 1953-1956, June 1983.

[26] T. C. L. G. Sollner, H. Q. Le, C. A. Correa, and W. D. Goodhue, "Persistent photoconductivity in quantum well resonators," *Appl. Phys. Lett.*, vol. 47, pp. 36-38, July 1985.

[27] M. Heiblum, D. C. Thomas, C. M. Knoedler, and M. I. Nathan, "Tunneling hot-electron transfer amplifier: A hot-electron GaAs device with gain," *Appl. Phys. Lett.*, vol. 47, pp. 1105-1107, Nov. 1985.

[28] H. L. Stormer, J. P. Eisenstein, A. C. Gossard, W. Wiegmann, and K. Baldwin, "Quantization of the Hall effect in an anisotropic three-dimensional electronic system," *Phys. Rev. Lett.*, vol. 56, pp. 85-88, Jan. 1986.

[29] R. J. Malik, T. R. Aucoin, R. L. Ross, K. Board, C. E. C. Wood, and L. F. Eastman, "Planar-doped barriers in GaAs by molecular beam epitaxy," *Electron. Lett.*, vol. 16, pp. 836-838, Oct. 1980.

[30] A. C. Gossard, W. Brown, C. L. Allyn, and W. Weigmann, "Molecular beam epitaxial growth and electrical transport of graded barriers for nonlinear current conduction," *J. Vac. Sci. Technol.*, vol. 20, pp. 694-700, Mar. 1982.

[31] A. F. J. Levi, J. R. Hayes, P. M. Platzman, and W. Wiegmann, "Injected hot-electron transport in GaAs," *Phys. Rev. Lett.*, vol. 55, pp. 2071-2074, 1985.

[32] M. A. Reed, R. T. Bate, K. Bradshaw, W. M. Duncan, W. R. Frensley, J. W. Lee, and H. D. Shih, "Spatial quantization in GaAs-AlGaAs multiple quantum dots," *J. Vac. Sci. Technol. B*, vol. 4, pp. 358-360, Jan./Feb. 1986.

[33] P. M. Petroff, A. C. Gossard, R. A. Logan, and W. Wiegmann, "Toward quantum well wires: Fabrication and optical properties," *Appl. Phys. Lett.*, vol. 41, pp. 635-638, Oct. 1, 1982.

[34] Y. Hirayama, Y. Suzuki, S. Tarucha, and H. Okamoto, "Compositional disordering of GaAs-Al$_x$Ga$_{1-x}$As superlattice by Ga focused ion beam implantation and its application to submicron structure fabrication," *J. Appl. Phys. (Japan)*, vol. 24, pp. L516-L518, July 1985.

[35] H. Sakaki, "Scattering suppression and high-mobility effect of size-quantized electrons in ultrafine semiconductor wire structures," *J. Appl. Phys. (Japan)*, vol. 19, pp. L735-L738, Dec. 1980.

[36] P. M. Petroff, A. C. Gossard, and W. Wiegmann, "Structure of AlAs-GaAs interfaces grown on (100) vicinal surfaces by molecular beam epitaxy," *Appl. Phys. Lett.*, vol. 45, pp. 620-622, Sept. 1984.

[37] W. I. Wang, "AlGaAs/GaAs heterostructures on any surfaces," *J. Vac. Sci. Technol. B*, to appear.

[38] D. L. Miller, "Lateral p-n junction formation in GaAs molecular beam epitaxy by crystal plane dependent doping," *Appl. Phys. Lett.*, vol. 47, pp. 1309-1311, Dec. 1985.

[39] G. C. Osbourn, P. L. Gourley, I. J. Fritz, R. M. Biefeld, L. R. Dawson, and T. E. Zipperian, "Principles and applications of semiconductor strained layer superlattices," in *Semiconductors and Semimetals Vol. 22*, R. Dingle, Ed. New York: Academic, to be published.

[40] R. Hull, J. C. Bean, F. Cerdeira, A. T. Fiory, and J. M. Gibson, "Stability of a semiconductor strained-layer superlattice," *Appl. Phys. Lett.*, vol. 48, pp. 56-58, Jan. 1986.

[41] K. A. Harris, S. Hwang, D. K. Blanks, J. W. Cook, Jr., J. F. Schetzina, N. Otsuka, J. P. Baukus, and A. T. Hunter, "Characterization study of a HgTe-CdTe superlattice by means of transmission electron microscopy and infrared photoluminescence," *Appl. Phys. Lett.*, vol. 48, pp. 396-399, Feb. 10, 1986.

[42] B. Abeles, L. Yang, P. D. Persans, H. S. Stasiewski, and W. Lanford, "Infrared spectroscopy of interfaces in amorphous hydrogenated silicon/silicon nitride superlattices," *Appl. Phys. Lett.*, vol. 48, pp. 168-170, Jan. 1986.

[43] W. I. Wang, "Molecular beam epitaxial growth and material properties of GaAs and AlGaAs on Si (100)," *Appl. Phys. Lett.*, vol. 44, pp. 1149-1151, June 1984.

[44] M. B. Panish, "Gas source molecular beam epitaxy of InP, GaInAs, and GaInAsP," *Prog. Cryst. Growth Characterization*, special issue, to appear.

[45] N. Vodjdani, A. Lamarchand, and M. Paradan, "Parametric studies of GaAs growth by metalorganic molecular beam epitaxy," *J. Phys. Colloq. C5*, vol. 43, pp. 339-349, 1982.

[46] W. T. Tsang, "Chemical beam epitaxy of InP and GaAs," *Appl. Phys. Lett.*, vol. 45, pp. 1234-1236, Dec. 1, 1984.

A. C. Gossard was born in Ottawa, IL, on June 18, 1935. He received the B.A. degree from Harvard University in 1956 and the Ph.D. degree from the University of California, Berkeley in 1960.

In 1960 he joined AT&T Bell Laboratories, Murray Hill, NJ, where he used magnetic resonance techniques in the study of magnetic, metallic, and superconducting materials. He is currently engaged in research on molecular beam epitaxy for production and study of finely layered semiconductor structures.

A Bird's-Eye View on the Evolution of Semiconductor Superlattices and Quantum Wells

LEO ESAKI, FELLOW, IEEE

(Invited Paper)

Abstract—Following the past seventeen-year developmental path in the research of semiconductor superlattices and quantum wells, significant milestones are presented with emphasis on experimental investigations in the device physics of reduced dimensionality performed in cooperation with the materials science of heteroepitaxial growth.

I. INTRODUCTION

IN 1969, research on semiconductor superlattices was initiated with a proposal by Esaki and Tsu [1], [2] for a one-dimensional potential structure "engineered" with epitaxy of alternating ultrathin layers. In anticipation of advancement in technology, two types of superlattices were envisioned: doping and compositional, as shown at the top and bottom of Fig. 1, respectively.

The superlattice idea occurred to us while examining the feasibility of structural formation by epitaxy for potential barriers and wells, thin enough to exhibit resonant electron tunneling through them [3]. Such resonant tunneling arises from the interaction of electron waves with potential barriers. If the thickness of potential wells is 50 Å, the calculated bound state energies of electrons in the wells are 0.08 eV for the ground state and 0.32 eV for the first excited state from the equation, $\lambda = h(2m^*E)^{-1/2}$ where λ is the deBroglie wavelength, h is Planck's constant and the effective mass m^* is assumed to be $0.1 m_o$. Thus, the corresponding voltages required for observation of resonant tunneling fell in a desirable range. Then, we attempted the formidable task of engineering such quantum structures which warranted serious effort. The superlattice was considered a natural extension of double- and multiple-barrier structures.

In general, if characteristic dimensions such as superlattice periods and well widths are reduced to less than the electron mean free path, the entire electron system will enter a quantum regime of reduced dimensionality in the presence of nearly ideal interfaces. Our effort for the semiconductor nanostructure [4] was intended to search for novel phenomena in such a regime with precisely engineered structures.

It was theoretically shown that superlattice structures

Manuscript received April 9, 1986; revised April 15, 1986. This work was supported in part by the United States Army Research Office.
The author is with the IBM Thomas J. Watson Research Center, Yorktown Heights, NY 10598.
IEEE Log Number 8609653.

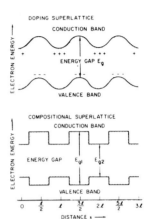

Fig. 1. Spatial variation of the conduction and valence bandedges in two types of superlattices. Doping (top) and compositional (bottom).

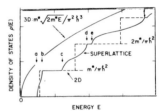

Fig. 2. Comparison of the density of states in a superlattice with those in the three-dimensional (3-D) and two-dimensional (2-D) electron systems.

possess unusual electronic properties of quasi-two-dimensional character [1], [2]. The introduction of the superlattice potential clearly perturbs the band structure of the host materials. Since the superlattice period is much longer than the original lattice constant, the Brillouin zone is divided into a series of minizones, giving rise to narrow subbands, separated by forbidden regions, analogous to the Kronig–Penney band model [5] for the conduction band or the valence band of the host crystal. Fig. 2 shows the density of states $\rho(E)$ for electrons in a superlattice in the energy range including the first three subbands: E_1 between a and b, E_2 between c and d, and E_3 between e and f (indicated by arrows in the figure). The parabolic

1612   IEEE JOURNAL OF QUANTUM ELECTRONICS, VOL. QE-22, NO. 9, SEPTEMBER 1986

curve for a three-dimensional electron system and the staircase-like density of states for a two-dimensional system are also shown for comparison.

The electron dynamics in the superlattice direction was analyzed for conduction electrons in narrow subbands of a highly perturbed energy-wave vector relationship with a simplified path integration method [6]. This calculation predicted an unusual current-voltage characteristic including a negative differential resistance, and even the occurrence of "Bloch oscillations [2]." The calculated Bloch frequency f is as high as 250 GHz from the equation $f = eFd/h$, for an applied field F and a superlattice period d of 10^3 V/cm and 100 Å, respectively.

In 1970, Esaki, Chang, and Tsu [7] reported an experimental result on a GaAs-GaAsP superlattice with a period of 200 Å synthesized with CVD (chemical vapor deposition) by Blakeslee and Aliotta [8]. Although transport measurements failed to show any predicted effect, this system probably constitutes the first strained-layer superlattice [9] having a relatively large lattice mismatch (1.8 percent) between GaAs and GaAs$_{0.5}$P$_{0.5}$. Early efforts for epitaxial growth of Ge$_{1-x}$Si$_x$ as well as Cd$_{1-x}$Hg$_x$Te superlattices, in our group were soon abandoned because of rather serious technological problems. However, the recent successful growth of such superlattices has received much attention because of their attractive properties.

In 1972, Esaki et al. [10] found that a MBE (molecular beam epitaxy)-grown GaAs-GaAlAs superlattice exhibited a negative resistance in its transport properties, which was, for the first time, interpreted in terms of the above-mentioned superlattice effect.

Our early efforts focussed on transport measurements. Nevertheless, Tsu and Esaki [11] calculated nonlinear respone of conduction electrons in a superlattice medium, leading to optical nonlinearity. It is worthwhile mentioning here that, in 1974, Gnutzmann and Clauseker [12] pointed out an interesting possibility; namely, the occurrence of a direct-gap superlattice made of indirect-gap host materials because of Brillouin-zone folding as a result of the introduction of the new superlattice periodicity. The idea suggests the synthesis of new optical materials.

The field of semiconductor superlattices and quantum wells in the interdisciplinary environment has proliferated widely since the first proposal and early experiments [13]. That makes a coherent, comprehensive review next to impossible. Here, I shall attempt to tread the developmental path and then survey achievements selected with emphasis on recent experimental studies in the physics of reduced dimensionality and the materials science of heteroepitaxial growth.

II. Epitaxy and Superlattice Growth

Heteroepitaxy is of fundamental interest for the superlattice growth. Steady improvements in growth techniques such as MBE [14] or MOCVD (metalorganic chemical vapor deposition) during the last decade have made possible high-quality heterostructures having de-

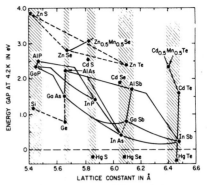

Fig. 3. Plot of energy gaps at 4.2 K versus lattice constants.

signed potential profiles and impurity distributions with dimensional control close to interatomic spacing and with virtually defect-free interfaces, particularly, in a lattice-matched case such as GaAs-Ga$_{1-x}$Al$_x$As. This great precision has cleared access to a quantum regime. The dynamics of the MBE growth process have been studied in detail [15]. The semiconductor superlattice structures have been grown with III-V, II-VI and IV-VI compounds, and elemental semiconductors as well as amorphous materials. In addition to MBE and MOCVD, new or unconventional techniques such as GS (gas source) MBE [16], LP (low pressure) MOCVD [17], CBE (chemical beam epitaxy) [18], HWE (hot wall epitaxy) [19], and ALE (atomic layer epitaxy) [20] have been explored for this purpose.

Fig. 3 shows the plot of energy gaps at 4.2 K versus lattice constants [21] for zinc-blende semiconductors together with Si and Ge. Joining lines represents ternary alloys except for Si-Ge, GaAs-Ge and InAs-GaSb. MnSe and MnTe are not shown here because their stable crystal structures are not zinc-blende. Superlattices and quantum wells or heterojunctions grown with pairs selected from those materials, include InAs-GaSb(-AlSb), InAlAs-InGaAs [22], InP-lattice matched alloys [23]-[26], Ge-GaAs [27], [28], CdTe-HgTe [29]-[31], PbTe-PbSnTe [32], [33], ZnS-ZnSe [19], [34], and ZnSe-ZnTe [35]. The introduction of II-VI compounds apparently extended the available range of energy gaps in both the high and the low direction: that of ZnS is as high as 3.8 eV and all the Hg compounds have a negative energy gap or can be called zero-gap semiconductors. The magnetic compounds, CdMnTe [36], [37] and ZnMnSe [38], are newcomers in the superlattice arena.

In Fig. 3, it should be noted that the energy gap generally decreases with an increase in the lattice constant or the atomic number [39], and also, that all the binary compounds fall into five distinct columns shown by the shaded areas, suggesting that the lattice constants are alike as long as the mean atomic-numbers of the binary constituents are the same. Following this rule, for instance, CdSe, InAs, GaSb, and ZnTe belong to the same column. AlSb and

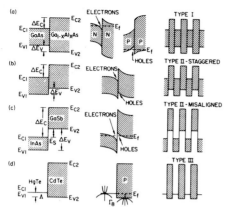

Fig. 4. Discontinuities of bandedges energies at four kinds of hetero-interfaces: band offsets (left), band bending and carrier confinement (middle), and superlattices (right).

HgSe are also added to this column since Ga and Cd can be substituted by Al and Hg, respectively, with no appreciable change in the lattice constant due to the fact that Ga and Al and also Cd and Hg have nearly equal values of tetrahedral radii.

Semiconductor hetero-interfaces exhibit an abrupt discontinuity in the local band structure, usually associated with a gradual band bending in its neighborhood which reflects space-charge effects. According to the character of such discontinuity, known hetero-interfaces can be classified into four kinds: Type I, Type II-staggered, Type II-misaligned, and Type III, as illustrated in Fig. 4(a)–(d): band offsets (left), band bending and carrier confinement (middle), and superlattices (right). The conduction band discontinuity ΔE_c is equal to the difference in the electron affinities of the two semiconductor. Case (a), called Type I, applies to the GaAs-AlAs, GaSb-AlSb, GaAs-GaP systems, etc., where their energy difference $\Delta E_g = \Delta E_c + \Delta E_v$. On the other hand, Cases (b) and (c), Type II, apply to pairs of InAs-GaSb, $(InAs)_{1-x}(GaAs)_x$-$(GaSb)_{1-y}(GaAs)_y$ [40], InP-$Al_{0.48}In_{0.52}As$ [26], etc., where their energy-gap difference $\Delta E_g = |\Delta E_c - \Delta E_v|$ and electrons and holes are confined in the different semiconductors at their heterojunctions and superlattices. Particularly, in Case (c), Type II-"misaligned," the top of the valence band in GaSb is located above the bottom of the conduction band in InAs by the amount of E_s, differing from Case (b), Type II-"staggered," as shown in Fig. 4. Type III in Case (d) is exemplified by HgTe-CdTe where one constituent is semimetallic. This type of superlattice can not be formed with III–V compounds.

The bandedge discontinuities at the hetero-interfaces obviously command all properties of quantum wells and superlattices, and thus constitute the most relevant parameters for device design [41]. Recently, considerable efforts have been made to understand the electronic structure at interfaces of heterojunctions [42]–[45]. Even in an ideal

situation, calculation of the discontinuity is a formidable theoretical task: propagating and evanescent Bloch waves should be matched across the interface, satisfying continuity conditions on the envelope wave functions [46], [47]. For the fundamental parameters ΔE_c and ΔE_v the predictive qualities of most of the theoretical models are not satisfactory and accurate experimental determination requires great care. In this regard, the GaAs-GaAlAs system is most extensively investigated with both spectroscopic and electrical measurements [48]. Recent experiments have revised early established values of $\Delta E_c/\Delta E_g$, 85 percent, and $\Delta E_v/\Delta E_g$, 15 percent, [49], [50] to about 60 percent and 40 percent, respectively [51]–[55]. The AlAs-AlGaAs system is quite unlike the above: Dawson et al. [56] recently presented optical evidence that the band alignment in AlAs-$Al_{0.37}Ga_{0.63}As$ quantum wells is indeed Type II-staggered, as shown in Fig. 4(b), because of the crossover between the direct Γ and indirect X minima.

Later, I shall mention studies of other superlattice types different from lattice-matched compositional superlattices, which include doping, amorphous, and strained-layer superlattices, and other structures.

III. RESONANT TUNNELING AND QUANTUM WELLS

Our superlattice concept arrived while seeking resonant tunneling. In 1973, Tsu and Esaki [57] computed the resonant transmission coefficient $T*T$ as a function of electron energy for double, triple, and quintuple barrier structures from the tunneling point of view, as shown in Fig. 5, leading to the derivation of the current-voltage characteristics. Note that the resonant energies for the triple-barrier case consist of a doublet, and those for the quintuple barrier are a quadruplet. In the double-well case, each single-well bound state is split into a symmetric combination and an asymmetric one. The superlattice band model previously presented, assumed an infinite periodic structure, whereas, in reality, not only a finite number of periods is prepared with alternating epitaxy, but also the electron mean free path is limited. Thus, this multibarrier tunneling model provided useful insight into the transport mechanism and laid the foundation for the following experiment.

In early 1974, Chang, Esaki and Tsu [58] observed resonant tunneling in double barriers, and subsequently, Esaki and Change [59] measured quantum transport properties for a superlattice having a tight-binding potential. The current and conductance versus voltage curves for a double-barrier with a well of 50 Å and two barriers of 80 Å made of $Ga_{0.3}Al_{0.7}As$ are shown in Fig. 6. The schematic energy diagram is shown in the inset where the two bound states, E_1 and E_2, are indicated. Resonance is achieved at such applied voltages as to align the Fermi level of the electrode with the bound states, as shown in Cases (a) and (c). The energies of the bound states can be obtained from such resonant curves: half of the voltages at the current peaks correspond to the bound energies. The measured values were in good agreement with the calcu-

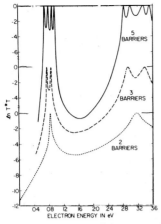

Fig. 5. Plot of ln T^*T (transmission coefficient) versus electron energy showing peaks at the energies of the bound states in the quantum wells. The curves labeled "2 barriers," "3 barriers," and "5 barriers" correspond to one, two, and four wells, respectively.

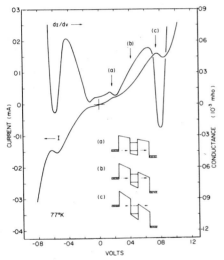

Fig. 6. Current-voltage and conductance-voltage characteristics of a double-barrier structure. Conditions at resonance (a), (c), and off-resonance (b), are indicated by arrows.

lated E_1 and E_2. This experiment, together with the quantum transport measurement [59], probably constitutes the first observation of man-made bound states in both single- and multiple-potential wells.

The technological advance in MBE for the last decade resulted in dramatically-improved resonant-tunneling characteristics [60] which renewed interest in such structures, possibly for applications. Recent reports are as follows: quantum well oscillators at frequencies up to 18 GHz [61]; room-temperature negative resistance [62]; persistent photocarriers in quantum well resonators [63]; calculations of two-body effects in tunneling currents [64];

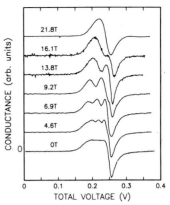

Fig. 7. Conductance versus applied voltage for various magnetic fields at 0.55 K. No difference was found at 4.2 K.

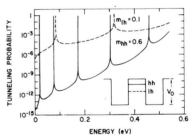

Fig. 8. Tunneling probability versus energy of a particle incident on a double-barrier structure. The calculation was done for a 50 Å –50 Å –50 Å structure with 0.55 eV potential barrier, for two different masses: 0.6 m_0 (continuous line) and 0.1 m_0 (discontinuous line). The inset shows the valence-band alignment for a AlAs-GaAs-AlAs heterostructure with the ground light-hole and heavy states sketched in the GaAs quantum states.

and a proposal for a three-terminal resonant-tunneling structure [65]. Capasso et al. [66] observed sequential resonant tunneling through a multiquantum well superlattice. Méndez et al. [67] reported resonance magnetotunneling up to 22 T at low temperatures where electron tunneling through Landau levels manifests itself as a periodic modulation of the conductance-voltage characteristics, as shown in Fig. 7, with the period proportional to the electron cyclotron energy. The observation of resonant tunneling in p type double barrier structures [68] revealed fine structure corresponding to each bound state of both heavy and light holes, confirming, in principle, the calculated tunneling probability shown in Fig. 8. In a specific configuration, Davies et al. [69] measured tunneling between coupled superlattices.

IV. OPTICAL ABSORPTION, PHOTOCURRENT SPECTROSCOPY, PHOTOLUMINESCENCE, AND STIMULATED EMISSION

Optical investigation on the man-made structures during the last decade has revealed the salient features of quantum confinement. Dingle et al. [49], [50] observed pronounced structure in the optical absorption spectrum, representing bound states in isolated [49] and double quantum wells [50]. For the former, GaAs well widths in the range between 70 Å and 500 Å were grown by MBE. The GaAs wells were separated by $Ga_{1-x}Al_xAs$ barriers which were normally thicker than 250 Å. In low-temperature measurements for such structures, several exciton peaks, associated with different bound-electron and bound-hole states, were resolved. For the latter study, a series of structures, with GaAs well widths in the range between 50 Å and 200 Å and $Ga_{1-x}Al_xAs(0.19 < x < 0.27)$ barrier widths between 12 Å and 18 Å, were grown on GaAs substrates. The spectra at low temperatures clearly indicate the evolution of resonantly split, discrete states into the lowest subband of a superlattice. From analysis of such spectra, the electron and hole well depths were mistakenly determined to be 85 percent and 15 percent of the total energy-gap difference, respectively. As mentioned before, recent photoluminescence measurements on both parabolic and square quantum wells by Miller *et al.* [51], [52] revised the band offsets in the GaAs-GaAlAs system.

Tsu *et al.* [70] made photocurrent measurements on GaAs-GaAlAs superlattices subject to an electric field perpendicular to the well plane via a semitransparent Schottky contact. It was confirmed that a series of peaks in the photocurrent spectrum correspond to transitions between quantum states in the valence and conduction bands. The photocurrent-voltage curve exhibited a pronounced negative resistance when the energy difference between the adjacent wells exceeded the superlattice bandwidth.

van der Ziel *et al.* [71] observed optically pumped laser oscillation from GaAs-GaAlAs quantum-well structures at 15 K. In 1978, Dupuis *et al.* [72] and Holonyak *et al.* [73] succeeded in room-temperature operation of quantum-well GaAlAs-GaAs laser diodes with a well width of 200 Å prepared by MOCVD. Recently, Tsang [74] succeeded in attaining a threshold current density Jth as low as 250 A/cm^2 in MBE-grown GaAlAs-GaAs laser diodes with a multiquantum-well structure. This was achieved as a result of utilizing the beneficial effects arising from the two-dimensional density of states of the confined carriers (Fig. 2). It is generally observed that, in multiquantum-well lasers, the beam width in the direction perpendicular to the junction plane, and also the temperature dependence of Jth, are significantly reduced in comparison with the regular double-heterostructure lasers. More recently, GaInAs-InP [75] and InGaAsP-InP [76] multiquantum-well laser diodes operating at 1.53 μm and 1.3 μm, respectively, were reported.

In undoped high-quality GaAs-GaAlAs quantum wells grown either by MBE [77], [78] or by MOCVD [79], the main photoluminescence peak was attributed to the excitonic transition between two-dimensional electrons and heavy holes. In 1982, Méndez *et al.* [80] studied the field-induced effect on the photoluminescence in such quantum wells: The electric field, for the first time in luminescence

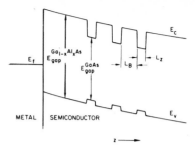

Fig. 9. Schematic energy-band diagram of a multiquantum-well Schottky junction.

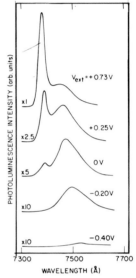

Fig. 10. Photoluminescence intensity for various applied voltages versus emission wavelength for GaAs quantum wells with $L_z = 35$ Å. The spectra were taken wih 1.96 eV excitation and 0.08 W/cm^2 power density.

measurements, was applied perpendicular to the well plane with the use of a Schottky-barrier configuration, as shown in Fig. 9 where the widths of L_B and L_z are 100 Å and 20 ~ 35 Å, respectively. Pronounced field-effects were discovered, as shown in Fig. 10: The spectra indicate two peaks associated, respectively, with exciton and impurity-related recombination; with increasing field, the peak position shifts to lower energies and the intensity decreases, with the excitons structure decreasing at a much faster rate, and becoming completely quenched at a field of a few tens of kV/cm. The results were interpreted as being caused by induced separation of confined carriers and modification of the quantum states. Miller and Gossard [81] studied similar field-effect in Be-doped quantum wells. Picosecond luminescence studies [82], [83] of quantum wells in such field-induced regime were reported.

Chemla et al. [84] observed a large shift of the excitonic absorption peak by applied electric fields, even at room temperature; Miller et al. [85] attempted to explain the phenomenon that the exciton resonances remain resolved for shifts much larger than the zero-field binding energy. Using such shifts of the absorption edge, an optical modulator [86] and a bistable optoelectronic device [87] were demonstrated with quantum wells in a p-i-n diode structure. DC photocurrent spectroscopy and the dynamics of photo-excited carriers were studied with applied fields by Polland et al. [88], Collins et al. [89], Matsumoto et al. [90], and Viña et al. [91], Alibert et al. [92] made electroreflectance measurements of field-induced energy-level shifts in quantum wells. The field-induced effects were investigated theoretically by several authors [93], [94]. In a different context, Capasso et al. [95] observed a large photocurrent gain in a forward-biased superlattice p-n junction.

V. RAMAN SCATTERING

Manuel et al. [96] reported the observation of enhancement in the Raman cross section for photon energies near electronic resonance in GaAs-Ga$_{1-x}$Al$_x$As superlattices of a variety of configurations. Both the energy positions and the general shape of the resonant curves agree with those derived theoretically, based on the two-dimensionality of the quantum states in such superlattices. Later, however, the significance of resonant inelastic light scattering as a spectroscopic tool was pointed out by Burstein et al. [97], claiming that the method yields separate spectra of single particle and collective excitations which will lead to the determination of electronic energy levels in quantum wells as well as Coulomb interactions. Subsequently, Abstreiter et al. [98] and Pinczuk et al. [99] observed light scattering by intersubband single particle excitations between discrete energy levels of two-dimensional carriers in GaAs-Ga$_{1-x}$Al$_x$As quantum wells. The technique also provided information on the dispersion of collective intrasubband as well as intersubband excitations [100] in such structures.

Meanwhile, Colvard et al. [101] reported the observation of Raman scattering from folded acoustic longitudinal phonons in a GaAs(13.6 Å)-AlAs(11.4 Å) superlattice. The superlattice periodicity leads to Brillouin zone folding (as previously mentioned), resulting in the appearance of gaps in the phonon spectrum for wave vectors satisfying the Bragg condition. Prior to this observation, Narayanamurti et al. [102] showed selective transmission of high-frequency phonons due to narrow band reflection determined by the superlattice period. Recently, Jusserand et al. [103] reported the folded acoustical zone-center gaps by Raman scattering measurements and their analysis. Furthermore, Raman scattering revealed confined optical phonons [104]-[106], interface vibrational modes [107], as well as resonant impurity states [108]-[110].

VI. MODULATION DOPING

It is usually the case that free carriers, electrons and holes created in a semiconductor by impurities inevitably

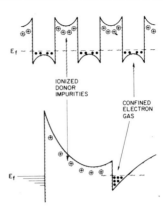

Fig. 11. Modulation doping for a superlattice (top) and a heterostructure with Schottky junction (bottom).

suffer from impurity scattering. There are a few exceptions, i.e., Si MOSFET's (metal–oxide–semiconductor field-effect transistor) where electrons or holes are induced by applied gate voltages. InAs-GaSb heterostructures are another example where electrons and holes are produced solely by electron transfer, as described later.

Now, in superlattices, it is possible to spatially separate free carriers and their parent impurity atoms by doping impurities in the region of the potential hills. Though this concept was expressed in the original article, [1] Dingle et al. [111] successfully implemented such a concept in modulation-doped GaAs-GaAlAs superlattices, as illustrated at the top of Fig. 11, achieving electron mobilities far exceeding the Brooks-Herring predictions. Modulation doping was performed by synchronizing the silicon (n-dopant) and aluminum fluxes in the MBE, so that the dopant was distributed only in the GaAlAs layers and was absent from the GaAs layers. Soon after, Störmer et al. [112] reported a high-mobility two-dimensional electron gas in modulation-doped GaAs-GaAlAs heterostructures. These heterostructures were used to fabricate a new high-speed field-effect transistor [113], [114] called MODFET (modulation-doped field-effect transistor) (its band energy diagram is shown at the bottom of Fig. 11). The device, if operated at 77 K, exhibited a performance three times faster than that of the conventional GaAs MESFET (metal–semiconductor field-effect transistor). Hall mobilities in the dark at 4.2 K for confined electrons in high-quality heterostructures exceeded 1 000 000 cm^2/V · s, [115]-[117] where a low-temperature persistent photoconductive effect was noticed [118], [119]. Such persistent photoconductivity was also reported in InGaAs-InP heterostructures [120]. Recently, an experiment of field-induced mobility modulation for a two-dimensional electron gas was reported [121].

Subsequently, a similar technique was used to form a two-dimensional hole gas at hetero-interface [122], resulting in p channel MODFET's [123]. Such a hole gas, however, was found to be involved in the complexity of band mixing as well as the effect of inversion symmetry [124]-[126]. Wang et al. [54] achieved a high-quality p channel and deduced a valence-band offset of 210 ± 30 meV for $Ga_{0.5}Al_{0.5}As$ − GaAs heterojunctions, corresponding to $\Delta E_c/\Delta E_g = 0.62 \pm 0.05$. More recently, high hole mobilities [127] at low temperatures were reported, reaching a value of $380\,000$ $cm^2/V \cdot s$ at 0.4 K. Theoretical calculations were made on the hole subbands [128], the effective masses [129], and band mixing [130].

VII. QUANTIZED HALL EFFECT AND FRACTIONAL FILLING

In 1980, Klitzing et al. [131], [132] demonstrated the interesting proposition that quantized Hall resistance could be used for precision determination of the fine structure constant α, using two-dimensional electrons in the inversion layer of a Si MOSFET. Subsequently, Tsui and Gossard [133] found modulation-doped GaAs-GaAlAs heterostructures desirable for this purpose, primarily because of their high electron mobilities, which led to the determination of α with great accuracy; i.e., $\alpha^{-1} = 137.035965(12)(0.089$ ppm$)$ [134].

The quantized Hall effect in a two-dimensional electron or hole [123] system is observable at such high magnetic fields and low temperatures as to locate the Fermi level in the localized states between the extended states. Under these conditions, the parallel component of resistance ρ_{xx} vanishes and the Hall resistance ρ_{xy} goes through plateaus. This surprising result can be understood by the argument that the localized states do not take part in quantum transport [135]. At the plateaus, the Hall resistance is given by $\rho_{xy} = h/e^2\nu = \mu_0 c/2\nu\alpha \approx 25,813\ \Omega/\nu$ where ν is the number of filled Landau levels; h Planck's constant; e the electronic charge; μ_0 the vacuum permeability; and c the speed of light in vacuum. Recently, the quantized Hall effect was also observed in a superlattice [136] which is not purely two-dimensional, as illustrated in Fig. 2.

In 1982, Tsui, Störmer, and Gossard [137] discovered a striking phenomenon: The existence of an anomalous quantized Hall effect, a Hall plateau in ρ_{xy}, and a dip in ρ_{xx}, at a fractional filling factor of $\frac{1}{3}$ in the extreme quantum limit at temperatures lower than 4.2 K. This discovery has spurred a large number of experimental and theoretical studies. Laughlin [138] explained such fractional filling by presenting variational ground- and excited-state wave functions to describe the condensation of a two-dimensional electron gas into a new state of matter, an incompressible quantum fluid. The elementary excitations of this quantum fluid are fractionally charged, and this elegant theory predicts a series of ground states characterized by the variational parameter m ($m = 3, 5 \cdots$), decreasing in density and terminating in a Wigner crystal [139]. Mendez et al. [140], [141] attempted to explore the extreme quantum limit with magnetotransport measurements up to 28 T for a dilute two-dimensional electron gas. Activation energies in the fractional quantum Hall effect [142], [143] were measured, being generally smaller than theoretical predictions [144]. The fractional quantum Hall effect was also found in a two-dimensional hole system [145].

To my knowledge, two experiments: AC conductance measurements [146] and magnetocapacitance [147] for GaAs-GaAlAs structures, claimed the observation of fractional quantization which does not depend on the measurement of the Hall effect in a two-dimensional carrier system.

VIII. InAs-GaSb SYSTEM

In 1977, while searching for a superlattice where the introduction of the periodic potential provides a greater modification to the host bandstructure than that in the GaAs-AlAs system, the InAs-GaSb system was the candidate selected because of its extraordinary bandedge relationship at the interface, called Type II-"misaligned" in Fig. 4(c). It was observed that, in the study of $(InAs)_{1-x}(GaAs)_x$-$(GaSb)_{1-y}(GaAs)_y$ p-n heterojunctions, [40] the rectifying characteristic changes to nonrectification as both x and y approach zero, implying the change-over from the "staggered" heterojunction to the "misaligned" one. Such unusual nonrectifying p-n junctions are the direct consequence of "interpenetration" between the GaSb valence band and the InAs conduction band. At the heterointerface, electrons which "flood" from the GaSb valence band to the InAs conduction band, leaving holes behind, produce a dipole layer consisting of two-dimensional electron and hole gases, as shown in the center of Fig. 4(c).

First, a one-dimensional calculation [148] and, subsequently, the LCAO band calculation [149] for InAs-GaSb superlattices were performed, indicating a strong dependence of the subband structure on the period. The semiconducting energy gap decreases when increasing the period, becoming zero at 170 Å, corresponding to a semiconductor-to-semimetal transition. In these calculations, the misaligned magnitude E_s [seen in Fig. 4(c)], was set at 0.15 eV; a value which had been derived from analysis of optical absorption [150]. Recently, Altarelli [151] performed self-consistent electronic structure calculations in the envelope-function approximation with a three-band $k \cdot p$ formalism for this superlattice.

The electron concentration in superlattices was measured as a function of InAs layer thickness [152]; it exhibited a sudden increase of an order-of-magnitude in the neighborhood of 100 Å. Such increase indicates the onset of electron transfer from GaSb to InAs which is in good agreement with theoretical prediction. Far-infrared magneto-absorption experiments [153], [154] were performed at 1.6 K for semimetallic superlattices which confirmed their negative energy-gap.

MBE-grown GaSb-InAs-GaSb quantum wells have been investigated where the unique bandedge relationship allows the coexistence of electrons and holes across the two interfaces, as shown in Fig. 12. Prior to experimental studies, Bastard et al. [155] performed self-consistent calculations for the electronic properties of such quantum wells, predicting the existence of a semiconductor-to-semimetal transition as a result of electron transfer from GaSb at the threshold thickness of InAs; this is somewhat

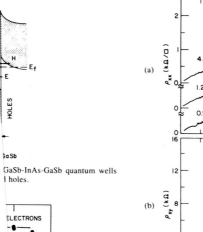

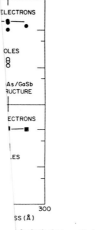

GaSb-InAs-GaSb quantum wells
holes.

for both electrons (in InAs)
well width at 4.2 K.

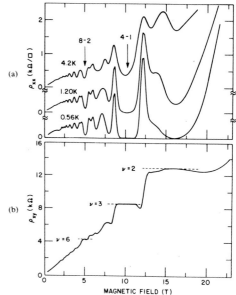

Fig. 14. (a) Magnetoresistivity at three representative temperatures. (b) Hall resistivity at 0.56 K, versus magnetic field.

s-GaSb superlattices.

termined by recent experiments carried out by Munekata *et al.* [156], as shown in Fig. 13: the threshold thickness was found to be 60 Å. In quantum wells when their InAs layer thickness exceeds 100 Å, the electron and hole mobilities are $1 \sim 3 \times 10^5$ cm²/V · s and $1 \sim 2 \times 10^4$ cm²/V · s and their corresponding densities are $\sim 10^{12}$ cm⁻² and 3×10^{11} cm⁻², respectively, at 4.2 K. Those mobilities are the highest ever reported for InAs and GaSb.

Since the electron and hole densities are not the same, probably because of the existence of some extrinsic electronic states, magnetotransport measurements [157], [158] show rather complex structure. Recent analysis [159] for such a two-dimensional electron-hole gas elucidated, for the first time, the fact that the quantum Hall effect is determined by the degree of uncompensation of the system. Fig. 14 shows (a) magnetoresistivity at three temperatures, and (b) Hall resistivity at 0.56 K, versus magnetic field. Although Shubnikov–de Haas oscillations due to electrons are shown at low fields in Fig. 14(a), the emphasis is on high fields, when both electrons and holes are in the quantum regime. The arrows indicate the fields at which the Fermi level is simultaneously between electron and hole magnetic levels and the labels above them give the corresponding filling factors. The broken lines in (b) indicate the theoretical values $h/\nu e^2$, for $\nu = 2, 3, 6$; the filling factor, determined by the quantum Hall effect, represents the "difference" between the electron and hole filling factors.

IX. OTHER SUPERLATTICES

The periodic structure called "n-i-p-i," an outgrowth of a doping superlattice in the original proposal [1], [2] was pursued by Döhler [160] and Ploog *et al.* [161]. As shown in Fig. 1, the periodic rise and fall of the bandedges is caused by a periodic variation of impurity doping. If this superlattice is illuminated, extra electrons and holes are attracted to minima in the conduction band and to maxima in the valence band, respectively. Thus, those extra carriers are spatially separated, resulting in anomalously long lifetimes. An interesting consequence of this fact is that the amplitude of the periodic potential is reduced by the extra carriers, leading to a crystal which has a variable energy gap [162]. Recently, Vojak *et al.* [163] attributed photopumped laser emission at low energies to donor-to-acceptor transitions that occur after a GaAs doping superlattice is excited to a flat-band condition. Schubert *et al.* [164] reported a new doping-superlattice injection laser. Doping superlattices were also grown with InP [165] and PbTe [166].

Amorphous materials are clearly not suitable for observation of electron confinement and other superlattice effects. Nevertheless, Abeles and Tiedje [167] pioneered the development of amorphous superlattices with thin layers of hydrogenated silicon, germanium, silicon nitride, and silicon carbide. In these structures, Tiedje *et al.* [168] recently found the enhancement of photoluminescence when the layer thickness is reduced. Santos *et al.* [169] observed folded-zone acoustical phonons by Raman scattering in amorphous superlattices.

Some degree of the lattice-mismatch, however small, at hetero-interfaces is inevitable because of the joining of two different semiconductors. It is certainly desirable to select a pair of materials closely lattice-matched in order to obtain defect- and stress-free interfaces. However, heterostructures lattice-mismatched to some extent, 1 or 2 percent, can be grown with essentially no misfit dilocations, if the layers are sufficiently thin because the mismatch is accommodated by uniform lattice strain [170]. On the basis of such premise, Osbourn [171] and his co-workers [172] prepared strained-layer superlattices from lattice-mismatched pairs, claiming their relatively high-quality suitable for some applications. Recent activities include the observation of the reversal of the heavy- and light-hole bands due to the strain effect [173], Raman scattering [174] and free-exciton luminescence [175] in GaSb-AlSb superlattices; quantum size effects [176] and lasing transitions [177] in GaAs-GaAsP; injection lasers [178] and optical investigation on energy-band configurations [179] in GaAs-GaInAs.

Kasper *et al.* [180] pioneered the MBE growth of Si-SiGe strained-layer superlattices, and Manasevit *et al.* [181] observed unusual mobility enhancement in such structures. The recent growth of high-quality Si-SiGe superlattices [182] attracted much interest in view of possible applications as well as scientific investigations. A large number of recent reports on such superlattices include Raman spectroscopy [183] for determination of built-in deformation, modulation doping [184], band alignments [185], confined electronic states [186], etc. There exist technological problems inherent to strained layers, i.e., critical layer thickness for degradation [187], stability [188], [189], and thermal relaxation [190].

Dilute-magnetic superlattices [191], such as CdTe-CdMnTe and ZnSe-ZnMnSe are recent additions to the superlattice family, and have already exhibited promising magneto-optical properties.

In 1976, Gossard *et al.* [192] achieved epitaxial structures with alternate atomic-layer composition modulation by MBE; these structures were characterized by transmission electron microscope as well as optical measurements. Recently, short-period superlattices of binary compounds were used for barrier of cladding layers, claiming the improved quality of quantum wells [175], [193], [194]. Ishibashi *et al.* [195] studied Raman scattering in such superlattices.

The superlattice synthesis so far has been limited to the dual-constituent system. Esaki *et al.* [196] proposed

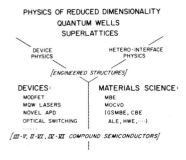

Fig. 15. Research in the interdisciplinary environment.

the introduction of a third constituent, AlSb in the InAs-GaSb system: this triple-constituent system offers an additional degree of freedom.

X. CONCLUSION

We have witnessed remarkable progress of an interdisciplinary nature on this subject. A variety of "engineered" structures exhibited extraordinary transport and optical properties; some of them, such as ultrahigh carrier mobilities, semimetallic coexistence of electrons and holes, and large electric field-induced effects on the optical properties, may not even exist in any "natural" crystal. Thus, this new degree of freedom offered in semiconductor research *through advanced material engineering* has inspired many ingenious experiments, resulting in observations of not only predicted effects but also totally unknown phenomenon such as fractional quantization which require novel interpretations. Activities in this new frontier of semiconductor physics, in turn, give immeasurable stimulus to device physics, leading to unprecedented transport and optoelectronic devices or provoking new ideas for applications. Figure 15 illustrates a pattern of such interdisciplinary research where beneficial cross fertilizations are prevalent.

I hope this article, which cannot possibly cover every landmark, provides some flavor of the excitement in this field. Finally, I would like to acknowledge many participants in and out of superlattice research for their contributions, T. P. Smith III for his critical reading, and the ARO's partial sponsorship from the very beginning of our investigation.

REFERENCES

[1] L. Esaki and R. Tsu, "Superlattice and negative conductivity in semiconductors," *IBM Res. Note*, RC-2418, Mar. 1969.
[2] ——, "Superlattice and negative differential conductivity in semiconductors," *IBM J. Res. Develop.*, pp. 61–65, Jan. 1970.
[3] D. Bohm, *Quantum Theory*. Englewood Cliffs, NJ: Prentice-Hall, 1951, p. 283.
[4] L. Esaki, "Long journey into tunneling," in *Les Prix Nobel en 1973*. Stockholm, Sweden: Imprimerie Royale, P. A. Norstedt & Söner, 1974, pp. 66–83.
[5] R. de L. Kronig and W. J. Penny, "Quantum mechanics of electrons in crystal lattices," *Proc. Roy. Soc.*, vol. A130, pp. 499–513, 1930.
[6] R. G. Chambers, "The kinetic formulation of conduction problems," in *Proc. Phys. Soc.*, London, vol. A65, pp. 458–459, 1952.

[7] L. Esaki, L. L. Chang, and R. Tsu, "A one-dimensional "superlattice" in semiconductors," in *Proc. 12th Int. Conf. Low Temp. Phys.*, Kyoto, Japan, 1970, pp. 551–553.

[8] A. E. Blakeslee and C. F Aliotta, "Man-made superlattice crystals," *IBM J. Res. Develop.*, vol. 14, pp. 686–688, Nov. 1970.

[9] See the review of strained-layer superlattices by G. C. Osbourn, P. L. Fourley, I. J. Fritz, R. M. Biefeld, L. R. Dawson, and T. E. Zipperian, in *Semiconductor and Semimetals*, R. K. Willardson and A. C. Beer, Eds. New York: Academic, to be published.

[10] L. Esaki, L. L. Chang, W. E. Howard, and V. L. Rideout, "Transport properties of a GaAs-GaAlAs superlattice," in *Proc. 11th Int. Conf. Phys. Semiconductors*, Warsaw, Poland, 1972, pp. 431–436.

[11] R. Tsu and L. Esaki, "Nonlinear optical response of conduction electrons in a superlattice," *Appl. Phys. Lett.*, vol. 19, pp. 246–248, Oct. 1971.

[12] U. Gnutzmann and K. Clauseker, "Theory of direct optical transitions in an optical indirect semiconductor with a superlattice structure," *Appl. Phys.*, vol. 3, pp. 9–14, 1974.

[13] L. Esaki, "Semiconductor superlattices and quantum wells," in *Proc. 17th Int. Conf. Semiconductors*, San Francisco, CA, Aug. 1984, pp. 473–483; "Semiconductor superlattices and quantum wells through development of molecular beam epitaxy," in *Molecular Beam Epitaxy and Heterostructures*, L. L. Chang and K. Ploog, Eds. Dordrecht, The Netherlands: Nijhoff, 1985, pp. 1–36; "History and perspective of semiconductor superlattices," in *Synthetic Modulated Structures*, by L. L. Chang and B. C. Giessen, Eds. Orlando, FL: Academic Press, 1985, pp. 3–41; "Compositional superlattices," in *The Technology and Physics of Molecular Beam Epitaxy*, E. H. C. Parker, Ed. New York: Plenum, 1985, pp. 143–184.

[14] A. Y. Cho, "Growth of periodic structures by the molecular-beam method," *Appl. Phys. Lett.*, vol. 19, pp. 467–468, 1971.

[15] J. H. Neave, B. A. Joyce, P. J. Dobson, and N. Norton, "Dynamics of film growth of GaAs by MBE from Rheed observations," *Appl. Phys.*, vol. A31, pp. 1–8, 1983.

[16] M. B. Panish and S. Sumski, *J. Appl. Phys.*, vol. 55, pp. 3571–3576, May 1984.

[17] M. Razeghi and J. P. Duchemin, "Recent advances in MOCVD growth of In$_x$Ga$_{1-x}$As$_y$P$_{1-y}$ alloys," *J. Cryst. Growth*, vol. 70, pp. 145–149, 1984.

[18] W. T. Tsang, "Chemical beam eiptaxy of InP and GaAs," *Appl. Phys. Lett.*, vol. 45, pp. 1234–1236, Dec. 1984.

[19] H. Fujiyasu, H. Takahashi, H. Shimizu, and A. Sasaki, "Optical properties of ZnS-ZnSe superlattices prepared by a HWE," in *Proc. 17th Int. Conf. Phys. Semiconductors*, San Francisco, CA, Aug. 1984, pp. 539–542.

[20] M. Pessa and O. Jylhä, "Growth of Cd$_{1-x}$Mn$_x$Te films with $0 < x < 0.9$ by atomic layer epitaxy," *Appl. Phys. Lett.*, vol. 45, pp. 646–648, Sept., 1984.

[21] Most numerical data obtained from *Landolt-Börnstein, New Series*, vol. 17, Berlin: Springer-Verlag.

[22] R. People, L. W. Wecht, K. Alavi, and A. Y. Cho. "Measurement of the conduction-band discontinuity of molecular beam epitaxial grown In$_{0.52}$Al$_{0.48}$As/In$_{0.53}$Ga$_{0.47}$As, N-n heterojunction by C-V profiling," *Appl. Phys. Lett.*, vol. 43, pp. 118–120, July 1983.

[23] K. Y. Cheng, A. Y. Cho, and W. R. Wagner, "Molecular-beam epitaxial growth of uniform Ga$_{0.47}$In$_{0.53}$As with a rotating sample holder," *Appl. Phys. Lett.*, vol. 39, pp. 607–609, Oct. 1981.

[24] M. Razeghi and J. P. Duchemin, "Low pressure-MOCVD growth of Ga$_{0.47}$In$_{0.53}$As-InP heterojunction and superlattices," *J. Vac. Sci. Technol. B*, vol. 1, pp. 262–265, Apr.–June 1983.

[25] M. Voos, "Electronic properties of MO-CVD grown InGaAs-InP heterojunctions and superlattices," *J. Vac. Sci. Technol. B*, vol. 1, pp. 404–408, Apr.–June 1983.

[26] E. J. Caine, S. Subbanna, H. Krömer, J. L. Merz, and A. Y. Cho, "Staggered-lineup heterojunctions as sources of tunable below-gap radiation: Experimental verification," *Appl. Phys. Lett.*, vol. 45, pp. 1123–1125, Nov. 1984.

[27] P. M. Petrof, A. C. Gossard, A. Savage, and W. Wiegmann, "Molecular beam epitaxy of Ge and Ga$_{1-x}$Al$_x$As ultra thin-film superlattices," *J. Cryst. Growth*, vol. 46, pp. 172–178, 1979.

[28] C.-A. Chang, A. Segmüller, L. L. Chang, and L. Esaki, "Ge-GaAs superlattices by molecular beam epitaxy," *Appl. Phys. Lett.*, vol. 38, pp. 912–914, June 1981.

[29] J. N. Schulman and T. C. McGill, "The CdTe/HgTe superlattice: Proposal for a new infrared material," *Appl. Phys. Lett.*, vol. 34, pp. 663–665, May 1979.

[30] G. Bastard, "Theoretical investigations of superlattice band struc-

ture in the envelope-function approximation," *Phys. Rev. B*, vol. 25, pp. 7584–7597, June 1982.

[31] J. P. Faurie, A. Million, and J. Piaguet, "CdTe-HgTe multilayers grown by molecular beam epitaxy," *Appl. Phys. Lett.*, vol. 41, pp. 713–715, Oct. 1982.

[32] H. Kinoshita and H. Fujiyasu, "PbTe-Pb$_{1-x}$Sn$_x$Te superlattices prepared by a hot wall technique," *J. Appl. Phys.*, vol. 51, pp. 5845–5846, Nov. 1980.

[33] E. F. Fantner and G. Bauer, *Two-Dimensional Systems, Heterostructures, and Superlattices*, G. Bauer, F. Kuchar and H. Heinrich, Eds. Berlin: Springer 1984, p. 207.

[34] S. Fujita, Y. Matsuda, and A. Sasaki, "Blue luminescence of a ZnSe-ZnS$_{0.1}$Se$_{0.9}$ strained-layer superlattice on a GaAs substrate grown by low-pressure organometallic vapor phase epitaxy," *Appl. Phys. Lett.*, vol. 47, pp. 955–957, Nov. 1985.

[35] M. Kobayashi, N. Mino, H. Katagiri, R. Kimura, M. Konagai, and K. Takahashi, "Growth of a ZnSe-ZnTe strained-layer superlattice on an InP substrate by molecular beam epitaxy," *Appl. Phys. Lett.*, vol. 48, pp. 296–297, Jan. 1986.

[36] R. N. Bicknell, R. W. Yanka, N. C. Giles-Taylor, D. K. Blanks, E. L. Buckland, and J. F. Schetzina, "Cd$_{1-x}$Mn$_x$Te-CdTe multilayers grown by molecular beam epitaxy," *Appl. Phys. Lett.*, vol. 45, pp. 92–94, July 1984.

[37] L. A. Kolodziejski, T. C. Bonsett, R. L. Gunshor, S. Datta, R. B. Bylsma, W. M. Becker, and N. Otsuka, "Molecular beam epitaxy of diluted magnetic semiconductor (Cd$_{1-x}$Mn$_x$Te) superlattices," *Appl. Phys. Lett.*, vol. 45, pp. 440–442, Aug., 1984.

[38] L. A. Kolodziejski, R. L. Gunshor, T. C. Bonsett, R. Venkatasubramanian, S. Datta, R. B. Bylsma, W. M. Becker, and N. Otsuka, "Wide gap II-VI superlattices of ZnSe-Zn$_{1-x}$Mn$_x$Se," *Appl. Phys. Lett.*, vol. 47, pp. 169–171, July 1985.

[39] R. H. Bube, *Electronic Properties of Crystalline Solids*. New York: Academic, 1974, p. 200.

[40] H. Sakaki, L. L. Chang, R. Ludeke, C.-A. Chang, G. A. Sai-Halasz, and L. Esaki, "In$_{1-x}$Ga$_x$As-GaSb$_{1-y}$As$_y$ heterojunctions by molecular beam epitaxy," *Appl. Phys. Lett.*, vol. 31, pp. 211–213, Aug. 1977.

[41] H. Krömer, "Barrier control and measurements: Abrupt semiconductor heterojunctions," *J. Vac. Sci. Technol. B*, vol. 2, pp. 433–439, July–Sept. 1984.

[42] M. L. Cohen, "Electrons at interfaces," in *Advances in Electronics and Electron Physics, Vol. 51*. New York: Academic, 1980, pp. 1–62.

[43] W. A. Harrison, "Elementary theory of heterojunctions," *J. Vac. Sci. Technol.*, vol. 14, pp. 1016–1021, July/Aug. 1977.

[44] W. R. Frensley and H. Krömer, "Theory of the energy-band lineup at an abrupt semiconductor heterojunction," *Phys. Rev. B*, vol. 16, pp. 2642–2652, Sept. 1977.

[45] J. Tersoff, "Theory of semiconductor heterojunctions: The role of quantum dipoles," *Phys. Rev. B*, vol. 30, pp. 4874–4877, Oct., 1984; "Schottky barrier heights and the continuum of GaP states," *Phys. Rev. Lett.*, vol. 52, pp. 465–468, 1984.

[46] G. Bastard, "Superlattice band structure in the envelope-function approximation," *Phys. Rev. B*, vol. 24, pp. 5693–5697, Nov. 1981.

[47] S. R. White and L. J. Sham, "Electronic properties of flat-band semiconductor heterostructures," *Phys. Rev. Lett.*, vol. 47, pp. 879–882, Sept. 1981.

[48] G. Duggan, "A critical review of heterojunction band offsets," *J. Vac. Sci. Technol. B*, vol. 3, pp. 1224–1230, July/Aug. 1985.

[49] R. Dingle, W. Wiegmann, and C. H. Henry, "Quantum states of confined carriers in very thin Al$_x$Ga$_{1-x}$As-GaAs-Al$_x$Ga$_{1-x}$As heterostructures," *Phys. Rev. Lett.*, vol. 33, pp. 827–830, Sept. 1974.

[50] R. Dingle, A. C. Gossard, and W. Wiegmann, "Direct observation of superlattice formation in a semiconductor heterostructure," *Phys. Rev. Lett.*, vol. 34, pp. 1327–1330, May 1975.

[51] R. C. Miller, A. C. Gossard, D. A. Kleinman, O. Munteanu, "Parabolic quantum wells with the GaAs-Al$_x$Ga$_{1-x}$As system," *Phys. Rev. B*, vol. 29, pp. 3740–3743, Mar. 1984.

[52] R. C. Miller, D. A. Kleinman, and A. C. Gossard, "Energy-gap discontinuities and effective masses for GaAs-Al$_x$Ga$_{1-x}$As quantum wells," *Phys. Rev. B*, vol. 29, pp. 7085–7087, June 1984.

[53] T. W. Hickmott, P. M. Solomon, R. Fischer, and H. Morkoç, "Negative charge, barrier heights, and the conduction-band discontinuity in Al$_x$Ga$_{1-x}$As capacitors," *J. Appl. Phys.*, vol. 57, pp. 2844–2853, Apr. 1985.

[54] W. I. Wang, E. E. Méndez, and F. Stern, "High mobility hole gas and valence-band offset in modulation-doped p-AlGaAs/GaAs heterojunctions," *Appl. Phys. Lett.*, vol. 45, pp. 639–641, Sept. 1984.

[55] W. I. Wang and F. Stern, "Valence band offset in AlAs/GaAs heterojunctions and the empirical relation for band alignment," *J. Vac. Sci. Technol. B*, vol. 3, pp. 1280–1284, Jul./Aug. 1985.

[56] P. Dawson, G. A. Wilson, C. W. Tu, and R. C. Miller, "Staggered band alignments in AlGaAs heterojunctions and the determination of valence-band offsets," *Appl. Phys. Lett.*, vol. 48, pp. 541–543, Feb. 1986.

[57] R. Tsu and L. Esaki, "Tunneling in a finite superlattice," *Appl. Phys. Lett.*, vol. 22, pp. 562–564, June 1973.

[58] L. L. Chang, L. Esaki, and R. Tsu, "Resonant tunneling in semiconductor double barriers," *Appl. Phys. Lett.*, vol. 24, pp. 593–595, June 1974.

[59] L. Esaki and L. L. Chang, "New transport phenomenon in a semiconductor 'superlattice,' " *Phys. Rev. Lett.*, vol. 33, pp. 495–497, Aug. 1974.

[60] T. C. L. G. Sollner, W. D. Goodhue, P. E. Tannenwald, C. D. Parker, and D. D. Peck, "Resonant tunneling through quantum wells at frequencies up to 2.5 THz," *Phys. Rev. Lett.*, vol. 43, pp. 558–590, Sept. 1983.

[61] T. C. L. G. Sollner, P. E. Tannenwald, D. D. Peck, and W. D. Goodhue, "Quantum well oscillators," *Appl. Phys. Lett.*, vol. 45, pp. 1319–1321, Dec. 1984.

[62] T. J. Shewchuk, P. C. Chapin, P. D. Coleman, W. Kopp, R. Fisher, and H. Morkoç, "Resonant tunneling oscillations in a GaAs-$Al_xGa_{1-x}As$ heterostructure at room temperature," *Appl. Phys. Lett.*, vol. 46, pp. 508–510, Mar. 1985.

[63] T. C. L. G. Sollner, H. Q. Le, C. A. Correa, and W. D. Goodhue, "Persistent photoconductivity in quantum well resonators," *Appl. Phys. Lett.*, vol. 47, pp. 36–39, July 1985.

[64] D. D. Coon and H. C. Liu, "Tunneling currents and two-body effects in quantum well and superlattice structures," *Appl. Phys. Lett.*, vol. 47, pp. 172–174, July 1985.

[65] S. Luryi and F. Capasso, "Resonant tunneling of two-dimensional electrons through a quantum wire: A negative transconductance device," *Appl. Phys. Lett.*, vol. 47, pp. 1347–1349, Dec. 1985.

[66] F. Capasso, K. Mohanned, and A. Y. Cho, "Sequential resonant tunneling through a multiquantum well superlattice," *Appl. Phys. Lett.*, vol. 48, pp. 478–480, Feb. 1986.

[67] E. E. Méndez, L. Esaki, and W. I. Wang, "Resonant magnetotunneling in GaAlAs-GaAs-GaAlAs heterostructures," *Phys. Rev. B*, vol. 33, pp. 2893–2896, Feb. 1986.

[68] E. E. Méndez, W. I. Wang, B. Ricco, and L. Esaki, *Appl. Phys. Lett.*, vol. 47, pp. 415–417, Aug. 1985.

[69] R. A. Davis, M. J. Kelly, and T. M. Kerr, "Tunneling between two strongly coupled superlattices," *Phys. Rev. Lett.*, vol. 55, pp. 1114–1116, Sept. 1985.

[70] R. Tsu, L. L. Chang, G. A. Sai-Halasz, and L. Esaki, "Effects of quantum states on the photocurrent in a superlattice," *Phys. Rev. Lett.*, vol. 34, pp. 1509–1511, June 1975.

[71] J. P. van der Ziel, R. Dingle, R. C. Miller, W. Wiegmann, and W. A. Nordland, Jr., "Laser oscillation from quantum states in very thin GaAs-$Al_{0.2}Ga_{0.8}As$ multilayer structures," *Appl. Phys. Lett.*, vol. 26, pp. 463–465, Apr. 1975.

[72] R. D. Dupuis, P. D. Dapkus, N. Holonyak, Jr., E. A. Rezek, and R. Chin, "Room-temperature laser operation quantum-well $Ga_{1-x}Al_xAs$-GaAs laser diodes grown by metalorganic chemical vapor deposition," *Appl. Phys. Lett*, vol. 32, pp. 295–297, Mar. 1978.

[73] N. Holonyak, Jr., R. M. Kolbas, E. A. Rezek, R. Chin, R. D. Dupuis, and P. D. Dapkus, "Bandfilling in metalorganic chemical vapor deposited $Al_xGa_{1-x}As$-GaAs-$Al_xGa_{1-x}As$ quantum-well heterostructure lasers," *J. Appl. Phys.*, vol. 49, pp. 5392–5397, Nov. 1978.

[74] W. T. Tsang, "Extremely low threshold (AlGa) As modified multiquantum well heterostructure lasers grown by molecular-beam epitaxy," *Appl. Phys. Lett.*, vol. 39, pp. 786–788, Nov. 1981.

[75] ——, "$Ga_{0.47}In_{0.53}As$/InP multiquantum well heterostructure lasers grown by molecular beam epitaxy operating at 1.53 µm," *Appl. Phys. Lett.*, vol. 44, pp. 288–290, Feb. 1984.

[76] N. K. Dutta, S. G. Napholtz, R. Yen, R. Wessel, T. M. Shen, and N. A. Olsson, "Long wavelength InGaAsP ($\lambda \sim$ 1.3 µm) modified multiquantum well laser," *Appl. Phys. Lett.*, vol. 46, pp. 1036–1039, June 1985.

[77] R. C. Miller, D. A. Kleinman, W. A. Nordland, Jr., and A. C. Gossard, "Luminescence studies of optically pumped quantum wells in GaAs-$Al_xGa_{1-x}As$ multilayer structures," *Phys. Rev. B*, vol. 22, pp. 863–871, July 1980.

[78] P. M. Petroff, C. Weisbuch, R. Dingle, A. C. Gossard, and W. Wiegmann, "Luminescence properties of GaAs-$Ga_{1-x}Al_xAs$ dou-

ble heterostructures and multiquantum-well superlattices grown by molecular beam epitaxy," *Appl. Phys. Lett.*, vol. 38, pp. 965–967, June 1981.

[79] B. A. Vojak, N. Holonyak, Jr., W. D. Laidig, K. Hess, J. J. Coleman, and P. D. Dapkus, "The exciton in recombination in $Al_xGa_{1-x}As$-GaAs quantum-well heterostructures," *Solid State Commun.*, vol. 35, pp. 477–481, 1980.

[80] E. E. Méndez, G. Bastard, L. L. Chang, and L. Esaki, "Effect of an electric field on the luminescence of GaAs quantum wells," *Phys. Rev. B*, vol. 26, pp. 7101–7103, Dec. 1982.

[81] R. C. Miller and A. C. Gossard, "Some effects of a longitudinal electric field on the photoluminescence of p-doped GaAs-$Al_xGa_{1-x}As$ quantum well heterostructures," *Appl. Phys. Lett.*, vol. 43, pp. 954–956, Nov. 1983.

[82] J. A. Kash, E. E. Méndez, and H. Morkoç, "Electric field induced decrease of photoluminescence lifetime in GaAs quantum wells," *Appl. Phys. Lett.*, vol. 46, pp. 173–175, Jan. 1985.

[83] H. J. Polland, L. Schultheis, J. Kuhl, E. O. Göbel, and C. W. Tu, "Lifetime enhancement of two-dimensional excitons by the quantum-confined Stark effect," *Phys. Rev. Lett.*, vol. 55, pp. 2610–2613, Dec. 1985.

[84] D. S. Chemla, T. C. Damen, C. A. B. Miller, A. C. Gossard, and W. Wiegmann, "Electroabsorption by Stark effect on room-temperature excitons in GaAs/GaAlAs multiple quantum well structures," *Appl. Phys. Lett.*, vol. 42, pp. 864–866, May 1983.

[85] D. A. B. Miller, D. S. Chemla, T. C. Damen, A. C. Gossard, W. Wiegmann, T. H. Wood, and C. A. Burrus, "Band-edge electroabsorption in quantum well structures: The quantum-confined Stark effect," *Phys. Rev. Lett.*, vol. 53, pp. 2173–2176, Nov. 1984; "Electric field dependence of optical absorption near the band gap of quantum-well structures," *Phys. Rev. B*, vol. 32, pp. 1043–1060, July 1985.

[86] T. H. Wood, C. A. Burrus, D. A. B. Miller, D. S. Chemla, T. C. Damen, A. C. Gossard, and W. Wiegmann, "High-speed optical modulation with GaAs/GaAlAs quantum wells in a p-i-n diode structure," *Appl. Phys. Lett.*, vol. 44, pp. 16–18, Jan. 1983.

[87] D. A. B. Miller, D. S. Chemla, T. C. Damen, A. C. Gossard, W. Wiegmann, T. H. Wood, and C. A. Burrus, "Novel hybrid optically bistable switch: The quantum well self-electro-optic effect device," *Appl. Phys. Lett.*, vol. 45, pp. 13–15, July 1984.

[88] H.-J. Polland, Y. Horikoshi, R. Höger, E. O. Göbel, J. Kuhl, and K. Ploog, "Influence of electric fields on the hot carrier kinetics in AlGaAs/GaAs quantum wells," in *Proc. 4th Int. Conf. Hot Electrons Semiconductors*, Innsbruck, Austria, July 1985, to be published in *Physica B*.

[89] R. T. Collins, K. von Klitzing, and K. Ploog, "Photocurrent spectroscopy of GaAs/AlGaAs quantum wells in an electric field," *Phys. Rev. B.*, vol. 33, pp. 4378–4381, Mar. 1986.

[90] Y. Matsumoto, S. Tarucha, and H. Okamoto, "Tunneling dynamics of photo-generated carriers in semiconductor superlattices," *Phys. Rev. B.*, vol. 33, pp. 5961–5964, Apr. 1986.

[91] L. Viña, R. T. Collins, E. E. Méndez, and W. I. Wang, "Electric field effects on GaAs/GaAlAs quantum wells measured by photoluminescence and photocurrent spectroscopy," *Phys. Rev. B.*, vol. 33, pp. 5939–5942, Apr. 1986.

[92] C. Alibert, S. Gaillard, J. A. Brum, G. Bastard, P. Frijlink, and M. Erman, "Measurements of electri-field-induced energy-level shifts in GaAs single-quantum-wells using electroreflectance," *Solid State Commun.*, vol. 53, pp. 457–460, 1985.

[93] G. Bastard, E. E. Méndez, L. L. Chang, and L. Esaki, "Variational calculations on a quantum well in an electric field," *Phys. Rev. B*, vol. 28, pp. 3241–3245, Sept. 1983.

[94] E. J. Austin and M. Jaros, "Electronic structure of an isolated GaAs-GaAlAs quantum well in a strong electric field," *Appl. Phys. Lett.*, vol. 31, pp. 5569–5572, Apr. 1985.

[95] F. Capasso, K. Mohammed, A. Y. Cho, R. Hull, and A. L. Hutchinson, "New quantum photoconductivity and large photocurrent gain be effective-mass filtering in a forward-biased superlattice p-n junction," *Phys. Rev. Lett.*, vol. 55, pp. 1152–1155, Sept. 1985.

[96] P. Manuel, G. A. Sai-Halasz, L. L. Chang, Chin-An Chang, and L. Esaki, "Resonant Raman scattering in a semiconductor superlattice," *Phys. Rev. Lett.*, vol. 37, pp. 1701–1704, Dec. 1976.

[97] E. Burstein, A. Pinczuk, and S. Buchner, "Resonance inelastic light scattering by charge carriers at semiconductor surfaces," in *Physics of Semiconductors 1978, Institute of Physics Conference Series*, no. 43. London: The Institute of Physics, 1979, pp. 1231–1234.

[98] G. Abstreiter and K. Ploog, "Inelastic light scattering from a quasi-two-dimensional electron system in GaAs-$Al_xGa_{1-x}As$ heterojunc-

tions," *Phys. Rev. Lett.*, vol. 42, pp. 1308-1311, May 1979.

[99] A. Pinczuk, H. L. Störmer, R. Dingle, J. M. Worlock, W. Wiegmann, and A. C. Gossard, "Observation of intersubband excitations in a multilayer two-dimensional electron gas," *Solid State Commun.*, vol. 32, pp. 1001-1003, 1979.

[100] R. Sooryakumar, A. Pinczuk, A. Gossard, and W. Wiegmann, "Dispersion of collective intersubband excitations in semiconductor superlattices," *Phys. Rev. B*, vol. 31, pp. 2578-2580, Feb. 1985.

[101] C. Colvard, R. Merlin, and M. V. Klein, and A. C. Gossard, "Observation of folded acoustic phonons in a semiconductor superlattice," *Phys. Rev. Lett.*, vol. 45, pp. 298-301, July 1980.

[102] V. Narayanamurti, H. L. Störmer, M. A. Chin, A. C. Gossard, and W. Wiegmann, "Selective transmission of high-frequency phonons by a superlattice: the 'Dielectric' phonon filter," *Phys. Rev. Lett.*, vol. 43, pp. 2012-2016, Dec. 1979.

[103] B. Jusserand, F. Alexandre, J. Dubard, and D. Paguet, "Raman scattering study of acoustical zone-center gaps in GaAs/AlAs superlattices," *Phys. Rev. B*, vol. 33, pp. 2897-2899, Feb. 1986.

[104] B. Jusserand, D. Paguet, and A. Regreny, "'Folded' optical phonons in GaAs/Ga$_{1-x}$Al$_x$As superlattices," *Phys. Rev. B*, vol. 30, pp. 6245-6247, Nov. 1984.

[105] A. K. Sood, J. Méndez, M. Cardona, and K. Ploog, "Resonance Raman scattering by confined LO and TO phonons in GaAs-AlAs superlattices," *Phys. Rev. Lett.*, vol. 54, pp. 2111-2114, May 1985.

[106] C. Colvard, T. A. Grant, and M. V. Klein, R. Melin, R. Fischer, H. Morkoç, and A. C. Gossard, "Folded acoustic and quantized optic phonons in (GaAl)As superlattices," *Phys. Rev. B*, vol. 31, pp. 2080-2091, Feb. 1985.

[107] A. K. Sood, J. Méndez, M. Cardona, and K. Ploog, "Interface vibrational modes in GaAs-AlAs superlattices," *Phys. Rev. Lett.*, vol. 54, pp. 2115-2118, May 1985; "Second-order Raman scattering by confined optical phonons and interface vibrational modes in GaAs-AlAs superlattices," *Phys. Rev. B*, vol. 32, pp. 1412-1414, July 1985.

[108] B. V. Shanabrook, J. Comas, T. A. Perry, and R. Merlin, "Raman scattering form electrons bound to shallow donors in GaAs-Al$_x$Ga$_{1-x}$As quantum-well structures," *Phys. Rev. B*, vol. 29, pp. 7096-7098, June 1984.

[109] T. A. Perry, R. Merlin, B. V. Shanabrook, and J. Comas, "Observation of resonant impurity states in semiconductor quantum-well structures," *Phys. Rev. Lett.*, vol. 54, pp. 2623-2626, June 1985.

[110] D. Gammon, R. Merlin, W. T. Masselink, and H. Morkoç, "Raman spectra of shallow acceptors in quantum-well structures," *Phys. Rev. B*, vol. 33, pp. 2919-2922, Feb. 1986.

[111] R. Dingle, H. L. Störmer, A. C. Gossard, and W. Wiegmann, "Electron mobilities in modulation-doped semiconductor heterojunction superlattices," *Appl. Phys. Lett.*, vol. 33, pp. 665-667, Oct. 1978.

[112] H. L. Störmer, A. C. Gossard, W. Wiegmann, and M. D. Sturge, "Two-dimensional electron gas at a semiconductor-semiconductor interface," *Solid State Commun.*, vol. 29, pp. 705-709, 1979.

[113] T. Mimura, S. Hiyamizu, T. Fujii, and K. Nanbu, "A new field-effect transistor with selectively doped GaAs/n-Al$_x$Ga$_{1-x}$As heterojunctions," *Jpn. J. Appl. Phys.*, vol. 19, pp. L225-L227, May 1980.

[114] P. Delescluse, M. Laviron, J. Chaplart, D. Delagebeaudeuf, and N. T. Linh, "Transport properties in GaAs-Al$_x$Ga$_{1-x}$As heterostructures and MESFET application," *Electron Lett.*, vol. 17, pp. 342-344, May 1981.

[115] M. Hieblum, E. E. Méndez, and F. Stern, "High mobility electron gas in selectively doped n : AlGaAs/GaAs heterojunctions," *Appl. Phys. Lett.*, vol. 44, pp. 1064-1066, June 1984.

[116] E. E. Méndez, P. J. Price, and M. Heiblum, "Temperature dependence of the electron mobility in GaAs-GaAlAs heterostructures," *Appl. Phys. Lett.*, vol. 45, pp. 294-296, Aug. 1984.

[117] B. J. F. Lin, D. C. Tsui, M. A. Paalanen, and A. C. Gossard, "Mobility of the two-dimensional electron gas in GaAs-Al$_x$Ga$_{1-x}$As heterostructures," *Appl. Phys. Lett.*, vol. 45, pp. 695-697, Sept. 1984.

[118] H. L. Störmer, A. C. Gossard, W. Wiegmann, and K. Baldwin, "Dependence of electron mobility in modulation-doped GaAs-(AlGa)As heterojunction interfaces on electron density and Al concentration," *Appl. Phys. Lett.*, vol. 39, pp. 912-914, Dec. 1981.

[119] T. J. Drummond, W. Kopp, R. Fischer, H. Morkoç, R. E. Thorne, and A. Y. Cho, "Photoconductivity effects in extremely high mobility modulation-doped (Al,Ga)As/GaAs heterostructures," *J. Appl. Phys.*, vol. 53, pp. 1238-1240, Feb. 1982.

[120] H. P. Wei, D. C. Tsui, and M. Razeghi, "Persistent photoconduc-

[121] K. Hirakawa, H. Sakaki, and J. Yoshino, "Mobility modulation of the two-dimensional electron gas via controlled deformation of the electron wave function in selectively doped AlGaAs-GaAs heterojunctions," *Phys. Rev. Lett.*, vol. 54, pp. 1279-1282, Mar. 1985.

[122] H. L. Störmer and W. T. Tsang, "Two-dimensional hole gas at a semiconductor heterojunction interface," *Appl. Phys. Lett.*, vol. 36, pp. 685-687, Apr. 1980.

[123] H. L. Störmer, K. Baldwin, A. C. Gossard, and W. Wiegmann, "Modulation-doped field-effect transistor based on a two-dimensional hole gas," *Phys. Rev. Lett.*, vol. 44, pp. 1062-1064, June 1984.

[124] H. L. Störmer, Z. Schlesinger, A. Chang, D. C. Tsui, A. C. Gossard, and W. Wiegmann, "Energy structure and quantized Hall effect of two-dimensional holes," *Phys. Rev. Lett.*, vol. 51, pp. 126-129, July 1983.

[125] J. P. Eisenstein, H. L. Störmer, V. Narayanamurti, A. C. Gossard, and W. Wiegman, "Effect of inversion symmetry on the band structure of semiconductor heterostructures," *Phys. Rev. Lett.*, vol. 53, pp. 2579-2581, Dec. 1984.

[126] Y. Iye, E. E. Méndez, W. I. Wang, and L. Esaki, "Magnetotransport properties and subband structure of the two-dimensional hole gas in GaAs-Ga$_{1-x}$Al$_x$As heterostructures," *Phys. Rev. B*, vol. 33, pp. 5854-5857, Apr. 1986.

[127] E. E. Mendez and W. I. Wang, "Temperature dependence of hole mobility in GaAs-Ga$_{1-x}$Al$_x$As heterojunctions," *Appl. Phys. Lett.*, vol. 46, pp. 1159-1161, June 1985.

[128] U. Ekenberg and M. Altarelli, "Calculation of hole subbands at the GaAs-Al$_x$Ga$_{1-x}$As interface," *Phys. Rev. B*, vol. 30, pp. 3569-3572, Sept. 1984.

[129] D. A. Broido and L. J. Sham, "Effective masses of holes at GaAs-AlGaAs heterojunctions," *Phys. Rev. B*, vol. 31, pp. 888-892, Jan 1986.

[130] J. N. Schulman and Y.-C. Chang, "Band mixing in semiconductor superlattices," *Phys. Rev. B*, vol. 31, pp. 2056-2068, Feb. 1985.

[131] K. von Klitzing, G. Doreda, and M. Pepper, "New method for high-accuracy determination of the fine-structure constant based on quantized hall resistance," *Phys. Rev. Lett.*, vol. 45, pp. 494-497, Aug. 1980.

[132] K. von Klitzing, "Two-dimensional systems: A method for the determination of the fine structure constant," *Surface Sci.*, vol. 113, pp. 1-6, Aug. 1981.

[133] D. C. Tsui and A. C. Gossard, "Resistance standard using quantization of the Hall resistance of GaAs-Al$_x$Ga$_{1-x}$As heterostructures," *Appl. Phys. Lett.*, vol. 38, pp. 550-552, Apr. 1981.

[134] D. C. Tsui, A. C. Gossard, B. F. Field, M. E. Cage, and R. F. Dziuba, "Determination of the fine-structure constant using GaAs-Al$_x$Ga$_{1-x}$As heterostructures," *Phys. Rev. Lett.*, vol. 48, pp. 3-6, Jan. 1982.

[135] T. Ando and Y. Uemura, "Theory of quantum transport in a two-dimensional electron system under magnetic fields," *J. Phys. Soc. Japan*, vol. 36, pp. 959-968, Apr. 1974.

[136] H. L. Störmer, J. P. Eisenstein, A. C. Gossard, W. Wiegmann, and K. Baldwin, "Quantization of the Hall effect in an anisotropic three-dimensional electronic system," *Phys. Rev. Lett.*, vol. 56, pp. 85-88, Jan. 1986.

[137] D. C. Tsui, H. L. Störmer, and A. C. Gossard, "Two-dimensional magnetotransport in the extreme quantum limit," *Phys. Rev. Lett.*, vol. 48, pp. 1559-1562, May 1982.

[138] R. B. Laughlin, "Anomalous quantum Hall effect: An incompressible quantum fluid with fractionally charged excitations," *Phys. Rev. Lett.*, vol. 50, pp. 1395-1398, May 1983.

[139] P. K. Lam and S. M. Girvin, "Liquid-solid transition and the fractional quantum Hall effect," *Phys. Rev. B*, vol. 30, pp. 473-475, July 1984.

[140] E. E. Méndez, M. Heiblum, L. L. Chang, and L. Esaki, "High-magnetic-field transport in a dilute two-dimensional electron gas," *Phys. Rev. B*, vol. 28, pp. 4886-4888, Oct. 1983.

[141] E. E. Méndez, L. L. Chang, M. Heiblum, L. Esaki, M. Naughton, K. Martin, and J. Brooks, "Fractionally quantized Hall effect in two-dimensional systems of extreme electron concentration," *Phys. Rev. B*, vol. 30, pp. 7310-7312, Dec. 1984.

[142] S. Kawaji, J. Wakabayashi, J. Yoshino, and H. Sakaki, "Activation energies of the $\frac{1}{3}$ and $\frac{2}{3}$ fractional quantum Hall effect in GaAs/Al$_x$Ga$_{1-x}$As heterostructures," *J. Phys. Soc. Japan*, vol. 53, pp. 1915-1918, June 1984.

[143] G. S. Boebinger, A. M. Chang, H. L. Störmer, and D. C. Tsui, "Magnetic field dependence of activation energies in the fractional quantum Hall effect," *Phys. Rev. Lett.*, vol. 55, pp. 1606-1609, Oct. 1985.

[144] S. M. Girvin, A. H. MacDonald, and P. M. Platzman, "Collective-excitation gap in the fractional quantum Hall effect," *Phys. Rev. Lett.*, vol. 54, pp. 581-583, Feb. 1985.

[145] E. E. Méndez, W. I. Wang, L. L. Chang, and L. Esaki, "Fractional quantum Hall effect in a two-dimensional hole system," *Phys. Rev. B*, vol. 30, pp. 1087-1089, July 1984.

[146] H. W. Hickmott, private communication.

[147] T. P. Smith, III, private communication.

[148] B. A. Sai-Halasz, R. Tsu, and L. Esaki, "A new semiconductor superlattice," *Appl. Phys. Lett.*, vol. 30, pp. 651-654, June 1977.

[149] G. A. Sai-Halasz, L. Esaki, and W. A. Harrison, "InAs-GaSb superlattice energy structure and its semiconductor-semimetal transition," *Appl. Phys. Lett.*, vol. 18, pp. 2812-2818, Sept. 1978.

[150] G. A. Sai-Halasz, L. L. Chang, J. M. Welter, C. A. Chang, and L. Esaki, "Optical absorption of In₁₋ₓGaₓAs-GaSb₁₋ₓAsₓ superlattices," *Solid State Commun.*, vol. 27, pp. 935-937, June 1978.

[151] M. Altarelli, "Electronic structure and semiconductor-semimetal transition in InAs-GaSb superlattice," *Phys. Rev. B*, vol. 28, pp. 842-845, July 1983.

[152] L. L. Chang, N. J. Kawai, G. A. Sai-Halasz, R. Ludeke, and L. Esaki, "Observation of semiconductor-semimetal transition in InAs-GaSb superlattices," *Appl. Phys. Lett.*, vol. 35, pp. 939-942, Dec. 1979.

[153] Y. Guldner, J. P. Vieren, P. Voisin, M. Voos, L. L. Chang, and L. Esaki, "Cyclotron resonance and far-infrared magneto-absorption experiments on semimetallic InAs-GaSb superlattices," *Phys. Rev. Lett.*, vol. 45, pp. 1719-1722, Nov. 1980.

[154] J. C. Mann, Y. Guldner, J. P. Vieren, P. Voisin, M. Voos, L. L. Chang, and L. Esaki, "Three-dimensional character of semimetallic InAs-GaSb superlattices," *Solid State Commun.*, vol. 39, pp. 683-686, 1981.

[155] G. Bastard, E. E. Méndez, L. L. Chang, L. Esaki, "Self-consistent calculations in InAs-GaSb heterojunctions," *J. Vac. Sci. Technol.*, vol. 21, pp. 531-534, July/Aug. 1982.

[156] H. Munekata, E. E. Méndez, Y. Iye, and L. Esaki, "Densities and mobilities of coexisting electrons and holes in MBE grown GaSb-InAs-GaSb quantum well," in *Proc. 2nd Int. Conf. Modulated Semiconductor Structures*, Kyoto, Japan, 1985; to be published in *Surface Sci.*

[157] E. E. Méndez, L. L. Chang, C.-A. Chang, L. F. Alexander and L. Esaki, "Quantized Hall effect in single quantum wells of InAs," *Surface Sci.*, vol. 142, pp. 215-219, 1984.

[158] S. Washburn, R. A. Webb, E. E. Méndez, L. L. Chang, and L. Esaki, "New Shubnikov-de Haas effects in a two-dimensional electron-hole system," *Phys. Rev. B*, vol. 31, pp. 1198-1201, Jan. 1985.

[159] E. E. Méndez, L. Esaki, and L. L. Chang, "Quantum Hall effect in a two-dimensional electron-hole gas," *Phys. Rev. Lett.*, vol. 55, pp. 2216-2219, Nov. 1985.

[160] G. H. Döhler, "Electron states in crystals with nipi-superstructure," *Phys. Status Solidi(b)*, vol. 52, pp. 79-91, 1972; "Electrical and optical properties of crystals with 'nipi-superstructures'," *Phys. Status Solidi(b)*, vol. 52, pp. 533-545, 1972.

[161] K. Ploog, A. Fischer, G. H. Döhler, and H. Künzel, in *Gallium Arsenide and Related Compounds 1980*, Institute of Physics Conference Series no. 56, H. W. Thim, Ed. London: Institute of Physics, 1981, p. 721.

[162] G. H. Döhler, H. Künzel, D. Olego, K. Ploog, P. Ruden, H. J. Stolz, and G. Abstreiter, "Observation of tunable band gap and two-dimensional subbands in a novel GaAs superlattice," *Phys. Rev. Lett.*, vol. 47, pp. 864-867, Sept. 1981.

[163] B. A. Vojak, G. W. Zajac, F. A. Chambers, J. M. Meese, and P. E. Chumbey, "Photopumped laser operation of GaAs doping superlattice," *Appl. Phys. Lett.*, vol. 48, pp. 251-253, Jan. 1986.

[164] E. F. Schuber, A. Fischer, Y. Horikoshi, and K. Ploog, "GaAs sawtooth superlattice laser emitting at wavelengths λ > 0.9 μm," *Appl. Phys. Lett.*, vol. 47, pp. 219-221, Aug. 1985.

[165] J. S. Yuan, M. Gal, P. C. Taylor, and G. B. Stringfellow, "Doping superlattices in organometallic vapor phase epitaxial InP," *Appl. Phys. Lett.*, vol. 47, pp. 405-407, Aug. 1985.

[166] W. Jantsch, G. Bauer, P. Pichler, and H. Clemens, "Anomalous transport in PbTe doping superlattices," *Appl. Phys. Rev.*, vol. 47, 738-740, Oct. 1985.

[167] A. Abeles and T. Tiedje, "Amorphous semiconductor superlattice," *Phys. Rev. Lett.*, vol. 51, pp. 2003-2006, Nov. 1983.

[168] T. Tiedje, B. Abeles, and B. G. Brooks, "Energy transport and size effects in the photoluminescence of amorphous-germanium/amorphous-silicon multilayer structure," *Phys. Rev. Lett.*, vol. 54, pp. 2545-2547, June 1985.

[169] P. Santos, M. Hundhausen, and L. Ley, "Observation of folded-zone acoustical phonons by Raman scattering in amorphous Si-SiNₓ superlattices," *Phys. Rev. B*, vol. 33, pp. 1516-1518, Jan. 1986.

[170] J. H. van der Merwe, "Crystal interfaces," *J. Appl. Phys.*, vol. 34, pp. 117-127, Jan 1963.

[171] G. C. Osbourn, "Strained-layer superlattices from lattice mismatched materials," *J. Appl. Phys.*, vol. 53, pp. 1586-1589, Mar. 1982.

[172] G. C. Osbourn, R. M. Biefeld and P. L. Gourley, "A GaAsₓP₁₋ₓ/GaP strained-layer superlattice," *J. Appl. Phys. Lett.*, vol. 41, pp. 172-174, July 1982.

[173] P. Voisin, C. Delalande, M. Voos, L. L. Chang, A. Segmüller, C. A. Chang, and L. Esaki, "Light and heavy valence subband reversal in GaSb-AlSb superlattices," *Phys. Rev. B*, vol. 30, pp. 2276-2278, Aug. 1984.

[174] B. Jusserand, P. Voisin, M. Voos, L. L. Chang, E. E. Méndez, and L. Esaki, "Raman scattering in GaSb-AlSb strained layer superlattices," *Appl. Phys. Lett.*, vol. 46, pp. 678-680, Apr. 1985.

[175] K. Ploog, Y. Ohmori, H. Okamoto, W. Stolz, and J. Wagner, *Appl. Phys. Lett.*, vol. 47, pp. 384-386, Aug. 1985.

[176] P. L. Gourley and R. M. Biefeld, "Quantum size effects in GaAs/GaAsₓP₁₋ₓ strained-layer superlattices," *Appl. Phys. Lett.*, vol. 45, pp. 749-751, Oct. 1984.

[177] P. L. Gourley, J. P. Hohimer, and R. M. Biefeld, "Lasing transitions in GaAs/GaAsₓP₁₋ₓ strained-layer superlattices with x = 0.1-0.5," *Appl. Phys. Lett.*, vol. 47, pp. 552-554, Sept. 1985.

[178] W. D. Laidig, P. J. Caldwell, Y. F. Lin, and C. K. Peng, "Strained-layer quantum-well injection laser," *Appl. Phys. Lett.*, vol. 44, pp. 653-655, Apr. 1984.

[179] J. Y. Marizn, M. N. Charasse, and B. Sermage, "Optical investigation of a new type of valence-band configuration in In₍Ga₁₋ₓ₎As GaAs strained superlattices," *Phys. Rev. B*, vol. 31, pp. 8298-8301, June 1985.

[180] E. Kasper, H. J. Herzog, and H. Kibbel, "A one-dimensional SiGe superlattice grown by UHV epitaxy," *Appl. Phys.*, vol. 8, pp. 199-205, 1975.

[181] H. M. Manasevit, I. S. Gergis, and A. B. Jones, "Electron mobility enhancement in epitaxial multilayer Si-Si₁₋ₓGe, alloy films on (100) Si," *Appl. Phys. Lett.*, vol. 4, pp. 464-466, Sept. 1982.

[182] J. C. Bean, L. C. Feldman, A. T. Fiory, S. Nakahara, and J. D. Robinson, "Ge₍Si₁₋ₓ₎/Si strained-layer superlattice grown by molecular beam epitaxy," *J. Vac. Sci. Technol.*, vol. A2, pp. 436-440, Apr./June 1984.

[183] F. Cerdeira, A. Pinzcuk, J. C. Bean, B. Batlogg, and B. A. Wilson, "Raman scattering from Ge₍Si₁₋ₓ₎/Si strained-layer superlattices," *Appl. Phys. Lett.*, vol. 45, pp. 1138-1140, Nov. 1984.

[184] R. People, J. C. Bean, D. V. Lang, A. M. Sergent, H. L. Störmer, K. W. Wecht, R. T. Lynch, and K. Baldwin, "Modulation doping in Ge₍Si₁₋ₓ₎/Si strained layer heterostructures," *Appl. Phys. Lett.*, vol. 45, pp. 1231-1232, Dec. 1984.

[185] R. People and J. C. Bean, "Band alignments of coherently strained Ge₍Si₁₋ₓ₎/Si heterostructures on ⟨001⟩ Ge₍Si₁₋ᵧ₎ substrates," *Appl. Phys. Lett.*, vol. 48, pp. 538-540, Feb. 1986.

[186] F. Cerdeira, A. Pinczuk, and J. C. Bean, "Observation of confined electronic states in Ge₍Si₁₋ₓ₎/Si strained-layer superlattices," *Phys. Rev. B*, vol. 31, pp. 1201-1204, Jan. 1985.

[187] I. J. Fritz, S. T. Picraux, L. R. Dawson, and T. J. Drummond, W. D. Laidig, and N. G. Anderson, "Dependence of critical layer thickness on strain for In₍Ga₁₋ₓ₎As/GaAs strained layer superlattices," *Appl. Phys. Lett.*, vol. 46, pp. 967-969, May 1985.

[188] R. Hull, J. C. Bean, F. Cerdiera, A. T. Fiory, and J. M. Gibson, "Stability of semiconductor strained-layer superlattices," *Appl. Phys. Lett.*, vol. 48, pp. 56-58, Jan. 1986.

[189] B. W. Dodson, "Stability of registry in strained-layer superlattice interfaces," *Phys. Rev. B*, vol. 30, pp. 3545-3546, Sept. 1984.

[190] A. T. Fiory, J. C. Bean, R. Hull, and S. Nakahara, "Thermal relaxation of metastable strained-layer Ge₍Si₁₋ₓ₎/Si epitaxy," *Phys. Rev. B*, vol. 31, pp. 4063-4065, Mar. 1985.

[191] R. L. Gunshor, L. A. Kolodziejski, N. Otsuka, and S. Datta, "ZnSe-ZnMn Se and CdTe-CdMnTe superlattices," in *Proc. 2nd Int. Conf. Modulated Semiconductors Structure*, Kyoto, Japan,

1985, to be published in *Surface Sci.*

[192] A. C. Gossard, P. M. Petroff, W. Weigmann, R. Dingle, and S. Savage, "Epitaxial structures with alternate-atomic-layer composition modulation," *Appl. Phys. Lett.*, vol. 29, pp. 323–325, Sept. 1976.

[193] K. Fujiwara and K. Ploog, "Photoluminescence of GaAs single quantum wells confined by short-period all-binary GaAs/AlAs superlattices," *Appl. Phys. Lett.*, vol. 45, pp. 1222–1224, Dec. 1984.

[194] H. Sakaki, M. Tsuchiya, and J. Yoshino, "Energy levels an electron functions in semiconductor quantum wells having superlattice alloy-like material (0.9 nm GaAs/0.9 nm AlGaAs) as barrier layers," *Appl. Phys. Lett.*, vol. 47, pp. 295–297, Aug. 1985.

[195] A. Ishibashi, M. Itabashi, Y. Mori, K. Kaneko, S. Kawado, and N. Watanabe, "Raman scattering from $(AlAs)_m (GaAs)_n$ ultrathin-layer superlattices," *Phys. Rev. B*, vol. 33, pp. 2887–2889, Feb. 1986.

[196] L. Esaki, L. L. Chang, and E. E. Méndez, "Polytype superlattice and multiheterojunctions," *J. Appl. Phys.*, Japan, vol. 20, pp. L529–L532, July 1981.

Leo Esaki (SM'60–F'62) received the B.S. and Ph.D. degrees in physics from the University of Tokyo, Tokyo, Japan, in 1947 and 1959, respectively.

He is an IBM Fellow and has been engaged in research at the IBM Thomas J. Watson Research Center, Yorktown Heights, NY, since 1960. Prior to joining IBM, he worked at the Sony Corporation where his research on heavily-doped Ge and Si resulted in the discovery of the tunnel diode. His major field is semiconductor physics. His current interest includes artificial semiconductor superlattices in search of predicted quantum mechanical effects.

Dr. Esaki was awarded the Nobel Prize in Physics in 1973 in recognition of his pioneering work on tunneling in solids and discovery of the tunnel diode. Other awards include the Nishina Memorial Award, the Asahi Press Award, an achievement award from the Tokyo Chapter of the U.S. Armed Forces Communications and Electronics Association, the Toyo Rayon Foundation Award for the Promotion of Science and Technology, the Morris N. Liebmann Memorial Prize, the Stuart Ballantine Medal, and the Japan Academy Award in 1965. He holds honorary degrees from Doshisha School, Japan, and the Universidad Politecnica de Madrid, Spain. From 1971 to 1975 he served as Councillor-at-Large of the American Physical Society, and as a Director of the American Vacuum Society from 1973 to 1975. Dr. Esaki is a Director of IBM-Japan, Ltd., on the Governing Board of the IBM-Japan Science Institute, and a Director of the Yamada Science Foundation. He serves on numerous international scientific advisory boards and committees. Currently, he is a Guest Editorial writer for the Yomiuri Press. He was chosen for the Order of Culture by the Japanese Government in 1974. He was elected a Fellow of the American Academy of Arts and Sciences in May 1974, a member of the Japan Academy on November 12, 1975, a Foreign Associate of the National Academy of Sciences (U.S.A.) on April 27, 1976, and a Foreign Associate of the National Academy of Engineering (U.S.A.) on April 1, 1977. On May 31, 1978, he was elected a Corresponding Member of the Academia Nacional De Ingenieria of Mexico. He was awarded the U.S.-Asia Institute 1983 Achievement Award. Dr. Esaki was awarded the IEEE Centennial Medal in December 1984. The American Physical Society 1985 International Prize for New Materials was awarded to him for his pioneering work in artificial semiconductor superlattices. Most recently he was chosen as a 1986 recipient of the Distinguished Foreign-Born Individual Award. He is a Fellow of the American Physical Society, and a member of the Physical Society of Japan and the Institute of Electrical Communication Engineers of Japan.

Ann. Rev. Mater. Sci. 1986. 16 : 263–91

COMPOSITIONALLY GRADED SEMICONDUCTORS AND THEIR DEVICE APPLICATIONS

Federico Capasso

AT&T Bell Laboratories, Murray Hill, New Jersey 07974

Introduction

This paper reviews the electronic transport properties of compositionally graded materials. Band gap grading is a powerful tool for engineering the energy band diagram of a device and thus modifying its electrical transport properties (band gap engineering) (1). The most interesting property, which has far reaching consequences for devices made of these materials, is that electrons and holes experience different electric forces so the transport properties of the two types of carriers can be independently tuned.

With the advent and rapid development of molecular beam epitaxy (MBE) (2) in recent years, graded-gap materials can presently be grown with controlled compositional variations over distances of $\lesssim 100$ Å. In addition, different functional forms of grading (linear, parabolic, etc.) can be obtained by accurately controlling the temperature and/or the opening of the cells in the MBE system. High electron velocities ($> 10^7$ cm/sec) have recently been measured in heavily doped p-type graded-gap $Al_xGa_{1-x}As$.

Ultrahigh-speed phototransistors and transistors with a graded-gap base have been reported. Heterojunction bipolar transistors (HBT's) with a graded emitter-base interface have also been studied in different material systems; for example, parabolic grading eliminates the collector-emitter offset voltage and maximizes the injection efficiency. Unipolar single and multiple sawtooth graded-gap structures have shown interesting physical properties and device applications. For example, because of the lack of reflection symmetry, sawtooth superlattices can be electrically polarized or used as rectifying elements. Grading of the high field region in avalanche photodiodes has been used to enhance the ionization rates ratio. One of the most exciting recent applications of graded materials is the "staircase"

263

0084–6600/86/0801–0263$02.00

potential, which can be used as a solid-state photomultiplier and a repeated velocity overshoot device.

Quasi-Electric Fields in Graded-Gap Materials

Kroemer (3) first considered the problem of transport in a graded-gap semiconductor. As a result of compositional grading, electrons and holes experience "quasi-electric" fields, F, of different intensities,

$$F_e = -\frac{dE_c}{dz}$$

$$F_h = +\frac{dE_v}{dz},$$ 1.

where $E_c(z)$ and $E_v(z)$ are the conduction and valence band edges. In addition, the forces resulting from these fields push electrons and holes in the same direction. This is illustrated in Figure 1a for the case of an intrinsic material. Such a graded material can be thought of as a stack of many isotype heterojunctions of progressively varying band gap. If the conduction and valence band edge discontinuities ΔE_c and ΔE_v of such heterojunctions are known and depend little on the alloy composition (as in the case of $Al_xGa_{1-x}As$ heterojunctions), then one can also expect that for the structure in Figure 1a the ratio of the quasi-electric fields F_e/F_h will be equal to $\Delta E_c/\Delta E_v$.

For a p-type graded-gap material the situation is different; the energy band diagram is shown in Figure 1b. The valence band edge is now horizontal so no effective field acts on the holes, while the effective field for the electrons is $F_e = -dE_g/dz$, which can be significantly greater than in the intrinsic case. In other words, all the band gap grading is transferred to the conduction band. This can be interpreted physically using the following heuristic argument.

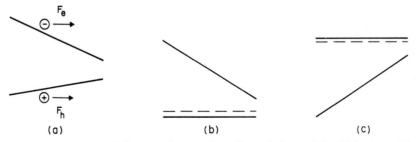

Figure 1 Energy band diagram of compositionally graded materials: (a) intrinsic, (b) p-type, and (c) n-type.

Consider the effect of p-type doping on an initially intrinsic material of the type in Figure 1a. The acceptor atoms will introduce holes, which under the action of the valence band quasi-electric field will be spatially separated from their negatively ionized parent acceptor atoms. This separation produces an electrostatic (space-charge) field. Holes accumulate (on the right-hand side of Figure 1b) until this space-charge field equals the magnitude of, and cancels, the hole quasi-electric field, F_h, thus achieving the thermodynamic equilibrium configuration (flat valence band) of Figure 1b. Note, however, that as a result of this process the equilibrium hole density is spatially nonuniform. The electrostatic field of magnitude $|dE_v/dz|$ produced by the separation of holes and acceptors adds instead to the conduction band quasi-electric field to give a total effective field acting on an electron of

$$F_e = - \left(\frac{dE_c}{dz} + \frac{dE_v}{dz} \right) = - \frac{dE_g}{dz}. \qquad 2.$$

Thus in a p-type material the conduction band field is made up of a nonelectrostatic (quasi-electric field) and an electrostatic (space-charge) contribution.

For an n-type material the same kind of argument can be applied to electrons; which yields the band diagram of Figure 1c and to the same effective field acting on the hole as given by Equation 2. Consider, for example, the case of an $Al_xGa_{1-x}As$ graded-gap p-type semiconductor. If we recall that in an $Al_xGa_{1-x}As$ heterojunction 62% of the band gap difference is in the conduction band (4), it follows that 62% of the effective conduction band field $F = - dE_g/dz$ will be quasi-electric in nature and the rest (38%) electrostatic. The opposite occurs in the case of an n-type $Al_xGa_{1-x}As$ graded material, where 62% of the valence band effective field is electrostatic in nature.

So far we have only considered quasi-electric fields arising from band gap grading. When the composition of the alloy is changed, however, the effective masses m_e^* and m_h^* of the carriers also change, giving rise to additional quasi-electric fields for electrons and holes. These are given, in the case of intrinsic materials (5), by

$$F_e = \frac{d}{dz} \left[\frac{3}{2} kT \ln (m_e^*/m_0) \right] \qquad 3.$$

$$F_h = \frac{d}{dz} \left[\frac{3}{2} kT \ln (m_h^*/m_0) \right] \qquad 4.$$

In the case of n- and p-type graded-gap materials one can repeat the previous reasoning and obtain the following expression for the quasi-field due to effective mass gradients:

$$F_{n,p} = \frac{d}{dz}\left[\frac{3kT}{4}\ln\left(m_e^* m_h^*/m_0^2\right)\right].$$ 5.

The quasi-electric fields in direct-band-gap-graded composition $Al_xGa_{1-x}As$ are primarily due to band gap grading; the quasi-electric fields due to the effective mass gradients are negligible in this case (5). However, effective mass gradients can make a substantial contribution to the quasi-electric field for $Al_xGa_{1-x}As$ graded materials in which the composition x is varied through the direct-indirect transition at $x = 0.45$. This is because the effective mass of the electron varies by about one order of magnitude in the direct-indirect transition region. Similar considerations are also thought to apply to other III–V alloys.

Electron Velocity Measurements

Quasi-electric fields are particularly important because they can be used to enhance the velocity of minority carriers that would otherwise move by diffusion (a relatively slow process) rather than by drift. In fact, Kroemer (3) first proposed the use of a graded-gap p-type layer (Figure 1b) for the base of a bipolar transistor to reduce the minority carrier (electron) transit time in the base. Recently Levine et al (6, 7), using an all-optical method, measured for the first time the electron velocity in a heavily p^+-doped compositionally graded $Al_xGa_{1-x}As$ layer grown by molecular beam epitaxy.

The energy band diagram of the sample is sketched in Figure 2 along with the principle of the experimental method. The measurement technique is a "pump and probe" scheme. The pump laser beam, transmitted through one of the AlGaAs window layers, is absorbed in the first few thousand angstroms of the graded layer. Optically generated electrons under the influence of the quasi-electric field drift towards the right in Figure 2 and accumulate at the end of the graded layer. This produces a change in the refractive index at the interface with the second window layer. This refractive index variation produces a reflectivity change that can be probed with the counter-propagating laser beam. This reflectivity change is measured as a function of the delay between pump and probe beams using phase-sensitive detection techniques. The reflectivity data (6) are shown in Figure 3 for a sample with a 1-μm-thick transport layer graded from $Al_{0.1}Ga_{0.9}As$ to GaAs and doped to $p \cong 2 \times 10^{18}$ cm^{-3}. This corresponds to a quasi-electric field of 1.2 kV/cm. The laser pulse

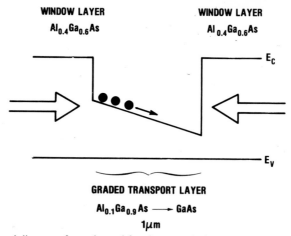

WINDOW LAYER
$Al_{0.4}Ga_{0.6}As$

WINDOW LAYER
$Al_{0.4}Ga_{0.6}As$

E_C

E_V

GRADED TRANSPORT LAYER

$Al_{0.1}Ga_{0.9}As \longrightarrow GaAs$

$1\mu m$

Figure 2 Band diagram of sample used for electron velocity measurements and schematic illustration of the pump-and-probe measurement technique.

width was 15 ps, and the time zero in Figure 3 represents the center of the pump pulse as determined by two photon absorption in a GaP crystal cemented near the sample. The approximate transit time is given by the shift of the half height of the reflectivity curve from zero, which is $\tau = 33$ ps. The drift length is taken as the thickness of the graded layer

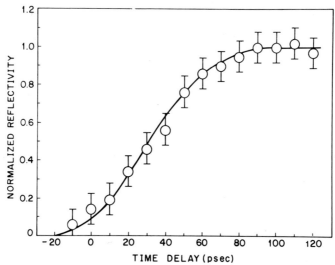

Figure 3 Normalized experimental results for pump-induced reflectivity change versus time delay obtained in 1-μm-thick, graded-gap, p$^+$ AlGaAs at a quasi-electric field $F = 1.2$ kV/cm. (From Reference 6.)

minus the absorption length of the pump beam $(1/\alpha \cong 2500 \text{ Å})$, and one finds a minority carrier velocity of $v \approx (L-\alpha^{-1})/\tau \approx 2.3 \times 10^6$ cm/sec. In this relatively thick sample carrier diffusion is important and causes a spread in the electron arrival time at the end of the sample, which is roughly the rise time of the reflectivity curve from 10 to 90%, i.e. 63 ps.

It is interesting to note that the drift mobility obtained from the measurement is $\mu_d = v_e/F = 1900$ cm^2/V sec, which is comparable with the usual mobility of 2200 cm^2/V sec at the doping level of the graded layer in GaAs.

Electron velocity measurements were also made in a 0.42-μm-thick, strongly graded ($F_e = 8.8$ kV/cm), highly doped ($p = 4 \times 10^{18}$ cm^{-3}) Al$_x$Ga$_{1-x}$As layer graded from Al$_{0.3}$Ga$_{0.7}$As to GaAs. A transit time of only 1.7 ps was measured, more than an order of magnitude shorter than that for $F = 1.2$ kV/cm, which corresponds to a velocity of roughly $v_e \cong 2.5 \times 10^7$ cm/sec (7). The velocity can be obtained rigorously and accurately ($\pm 10\%$ error) from the reflectivity data, by solving the drift diffusion equation and taking into account the effects of the pump absorption length (especially important in the thin sample) and the partial penetration of the probe beam into the graded material. If one includes all these effects, one finds that the reflectivity data can be fitted using only one adjustable parameter, the electron drift velocity (7). This velocity is $v_e = 2.8 \times 10^6$ cm/sec for $F = 1.2$ kV/cm and $p = 2 \times 10^{18}$ cm^{-3}, and $v_e = 1.8 \times 10^7$ cm/sec for $F = 8.8$ kV/cm and $p = 4 \times 10^{18}$ cm^{-3}.

We see that when we increase the quasi-electric field from 1.2 to 8.8 kV/cm (a factor of 7.3) the velocity increases from 2.8×10^6 to 1.8×10^7 cm/s (a factor of 6.4). That is, we observe the approximate validity of the relation $v = \mu F$. Using $\mu = 1700$ cm^2/Vs (for $p = 4 \times 10^{18}$ cm^{-3}) we calculate $v = 1.5 \times 10^7$ cm/s for $F = 8.8$ kV/cm, which is in reasonable agreement with the experimental results. The measured velocity of 1.8×10^7 cm/s (in the quasi-electric field) is significantly larger than that of undoped GaAs, in which $v = 1.2 \times 10^7$ cm/s for an ordinary electric field of $F = 8.8$ kV/cm. Our measured high velocity is comparable to the peak velocity reached in GaAs for $F = 3.5$ kV/cm before the transfer from the Γ to the L valley occurs. Our measured velocity is also comparable to the maximum possible phonon-limited velocity in the Γ minimum of GaAs. This is given by $v_{max} = [(E_p/m^*)\tanh(E_p/2kT)]^{1/2} = 2.3 \times 10^7$ cm/s, where $E_p = 35$ meV is the optical phonon energy and the effective mass $m^* = 0.067 \, m_0$.

This high velocity can be understood without reference to transient effects because the transit time is much larger than the momentum relaxation time of 0.3 ps. The high velocity results from the fact that the electrons spend most of their time in the high velocity central Γ valley rather than in the low velocity L valley. This may result from the injected electron density being so much less than the hole doping density that strong hole scattering can rapidly cool the electrons without excessively heating the holes. Furthermore, the electrons remain in the Γ valley throughout their transit across the graded layer since the total conduction band edge drop ($\Delta E_\mathrm{g} = 0.37$ eV) is comparable to the GaAs Γ–L separation ($\Delta E_{\Gamma\mathrm{L}} = 0.33$ eV), and therefore they do not have sufficient excess energy for significant transfer to the L valley.

High-Speed Graded-Base Transistors

The first device to utilize the high electron velocity found in p-type graded materials was reported by Capasso et al (8). The structure was a photo-transistor with an AlGaAs graded-gap base with a quasi-electric field of about 10^4 kV/cm. The device was grown by molecular beam epitaxy on a Si-doped ($\approx 4 \times 10^{18}$ cm^{-3}), n$^+$-type GaAs substrate. A buffer layer of n$^+$-type GaAs was grown first, followed by a Sn-doped, n-type ($\approx 10^{15}$ cm^{-3}), 1.5-μm-thick GaAs collector layer. The 0.45-μm-thick base layer was compositionally graded from GaAs (on the collector layer side) to Al$_{0.20}$Ga$_{0.80}$As ($E_\mathrm{g} = 1.8$ eV) and was heavily doped with Be ($p^+ \cong 5 \times 10^{18}$ cm^{-3}). The abrupt wide gap emitter consisted of an Al$_{0.45}$Ga$_{0.55}$As ($E_\mathrm{g} = 2.0$ eV), 1.5-μm-thick, window layer n-doped with Sn in the range of 2×10^{15} to 5×10^{15} cm^{-3}. Figure 4b shows the energy band diagram of the phototransistor.

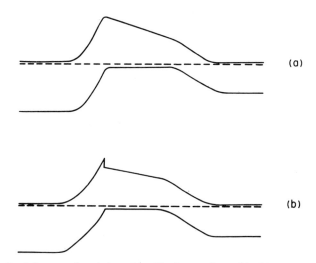

Figure 4   Band diagram of graded-gap base bipolar transistor: (*a*) with graded emitter-base interface, and (*b*) with ballistic launching ramp for even higher velocity in the base.

To study the effect of grading in the base on the speed of the device, 4-ps laser pulses were used. The wavelength ($\lambda = 6400$ Å) was chosen so that the light could only be absorbed in the base layer. The incident power was kept relatively high (100 mW) to minimize the effective emitter charging time. Under these conditions the speed-limiting factors are the RC time constant and the base transit time. Figure 5 shows the pulse response of the device as monitored by a fast sampling scope. In the lower part of Figure 5 the response was signal averaged; note the symmetrical rise and fall time and the absence of long tails, which are normally very difficult to achieve in picosecond photodetectors. From the observed 10–90% response time of 30 ps, a sum-of-squares approximation was used to estimate an intrinsic detector response time of about 20 ps. In the absence of a quasi-electric field in the heavily doped p^+ base, a broadened response followed by a tail with a square root of time dependence (due to slow diffusion) is expected. The diffusion time t_D is given by $W^2/2D$, where W is the base thickness and D is the diffusion coefficient. For a GaAs phototransistor with a base of $p^+ = 10^{18}$ cm^{-3} D is approximately 16 cm^2/s. In our structure D is likely to be smaller because AlGaAs has a lower mobility than GaAs, and because of the higher doping. For our structure $t_D \gtrsim 50$ ps. The fact that the expected broadening is not observed indicates that the quasi-electric field in the base sweeps out the electrons in a time much shorter than the diffusion time. From the velocity measurements previously discussed we know that the base transit time is about 2 ps, which is indeed much less than t_D. Thus the pulse response of this device is consistent with Kroemer's prediction (3) and is the first experimental verification of this effect (8).

Finally, the combination of the graded-gap base and the abrupt wide gap emitter (Figure 4*b*) suggests a new high-speed ballistic transistor (8, 9). In fact, the conduction band discontinuity can be used to ballistically launch electrons into the base with a high initial velocity; the quasi-electric field in the base will maintain an average velocity substantially higher than 10^7 cm/s. If no electric field were introduced in the base, ballistic launching alone, using the abrupt base emitter heterojunction, would not be sufficient to achieve a very high velocity in the base because collisions with plasmons or coupled plasmon-phonon modes in the heavily doped base would rapidly relax the initial forward momentum and velocity. It is sufficient for an initial high velocity that the conduction band discontinuity used for the launching be a few kT (typically 50 mV at 300 K).

Recently the first bipolar transistor with a compositionally graded base was reported (10, 11). Incorporation of a graded-gap base gives much faster base transit times because of the induced quasi-electric field for electrons, which allows a valuable tradeoff against the base resistance. To understand this last point consider a base of width W linearly graded from

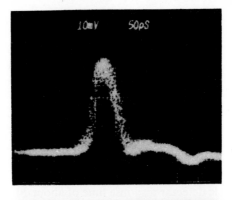

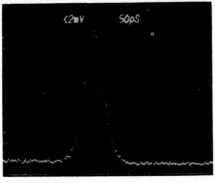

Figure 5 Pulse response of graded-gap base AlGaAs/GaAs phototransistor to a 4-ps laser pulse displayed on a sampling scope (*top*), and after signal averaging the sampling scope signal (*bottom*). (From Reference 8.)

one alloy with a band gap of E_{g1} to another with a band gap of E_{g2}. The quasi-electric field for electrons $(E_{g1} - E_{g2})/eW$ results in a base transit time (neglecting diffusion effects) of

$$\tau_b' = \frac{eW^2}{\mu(E_{g1} - E_{g2})}. \qquad \qquad 6.$$

We have made use of the experimental fact that the velocity in the graded base nearly equals μF_e, where F_e is the quasi-electric field (7). This time must be compared with the diffusion-limited base transit time of a transistor with an ungraded GaAs base of the same thickness and doping level

$$\tau_b = \frac{W^2}{2D}, \qquad \qquad 7.$$

where D is the ambipolar diffusion coefficient. If we compare Equations 6 and 7 and use the Einstein relationship $D = \mu kT/e$, we find that the base transit time is shortened by the factor

$$\frac{\tau_b}{\tau_b'} = \frac{E_{g1} - E_{g2}}{2kT},$$

8.

using a graded-gap base. Although Equation 8 is rigorous only in the limit $E_{g1} - E_{g2} \gg kT$, it can be employed as a useful "rule of thumb" in cases where $E_{g1} - E_{g2}$ is several times kT. Thus the band gap difference must be made as large as possible without exceeding the intervalley energy separation $\Delta E_{\Gamma L}$, which would greatly reduce the electron velocity. Using $E_{g1} - E_{g2} = 0.2$ eV, the transit time is reduced by a factor of about four at 300 K relative to an ungraded base of the same thickness. This allows a valuable tradeoff against the base resistance (R_b), since the base thickness can be increased to reduce R_b, while still keeping a reasonable base transit time. Finally, an added advantage of the quasi-electric field is the increased base transport factor that comes about because the short transit time reduces minority carrier recombination in the base.

Devices (10) grown by MBE on an n^+ substrate had a 1.5-μm GaAs buffer layer followed by a 5000-Å-thick collector doped to $n \cong 5 \times 10^{16}$ cm^{-3}. The p-type (2×10^{18} cm^{-3}) base was graded from Ga$_{0.98}$Al$_{0.02}$As to Ga$_{0.8}$Al$_{0.2}$As over 4000 Å. This grading corresponds to a field of about 5.6 kV/cm. The lightly doped ($n \cong 2 \times 10^{16}$ cm^{-3}), wide gap emitter consisted of an Al$_{0.35}$Ga$_{0.65}$As layer 3000 Å thick and a region adjacent to the base graded from Ga$_{0.8}$Al$_{0.2}$As to Ga$_{0.65}$Al$_{0.35}$As over 500 Å. This corresponds to a base/emitter energy gap difference of approximately 0.18 eV. This grading removes a large part of the conduction band spike, allowing most of the band gap difference to fall across the valence band and blocking the unwanted injection of holes from the base (12). Figure 4a shows the energy band diagram of the structure in the equilibrium (unbiased) configuration.

These devices had a current gain of 35 at a base current of 1.6 mA, and the collector characteristics were nearly flat with minimum collector-emitter offset voltage. More recently, high current gain, graded base bipolars with good high-frequency performance have been reported (Malik et al 12). The base layer was linearly graded over 1800 Å, from $x = 0$ to $x = 0.1$, which resulted in a quasi-electric field of 5.6 kV/cm, and was doped with Be to $p = 5 \times 10^{18}$ cm^{-3}. The emitter-base junction was graded over 500 Å from $x = 0.1$ to $x = 0.25$ to enhance hole confinement in the base. The 0.2-μm-thick Al$_{0.25}$Ga$_{0.75}$As emitter and the 0.5-μm-thick collector were doped n-type at 2×10^{17} cm^{-3} and 2×10^{16} cm^{-3}, respectively. The Al$_x$Ga$_{1-x}$As layers were grown at a substrate temperature of 700°C. It was found that this high growth temperature resulted in better $Al_xGa_{1-x}As$ quality, as determined by photoluminescence. However, it is known that significant Be diffusion occurs during MBE growth at high substrate temperatures and at high doping levels ($p > 10^{18}$ cm^{-3}). Secondary ion mass spectrometry (SIMS) data also indicated that the p-n junction was misplaced into the wide band gap emitter at 700°C. It was determined empirically that the insertion of an undoped setback layer of 200–500 Å between the base and emitter compensated for the Be diffusion and resulted in significantly increased current gains. Zn diffusion was used to contact the base and provided a low base contact resistance.

With a dopant setback layer in the base of 300 Å the maximum differential dc current gain was 1150, obtained at a collector current density of $J_c = 1.1 \times 10^3$ Acm^{-2}, a higher gain than previously reported for graded-gap base HBT's. These gains can be compared with those found in previous work, which were consistently <100 in HBT's without the set-back layer (10, 11). Several transistor wafers were processed with undoped setback layers in the base of 200–500 Å, and all exhibited greater current gains.

Graded-gap base HBT's were fabricated for high-frequency evaluation using the Zn diffusion process. A single 5-μm-wide emitter strip contact with dual adjacent base contacts was used. The areas of the emitter and collector junctions were approximately 2.3×10^{-6} cm^2 and 1.8×10^{-5} cm^2, respectively. The transistors were wire bonded in a microwave package, and automated s-parameter measurements were made with an HP 8409 network analyzer. The transistor has a current gain cutoff frequency of $f_T \approx 5$ GHz and a maximum oscillation frequency of $f_{max} \approx 2.5$ GHz. Large signal pulse measurements indicated rise times of $\tau_r \simeq 150$ ps and pulsed collector currents of $I_c > 100$ mA, suitable for high-current laser driver applications.

Emitter Grading in Heterojunction Bipolar Transistors

The essential feature of the heterojunction bipolar transistor is the use of part of the energy band gap difference between the wide band gap emitter and the base to suppress hole injection. This allows the base to be more heavily doped than the emitter, which leads to the low base resistance and low emitter-base capacitance necessary for high-frequency operation while still maintaining a high emitter injection efficiency (9). In this section we discuss in detail the grading problem in heterojunction bipolars. The performances of recently developed $Al_{0.48}In_{0.52}As/Ga_{0.47}In_{0.53}As$ bipolars with graded and ungraded emitters are compared (13), and the optimum grading of the emitter is discussed.

Most of the work on MBE-grown heterojunction bipolar transistors has concentrated on the AlGaAs/GaAs system. Recently the first vertical Npn $Al_{0.48}In_{0.52}As/Ga_{0.47}In_{0.53}As$ heterojunction bipolar transistors grown by MBE with high current gain have been reported by Malik et al (13). The (Al,In)As/(Ga,In)As layers were grown by molecular beam epitaxy (MBE) lattice-matched to an Fe-doped semi-insulating InP substrate. Two HBT structures were grown: the first with an abrupt emitter of $Al_{0.48}In_{0.52}As$ on a $Ga_{0.47}In_{0.53}As$ base, and a second with a graded emitter comprising a quaternary layer of AlGaInAs 600 Å wide and linearly graded between the two ternary layers. Grading from $Ga_{0.47}In_{0.53}As$ to $Al_{0.48}In_{0.52}As$ was achieved by simultaneously lowering the Ga and raising the Al oven temperatures in such a manner as to keep the total Group III flux constant during the transition. It should be noted that this is the first use of a graded quaternary alloy in a device structure.

The energy band diagram for the abrupt and graded emitter transistors are shown in Figure 6a and b, respectively. The effect of the grading is to eliminate the conduction band notch in the emitter junction. This in turn

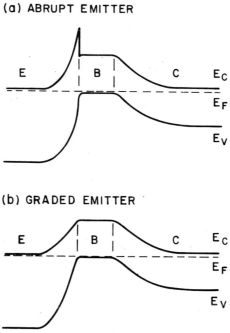

Figure 6   Band diagrams under equilibrium of heterojunction bipolar with (*a*) an abrupt emitter and (*b*) a graded emitter. Note that the conduction band notch is eliminated through the use of a graded emitter and the increase of the emitter-base valence band barrier. (From Reference 13.)

leads to a larger emitter-base valence band difference under forward bias injection. The following material parameters were used in both types of transistors. The $Al_{0.48}In_{0.52}As$ emitter and $Ga_{0.47}In_{0.53}As$ collector were doped n-type with Sn at levels of 5×10^{17} cm^{-3} and 5×10^{16} cm^{-3}, respectively. The $Ga_{0.47}In_{0.53}As$ base was doped p-type with Be to a level of 5×10^{18} cm^{-3}. Recent experimental determination of the band edge discontinuities in the $Al_{0.48}In_{0.52}As/Ga_{0.47}In_{0.53}As$ heterojunction indicates $\Delta E_c \cong 0.50$ eV and $\Delta E_v \cong 0.20$ eV (14). This value of ΔE_v is large enough to allow the use of an abrupt $Al_{0.48}In_{0.52}As/Ga_{0.47}In_{0.53}As$ emitter at 300 K. Nevertheless, a current gain increase by a factor of two (from $\beta = 200$ to $\beta = 400$) is achieved through the use of the graded-gap emitter, which is attributed to a larger valence band difference between the emitter and base under forward bias injection.

The common emitter characteristic of HBT's exhibits a relatively large collector-emitter offset voltage. This voltage is equal to the difference between the built-in potential for the emitter-base p-n junction and that of the base-collector p-n junction. Therefore no such offset is present in homojunction Si bipolars.

We have recently shown that by appropriately grading the emitter near the interface with the base such offset can be reduced and even totally eliminated (15). The other advantage of grading the emitter is of course that the potential spike in the conduction band can be reduced, thus increasing the injection efficiency. The conduction band potential has two components: the electrostatic potential ϕ_{es} equal to V_{bi} (the built-in potential) $- V_{be}$ (the base/emitter voltage), which varies parabolically, and the grading potential ϕ_g. If linear grading is used there is always unwanted structure in the conduction band (spikes or notches, see Figure 7). The "notches" can reduce the injection efficiency by promoting carrier recombination. It has now become obvious that such structures can be eliminated by grading with the complementary function of the electrostatic potential in the emitter region ($1-\phi_{es}$) over the depletion layer width at a forward bias equivalent to the base band gap (Figure 8). In this case if the base emitter junction is forward biased at 1.42 eV, the two potentials (grading and electrostatic) give rise to a smooth conduction band edge and one attains the flat band condition with a built-in voltage for the base emitter equivalent to the band gap in the base (1.42 eV).

A HBT with such a parabolic grading has been fabricated, using MBE, with a $Ga_{0.7}Al_{0.3}As$ emitter and a GaAs base and collector (15). The emitter/base junction was graded from $x = 0$ to $x = 0.3$ on the emitter side over a distance of 600 Å; the parabolic grading function was approximated by linear grading over nine regions. It was found that collector-emitter offset voltage is very small (about 0.03 V).

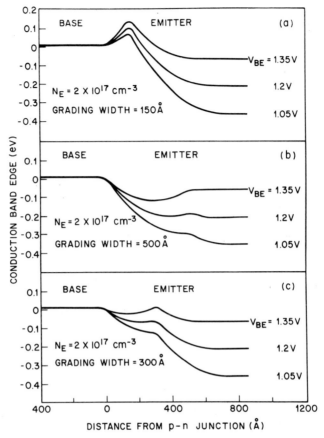

Figure 7 Conduction band edge versus distance from the p⁺-n base-emitter junction for three different linear grading widths at different base emitter forward bias voltages. (From Reference 15.)

Graded-Gap Lasers, Solar Cells, and Avalanche Photodiodes

If the active layer of a conventional double heterostructure laser is reduced to a thickness where the quantum size effect becomes important, i.e. less than 400 Å, the structure lases at a very high threshold current density because the overlap of the photon and electron populations is very small. To increase this overlap a separate optical cavity with graded composition is grown around the quantum well active layer. Here the electrons are "funnelled" into the quantum well region by the quasi-electric field of the

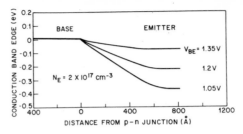

Figure 8 Conduction band edge versus distance from the p⁺-n junction, using a parabolically graded layer 500 Å wide at different forward bias voltages. (From Reference 15.)

graded composition layers and therefore have a higher probability of
capture by the quantum well. This device is known as the GRINSCH
(graded-index separate confinement heterostructure) laser after Tsang (16),
who first combined the quantum well laser (17) with the graded optical
confinement region, which had been proposed by Kazarinov & Tsarenkov
(18).

In solar cells, band gap grading in the top layer has been used to
efficiently collect carriers optically generated near the surface, before they
recombine through surface states. High-efficiency $Al_xGa_{1-x}As$-GaAs solar
cells have been fabricated by Woodall & Hovel (19) using this scheme.

Graded gaps can also be used to enhance the ratio of ionization
coefficients (α/β) in avalanche photodiodes (20). The ionization coefficient
$\alpha(\beta)$ is defined as the number of secondary pairs created per unit length
along the direction of the field by an electron (hole) by impact ionization.
The value α/β plays a crucial role in the signal-to-noise ratio of an avalanche photodiode (21). The α/β ratio must be either very large or very
small to minimize the avalanche excess noise.

Recently Capasso et al (20) proposed and demonstrated experimentally
a new avalanche detector in which α/β is enhanced with respect to the bulk
value of the alloys constituting the graded-gap material. The structure, grown by MBE, consists of a nominally intrinsic region, graded
from GaAs to $Al_{0.45}Ga_{0.55}As$ ($E_g = 2.0$ eV) and sandwiched between n⁺
and p⁺ regions. We demonstrated that the effective α/β ratio of this
structure is significantly increased over 1 ($\alpha/\beta \cong 5$–7.5) if the width of the
graded layer is $\lesssim 0.5 \ \mu m$. The principle of the device is illustrated in Figure
9. The electrons have a lower ionization energy than the holes because
they are moving towards lower gap regions. The effect is a significant
increase in the α/β ratio when the grading exceeds 1 eV/μm. Another
important property of this structure is the soft breakdown, which can lead
to much greater gain stability compared to ungraded diodes. This is due
to the fact that avalanche multiplication is always initiated in the low gap

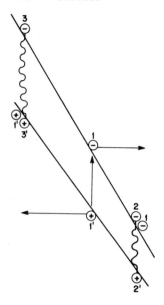

Figure 9   Band diagram of graded band gap avalanche photodiode showing impact ionization by the initial electron-hole pair 1–1'. (From Reference 20.)

regions and then spreads out to the higher gap regions as the electric field is increased.

Multilayer Sawtooth Materials

In this section we examine the electronic transport properties of sawtooth structures obtained by periodically varying the composition of the alloy in an asymmetric fashion. The key feature of such structures is the lack of reflection symmetry (22). This has several important consequences; for example, these devices can be used as rectifying elements or, under suitable conditions, one can optically generate in these structures a macroscopic electrical polarization that gives rise to a cumulative photovoltage across the uniformly doped sawtooth material. In addition, under appropriate bias they give rise to a staircase potential which has several intriguing applications.

RECTIFIERS   The basic principle of sawtooth rectifiers, recently demonstrated by Allyn et al (23, 24), is shown in Figure 10. A sawtooth-shaped potential barrier is created by growing a semiconductor layer of graded chemical composition followed by an abrupt composition discontinuity. The adjoining layers, with which contact is made, are of the same conductivity type. In the present case, the barrier material is aluminum gallium arsenide ($Al_xGa_{1-x}As$) in which the aluminum content is graded and the adjoining layers are n-type GaAs. Near zero bias, conduction in the direction perpendicular to the layer is inhibited by the barrier. When the device is biased in the forward direction (as shown in Figure 10c) the voltage drop initially occurs across the graded layer, reducing the slope of the potential barrier, and allowing increased thermionic emission over the reduced barrier. When the applied voltage exceeds the barrier height, the device will conduct completely, as in the case of a Schottky barrier. In the reverse direction (Figure 10d) electrons will be attracted to, but inhibited from passing through, the abrupt potential discontinuity at the sharp edge of the sawtooth. The width of the interface and potential discontinuity is known to be only 5–10 Å. Thus, the primary reverse current-carrying mechanism will be tunnelling. The barrier can be either doped or undoped, although depletion of carriers from within the barrier (in the case of doped barriers) leads to band bending, which reduces the equilibrium height and width of the barrier. Multiple sawtooth barriers with five periods were also fabricated (23). These showed a turn-on

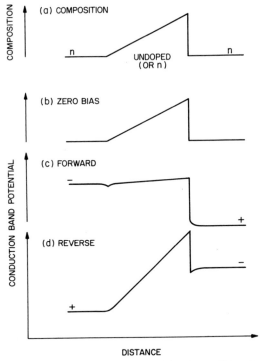

Figure 10 (a) Compositional structure of a sawtooth barrier rectifying structure, (b) potential distribution for band-edge conduction electrons at zero bias (undoped barrier case), (c) potential distribution under forward bias, and (d) potential distribution under reverse bias. (From Reference 23.)

voltage equal to five times that of the single barrier, thus demonstrating
the additivity of the technique.

ELECTRICAL POLARIZATION EFFECTS IN SAWTOOTH SUPERLATTICES   The lack
of planes of symmetry in sawtooth superlattice material, compared to
conventional superlattices with rectangular wells and barriers, can lead to
electrical polarization effects. Recently Capasso et al (25) reported for the
first time on the generation of a transient macroscopic electrical polarization extending over many periods of the superlattice. This effect is a
direct consequence of the above-mentioned lack of reflection symmetry in
these structures.

The energy band diagram of a sawtooth p-type superlattice is sketched
in Figure 11a, in which we have assumed a negligible valence band offset.
The layer thicknesses are typically a few hundred angstroms, and a suitable
material is graded-gap $Al_xGa_{1-x}As$. The superlattice is sandwiched
between two highly doped p^+ contact regions.

Let us assume that electron-hole pairs are excited by a very short light
pulse, as shown in Figure 11a. Due to the grading, electrons experience a
higher quasi-electric field than do holes. For this reason, and because

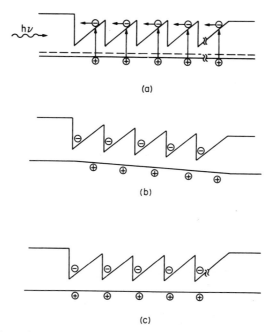

Figure 11   Formation and decay of the macroscopic electrical polarization in a sawtooth
superlattice. (From Reference 25.)

of their much higher velocity, electrons separate from holes and reach the low gap side in a subpicosecond time ($\simeq 10^{-13}$ sec). This sets up an electrical polarization in the sawtooth structure, which results in the appearance of a photovoltage across the device terminals (Figure 11b). The macroscopic dipole moment and its associated voltage subsequently decay in time by a combination of (a) dielectric relaxation and (b) hole drift.

The excess hole density decays by dielectric relaxation to restore a flat valence band (equipotential) condition, as illustrated in Figure 11c. Note that in this final configuration holes have redistributed to neutralize the electrons at the bottom of the wells. Thus the net negative charge density on the low gap side of the wells decreases with the same time constant as the positive charge packet (the dielectric relaxation time). The other mechanism by which the polarization decays is hole drift caused by the electric field created by the initial spatial separation of electrons and holes.

The graded-gap superlattice structure shown in Figure 11 and the underlying p$^+$-GaAs buffer layer were grown by MBE. A total of ten graded periods were grown with a period of ≈ 500 Å. The layers are graded from GaAs to $Al_{0.2}Ga_{0.8}As$. A heavily doped GaAs contact layer of ≈ 700 Å was grown on top of the 1-μm-thick $Al_{0.45}Ga_{0.55}$ (p $\approx 5 \times 10^{18}$ cm^{-3}) window layer. Unbiased devices were mounted in a microwave stripline and illuminated with short light pulses (4 ps) of wavelength $\lambda = 6400$ Å. The absorption length is ≈ 3500 Å. In this particular wafer the carrier concentration was 10^{16} cm^{-3}. It was found that the rise time of the pulse response is ≤ 25 ps, while the fall time (at the $1/e$ point) is $\cong 200$ ps. Unlike conventional detectors, the current carried in this photodetector is of a displacement rather than a conduction nature since it is associated with a time-varying polarization. This current, by continuity, equals the conduction current in the external load.

STAIRCASE STRUCTURES   Recently Capasso et al introduced the concept of a staircase potential (26–30). This innovative structure has several interesting applications. We shall concentrate on the staircase avalanche photodiode (APD) (1, 26–30) and on the repeated velocity overshoot device (31).

Staircase solid-state photomultipliers and avalanche photodiodes   Figure 12a shows the band diagram of the graded-gap multilayer material (assumed intrinsic) at zero applied field. Each stage is linearly graded in composition from a low (E_{g1}) to a high (E_{g2}) band gap, with an abrupt step back to low band gap material. The conduction band discontinuity shown accounts for most of the band gap difference, as is typical of many III–V heterojunctions. The materials are chosen for a conduction band discontinuity comparable to or greater than the electron ionization energy E_{ie} in the low gap material following the step. The biased detector is shown in Figure 12b. Consider a photoelectron generated near the p$^+$ contact: The electron does not impact ionize in the graded region before the conduction band step because the net electric field is too low. At the step, however, the electron ionizes and the process is repeated at every stage. Note that the steps correspond to the dynodes of a phototube. Holes created by electron impact ionization at the steps do not impact ionize, since the valence band steps are of the wrong sign to assist ionization and the electric field in the valence band is too low to cause hole-initiated ionization. Obviously holes multiply since at every step both an electron and a hole are created. The gain is $M = (2-\delta)^N$ where δ is the fraction of electrons that do not ionize per stage. The noise per unit bandwidth on the output signal, neglecting dark current, is given by $\langle i^2 \rangle = 2eI_{ph}M^2F$, where I_{ph} is the primary photocurrent and F the avalanche excess noise factor. For the staircase APD F is given by (30)

$$F = 1 + \frac{\delta[1-(2-\delta)^{-N}]}{2-\delta} \qquad\qquad 9.$$

(a)

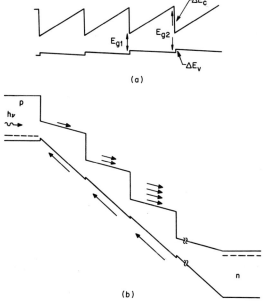

(b)

Figure 12   Band diagram of staircase solid-state photomultiplier. The arrows in the valence band simply indicate that holes do not impact ionize. (From Reference 30.)

Note that for small δ, $F \cong 1$ and is practically independent of the number of stages. Thus, the multiplication process is essentially noise free. It is interesting to note that the excess noise of this structure does not follow the McIntyre theory of conventional APD's (21). In a conventional APD the minimum excess noise factor at high gain (> 10) is two if one of the ionization rates is zero. The fact that in the staircase APD the avalanche noise is lower than in the best conventional APD's ($\alpha/\beta = \infty$) can be understood as follows: In a conventional APD the avalanche is more random because carriers can ionize everywhere in the avalanche region, while in the staircase APD electrons ionize at well-defined positions in space (i.e. the multiplication process is more deterministic). Note that, similarly, in a photomultiplier tube the avalanche is essentially noise free ($F \cong 1$).

Finally, the low voltage operation of this device with respect to conventional APD's should be mentioned. For a five stage detector and $\Delta E_c \cong E_{g1} \cong 1$ eV, the applied voltage required to achieve a gain of about 32 is slightly greater than 5 V. Possible material systems for the implementation of the device in the 1.3–1.6 μm region are AlGaAs/GaSb and HgCdTe. In a practical structure one should always leave an ungraded layer immediately after the step having a thickness of the order of a few ionization mean free paths ($\lambda_i \cong 50$–100 Å) to ensure that most electrons ionize near the step.

In progress toward the staircase APD, which has not yet been implemented, recently Capasso et al (32) demonstrated experimentally an enhancement of the α/β ($\cong 8$) in an AlGaAs/GaAs quantum well superlattice. The effect has been attributed to the difference between the conduction and valence band discontinuities ($\Delta E_c > \Delta E_v$). Thus electrons enter the well with a higher kinetic energy than holes and have a higher probability of ionizing. Note that the staircase APD is the limiting case of this detector since the whole ionization energy is gained at the band discontinuity. The staircase devices are probably the best example of the band gap engineering concept.

Repeated velocity overshoot devices   Another interesting application of staircase potentials has been proposed, the repeated velocity overshoot device (31). This structure offers the potential for achieving average drift velocities well in excess of the maximum steady-state velocity over distances greater than 1 μm. Figure 13a shows a general type of staircase potential structure. The corresponding electric field, shown in Figure 13b, consists of a series of high field regions of value E_1 and width d superimposed upon a background field E_0. To illustrate the electrical behavior and design considerations for a specific case, we consider electrons in the

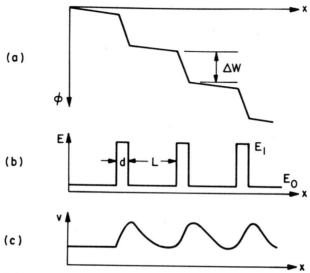

Figure 13 Principle of repeated velocity overshoot staircase potential and corresponding electric field. The ensemble velocity as a function of position is also illustrated. (From Reference 31.)

central valley of GaAs. The background field E_0 is chosen so that the steady-state electron energy distribution is not excessively broadened beyond its thermal equilibrium value, but at the same time the average drift velocity is still relatively high. For GaAs, an appropriate value would be around 2.5 kV/cm. At this field, the steady-state drift velocity is 1.8×10^7 cm/s and fewer than 2% of the electrons reside in the satellite valley. The electron distribution immediately downstream from the high field region is shifted to higher energy by an amount $\Delta W = E_1 d$. (Note that while the distribution is shifted uniformly in energy, it is compressed in momentum in the direction of transport.) We choose d so that the transit time across the high field region is shorter than the mean phonon scattering time, which is about 0.13 ps in GaAs. The energy step ΔW is chosen to maximize the average velocity of the distribution after the step while still keeping most of the distribution below the threshold energy for transfer to the satellite valley. In GaAs, the intervalley separation is about 0.3 eV, so an appropriate value of ΔW would be about 0.2 eV, resulting in an average velocity of approximately 1×10^8 cm/s immediately after the step. The momentum decays rapidly beyond the step due to scattering by polar optical phonons, with the result that the velocity decreases roughly linearly with distance, as shown in Figure 13c. During this time the distribution is broadened considerably in momentum. After the momen-

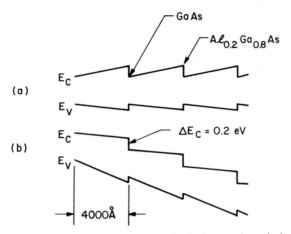

Figure 14   Band diagram of a graded-gap repeated velocity overshoot device. (From Reference 31.)

tum (and velocity) have relaxed, the distribution requires additional time to relax to its original energy. Thus, the spacing L between high field regions must be large enough to allow sufficient cooling of the electron distribution before another overshoot can be attempted. This is necessary to avoid populating the high mass satellite valleys. The effect of the resulting repeated velocity overshoot shown in Figure 13c is that average drift velocities greater than the maximum steady-state velocity can be maintained over very long distances. A practical way to achieve this device with graded-gap materials is shown in Figure 14.

Superlattice Band-Gap Grading and Pseudo-Quaternary Alloys

The growth of graded-gap structures of very short period represents a real challenge for the MBE crystal grower. In addition to a computer-controlled MBE system, new techniques to achieve very short distance compositional grading are necessary. One such technique is the recently introduced pulsed-beam method (33). This technique can be used, for example, to grow a variable gap alloy by alternately opening and closing the shutters of aluminum and gallium ovens. The result is an AlAs/GaAs superlattice with an ultrathin fixed period (≈ 20 Å) but a varying ratio of AlAs to GaAs layer thicknesses. The local band gap is therefore that of the alloy corresponding to the local average composition, determined by the thicknesses of the AlAs and the GaAs. Since the period of the superlattice is much smaller than the de Broglie wavelengths of the carriers, the material behaves basically like a *variable gap ordered alloy*. Such techniques have been used recently to grow parabolic quantum wells (34) (Figure 15).

Another interesting example of superlattice alloys are the pseudo-quaternary materials introduced by Capasso et al (35). Such artificial structures are capable of conveniently replacing GaInAsP semiconductors in a variety of applications. The concept of a pseudo-quaternary GaInAsP semiconductor is easily explained. Consider a multilayer structure of alternated $Ga_{0.47}In_{0.53}As$ and InP. If the layer thicknesses are sufficiently thin (typically a few tens of angstroms) one is in the superlattice regime. As a result, this novel material has its own band gap, intermediate between that of $Ga_{0.47}In_{0.53}As$ and InP. In the limit of layer thicknesses of the order of a few monolayers the energy band gap can be approximated by the expression

$$E_g = \frac{E_g(Ga_{0.47}In_{0.53}As)L(Ga_{0.47}In_{0.53}As) + E_g(InP)L(InP)}{L(Ga_{0.47}In_{0.53}As) + L(InP)}, \qquad 10.$$

where the L's are the layer thicknesses.

These superlattices can be regarded as novel pseudo-quaternary GaInAsP semiconductors. In fact, like $Ga_{1-x}In_xAs_{1-y}P_y$ alloys, they are grown lattice-matched to InP and their band gap can be varied between that of InP and that of $Ga_{0.47}In_{0.53}As$. The latter is done by adjusting the ratio of the $Ga_{0.47}In_{0.53}As$ and InP layer thicknesses. Pseudo-quaternary GaInAsP is particularly suited to replace variable gap $Ga_{1-x}In_xAs_{1-y}P_y$. Such alloys are very difficult to grow since the mode fraction x (or y) must be continuously varied while maintaining lattice matching to InP.

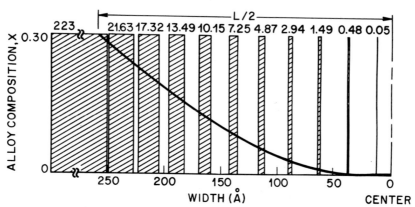

Figure 15 Compositional structure of parabolic quantum well versus distance from the well center (only half of the well is shown). The parabolic compositional profile (solid line) is obtained by growing a superlattice of alternated $Al_{0.3}Ga_{0.70}As$ and GaAs layers (dashed and white regions respectively) of varying thicknesses. The numbers at the top of the figure are the thicknesses of the $Al_{0.3}Ga_{0.70}As$ layers. (Courtesy of R. C. Miller.)

Figure 16a shows a schematic of the energy-band diagram of undoped (nominally intrinsic) graded-gap pseudo-quaternary GaInAsP. The structure consisted of alternated ultrathin layers of InP and $Ga_{0.47}In_{0.53}As$ and was grown by a new vapor phase epitaxial growth technique, levitation epitaxy (36). Other techniques, such as molecular beam epitaxy or metallorganic chemical vapor deposition, may also be suitable to grow such superlattices. From Figure 16a it is clear that the duty factor of the InP and $Ga_{0.47}In_{0.53}As$ layer is gradually varied, while the period of the superlattice is kept constant. As a result the average composition and band gap (dashed lines in Figure 16a) of the material are also spatially graded between the two extreme points (InP and $Ga_{0.47}In_{0.53}As$). In our structure both ten and twenty periods (1 period = 60 Å) were used. The InP layer thickness was decreased linearly with distance, from $\cong 50$ Å to $\cong 5$ Å,

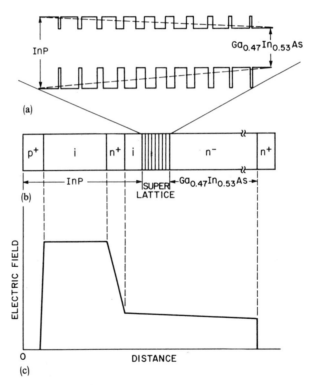

Figure 16   (*a*) Band diagram of a pseudo-quaternary graded-gap semiconductor. The dashed lines represent the average band gap seen by the carriers; (*b*) and (*c*) are the schematic structure and the electric field profile of a high-low avalanche photodiode using the pseudo-quaternary layer to achieve high speed. (From Reference 35.)

while the corresponding $Ga_{0.47}In_{0.53}As$ thickness was increased to keep the superlattice period constant ($= 60$ Å).

The graded-gap superlattice was incorporated in a long-wavelength $InP/Ga_{0.47}In_{0.53}As$ avalanche photodiode, as shown in Figure 16b. This device is basically a photodetector with separate absorption ($Ga_{0.47}In_{0.53}As$) and multiplication (InP) layers and a high-low electric field profile (HI-LO SAM APD). This profile (Figure 16c) is achieved by a thin doping spike in the ultralow doped InP layer and considerably improves the device performance compared to conventional SAM APD's (37). The $Ga_{0.47}In_{0.53}As$ absorption layer is undoped ($n \approx 1 \times 10^{15}$ cm^{-3}) and 2.5 μm thick. The n^+ doping spike thickness and carrier concentration were varied between 500 and 200 Å and 1×10^{17} to 5×10^{17} cm^{-3}, respectively (depending on the wafer), while maintaining the same carrier sheet density ($\cong 2.5 \times 10^{12}$ cm^{-2}). The n^+ spike was separated from the superlattice by an undoped 700 to 1000-Å-thick InP spacer layer. The p^+ region was defined by Zn diffusion in the 3-μm-thick low carrier density ($n^- \approx 10^{14}$ cm^{-3}) InP layer. The junction depth was varied from 0.8 to 2.5 μm. Similar devices without the superlattice region were also grown.

Previous pulse response studies of conventional SAM APD's with abrupt $InP/Ga_{0.47}In_{0.53}As$ heterojunctions found a long (> 10 ns) tail in the fall time of the detector due to the piling up of holes at the hetero-interface (38). This is caused by the large valence band discontinuity ($\cong 0.45$ eV). It has been proposed that this problem can be eliminated by inserting between the InP and $Ga_{0.47}In_{0.53}As$ region a $Ga_{1-x}In_xAs_{1-y}P_y$ layer of intermediate band gap (39). This quaternary layer is replaced in our structure by the $InP/Ga_{0.47}In_{0.53}As$ variable gap superlattice. This not only offers the advantage of avoiding the growth of the critical, inde-pendently lattice-matched GaInAsP quaternary layer, but also may lead to an optimum "smoothing out" of the valence band barrier for reproducible high-speed operation. This feature is essential for HI-LO SAM APD's since the heterointerface electric field is lower than in conventional SAM devices.

For the HI-LO SAM APD pulse response measurement we used a 1.55-μm GaInAsP device driven by a pulse pattern generator. Figure 17 shows the response to a 2-ns laser pulse with (a) and without (b) a 1300-Å-thick superlattice. Both devices had similar doping profiles and breakdown voltage ($\cong 80$ V) and were biased at -65.5 V. At this volt-age the ternary layer was completely depleted in both devices, and the measured external quantum efficiency was about 70%. The results of Figure 17 were reproduced in many devices on several wafers. The long tail in Figure 17b is due to the pile-up effect of holes, which is associated with the abruptness of the heterointerface. In devices with the graded-gap superlattice (Figure 17a) there are no long tails. In these cases the height

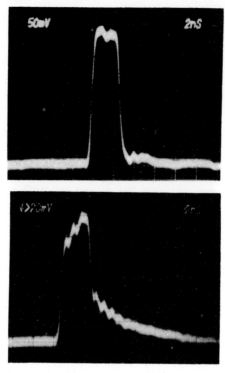

Figure 17   Pulse response of a high-low avalanche detector with pseudo-quaternary layer (*top*), and without (*bottom*), to a 2-ns, $\lambda = 1.55$-μm laser pulse. The bias voltage is -65.5 V for both devices; the time scale is 2 ns/div. (From Reference 35.)

of the barrier seen by the holes is no longer the valence band discontinuity ΔE_v, but

$$\Delta E = \Delta E_v - e\varepsilon_1 L,$$
11.

where ε_1 is the value of the electric field at the InP-superlattice interface and L is the thickness of the pseudo-quaternary layer. The devices are biased at voltage such that $\varepsilon_1 > \Delta E_v/eL$ so that $\Delta E = 0$ and no trapping occurs. In the devices with no superlattice instead ΔE is equal to ΔE_v for every ε_1, so long tails in the pulse response are observed at all voltages.

Conclusions

The previous discussion has illustrated the tremendous flexibility introduced by graded-gap material in heterostructure design. Band gap grading allows one to literally design and tailor the important transport properties to a given application (1). The most important characteristic of this band gap engineering approach is that the electron and hole transport properties can be varied independently. Another characteristic is that the energy band diagram and the associated transport properties can be varied continuously. Thus the band gap can be considered, just as the doping or the layer thicknesses, as an independent design variable.

ACKNOWLEDGMENTS

It is a pleasure to acknowledge the many colleagues who have collaborated with the author: A. Y. Cho, A. C. Gossard, W. T. Tsang, H. M. Cox, R. J. Malik, B. F. Levine, J. A. Cooper, K. K. Thornber, S. Luryi, G. F. Williams, C. G. Bethea, A. L. Hutchinson, and R. A. Kiehl. R. C. Miller kindly supplied Figure 15. I am also grateful to J. A. Hutchby for enlightening discussions.

Literature Cited

1. Capasso, F. 1983. *J. Vac. Sci. Technol.* B 1: 457–61
2. Cho, A. Y., Arthur, J. R. 1975. *Progr. Solid State Chem.* 10: 157
3. Kroemer, H. 1957. *RCA Rev.* 18: 332
4. Miller, R. C., Kleinman, D. A., Gossard, A. C., Munteanu, O. 1984. *Phys. Rev. B* 29: 7085
5. Hutchby, J. A. 1978. *J. Appl. Phys.* 49: 4041–46
6. Levine, B. F., Tsang, W. T., Bethea, C. G., Capasso, F. 1982. *Appl. Phys. Lett.* 41: 470–72
7. Levine, B. F., Bethea, C. G., Tsang, W. T., Capasso, F., Thornber, K. K., et al. 1983. *Appl. Phys. Lett.* 42: 769–71
8. Capasso, F., Tsang, W. T., Bethea, C. G., Hutchinson, A. L., Levine, B. F. 1983. *Appl. Phys. Lett.* 42: 93–95
9. Kroemer, H. 1983. *J. Vac. Sci. Technol.* B 1: 126–29
10. Hayes, J. R., Capasso, F., Gossard, A. C., Malik, R. J., Wiegmann, W. 1983. *Electron. Lett.* 19: 410–11
11. Miller, D. L., Asbeck, P. M., Anderson, R. J., Eisen, F. H. 1983. *Electron. Lett.* 19: 367–68
12. Malik, R. J., Capasso, F., Stall, R. A., Kiehl, R. A., Wunder, R., Bethea, C. G. 1985. *Appl. Phys. Lett.* 46: 600–2
13. Malik, R. J., Hayes, J. R., Capasso, F., Alavi, K., Cho, A. Y. 1983. *IEEE Electron. Devices Lett.* 4: 383–85
14. People, R., Wecht, K. W., Alavi, K., Cho, A. Y. 1983. *Appl. Phys. Lett.* 43: 118–20
15. Hayes, J. R., Capasso, F., Malik, R. J., Gossard, A. C., Wiegmann, W. 1983. *Appl. Phys. Lett.* 43: 949–51
16. Tsang, W. T. 1981. *Appl. Phys. Lett.* 39: 134–37
17. Holonyak, N. Jr., Kolbas, R. M., Dupuis, R. D., Dapkus, P. D. 1980. *IEEE J. Quantum Electron.* 16: 134–37
18. Kazarinov, R. F., Tsarenkov, G. V. 1976. *Sov. Phys. Semicond.* 10: 178–82
19. Woodall, J. M., Hovel, H. J. 1977. *Appl. Phys. Lett.* 30: 492–93
20. Capasso, F., Tsang, W. T., Hutchinson, A. L., Foy, P. W. 1982. *Proc. 1981 Symp. GaAs and Related Compounds, Oiso,* Inst. Phys. Conf. Ser. 63, pp. 473–78. London: Inst. Phys.
21. McIntyre, R. J. 1966. *IEEE Trans. Electron. Devices* 13: 164–68
22. Price, P. J. 1981. *IEEE Trans. Electron. Devices* 28: 911–14
23. Allyn, C. L., Gossard, A. C., Wiegmann, W. 1980. *Appl. Phys. Lett.* 36: 373–76
24. Gossard, A. C., Brown, W., Allyn, C. L., Wiegmann, W. 1982. *J. Vac. Sci. Technol.* 20: 694–700
25. Capasso, F., Luryi, S., Tsang, W. T., Bethea, C. G., Levine, B. F. 1983. *Phys. Rev. Lett.* 51: 2318–21
26. Capasso, F., Williams, G. F., Tsang, W. T. 1982. *Tech. Digest IEEE Specialist Conf. on Light Emitting Diodes and Photodetectors, Ottawa,* Hull, pp. 166–67
27. Capasso, F., Tsang, W. T. 1982. *Tech. Digest Intern. Electron. Devices Meet., Washington, D.C.,* pp. 334–37
28. Capasso, F. 1983. *IEEE Trans. Nucl. Sci.* 30: 424–28
29. Capasso, F. 1983. *Surf. Sci.* 132: 527–39
30. Capasso, F., Tsang, W. T., Williams, G.

F. 1983. *IEEE Trans. Electron. Devices* 30 : 381–90
31. Cooper, J. A. Jr., Capasso, F., Thornber, K. K. 1982. *IEEE Electron. Devices Lett.* 3 : 402–8
32. Capasso, F., Tsang, W. T., Hutchinson, A. L., Williams, G. F. 1982. *Appl. Phys. Lett.* 40 : 38–40
33. Kawabe, M., Matsuuza, N., Inuzuka, H. 1982. *Jpn J. Appl. Phys.* 21 : L447–48
34. Miller, R. C., Kleinman, D. A., Gossard, A. C. 1984. *Phys. Rev. B* 29 : 7085
35. Capasso, F., Cox, H. M., Hutchinson, A. L., Olsson, N. A., Hummel, S. G. 1984. *Appl. Phys. Lett.* 45 : 1193–95
36. Cox, H. M. 1984. *J. Cryst. Growth* 69 : 641–42
37. Capasso, F., Cho, A. Y., Foy, P. W. 1984. *Electron. Lett.* 20 : 635–37
38. Forrest, S. R., Kim, O. K., Smith, R. G. 1982. *Appl. Phys. Lett.* 41 : 95–97
39. Campbell, J., Dentai, A. G., Holden, W. S., Kasper, B. L. 1983. *Electron. Lett.* 19 : 818–20

Resonant Tunneling Through Double Barriers, Perpendicular Quantum Transport Phenomena in Superlattices, and Their Device Applications

FEDERICO CAPASSO, SENIOR MEMBER, IEEE, KHALID MOHAMMED AND
ALFRED Y. CHO, FELLOW, IEEE

(Invited Paper)

Abstract—New results on the physics of tunneling in quantum well heterostructures and its device applications are discussed. Following a general review of the field in the Introduction, in the second section resonant tunneling through double barriers is investigated. Recent conflicting interpretations of this effect are reconciled via an analysis of scattering. It is shown that the ratio of the intrinsic resonance width to the total scattering width (collision broadening) determines which of the two mechanisms controls resonant tunneling. The role of symmetry is quantitatively analyzed and two recently proposed resonant tunneling transistor structures are discussed. The third section deals with perpendicular transport in superlattices. A simple expression for the low field mobility in the miniband conduction regime is derived; localization effects, hopping conduction, and effective mass filtering are discussed. In the following section, experimental results on tunneling superlattice photoconductors based on effective mass filtering are presented. In the fifth section, negative differential resistance resulting from localization in a high electric field is discussed. In the last section, we report the observation of sequential resonant tunneling in superlattices. We point out a remarkable analogy between this phenomenon and paramagnetic spin resonance. New tunable infrared semiconductor lasers and wavelength selective detectors based on this effect are discussed.

Introduction

HETEROJUNCTION superlattices and their transport properties were first investigated by Esaki and Tsu in 1970 [1]. They predicted negative conductance associated with electron transfer into the negative mass regions of the minizone and Bloch oscillations. In 1971, Kazarinov and Suris theoretically studied the current–voltage (I–V) characteristic of multiquantum well structures with weak coupling between wells (tight-binding superlattices) and predicted the existence of peaks corresponding to resonant tunneling (RT) between the ground and excited states of adjacent wells [2], [3]. Calculations of RT through multiple barriers were also presented by Tsu and Esaki [4], followed in 1974 by the observation of RT through a double barrier [5]. In 1974, Esaki and Chang observed oscillatory conductance along the superlattice axis in an

AlAs/GaAs multilayer unipolar structure [6]. The voltage period of the oscillations was comparable to the energy separation between the first two conduction minibands. This effect was interpreted in terms of RT between adjacent quantum wells occurring within an expanding high-field domain. The following year, Dohler and Tsu [7], [8] predicted the existence of a new type of negative differential resistance (NDR) in a superlattice which occurs when the potential drop across the superlattice period exceeds the miniband width and the transport mechanism accordingly changes from miniband conduction to phonon-assisted tunneling (hopping). Preliminary experimental evidence of this effect was reported shortly after by Tsu et al. [9]. Tunneling injection of minority carriers (electrons) into the resonances of a quantum well and a superlattice were subsequently observed by Rezek et al. [10] and by Vojak et al. [11], respectively.

In recent years, there has been a revival of interest in RT and perpendicular quantum transport in superlattices, in large part motivated by the impressive progress achieved in molecular beam epitaxy (MBE). Low interface states densities ($\ll 10^{11}/cm^2$) can now be routinely achieved as demonstrated by high-quality modulation-doped heterojunctions exhibiting ultrahigh mobility [12]. Heterointerface abruptness is another important factor in resonant tunneling heterostructures. The interface width or abruptness (typically a monolayer) of MBE-grown heterostructures was shown to be limited by intralayer thickness fluctuations in a pioneering paper by Weisbuch et al. [13]. Recently, there has been an important breakthrough in this area. Madhukar et al. and Sakaki et al. [14], [15] demonstrated that the interruption of growth between deposition of layers can improve the morphological quality of the interfaces by allowing the surface kinetic processes to relax the growth front step density towards the generally lower step densities found for no growth III–V compound semiconductor surfaces. Essentially, the interruption of the growth for a few seconds to a few tens of seconds allows one to reduce the density of monolayer terraces in the plane of the heterointerfaces. Intralayer thickness fluctuations caused by such terraces can have a detrimental effect on resonant tunneling by weakening the coherence of the interfering electron waves reflected by

Manuscript received April 28, 1986.

F. Capasso and A. Y. Cho are with AT&T Bell Laboratories, Murray Hill, NJ 07974.

K. Mohammed was with AT&T Bell Laboratories, Murray Hill, NJ 07974. He is now with Philips Laboratories, Briarcliff Manor, NY 10510.

IEEE Log Number 8609504.

the two barriers. Intralayer fluctuations, in addition, cause fluctuations of the energy levels of the wells. As a result, in a superlattice, the overlap between the states of neighboring wells can be strongly reduced, leading to localization of the states and to hopping rather than miniband conduction. Such interrupted growth techniques appear, therefore, to have considerable potential for quantum devices and are presently widely explored in many laboratories.

The static and high-frequency transport characteristics of double barrier resonant tunneling diodes recently have been under intense experimental [16]–[22] and theoretical investigation [23]–[28] following the microwave experiments of Sollner [16], [17]; NDR in chirped superlattices [29] and in coupled superlattices [30], [31] has also been reported.

In this paper, we discuss primarily our recent work in the area of RT and perpendicular transport in superlattices.

In the second section, we discuss the physics of RT. The distinction between coherent (Fabry–Perot type) and incoherent (sequential) RT, the role of symmetry and scattering in determining which of the two RT mechanisms is operational are analyzed quantitatively and several RT transistor [32], [33] structures are presented. In the third section, the different modes of perpendicular transport in superlattices and the concept of effective mass filtering [34], [35] are analyzed, along with a discussion of localization effects. Experimental results on effective mass filtering are presented and the performance of detectors utilizing this phenomenon are discussed in the fourth section. The last section deals with the observation of sequential resonant tunneling [36] in superlattices and its applications to lasers and detectors.

II. RESONANT TUNNELING THROUGH DOUBLE BARRIERS

A. The Origin of Negative Differential Resistance

Resonant tunneling through a double barrier occurs when the energy of an incident electron in the emitter matches that of an unoccupied state in the quantum well corresponding to the same lateral momentum. Negative differential resistance arises simply from momentum and energy conservation considerations and does not require the presence of a Fabry–Perot effect. This has been clarified recently by Luryi [25] and is illustrated in Fig. 1.

Consider the Fermi sea of electrons in the degenerately doped emitter. Assuming that the barriers are free of impurities and inhomogeneities, the lateral electron momentum (k_x, k_y) is conserved in tunneling. This means that for $E_C < E_0 < E_F$ (where E_C is the bottom of the conduction band in the emitter and E_0 is the bottom of the subband in the QW), tunneling is possible only for electrons whose momenta lie in a disk corresponding to $k_z = k_0$ (shaded disk in the figure) where $\hbar^2 k_0^2 / 2m^* = E_0 - E_C$. Only those electrons have isoenergetic states in the QW with the same k_x and k_y. This is a general feature of tunneling into a two-dimensional system of states. As the emitter-base potential rises, so does the number of electrons which can tun-

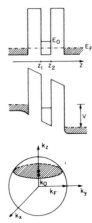

Fig. 1. Illustration of the operation of a double-barrier resonant-tunneling diode. The top part shows the electron energy diagram in equilibrium. The middle displays the band diagram for an applied bias V when the energy of certain electrons in the emitter matches unoccupied levels of the lowest subband E_0 in the quantum well. The bottom illustrates the Fermi surface for a degenerately doped emitter. Assuming conservation of the lateral momentum during tunneling, only those emitter electrons whose moments lie on a disk $k_z = k_0$ (shaded disk) are resonant. The energy separation between E_0 and the bottom of the conduction band in the emitter is given by $\hbar^2 k_0^2 / 2m^*$. In an ideal diode at zero temperature, the resonant tunneling occurs in a voltage range during which the shaded disk moves down from the pole to the equatorial plane of the emitter Fermi sphere. At higher V (when $k_0^2 < 0$), resonant electrons no longer exist. (From Luryi [25].)

nel: the shaded disk moves downward to the equatorial plane of the Fermi sphere. For $k_0 = 0$, the number of tunneling electrons per unit area equals $m^* E_F / \pi \hbar^2$. When E_C rises above E_0, then at $T = 0$ temperature, there are no electrons in the emitter which can tunnel into the QW while conserving their lateral momentum. Therefore, one can expect an abrupt drop in the tunneling current. Of course, similar arguments of conservation of lateral momentum and energy leading to NDR apply also to systems of lower dimensionality, e.g., to tunneling of two-dimensional electrons through a quantum wire and to resonant tunneling in one dimension.

B. Coherent (Fabry–Perot Type) Resonant Tunneling

Let us now consider the Fabry–Perot effect. In the presence of negligible scattering of the electrons in the well, the above NDR effect is accompanied by a resonant enhancement of the transmission identical to that occurring in an optical Fabry–Perot. Physically, what happens is that the amplitude of the resonant modes builds up in the quantum well to the extent that the electron waves leaking out in both directions cancel the reflected waves and enhance the transmitted ones. This can lead to much higher peak currents than in the case when phase coherence of the electrons waves is destroyed by scattering. In the latter case, collisions in the double barrier region randomize the phase of the electron waves and prevent the build up of the amplitude of the wave function in the well by multiple reflections. No resonant enhancement of the trans-

TABLE I
PEAK TRANSMISSION OF A DOUBLE BARRIER $Al_{0.30}Ga_{0.70}As/GaAs$ RT
DIODE, BIASED AT RESONANCE, FOR DIFFERENT EXIT BARRIER THICKNESSES
(L_{BR})

L_{BL} (Å)	L_W (Å)	L_{BR} (Å)	T_R
50	50	60	0.61
50	50	70	0.85
50	50	80	0.994
50	50	100	0.65

mission is then possible, and the electrons must be viewed as tunneling into and out of the well *sequentially* without preserving the phase coherence of the incident wave.

One should distinguish, therefore, coherent (Fabry–Perot like) RT from incoherent (sequential) RT. In the case of coherent resonant tunneling, the peak transmission at resonance is equal to T_{min}/T_{max} where T_{min} is the smallest among the transmission coefficients of the two barriers and T_{max} is the largest [23]. It is clear, therefore, that to achieve unity transmission at the resonance peak, the transmission of the left and right barriers must be equal, just like in an optical Fabry–Perot. This crucial role of symmetry has been discussed in detail by Ricco and Azbel [23]. Application of an electric field to a symmetric double barrier introduces a difference between the transmission of the two barriers, thus significantly decreasing below unity the overall transmission at the resonance peaks. Unity transmission can be restored if the two barriers have different and appropriately chosen thicknesses; obviously, with this procedure, one can only optimize the transmission of one of the resonance peaks. We have theoretically investigated this optimization (see Table I) in the case of resonant tunneling through the ground state resonance of a structure consisting of a 50 Å GaAs quantum well sandwiched between two $Al_{0.3}Ga_{0.7}As$ barriers. The barrier height in this case is 0.2 eV (taking $\Delta E_c = 0.57\Delta E_g$ for the conduction band discontinuity). The ground state resonance lies at 69.04 meV from the bottom of the quantum well. Consider electrons at the bottom of the conduction band in the emitter layer. When a bias $\simeq 2E_1/e$ is applied to the double barrier, electrons can resonantly tunnel through the double barrier. The peak transmission (as a function of voltage) is not unity, but 0.343. This is because the exit barrier becomes lower, and therefore has a higher transmission, under application of the electric field, than the input barrier. Unity transmission can be restored by making the exit barrier thicker as illustrated in Table I. Note that the transmission at resonance increases as the exit barrier is made thicker and reaches unity when the thickness is $\simeq 80$ Å.

C. The Role of Scattering: Incoherent (Sequential) Resonant Tunneling

RT through a double barrier has been investigated experimentally by many researchers [5], [16]–[22]. All of these investigations assumed that a Fabry–Perot type enhancement of the transmission was operational in such structures. However, as previously discussed, the observation of NDR does not imply a Fabry–Perot mechanism. Other types of tests are necessary to show the presence of a resonant enhancement of the transmission, such as the dependence of the peak current on the thickness of the exit barrier discussed in the previous section.

Scattering can considerably weaken the above enhancement of the transmission. A lucid discussion of this point has been recently given by Stone and Lee [37] in the context of RT through an impurity center. Unfortunately, their work has gone unnoticed among workers in the area of quantum well structures. Their conclusions can also be applied to the case of RT through quantum wells and we shall discuss them in this context.

To achieve the resonant enhancement of the transmission (Fabry–Perot effect), the electron probability density must be peaked in the well. Therefore, it takes a certain time constant to build up the steady-state resonant probability density in the well, i.e., to achieve high transmission at resonance. This time constant τ_0 is on the order of h/Γ_r where Γ_r is the full width at half maximum of the transmission peak. Collisions in the double barrier tend to destroy the coherence of the wavefunction, and therefore the electronic density in the well will never be able to build up to its full resonant value. If the scattering time τ is much shorter than τ_0, the peak transmission at resonance is expected to be decreased by the ratio τ_0/τ. The scattering time τ is simply the reciprocal of the *total* scattering rate, and thus includes both elastic scattering by carriers, impurities, and nonhomogeneities in the layer thicknesses (terraces) and inelastic scattering by phonons. In their treatment of one-dimensional resonant tunneling, Lee and Stone [37] only considered the effect of inelastic collisions on the transmission. It should be clear that their main physical conclusions are also valid if elastic collisions are added since every type of collision tends to prevent the resonant build up of the wavefunction in the well. The principal effects of collisions are to decrease the peak transmission by the ratio $\tau_0/(\tau_0 + \tau)$ and to broaden the resonance. In addition, the ratio of the number of electrons that resonantly tunnel without undergoing collisions to the number that tunnel after undergoing collisions is equal to τ/τ_0 [37]. To summarize, coherent resonant tunneling is observable when the intrinsic resonance width $(\simeq h/\tau_0)$ exceeds or equals the collision broadening $(\simeq h/\tau)$. In the other limit, electrons will always tunnel through one of the intermediate states of the well, but they will do it incoherently without resonant enhancement of the transmission. We shall apply now the above criterion to RT through AlGaAs/GaAs double barrier recently investigated in many experiments.

Consider a 50 Å thick GaAs well sandwiched between two $Al_{0.30}Ga_{0.70}As$ barriers. Table II shows the ground state resonance widths Γ_r (full width at half maximum of the transmission curve) calculated for different values of the barrier thicknesses L_B (assumed equal). Note the strong dependence of Γ_r on L_B. This is due to the fact that Γ_r is proportional to the transmission coefficient of the individual barriers which decreases exponentially with increasing L_B. The case $L_B = 50$ Å corresponds to the microwave oscillator recently reported by Sollner [17].

Because of dimensional confinement in the wells and because the wells are undoped, one can obtain a good estimate of the scattering time of electrons in the wells from the mobility of the two-dimensional electron gas (in the plane of the layers), measured in selectively doped AlGaAs/GaAs heterojunctions [12]. For state-of-the-art selectively doped AlGaAs/GaAs heterojunctions, the electron mobility at 300 K is ≈ 7000 cm^2/s · V. From this value, we can infer an average scattering time $\approx 3 \times 10^{-13}$ s which corresponds to a broadening of ≈ 2 meV. In Table II, we have also plotted the ratio of the resonance width Γ_r to the collision broadening Γ_c. For the 50 and 70 Å barrier case, the resonance width is much smaller than the collision broadening so that, by the previously discussed criterion, there is very little resonant enhancement of the transmission via the Fabry–Perot mechanism at 300 K. However, the latter effect should become visible in structures with thinner barriers < 30 Å, as seen from Table II. Consider now a temperature of 200 K; from the mobility ($\approx 2 \times 10^4$ cm^2/s · V) [12], one deduces $\tau \approx 1$ ps which corresponds to a broadening of ≈ 0.67 meV. This value is comparable to the resonance width for a barrier width of 50 Å. This implies that in Sollner's microwave oscillators [17] (which operated at 200 K), coherent resonant tunneling effects were probably present. This is definitely not the case for the mixing and detection experiments performed up to terahertz frequencies in double barrier RT structures with $L_w = L_B = 50$ Å, $x = 0.25$–0.30 at a temperature of 25 K. In this case, the well was intentionally doped to $\approx 10^{17}$ cm^{-3} which would correspond to a mobility of ≈ 3000 cm^2/s · V which gives a collision broadening of 4 meV which is significantly larger than the resonance width. Thus, in this case, electrons are tunneling incoherently (i.e., sequentially) through the double barrier. A similar conclusion has been reached by Luryi [25] based on a calculation of the intrinsic RC time constant of RT double barriers.

Finally, in Table II, we have estimated Γ_r/Γ_c for a temperature of 77 K. State-of-the-art mobilities in selectively doped interfaces exceed 10^5 cm^2/s · V so that scattering times are typically longer than 1 ps and the broadenings are less than 0.5 meV. Thus, coherent RT will significantly contribute to the current for barrier widths ≤ 70 Å and dominate for $L_B \leq 30$ Å. The values of Γ_r/Γ_c at 70 K in Table II were obtained using a mobility of 3×10^5 cm^2/s · V [12].

The situation appears to be different in the case of AlAs/GaAs double barrier with well widths of 50 Å. The confining barriers in this case are much higher (≈ 1.35 eV) [21], and for barrier thicknesses in the 30–70 Å range, the resonance widths are $\leq 10^{-2}$ meV. Thus, coherent RT is negligible at room temperature, but is expected to be-

TABLE II
RESONANCE AND COLLISION WIDTHS OF Al$_{0.30}$Ga$_{0.70}$As/GaAs RT DIODE (AT ZERO BIAS) FOR DIFFERENT BARRIER THICKNESSES

L_w (Å)	L_B (Å)	Γ_r (meV)	Γ_r/Γ_c (≡ RESONANCE WIDTH/COLLISION BROADENING)		
			300 K	200 K	70 K
50	70	1.28×10^{-2}	6×10^{-3}	1.93×10^{-2}	2.6×10^{-1}
50	50	1.5×10^{-1}	7.5×10^{-2}	2.26×10^{-1}	3.08
50	30	1.76	8.8×10^{-1}	1.32	3.62
50	20	6.03	3.02	4.56	124.02

come dominant at 70 K for $L_B \leq 70$ Å in high-quality double barriers.

D. Resonant Tunneling Transistors

From the considerations previously developed, it is clear that in order to achieve near unity transmission at all the resonance peaks, the transmission of the left and right barriers must be equal at all the quasi-eigenstate energies and the collisional broadening must be much smaller than the intrinsic resonance widths of all resonances. Let us assume that the latter condition is satisfied by appropriately choosing the double barrier height and dimensions and the operating temperature. The first condition, on the other hand, can never be satisfied if the tunneling is induced by applying a field to the double barrier, as previously discussed. To overcome this problem, recently Capasso and Kiehl [32] have proposed a new class of structures where resonant tunneling through a symmetric double barrier is achieved not by applying an electric field to the barriers, but by minority carrier high-energy or ballistic injection. This method does not alter the transmission of the two barriers, and therefore should lead to near unity transmission at all resonance peaks and to larger negative conductance and peak-to-valley ratios than conventional resonant tunneling structures.

Fig. 2 shows the band diagram of one of the devices. The structure is a heterojunction bipolar transistor with a degenerately doped tunneling emitter and a symmetric double barrier in the base. The collector current as a function of the base–emitter voltage V_{BE} exhibits a series of peaks corresponding to resonant tunneling through the various quasi-stationary states of the well. Multiple negative conductance in the collector circuit can therefore be achieved.

An alternative injection method is the abrupt or nearly abrupt emitter which can be used to ballistically launch electrons into the quasi-eigenstates with high momentum coherence. As V_{BE} is increased, the top of the launching ramp eventually reaches the same energy of the quasi-eigenstates so that electrons can be ballistically launched into the resonant states [Fig. 3(a)].

To achieve equally spaced resonances in the collector current, the rectangular quantum well in the base should be replaced by a parabolic one [Fig. 3(b)]. Parabolic

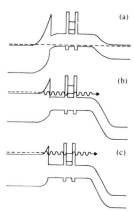

Fig. 2. Band diagram of resonant tunneling transistor (RTT) with tunneling emitter under different bias conditions: (a) in equilibrium, (b) resonant tunneling through the first level in the well, (c) resonant tunneling through the second level. (Not to scale.)

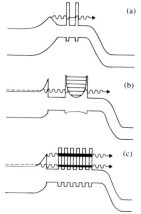

Fig. 3. (a) Band diagram of RTT with graded emitter (at resonance). Electrons are ballistically launched into the first quasi-eigenstate of the well. (b) RTT with parabolic quantum well in the base and tunneling emitter. A ballistic emitter can also be used. (c) RTT with superlattice base. (Not to scale.)

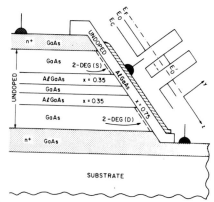

Fig. 4. Schematic cross section of the proposed surface resonant tunneling device structure and band diagram along the z direction. Thicknesses of the two undoped GaAs layers outside the double-barrier region should be sufficiently large (≥ 1000 Å) to prevent the creation of a parallel conduction path by the conventional (bulk) resonant tunneling. E_0 is the bottom of the 2D subband separated from the classical conduction band minimum by the energy of the zero-point motion in the y direction; E_0' is the bottom of the 1D subband in the quantum wire, separated from E_0 by the confinement energy in the z direction. In the operating regime, the Fermi level E_f lies between E_0 and E_0'.

quantum wells have been recently realized in the AlGaAs system [38]. Assuming the depth of the parabolic well in the conduction band to be 0.34 eV (corresponding to grading from $Al_{0.45}Ga_{0.55}As$ to GaAs) and its width to be 200 Å, one finds that the first state is at an energy of 32 meV from the bottom of the well and that the resonant states are separated by $\simeq 64$ meV. This gives a total of five states in the well.

Finally, in Fig. 3(c), we illustrate another application, that of studying high-energy injection and transport in the minibands of a superlattice, using ballistic launching or tunnel injection.

These new functional devices, because of their multiple resonant characteristic, can have potential for multiple-valued logic applications. In addition, by combining a number of these transistors in a parallel array, an ultra-high speed (~ 20 GHz) analog-to-digital converter could be realized [39].

Recently, Yokohama *et al.* [40] have demonstrated a unipolar resonant tunneling hot electron transistor in which the double barrier is placed in the emitter of an unipolar structure. Negative conductance has been achieved at 77 K, controlled by the base–emitter voltage.

Luryi and Capasso [33] described another type of RT transistor. The main difference compared to the previous bipolar device is that the structure is unipolar. In addition, the QW is linear rather than planar and the tunneling is of 2D electrons into a 1D density of states. Fig. 4 shows the schematic cross section of the proposed device. It consists of an epitaxially grown undoped planar QW and a double AlGaAs barrier sandwiched between two undoped GaAs layers and heavily doped GaAs contact layers. The working surface defined by a V-groove etching is subsequently overgrown epitaxially with a thin AlGaAs layer and is gated. Application of a positive gate voltage V_G induces 2D electron gases at the two interfaces with the edges of undoped GaAs layers outside the QW. These gases will act as the source (S) and drain (D) electrodes. At the same time, there is a range of V_G in which electrons are not yet induced in the "quantum wire" region (which is the edge of the QW layer) because of the additional dimensional quantization. The operating regime of our device is in this range. Application of a positive drain voltage V_D brings about the resonant tunneling condition, and one expects an NDR in the dependence $I(V_D)$. What is more interesting is that this condition is also controlled by V_G. The control is affected by fringing electric fields: in the operating regime, an increasing $V_G > 0$ *lowers* the electrostatic potential energy in the base with respect to the emitter—nearly as effectively as does the increasing V_D (this has been confirmed by solving the corresponding electrostatic problem exactly with the help of suitable conformal mappings). At a fixed V_G having established the peak of $I(V_D)$, we can then quench the tunneling current by increasing V_G. This implies the possibility of achieving *negative transconductance*—an entirely novel feature in a unipolar device. A negative-transconductance transistor can perform the functions of a complementary device analogous to a p-channel transistor in the silicon CMOS logic. A circuit formed by a conventional n-channel field-effect transistor and our device can act like a low-power inverter in which a significant current flows only during switching. This feature can find applications in logic circuits.

III. PERPENDICULAR TRANSPORT IN A SUPERLATTICE: MINIBAND CONDUCTION, LOCALIZATION, AND HOPPING

In a superlattice, the barrier thicknesses become comparable to the carrier de Broglie wavelength; thus, the wavefunctions of the individual wells tend to overlap due to tunneling and an energy miniband of width 2Δ is formed [1]. The width 2Δ is proportional to the tunneling probability through the barriers which, for rectangular barriers and not too strong coupling between wells, can be approximated by

$$T_e \simeq \exp\left[-\sqrt{\frac{8m_e^*}{\hbar^2} (\Delta E_c - E_{i,e})} L_B \right] \quad (1)$$

in the case of electrons and

$$T_{hh} \simeq \exp\left[-\sqrt{\frac{8m_{hh}^*}{\hbar^2} (\Delta E_v - E_{i,hh})} L_B \right] \quad (2)$$

in the case of heavy holes where m_e^* and m_{hh}^* are the electron and heavy hole effective masses, ΔE_c and ΔE_v are the conduction and valence band discontinuities, $E_{i,e}$ and $E_{i,hh}$ are the bottom of the ground state electron and heavy hole minibands, and L_B is the barrier thickness. The miniband width and, of course, the miniband energies can be calculated rigorously by solving Schroedinger's equation [41].

The above picture assumes a perfect superlattice, with no thickness or potential fluctuations and no scattering by either impurities or phonons. In reality, one must contend with such fluctuations and with the unavoidable presence of scattering. Such effects tend to disturb the coherence of the wavefunction and the formation of extended Bloch states which give rise to the miniband picture and have profound effects on perpendicular transport.

Consider first the weak electric field limit in which the potential energy drop across the superlattice period is smaller than the miniband width. Transport then proceeds by miniband conduction if the low-field mean free path of the carriers appreciably exceeds the superlattice period

[1], [6]. Palmier and Chomette [42] have studied this transport regime and calculated scattering rates for different scattering mechanisms, but did not give any simple analytical expression for the mobility.

It is easy to derive a phenomenological expression of the mobility $\mu_\parallel$ along the superlattice axis. Let us consider for simplicity a one-dimensional model and describe the band structure along the superlattice axis by the energy dispersion relationship [1]

$$E(k) = \Delta[1 - \cos k_\parallel d] \quad (3)$$

in which $k_\parallel$ is the component of the wavevector parallel to the superlattice axis and d is the superlattice period.

The average group velocity along the superlattice axis (drift velocity) is obtained from (3):

$$v_d = \left(\frac{1}{\hbar} \frac{dE}{dk_\parallel}\right)_{k=\bar{k}\parallel} = \frac{\Delta d}{\hbar} \sin(\bar{k}_\parallel d) \quad (4)$$

where $\bar{k}_\parallel$ is the steady-state average wavevector obtained from the momentum rate equation

$$\hbar \frac{d\bar{k}_\parallel}{dt} = eF - \frac{\hbar \bar{k}_\parallel}{\tau} = 0 \quad (5)$$

where F is the electric field and τ is the relaxation time for the momentum $p_\parallel = \hbar k_\parallel$. Substituting (3) and (5) in (4), one obtains

$$v_d = \frac{\Delta d}{\hbar} \sin\left(\frac{eF\tau}{\hbar} d\right). \quad (6)$$

For small electric fields (mobility regime), (6) reduces to

$$v_d = \frac{e\Delta d^2 \tau}{\hbar^2} F. \quad (7)$$

The mobility $\mu_\parallel$ is then

$$\mu_\parallel = \frac{e\Delta d^2}{\hbar^2} \tau. \quad (8)$$

Note that $\hbar^2/\Delta d^2$ represents the band-edge effective mass in the direction parallel to the superlattice axis, as can be seen using (3) and the definition

$$m_\parallel^* = \frac{\left(\dfrac{\hbar^2}{d^2 E}\right)}{\left(\dfrac{}{dk_\parallel^2}\right)_{k_\parallel = 0}}. \quad (9)$$

There are several important conclusions to be drawn from (8). Since the mobility is proportional to the miniband width, and the latter is proportional to the transmission coefficient of the superlattice, $\mu_\parallel$ decreases strongly with increasing barrier and well thicknesses [42]. It follows that $\mu_\parallel$ can be varied over a wide range by slight variations of the barrier and well layer thicknesses. Furthermore, for superlattices in which the barriers seen by holes are not much lower than those seen by electrons (which occurs in many heterojunctions), the electron mobility $\mu_{e\parallel}$ can be made much greater than $\mu_{hh\parallel}$ since the tunneling probability depends exponentially on the effective mass [see (1) and (2)]. This implies that the superlattice can act as a filter for effective masses [34], [35] by easily transmitting the light carriers (electrons) and effectively slowing down the heavy carrier (heavy hole). In fact, it will soon be clear that heavy holes in most practical cases remain localized in the quantum wells.

Once τ is known, one can obtain $\mu_{\parallel}$ from (8). One can get, nevertheless, a rough estimate of $\mu_{\parallel}$ by assuming that τ in (8) is not too different from τ in an alloy with a composition equal to the average composition of the superlattice. The smaller the superlattice period, the better this approximation is. If the mass and the mobility μ_{all} of this alloy are known, one can then obtain an estimate of $\mu_{\parallel}$ from

$$\mu_{\parallel} = m_{all}^* \frac{\Delta d^2}{\hbar^2} \mu_{all}. \qquad (10)$$

Consider the case of an $Al_{0.48}In_{0.52}As/Ga_{0.47}In_{0.53}As$ superlattice with 35 Å wells and 35 Å barriers. The width of the first miniband is 17 meV [35] and $d \simeq 70$ Å. For $m_{all\,of\,the\,electrons}^*$, we can take the mean between the masses of the two constituents [42], i.e., $\simeq 0.06 m_0$, and for μ_{all} estimate ≈ 3000 cm^2/s · V at 300 K. One then obtains $\mu_{\parallel} \simeq 990$ cm^2/s · V for the electron mobility.

As the barrier thickness increases, the miniband width 2Δ decreases exponentially. The maximum group velocity in the miniband ($v_{max} = \Delta d/\hbar$) and the mobility decrease proportionally with the bandwidth. The relaxation time, however, is practically independent of d since it is dominated by intralayer processes. Eventually, $v_{max}\tau$, which is always greater than the mean free path λ, becomes smaller than d even for an ideal superlattice without layer thickness or compositional fluctuations [7]. This can be written approximately as

$$\frac{\Delta d}{\hbar} \tau < d \qquad (11)$$

from which it follows

$$\hbar/\tau > \Delta, \qquad (12)$$

i.e., the collision broadening is greater than the miniband width, which also implies that $\lambda < d$. If this condition or the one on the mean free path is satisfied, the states of the superlattice are no longer Bloch waves, but are localized in the wells along the direction perpendicular to the layer (the states, however, will be, in general, always delocalized in the plane of the layers). As discussed in the Introduction, localization may typically arise as a result of intralayer and interlayer thickness fluctuations and alloy disorder which cause fluctuations of the energies of the quasi-eigenstates of the wells. If this nonhomogeneous broadening exceeds the intrinsic miniband width, there are no Bloch states, and again the wavefunction becomes localized in the wells [34], [35]. From the previous discussion, it should be clear that phonon scattering alone, if sufficiently strong, can induce localization.

It is important to note that this type of localization occurring in superlattices is of the Anderson type [43] and has profound effects on perpendicular transport. In fact, in this case, conduction proceeds by phonon-assisted tunneling (hopping) between adjacent layers. For superlattices of III–V materials such as $Al_{0.48}In_{0.52}As/Ga_{0.47}In_{0.53}As$, $AlGa_{1-x}As/GaAs$, $InP/Ga_{0.47}In_{0.53}As$ and equal barrier and well thicknesses, it is easily shown, using the above criterion, that localization of the electrons occurs when d is on the order of or exceeds 100 Å.

Heavy holes, on the other hand, in these and most other superlattices become localized for much smaller d (≈ 15 Å) due to their much larger effective mass. This implies a much smaller miniband width than electrons, so that the localization criterion is very easily satisfied. A superlattice therefore tends to selectively localize carriers, acting as an effective mass filter.

The mobility perpendicular to the layers then cannot be described by (8) and becomes very small. Several authors have investigated this case. In the limit of strong localization where one can neglect transitions other than those between adjacent wells, the mobility is given by [44]

$$\mu \simeq \frac{ed^2}{kT} \langle W \rangle \qquad (13)$$

where $\langle W \rangle$ is the thermodynamically averaged phonon-assisted tunneling rate between adjacent wells, which is proportional to the tunneling probability.

The concept of effective mass filtering is, of course, also valid in the case where conduction occurs by phonon-assisted tunneling since the electron hopping mobility is much larger, in general, than that of heavy holes.

At high electric fields (approaching $\sim 10^5$ V/cm), heavy holes, however, tend to be much less localized than at low fields, due to hot carrier effects and barrier lowering (enhanced thermionic emission) which dramatically increase the tunneling probability and reduce the filtering effect of the superlattice. This phenomenon was observed by Capasso et al. [45] in transport through a graded gap chirped superlattice and was theoretically investigated by Weil and Winter [46].

IV. EFFECTIVE MASS FILTERING: TUNNELING PHOTOCONDUCTORS

The effective mass filtering effect associated with the large difference between the tunneling rates of electrons and holes gives rise to a new type of photoconductivity of quantum mechanical origin [34], [35]. The underlying mechanism of this effect is illustrated in Fig. 5 which shows the band diagram of the superlattice photoconductor with applied bias. Photogenerated (heavy) holes remain relatively localized (their hopping probability is negligible), while photoelectrons and those injected by the n^+ contacts are transported through the superlattice by phonon-assisted tunneling [Fig. 5(a)] or miniband conduction [Fig. 5(b)], depending on whether the electronic states are localized or not. This effective mass filtering effect produces a photoconductive gain given by the ratio

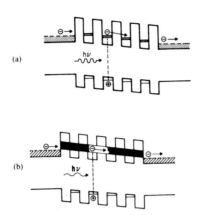

(a)

(b)

Fig. 5. Band diagram with applied bias of the superlattice detector and schematic illustration of effective mass filtering in the case of (a) phonon-assisted tunneling between the wells and (b) miniband conduction.

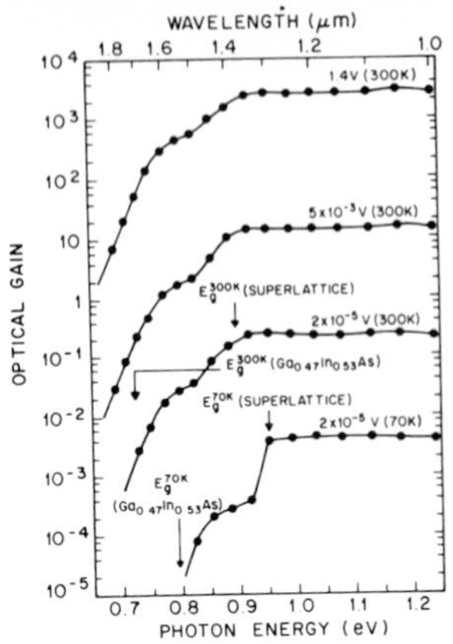

Fig. 6. Spectral response of the effective mass filter at different biases with positive bias polarity with respect to the substrate. The arrows indicate the bandgaps of the bulk and superlattice layers of the detector.

of the electron lifetime τ (which defines the response time of a photoconductor) to the electron transit time ($t_e = L/\mu_e F$). The latter increases exponentially with the barrier thickness and is also sensitive to the well thickness [see (8) and (13)]. It follows that the gain ($=\tau/t_e$) and gain-bandwidth product ($=1/t_e$) of these novel photodetectors can be easily tuned over a very wide range by varying the superlattice period and/or the duty factor, a unique feature not available in conventional photoconductors. The advantage of the scheme of Fig. 5(b) over that of Fig. 5(a) is that electrons can attain much higher mobilities if the miniband is sufficiently wide. Much shorter transit times and greater gain–bandwidth products (several gigahertz) should therefore be attainable.

In conventional photoconductors, the current gain is given by the ratio of the electron and hole velocities if the lifetime exceeds the hole transit time. In the opposite case, instead, the gain is given by the ratio of the lifetime to the electron transit time. This means that the current gain and also the speed of the photoconductor are controlled by bulk material properties such as mobilities and lifetimes. In superlattice quantum photoconductors, instead, the gain is in general controlled by the lifetime-to-electron transit time ratio since holes are localized in the wells.

Effective mass filtering was recently demonstrated by Capasso et al. [34], [35]. The structures were grown on $\langle 100 \rangle n^+$-InP and consist of an undoped ($n \approx 5 \times 10^{15}$ cm^{-3}) Al$_{0.48}$In$_{0.52}$As (35 Å)/Ga$_{0.47}$In$_{0.53}$As (35 Å) 100 period superlattice sandwiched between two degenerately doped n^+ (2×10^{18}/cm^3) 0.45 μm thick Ga$_{0.47}$In$_{0.53}$As layers.

Fig. 6 illustrates the spectral dependence of the optical gain at different bias voltages, with the top of the device positively biased with respect to the substrate. The optical gain is given by $G_0 = (h\nu/e)\ (I/P_0)$ where I is the photocurrent, P_0 is the incident optical power, I/P_0 is the responsivity, $h\nu$ is the photon energy, and e is the electronic charge. G_0 can also be expressed as the product of the

external quantum efficiency η and the current gain. The optical gain is extremely sensitive to the applied bias and increases by orders of magnitude as the voltage is increased by a few millivolts, reaching values well in excess of 10^3 at 1.4 V. Note that at bias as low as 5 mV, corresponding to an electric field of 66 V/cm, there is already a sizable current gain since the optical gain ≈ 10 at $\lambda < 1.4$ μm. The external quantum efficiency can be estimated from the layer thicknesses and absorption constants to be ≈ 0.14 at $\lambda \approx 1.2$ μm. Thus, the room-temperature spectral response curve at 2×10^{-5} V corresponds to the onset of current gain. At photon energies >0.82 eV at 300 K, there is a step-like increase corresponding to the onset of the absorption in the superlattice, i.e., to photoexcitation from the heavy hole miniband to the ground-state electron miniband. This transition defines the superlattice bandgap. Its theoretical value, indicated by the arrows, was obtained by adding to the Ga$_{0.47}$In$_{0.53}$As bulk bandgap the energies of the bottom of the ground-state electron and hole minibands (0.137 and 0.034 eV, respectively) [34]. For the band offsets, the experimental values $\Delta E_c = 0.5$ eV and $\Delta E_v = 0.23$ eV were used. Good agreement with the experiment is observed. The low-energy portion of the photoresponse curve between 0.7 and 0.82 eV is due to photocarriers which are photogenerated in the top and bottom n^+-Ga$_{0.47}$In$_{0.53}$As layers ($E_g = 0.73$ eV at 300 K) and diffuse to the superlattice region where they are collected. The low-temperature curve (70 K) reveals more clearly the superlattice effects.

In Fig. 7, we have plotted the measured responsivity (I/P_0) versus voltage for opposite bias polarities (with respect to the substrate) at $\lambda = 1.2$ μm on a semilog scale and on an expanded linear scale at very low biases in the inset. The responsivity is linear with voltage up to ≈ 0.2 V. Above 0.2 V, the asymmetry with opposite bias polarity becomes significant. This is due to the nonuniform photoexcitation of the superlattice and in part also to the

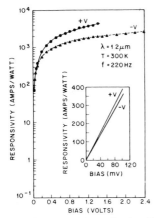

Fig. 7. Effective mass filter responsivity as a function of bias (for opposite polarities with respect to the substrate) at $\lambda = 1.2 \ \mu m$. The inset shows the responsivity at very low voltages on an expanded scale.

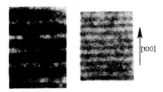

Fig. 8. Transmission electron micrograph of two superlattices used as effective mass filters: $Al_{0.48}In_{0.52}As$ (23 Å)/$Ga_{0.47}In_{0.53}As$ (49 Å) (left); $Al_{0.48}In_{0.52}As$ (35 Å)/$Ga_{0.47}In_{0.53}As$ (35 Å) (right).

observed asymmetry in the dark *I–V*. The latter is attributed to microscopic differences, observed by transmission electron microscopy, between the top and bottom superlattice/$Ga_{0.47}In_{0.53}As$ interfaces. The responsivity is $\approx 10^3$ A/W at 0.3 V and then tends to saturate at $\approx 4 \times 10^3$ A/W for positive polarity. This corresponds to an internal current gain of $\approx 2 \times 10^4$. Note that significant responsivities (≈ 50 A/W) are obtained at voltages as low as 20 mV. From measurements of the responsivity as a function of the light modulation frequency, we determined a response time $\tau \approx 10^{-3}$ s.

We also found that the responsivity decreased nearly exponentially with temperature in the range 300–70 K. In addition, the optical gain strongly decreased in structures with thicker barriers (70 Å) and no current gain was observed for barrier thicknesses ≥ 100 Å. These features clearly indicate that transport is controlled by phonon-assisted tunneling. The wells are coupled because of the small layer thickness (35 Å) so that the quantum states tend to form minibands. The calculated widths of the ground-state electron and hole minibands are, respectively, 17 and 0.23 meV [34]. Transmission electron microscopy studies in the present superlattice [Fig. 8(a)] indicate that the interfaces are abrupt within two monolayers, thus causing estimated fluctuations of the energy of the ground-state minibands on the order of (in the case

of electrons) or greater (in the case of holes) than the miniband widths. If one also considers compositional fluctuations and collisional broadening (≈ 5 meV), it can easily be shown that in our superlattices, the previously discussed localization criteria are satisfied for both the electron and hole states.

The photocurrent–voltage characteristic is described by

$$I = (e\eta P_0/h\nu) \ (\mu_e \tau V/L^2). \qquad (14)$$

The second factor in parentheses is the photoconductive gain. Fitting the linear part of the experimental responsivity (I/P_0) curve (inset of Fig. 7) with (14), we obtain for the mobility $\mu_e = 0.15$ cm^2/s $\cdot$ V. Such low mobility is expected for phonon-assisted tunneling conduction. Recent calculations [44] and conductivity measurements [47] in AlGaAs/GaAs superlattices find a mobility comparable to ours for a comparable superlattice period. Additional strong evidence for hopping electron conduction comes from the observed temperature dependence of the quantum efficiency. This decreases strongly with decreasing temperature since the phonon-assisted tunneling probability decreases with the number of available phonons. The electron hopping rate by thermionic emission across the barriers is negligible compared to the tunneling rate, due to the large $\Delta E_c/kT$.

There are several mechanisms that in a superlattice can lead to a large enhancement of the lifetime with values of τ on the order of those deduced in our structures. A reduced spatial overlap between electron and hole states can be, in this respect, an important factor. For example, the slow carrier (hole) can be captured by a defect state in the wide gap barriers. In particular, AlGaAs and $Al_{0.48}In_{0.52}As$ layers may contain relatively large densities of defects. Recombination with an electron will occur then through phonon-assisted electron tunneling into the barrier. This indirect recombination process has a very small probability, leading to long lifetimes [34].

The AlInAs/GaInAs photoconductors were also characterized from the point of view of noise performance. The noise equivalent power (NEP) was found to have a minimum at -0.2 V bias. This corresponds to $\approx 1.4 \times 10^{-13}$ W. The corresponding detectivity D^* at 1 kHz modulation frequency and $\lambda = 1.3 \ \mu m$ is $\approx 10^{11}$ (cm $\cdot$ Hz$^{1/2}$/W). These are the highest gain and lowest noise photoconductors achieved at such low bias.

We have also observed effective mass filtering in $Al_{0.35}Ga_{0.75}As$/GaAs MBE-grown photoconductors having similar layer thickness. The current gain and the response time are comparable to those measured in the $Al_{0.48}In_{0.52}As$/$Ga_{0.47}In_{0.53}As$ photoconductors.

Effective mass filtering associated with miniband conduction of electrons was observed recently by us in a forward-biased superlattice p-n junction [35]. The energy band diagram is illustrated in Fig. 9(a). The undoped n-type, 7200 Å thick, $Al_{0.48}In_{0.52}As$/$Ga_{0.47}In_{0.53}As$ superlattice had 23 Å thick barriers and 49 Å thick wells [Fig. 8(a)] and is sandwiched between a p$^+$ and an n$^+$

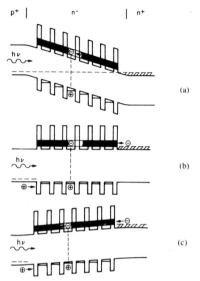

Fig. 9. Schematic energy band diagram of the superlattice p-n junction in equilibrium (a), at a forward bias voltage equal to the built-in potential (flat band) (b), beyond flat band (c). Shown also is the effective mass filtering mechanism.

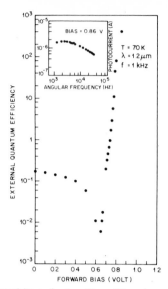

Fig. 10. Responsivity as a function of forward bias voltage at $\lambda = 1.2\ \mu m$ and $T = 70$ K of superlattice effective mass filter p-n junction. The inset shows the frequency response of the structure in the high photoconductive gain region (0.86 V bias).

$Ga_{0.47}In_{0.53}As$ layer. Rigorous calculations, which include nonparabolicities, show that the energies of the bottom of the ground state electron and hole minibands are 90 and 20 meV, respectively [35]. The calculated width of the electron ground-state miniband is 30 meV and is greater than the combined compositional nonhomogeneous broadening due to the fluctuations and collision broadening due to phonons ($\simeq 10$ meV). Electron transport perpendicular to the layers occurs, therefore, by miniband conduction.

The situation is very different for holes. The ground-state heavy-hole miniband is only 0.7 meV wide, which is much smaller than nonhomogeneous and collisional broadening. Thus, holes are localized perpendicular to the layers and are transported by hopping between adjacent wells.

For forward bias voltages smaller than the built-in voltage ($\simeq 0.65$ V), the p-n junction acts like a photodiode and, as the forward bias is increased, the quantum efficiency is expected to decrease and reach a minimum near flat band conditions [Fig. 9(b)]. As the forward bias is further increased, the electric field inside the device changes sign and the direction of motion of the photocarriers is reversed [Fig. 9(c)]. Now electrons (holes) drift in the same direction as the electrons (holes) injected from the contact regions; in other words, the photocurrent has the same direction as the dark current. Thus, one can observe photoconductivity and photoconductive current gain by effective mass filtering.

This is precisely what is found experimentally (Fig. 10). As the forward bias is increased, the responsivity decreases and reaches a minimum at $V = V_{bi}$. For $V > V_{bi}$,

the photocurrent changes sign and the responsivity increases by orders of magnitudes. The large value of the responsivity clearly indicates the presence of high photoconductive gain. The inset of Fig. 10 shows the frequency response of the photoconductor at +0.86 V bias.

Detailed studies of the photocurrent and of the spectral response at different temperatures and in samples with different superlattice barrier thicknesses show very clearly that the mechanism responsible for photoconductive gain is effective mass filtering. The current gain was found to increase with decreasing temperature (opposite to what occurs in the samples with phonon-assisted tunneling electron conduction). This is a manifestation of band-like electron transport described by a relaxive mobility (miniband conduction) discussed in a previous section [see (8)].

V. Negative Differential Resistance in the Transition from Miniband Conduction to Hopping

So far we have only considered situations in which the voltage drop across the superlattice period is smaller than the miniband width. In this case, whether the electronic states are localized or not depends on the magnitude of the broadening (collisional plus that due to disorder) relative to the miniband width. Localization, nevertheless, can also occur in a structurally perfect superlattice if the energy potential drop across the superlattice period exceeds the width of the miniband. This is because the overlap between the states of neighboring wells which produces a band is greatly reduced when this condition is met. When this occurs, there is a corresponding transition from band-like conduction to phonon-assisted tunneling (hopping) conduction between the localized states of the

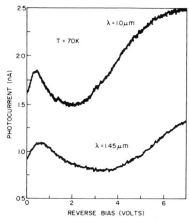

Fig. 11. Photocurrent versus reverse bias in a p⁺in⁺ device with superlattice in the i layer. Negative differential resistance occurs when the potential drop across the superlattice period exceeds the miniband width.

wells. This gives rise to NDR since, with increasing electric field, the spatial overlap between the states of neighboring wells decreases, thus decreasing the mobility and the current [7], [8].

The same mechanism can, of course, give rise to NDR in the photocurrent versus voltage characteristic of reverse-biased p-n junctions or Schottky diodes with a superlattice in the high field region, provided the recombination rate of carriers in the superlattice is not negligible compared to the photogeneration rate. The photocurrent can then be written as $I_{ph} = eAGv\tau$ where A is the sample area, v is the drift velocity, G is the generation rate, and τ is the lifetime and where we have assumed for simplicity that conduction is dominated by one type of carrier. The drift velocity v decreases strongly with increasing field F when the previously stated condition $eFd > 2\Delta$ is satisfied, giving rise to NDR in the photocurrent-voltage characteristic.

The first evidence of this effect was given by Tsu *et al.* [9] who observed negative conductance in the photocurrent of a Schottky barrier on AlAs/GaAs superlattices. We have also found this effect in the AlInAs/GaInAs superlattice p-n junctions discussed in the previous section and in other structures of the same material with ultrathin barriers [48].

The photocurrent versus reverse bias voltage was measured at different temperatures. The incident optical power was kept low ($\lesssim 5$ nW). No NDR is observed at temperatures between 150 and 300 K. Below 150 K, a distinct peak develops in the photocurrent versus voltage characteristic. Fig. 11 shows representative curves at 70 K at $\lambda = 10$ μm and 1.43 μm. The superlattice is transparent to these wavelengths so that light is absorbed only in the two contact regions. Photogenerated carriers diffuse to the superlattice and enter the high field region with thermal energies (i.e., at the band bottom). In the discussion that follows, we shall only consider electrons; the hole states

are localized in the wells, even at the lowest fields, and the hole hopping mobility is orders of magnitude smaller than the electron mobility so that their contribution to the transport phenomena discussed here is negligible. From C-V and doping profiling measurements, we were able to accurately calculate the field profile inside the superlattice at every voltage. The superlattice is depleted at $\cong 2$ V. The onset of NDR occurs at a voltage such that the average potential drop across the superlattice period $= 30.5$ meV. This value corresponds well with the threshold predicted theoretically $eFd \sim 2\Delta$ where 2Δ is the ground state miniband width calculated to be 34.1 meV. That the above NDR effect is only related to electrons was conclusively shown by achieving pure hole injection. This was done by back illuminating the sample with light strongly absorbed in the substrate. No NDR was observed.

VI. SEQUENTIAL RESONANT TUNNELING AND DEVICE APPLICATIONS

Another very interesting phenomenon occurs at even higher electric fields when $eFd = E_2 - E_1 \equiv \Delta_1$, i.e., when the ground state in the nth well becomes degenerate with the first excited state in the $(n + 1)$th well having the same transverse momentum $p_\perp$ (i.e., the momentum in the x, y plane perpendicular to the superlattice axis). Under these conditions, the current is due to RT; an electron from the ground state at the nth site tunnels to the vacant excited state at the $(n + 1)$th site, followed by nonradiative relaxation to the ground state at the $(n + 1)$th site [Fig. 12(a)]. Because of the resonant nature of this process, its probability is high, although the corresponding matrix elements are small. This gives rise to a peak in the current at fields corresponding to the above degeneracy, as first described by Kazarinov and Suris [2], [3]. A second peak is, of course, expected at fields such that the bottom of the ground state subband of the nth well is degenerate with the bottom of the second excited subband in the $(n + 1)$th well [Fig. 12(b)], i.e., $eFd = E_3 - E_1 \equiv \Delta_2$. Fig. 13 illustrates the same phenomenon by a means of a momentum space band diagram. It should be clear from this figure that NDR arises from conservation of transverse momentum and energy, as discussed in Section II-A.

This effect could not be studied in the structures discussed in the previous section. In those superlattices with thin wells, $\Delta_1 = 230$ meV so that the electric field required for the observation of sequential resonant tunneling is $\simeq 3 \times 10^5$ V/cm. The devices broke down before reaching that field. Thicker wells are required to observe sequential RT at lower fields. p⁺in⁺ diodes with the i layer consisting of a 35 period, undoped (n⁻ $\leq 1 \times 10^{14}$ cm⁻³) Al$_{0.48}$In$_{0.52}$As (139 Å), Ga$_{0.47}$In$_{0.53}$As (139 Å) superlattice were grown by MBE on an n⁺ ($= 10^{17}$ cm⁻³) Al$_{0.48}$In$_{0.52}$As buffer layer lattice matched to an n⁺⟨100⟩ InP substrate. The top p⁺ Al$_{0.48}$In$_{0.52}$As window layer is 1 μm thick and doped to $\cong 2 \times 10^{18}$ cm⁻³ and is followed by a 150 Å highly doped Ga$_{0.47}$In$_{0.53}$As p⁺ layer for con-

109

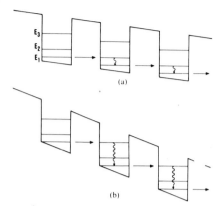

(a)

(b)

Fig. 12. Schematic illustration of sequential resonant tunneling of electrons for a potential energy drop across the superlattice period equal, respectively, to the energy difference between the first excited state and the ground state of the wells (a) and to the energy difference between the second excited state and the ground state of the wells (b).

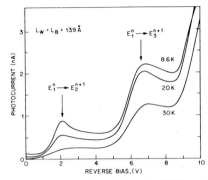

Fig. 14. Photocurrent-voltage characteristic at $\lambda = 0.6328 \mu m$ (pure electron injection) for a superlattice with 139 Å thick wells and barriers and 35 periods. The arrows indicate that the peaks correspond to resonant tunneling between the ground state of the nth well and the first two excited states of the $(n + 1)$th well.

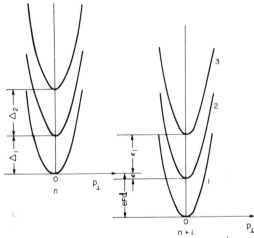

Fig. 13. Momentum space representation of resonant tunneling between two adjacent wells (n and $n + 1$) in a superlattice. Shown is the subband structure; $p_{\perp}$ is the momentum in the plane of the layers. eFd is the potential drop across the superlattice period; ϵ and ϵ_1 are the energy detunings from resonance [see (15)].

tact purposes. The area of the finished mesa etched diodes in 1.3×10^{-4} cm^2. Capacitance–voltage measurements in the temperature range between 300 and 8 K indicate that the i layer is completely depleted at zero bias. The photocurrent was measured as a function of reverse bias voltages at different temperatures [36].

In the temperature range where RT was observed (< 50 K), the reverse dark current of the diodes was below or at most comparable to the detection limits ($\leq 10^{-13}$ A) of our measuring apparatus and orders of magnitude smaller than the photocurrent.

To achieve pure electron injection into the superlattice and to ensure electron transport through the entire length of the multilayer region, the well-known method of mi-

nority-carrier injection was used. Suitably attenuated visible light from the He–Ne laser was shined on the p$^+$ layer where it is completely absorbed. In this way, only photogenerated minority carriers (electrons) which have diffused towards the i region are collected by the reverse-biased junction. The electron photocurrent was measured as a function of reverse bias voltage at different temperatures. Above 50 K, no negative conductance (NC) was observed. Below this temperature, two NC regions start to appear as shown by the peaks in Fig. 14 [36]. Similar results were obtained by varying the incident wavelength from 0.85 to 1.55 μm. This corresponds to mixed injection of electrons and holes within the superlattice region. The position of the peaks of the I–V and their shape did not vary as the photocurrent level was varied from 1 nA to 10 μA by changing the incident power, indicating that nonuniformities of the electric field induced by space-charge effects are negligible. To achieve instead pure hole injection, the He–Ne laser was shined on the substrate side (n$^+$) of the diodes. No NC was observed in this case. The above results prove that only electrons participate in the observed NC phenomenon, which is a manifestation of RT. RT of holes is too weak in our tight-binding superlattices to be observable.

The difference between the bias voltages corresponding to the two peaks divided by the number of superlattice periods ($= 140$ mV) is in excellent agreement with the calculated energy difference between the second and first excited states of the wells ($E_3 - E_2 \cong 143$ meV) [36]. A slightly less direct and accurate, but equivalent, comparison is obtained by adding to the applied voltages at the peaks the estimated built-in voltage drop across the superlattice (≈ 0.8 V) and dividing by the number of periods. These values are in good agreement with the calculated subband energy differences $E_2 - E_1$ and $E_3 - E_1$ [36]. This represents direct evidence of sequential RT through the entire superlattice. Note that in these superlattices with relatively thick barriers, the states are localized, even at zero electric field.

Above 10 V reverse bias, the photocurrent flattens out, implying that all electrons are collected from the wells (no recombination). Some of the electrons in the wells, for bias voltages between -2 and -10 V, are already hot and are therefore transported by thermionic emission across the barriers rather than by tunneling. The associated thermionic current will provide a raising background with increasing voltage (clearly seen in Fig. 14) which explains the asymmetry of the peaks.

Calculations give the following expression for the dependence of the current on the field near the first peak [2]:

$$J = edN(1 - e^{-\Delta_1/kT}) \frac{2|\Omega|^2 \tau_\perp}{1 + \epsilon^2 \tau_\perp^2 + 4|\Omega|^2 \tau_\parallel \tau_\perp} \quad (15)$$

where $h\epsilon = edF - \Delta_1$ is the energy detuning from resonance and Ω is the matrix element of the Hamiltonian between the ground state of the nth well and the first excited state of the $(n + 1)$th well divided by h. N is the carrier concentration in the wells.

The time constant $\tau_\parallel$ represents the energy relaxation time for electronic transitions from the excited to the ground state; $\tau_\perp$ is the relaxation time for the transverse momentum $p_\perp$. The Boltzmann factor in (15) describes the finite population of the excited state.

The half width at half maximum in the field dependence of the current density is given from (15) by

$$\delta F \cong h/ed\tau_\perp \quad (16)$$

where we have neglected $4|\Omega|^2 \tau_\parallel \tau_\perp$ in (15) which is a very good approximation in our tight-binding superlattices [2]. From the half width at half maximum of the first current peak at 8.6 K (Fig. 14), after subtraction of the broadening due to intralayer thickness fluctuations, we estimate using (16) a transverse momentum relaxation time $\tau_\perp \cong 10^{-13}$ s. This time describes the relaxation of the phase difference between the states involved in the RT process due to momentum relaxing collisions. To clarify this concept, let us consider the idealized case of RT of an electron, initially in the ground state of a well, into the first excited state of the adjacent well, in the absence of collisions and intrawell relaxation. No current is carried in this case and a straightforward solution of Schroedinger's equation shows that the peak of the electron probability density oscillates back and forth between the two wells with a frequency equal to Ω. Collisions that randomize the transverse momentum (described by $\tau_\perp$) tend to destroy the phase coherence of this oscillation since RT occurs between states of the same transverse momentum. Collisions produce a net current flow between wells. The time $\tau_\perp$ is comparable to the relaxation time obtained from mobility for motion parallel to the layers [2].

It is worth pointing out some remarkable analogies between sequential RT and paramagnetic spin resonance. The Liouville–Von Neumann equations for the diagonal and off-diagonal elements of the density matrix describing tunneling between adjacent wells are formally equivalent to the Bloch equations for paramagnetic spin resonance

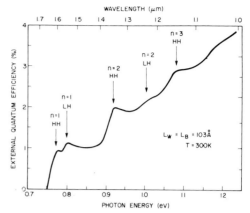

Fig. 15. Spectral response at room temperature of quantum well p-i-n diode. Note the exciton structure and the plateaus associated with the step like density of states of the wells.

[2], [3]. The analogy is more than simply formal since the two phenomena share some profound physical similarities. In both cases, we are dealing with effectively a two-level system. In spin resonance, the population difference oscillates coherently (in the absence of collisions) between the spin-up and the spin-down states with a period given by the reciprocal of the Rabi frequency $\nu_R = \mu_{12}E/h$ where μ_{12} is the matrix element of the dipole operator between the two states and E is the amplitude of the RF field. In RT (in the absence of collisions), the electron density transfers back and forth between the ground and excited state of adjacent wells with a period given by the reciprocal of $\nu = (edF/h) T_B$ where T_B is the barrier transmission (ν is related to Ω in (15) by $\nu = \Omega/2\pi$). Note the similarity between ν and ν_R, with μ_{12} corresponding to ed and E to F. In spin resonance, the dephasing effect of spin–spin interactions is described by the T_2 relaxation time; the corresponding time in sequential RT is, from the discussion in the previous paragraph, $\tau_\perp$. Finally, the relaxation time for the population difference T_1 in spin resonance corresponds to the energy relaxation time $\tau_\parallel$ in sequential RT.

The observation of two transport routes via sequential resonant tunneling through a series of 35 periods is direct evidence of the high quality of our superlattices. The remarkable quality of these superlattices and their suitability to study quantum effects on transport is further confirmed by photocurrent measurement versus photon energy, which show very clearly quantum size and excitonic effects at room temperature. Fig. 15 shows the room-temperature external quantum efficiency at zero bias voltage in a p^+in^+ superlattice diode. The i μm thick region is a 50 period $Al_{0.48}In_{0.52}As$ (103 Å)/$Ga_{0.48}In_{0.53}As$ (103 Å) superlattice and the p^+ and n^+ layers consist of 1 μm thick $Al_{0.48}In_{0.52}As$. The plateaus are due to the step-like density of states. Three plateaus are clearly identified which correspond to the transitions between the $n = 1, 2$, and 3 valence and conduction subbands. The light-hole

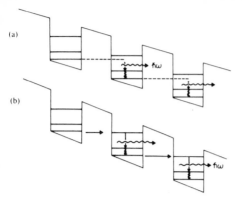

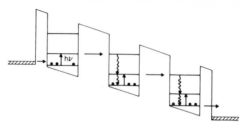

Fig. 17. Band diagram of sequential resonant tunneling photoconductive detector. Electrons in the doped wells are photoexcited to the first excited state of the well from where they tunnel into the neighboring well, followed by energy relaxation via phonons.

Fig. 16. Band diagram of far infrared laser using sequential resonant tunneling. (a) Laser photon is emitted during an *interwell* photon-assisted tunneling transition connecting the ground state of a well with one of the excited states of the adjacent well. (b) Laser photon is emitted during an *intrawell* transition between excited states, following resonant tunneling between wells. Relaxation to the ground state, following photon emission, is via phonons in both cases.

and heavy-hole $n = 1, 2$ excitons clearly emerge from the plateaus of the $n = 1$ and $n = 2$ intersubband transitions. The presence of excitonic effects at room temperature in quantum wells, due to their increased binding energy, is well understood [49]. Recently, Weiner et al. [50] have reported clear evidence of room-temperature excitons in the absorption spectra of $Al_{0.48}In_{0.52}As/Ga_{0.47}In_{0.53}As$ $p^{+}in^{+}$ quantum well structures with layer thicknesses of 110 Å.

There are some very interesting device applications of sequential RT to optoelectronic devices. For example, in 1970, Kazarinov and Suris [2] proposed a new type of infrared laser amplifier (Fig. 16). If $eFd > \Delta_1$, the bottom of the first excited quantum subband in the $(n + 1)$th well lies below the bottom of the ground state subband in the nth well. Under these conditions, we may observe the amplification (or laser action) of radiation with frequency $\nu = (eFd - \Delta_1)/h$ whose electric field is parallel to the external static field. This occurs via a photon-assisted tunneling transition whereby an electron from the ground state in the nth well tunnels to an excited state in the $(n + 1)$th well with the *simultaneous* emission of a photon of frequency ν [Fig. 16(a)]. The conditions for amplification are automatically satisfied since the initial state of the transition, being the ground state of the nth well, is more populated than the first excited state of the adjacent well. Clearly, the gain is proportional to the transmission of the barrier between the wells in question. An important property of the amplification is that the amplified frequency (or the laser frequency) can be varied over a wide range by varying the electric field.

As the second excited state in the $(n + 1)$ well approaches the ground state in the well n, the probability of the photon-assisted tunneling transition from the ground state in the cell nth well to the first excited state in the $n + 1$th well increases considerably. This increase in the

transition probability is due to the fact that the energy denominator involving the difference between the energies of the initial and the intermediate virtual state becomes small (resonant enhancement of a second-order process).

Consider finally the situation in which one is exactly at resonance. This situation is very different since the second excited state is populated by tunneling, and therefore stimulated emission is now a direct intrawell transition from the second to the first excited state of the wells. To achieve, therefore, laser amplification at the frequency $(E_3 - E_2)/h$, there must be a population inversion between states 3 and 2. This requires that the lifetime of the electron at the bottom of the third subband be larger than the electron lifetime at the bottom of the second subband. These lifetimes are primarily controlled by intersubband scattering by optical phonons if the intersubband separation is greater than the optical phonon energy. The scattering time for this process is on the order of 10^{-13} s. Consider an electron in the $n = 3$ subband at $k = 0$. Because of the dipole selection rule ($\Delta n = 1$), the only allowed intersubband transition via polar optical phonons is to the $n = 2$ subband. The scattering time (τ_{32}) for this transition is smaller than the time for the corresponding transition from the second subband (at $k = 0$) to the ground state. This is because the momentum transferred $h\Delta k$ in the $3 \rightarrow 2$ transition is smaller. (The polar phonon scattering matrix element is inversely proportional to $|\Delta k|$ and the density of final states is equal for the above two transitions because of the two dimensionality.) This implies that it is difficult to achieve a population inversion between the $n = 3$ and the $n = 2$ states unless the separation between the two subbands is chosen smaller than the optical phonon energy. Thus, the most promising scheme for achieving laser action is the one depicted in Fig. 16(a), i.e., photon-assisted tunneling. Of course, to achieve the pumping tunneling current density required for lasing, the barriers should be made relatively thin (20–50 Å). Such lasers, depending on the well thickness and the value of the electric field, can be made to emit in the 5–15 μm wavelength range and may be used as local oscillators in IR heterodyne detection. The emitted radiation is polarized normal to the plane of the layers.

Another interesting application of sequential RT is an infrared detector with high wavelength selectivity in the same spectral range. This scheme is illustrated in Fig. 17. The quantum wells are doped n type in the range 10^{17}–10^{18} cm^{-3} and have thicknesses in the 100 Å range. (Alternatively, one could use a modulation-doped geometry by doping the barriers to the same level.) The undoped barriers are typically in the 20–50 Å range and the whole structure is sandwiched between heavily doped n$^+$ contacts.

The bias across the device should be such that the second level in the nth well is degenerate with the third level in the $(n + 1)$th well. Consider now infrared radiation incident on the device, having a component of the electric field perpendicular to the plane of the well. It will be absorbed strongly only if the photon energy is equal or very near to $(E_2 - E_1)/h$ because of momentum conservation considerations. This is a dipole transition with a large oscillator strength. It has been recently observed by West and Eglash [51] in GaAs quantum wells with 65 and 82 Å thickness. The above transition in these structures exhibited resonant energies of 152 and 121 meV, respectively, and half maximum linewidths at room temperature of 10 meV. Thus, excellent wavelength selectively is ensured.

Once electrons have made the optical transition to the first excited state, they can resonantly tunnel to the bottom of the second excited state of the adjacent well. From here, the most likely route is intersubband scattering to the $n = 2$ level, followed by either intraband relaxation and resonant tunneling to the adjacent well or scattering to the $n = 1$ subband with final relaxation to the bottom of the well. The net effect of the absorption, tunneling, and relaxation processes is a photocurrent. To maximize the quantum efficiency of this detector, the barriers should be made sufficiently thin that the tunneling time from the $n = 2$ to the $n = 3$ state is smaller than the scattering time to the $n = 1$ subband. This is because to give rise to a photocurrent, the electron must relax in a well different from the one in which it has been photoexcited. The tunneling time can be estimated from the formula

$$t_0 \approx \frac{h}{2eFd T_B} \quad (17)$$

where T_B is the barrier transmission for an electron at the bottom of the second subband. For Al$_{0.48}$In$_{0.52}$As/Ga$_{0.47}$In$_{0.53}$As structures with 140 Å wells, the above condition implies that the barrier thickness should not exceed 40 Å.

In conclusion, we have presented a comprehensive discussion of recent experimental and theoretical results obtained in tunneling and resonant tunneling heterostructures. Many interesting device applications are possible, ranging from novel transistors to lasers and detectors.

Note Added in Proof: Following completion of the paper, several other papers pertinent to localization and tunneling in superlattices have been brought to our attention. These are included at the end of the list of References [52]–[55].

ACKNOWLEDGMENT

It is a pleasure to acknowledge fruitful collaborations and discussions with S. Luryi, R. Kazarinov, R. Kiehl, R. Hull, A. L. Hutchinson, R. C. Miller, D. A. Miller, J. Weiner, G. Derkits, J. Shah, and A. Pinczuk. A. C. Gossard kindly supplied the resonant tunneling data of Tables I and II.

REFERENCES

[1] L. Esaki and R. Tsu, "Superlattice and negative differential conductivity in semiconductors," *IBM J. Res. Develop.*, vol. 14, pp. 61–65, 1970.
[2] R. F. Kazarinov and R. A. Suris, "Possibility of amplification of electromagnetic waves in a semiconductor with a superlattice," *Fiz. Tekh. Poluprov.*, vol. 5, pp. 797–800, 1971; transl. in *Sov. Phys. Semicond.*, vol. 5, pp. 707–709, 1971.
[3] ——, "Electric and electromagnetic properties of semiconductors with a superlattice," *Fiz. Tekh. Poluprov.*, vol. 6, pp. 148–62, 1972; transl. in *Sov. Phys.—Semiconductors*, vol. 6, pp. 120–131, 1972.
[4] R. Tsu and L. Esaki, "Tunneling in a finite superlattice," *Appl. Phys. Lett.*, vol. 22, pp. 562–564, 1973.
[5] L. L. Chang, L. Esaki, and R. Tsu, "Resonant tunneling in semiconductor double barriers," *Appl. Phys. Lett.*, vol. 24, pp. 593–595, 1974.
[6] L. Esaki and L. L. Chang, "New transport phenomenon in a semiconductor 'superlattice,'" *Phys. Rev. Lett.*, vol. 33, pp. 495–498, 1974.
[7] G. H. Dohler, R. Tsu, and L. Esaki, "A new mechanism for negative differential conductivity in superlattices," *Solid State Commun.*, vol. 17, pp. 317–320, 1975.
[8] R. Tsu and G. Dohler, "Hopping conduction in a 'superlattice,'" *Phys. Rev. B.*, vol. 12, pp. 680–686, 1975.
[9] R. Tsu, L. L. Chang, G. A. Sai-Halasz, and L. Esaki, "Effects of quantum states on the photocurrent in a 'superlattice,'" *Phys. Rev. Lett.*, vol. 34, pp. 1509–1512, 1975.
[10] E. A. Rezek, N. Holonyak, Jr., B. A. Vojak, and H. Shichijo, "Tunnel injection into the confined-particle states of an In$_{1-x}$Ga$_x$P$_{1-z}$As$_z$ well in InP," *Appl. Phys. Lett.*, vol. 31, pp. 703–705, 1977.
[11] B. A. Vojak, N. Holonyak, Jr., R. Chin, E. A. Rezek, R. D. Dupuis, and P. D. Dapkus, "Tunnel injection and phonon-assisted recombination in multiple quantum-well Al$_x$Ga$_{1-x}$As/GaAs heterostructure lasers grown by metalorganic chemical vapor deposition," *J. Appl. Phys.*, vol. 50, pp. 5835–5840, 1979.
[12] C. W. Tu, R. Hendel, and R. Dingle, "Molecular beam epitaxy and the technology of selectively doped heterostructure transistors," in *Gallium Arsenide Technology*, D. K. Ferry, Ed. Indianapolis, IN: Howard & Sams, 1985, pp. 107–146.
[13] C. Weisbuch, R. Dingle, A. C. Gossard, and W. Wiegmann, "Optical properties and interface disorder of GaAs-Al$_x$Ga$_{1-x}$As multi-quantum well structures," in *Proc. 8th Int. Symp. GaAs and Related Compounds*, Vienna, Austria, Sept. 1980, Inst. Phys. Conf. Ser. 56, pp. 711–720, 1981.
[14] A. Madhukar, T. C. Lee, M. Y. Yen, P. Chen, J. Y. Kim, S. V. Ghaisas, and P. G. Newman, "Role of surface kinetics and interrupted growth during molecular beam epitaxial growth of normal and inverted GaAs/AlGaAs(100) interfaces: A reflection high-energy electron diffraction intensity dynamics study," *Appl. Phys. Lett.*, vol. 46, pp. 1148–1150, 1985.
[15] H. Sakaki, M. Tanaka, and J. Yoshino, "One atomic layer heterointerface fluctuations in GaAs-AlAs quantum well structures and their suppression by insertion of smoothing period in molecular beam epitaxy," *Japan. J. Appl. Phys.*, part 2, vol. 24, pp. L417–L420, 1985.
[16] T. C. L. G. Sollner, W. D. Goodhue, P. E. Tannenwald, C. D. Parker, and D. D. Peck, "Resonant tunneling through Quantum wells at frequencies up to 2.5 THz," *Appl. Phys. Lett.*, vol. 43, pp. 588–590, 1983.
[17] T. C. L. G. Sollner, P. E. Tannenwald, D. D. Peck, and W. D. Goodhue, "Quantum well oscillators," *Appl. Phys. Lett.*, vol. 45, pp. 1319–1321, 1984.
[18] P. Gavrilovic, J. M. Brown, R. W. Kaliski, N. Holonyak, Jr., K. Hess, M. J. Ludowise, W. T. Dietze, and C. R. Lewis, "Resonant tunneling in a GaAs$_{1-x}$P$_x$-GaAs strained-layer quantum-well heterostructure," *Solid State Commun.*, vol. 52, pp. 237–239, 1984.

1868 IEEE JOURNAL OF QUANTUM ELECTRONICS, VOL. QE-22, NO. 9, SEPTEMBER 1986

[19] A. R. Bonnefoi, R. T. Collins, T. C. McGill, R. D. Burnham, and F. A. Ponce, "Resonant tunneling in GaAs/AlAs heterostructures grown by metalorganic chemical vapor deposition," *Appl. Phys. Lett.*, vol. 46, pp. 285-287, 1985.

[20] T. J. Shewchuk, P. C. Chapin, P. D. Coleman, W. Kopp, R. Fischer, and H. Morkoc, "Resonant tunneling oscillations in a GaAs-Al$_x$Ga$_{1-x}$As heterostructure at room temperature," *Appl. Phys. Lett.*, vol. 46, pp. 508-510, 1985.

[21] M. Tsuchiya, H. Sakaki, and J. Yoshino, "Room temperature observation of differential negative resistance in an AlAs/GaAs/AlAs resonant tunneling diode," *Japan. J. Appl. Phys.*, part 2, vol. 24, pp. L466-L468, 1985.

[22] E. E. Mendez, W. I. Wang, B. Ricco, and L. Esaki, "Resonant tunneling of holes in AlAs-GaAs-AlAs heterostructures," *Appl. Phys. Lett.*, vol. 47, pp. 415-417, 1985.

[23] B. Ricco and M. Ya. Azbel, "Physics of resonant tunneling. The one dimensional double-barrier case," *Phys. Rev. B*, vol. 29, pp. 1970-1981, 1984.

[24] J. R. Barker, S. Collins, D. Lowe, and S. Murray, "Theory of transient quantum transport in heterostructures," in *Proc. 17th Int. Conf. Phys. Semicond.*, 1984, pp. 449-452, J. D. Chadi and W. A. Harrison, Eds. New York: Springer.

[25] S. Luryi, "Frequency limit of double-barrier resonant-tunneling oscillators," *Appl. Phys. Lett.*, vol. 47, pp. 490-492, 1985.

[26] B. Jogai and K. L. Wang, "Dependence of tunneling current on structural variations of superlattice devices," *Appl. Phys. Lett.*, vol. 46, pp. 167-168, 1985.

[27] W. R. Frensley, "Simulation of resonant-tunneling heterostructure devices," *J. Vac. Sci. Technol. B*, vol. 3, pp. 1261-1266, 1985; *Proc. 12th Annu. Conf. Phys. and Chem. of Semiconductor Interfaces*, Tempe, AZ, Jan. 1985.

[28] U. Ravaioli, M. A. Osman, W. Potz, N. Kluksdahl, and D. K. Ferry, "Investigation of ballistic transport through resonant-tunneling quantum wells using Wigner function approach," *Physica B & C*, vol. 134(B & C), pp. 36-40, 1985; *Proc. 4th Int. Conf. Hot Electrons in Semiconductors*, Innsbruck, Austria, July 1985.

[29] T. Nakagawa, H. Imamoto, T. Sakamoto, T. Kojima, K. Ohta, and N. J. Kawai, "Observation of negative differential resistance in CHIRP superlattices," *Electron. Lett.*, vol. 21, pp. 882-884, 1985.

[30] R. A. Davies, M. J. Kelly, and T. M. Kerr, "Tunneling between two strongly coupled superlattices," *Phys. Rev. Lett.*, vol. 55, pp. 1114-1116, 1985.

[31] ——, "Room-temperature oscillations in a superlattice structure," *Electron. Lett.*, vol. 22, pp. 131-133, 1986.

[32] F. Capasso and R. A. Kiehl, "Resonant tunneling transistor with quantum well base and high-energy injection: A new negative differential resistance device," *J. Appl. Phys.*, vol. 58, pp. 1366-1368, 1985.

[33] S. Luryi and F. Capasso, "Resonant tunneling of two-dimensional electrons through a quantum wire: A negative transconductance device," *Appl. Phys. Lett.*, vol. 47, pp. 1347-1349, 1985.

[34] F. Capasso, K. Mohammed, A. Y. Cho, R. Hull, and A. L. Hutchinson, "New quantum photoconductivity and large photocurrent gain by effective-mass filtering in a forward-biased superlattice p-n junction," *Phys. Rev. Lett.*, vol. 55, pp. 1152-1155, 1985.

[35] ——, "Effective mass filtering: Giant quantum amplification of the photocurrent in a semiconductor superlattice," *Appl. Phys. Lett.*, vol. 47, pp. 420-422, 1985.

[36] F. Capasso, K. Mohammed, and A. Y. Cho, "Sequential resonant tunneling through a multiquantum-well superlattice," *Appl. Phys. Lett.*, vol. 48, pp. 478-480, 1986.

[37] A. D. Stone and P. A. Lee, "Effect of inelastic processes on resonant tunneling in one dimension," *Phys. Rev. Lett.*, vol. 54, pp. 1196-1199, 1985.

[38] R. C. Miller, A. C. Gossard, D. A. Kleinman, and O. Munteanu, "Parabolic quantum wells with the GaAs-Al$_x$Ga$_{1-x}$As system," *Phys. Rev. B*, vol. 29, pp. 3740-3743, 1984.

[39] F. Capasso, "New high-speed quantum well and variable gap superlattice devices," in *Picosecond Electronics and Optoelectronics*, G. A. Mourou, D. M. Bloom, and C. H. Lee, Eds., Springer Ser. in Electrophys. 21. Berlin, Heidelberg: Berlin, Springer-Verlag, 1985, pp. 112-130.

[40] N. Yokoyama, K. Imanura, S. Muto, S. Hiyamizu, and H. Nishi, "A new functional, resonant-tunneling hot electron transistor (RHET)," *Japan. J. Appl. Phys.*, part 2, vol. 24, pp. L853-854, 1985.

[41] See, for example, G. Bastard, this issue, pp. 000-000.

[42] J. F. Palmier and A. Chomette, "Phonon-limited near equilibrium transport in a semiconductor superlattice," *J. Phys. (France)*, vol. 43, pp. 381-391, 1982.

[43] P. W. Anderson, "Absence of diffusion in certain random lattices," *Phys. Rev.*, vol. 109, pp. 1492-1505, 1958.

[44] D. Calecki, J. F. Palmier, and A. Chomette, "Hopping conduction in multiquantum well structures," *J. Phys. C*, vol. 17, pp. 5017-5030, 1984.

[45] F. Capasso, H. M. Cox, A. L. Hutchinson, N. A. Olsson, and S. G. Hummel, "Pseudo-quaternary GaInAsP semiconductors: A new Ga$_{0.47}$In$_{0.53}$As/InP graded gap superlattice and its applications to avalanche photodiodes," *Appl. Phys. Lett.*, vol. 45, pp. 1193-1195, 1984.

[46] T. Weil and B. Vinter, "Calculation of carrier transport in pseudo-quaternary alloys," *Surface Sci.*, 1986.

[47] J. F. Palmier, H. Le Person, C. Minot, A. Chomette, A. Regreny, and D. Calecki, "Hopping mobility in semiconductor superlattices," *Superlattices and Microstruct.*, vol. 1, pp. 67-72, 1985.

[48] F. Capasso, K. Mohammed, and A. Y. Cho, "Quantum photoconductive gain by effective mass filtering and negative conductance in superlattice p-n junctions," *Physica B&C*, vol. 134B, pp. 487-493, 1985. Extensive measurements by us after publication of this work have shown that the second peak in the photocurrent *I-V* reported in this paper is not always reproducible.

[49] R. Dingle, "Confined carrier quantum states in ultrathin semiconductor structures," in *Festkörperprobleme XV*, H. J. Queisser, Ed. Braunschweig: Pergamon/Vieweg, 1975, pp. 21-48.

[50] J. S. Weiner, D. S. Chemla, D. A. B. Miller, T. H. Wood, D. Sivco, and A. Y. Cho, "Room temperature excitons in 1.6 μm band-gap GaInAs/AlInAs quantum wells," *Appl. Phys. Lett.*, vol. 46, pp. 619-621, 1985.

[51] L. C. West and S. J. Eglash, "First observation of an extremely large dipole infrared transition within the conduction band of a GaAs quantum well," *Appl. Phys. Lett.*, vol. 46, pp. 1156-1158, 1985.

[52] R. Lang and K. Nishi, "Electronic state localization in semiconductor superlattices," *Appl. Phys. Lett.*, vol. 45, pp. 98-100, 1984.

[53] R. K. Littleton and R. E. Camley, "Investigation of localization in a 10-well superlattice," *J. Appl. Phys.*, vol. 59, pp. 2817-2820, 1986.

[54] T. Furuta, K. Hirakawa, J. Yoshino, and H. Sakaki, "Splitting of photoluminescence spectra and negative differential resistance caused by the electric field induced resonant coupling of quantized levels in GaAs-AlGaAs multi quantum well structures," *Japn. J. Appl. Phys.*, Feb. 1986.

[54] C. J. Summers and K. F. Bremran, "Variably spaced superlattice energy filter, A new device design concept for high-energy electron injection," *Appl. Phys. Lett.*, vol. 48, pp. 806-808, 1986.

Federico Capasso (M'79-SM'85) received the Doctor of Physics degree (summa cum Laude) from the University of Rome, Rome, Italy in 1973, with a thesis on stimulated Raman spectroscopy.

He was a Research Scientist at Fondazione Bardoni from 1974 to 1976, working on the theory of nonlinear phenomena in optical fibers and on dye lasers. He joined AT&T Bell Labs., Murray Hill, NJ in 1976 where he has since engaged in research on impact ionization, 1.3-1.6 μm detectors, liquid-phase epitaxy of III-V materials, surface passivation studies, deep-level spectroscopy, high-field transport in semiconductors, novel avalanche photodiode structures, bipolar transistors, resonant tunneling, and superlattices. He pioneered the technique of "band-gap engineering" and has used it extensively in the design of a new class of superlattice and variable-gap heterostructures, including new photodetectors and bipolar transistors. He has given 35 invited talks at international conferences and has co-authored over 90 papers.

In 1984, Dr. Capasso received the AT&T Bell Labs Distinguished Member of Technical Staff Award, and the Award of Excellence of the Society for Technical Communication. He is a member of the American Physics Society, the New York Academy of Sciences, and the Optical Society of America. He has been serving on the program committees of the Hot Electron Conference, Integrated and Guided Wave Optics Conference, Optical Society of America Meeting, High Speed Electronics Conference, Fourth ICTP-IUPAP Semiconductors Symposium on Shallow Centers, and

IEEE International Electron Devices Meeting. He is the program co-Chairman of the 1987 Topical Meeting on Picosecond Optoelectronics and Electronics and Chairman of the Solid State Detector Technical Group of the Optical Society of America.

Khalid Mohammed was born in Pakistan on January 2, 1957. He received the B.Sc. degree in physics from the University of Liverpool, Liverpool, England and the Ph.D. degree in solid-state physics from King's College, University of London, London, England, in 1982.

From 1982 to 1984, he was a Postdoctoral Fellow in the Solid-State Electronics laboratory at the University of California, Santa Barbara, and then moved on to the Optical-Electronics Research Department at AT&T Bell Laboratories, Murray Hill, NJ, as a member of the Technical Staff until February 1986. There he worked on, among other topics, resonant tunneling phenomenon in heterostructure superlattices. Currently, he is employed by North American Philips Corporation, Briarcliff Manor, NY, where he is working on the optical properties of II–VI heterostructures.

Alfred Y. Cho (S'57–M'60–SM'79–F'81) is Head of the Electronics and Photonics Materials Research Department of the Solid State Electronics Research Laboratory. He joined AT&T Bell Laboratories, Murray Hill, NJ, in 1968 where he developed a crystal growth technology called molecular beam epitaxy. He is also an Adjunct Professor in the Department of Electrical Engineering, at the University of Illinois, Urbana.

Dr. Cho is an Associate Editor of the *Journal of Crystal Growth* and *Journal of Vacuum Science and Technology*. He is a recipient of the Electronics Division Award of the Electrochemical Society (1977), the American Physical Society International Prize for New Materials (1982), the IEEE Morris N. Liebmann Award (1982), the AT&T Bell Laboratories Distinguished Technical Staff Award (1982), the University of Illinois Electrical and Computer Engineering Distinguished Alumnus Award (1985), and the Chinese Institute of Engineers U.S.A. Distinguished Achievement Award (1985). He is a fellow of the American Physical Society. He is a member of the New York Academy of Sciences, the American Association for the Advancement of Science, the American Vacuum Society, the Electrochemical Society, the Materials Research Society, the National Academy of Engineering, and the National Academy of Sciences.

Surface Science 132 (1983) 543–576
North-Holland Publishing Company

HETEROSTRUCTURE DEVICES: A DEVICE PHYSICIST LOOKS AT INTERFACES

Herbert KROEMER

Department of Electrical and Computer Engineering, University of California, Santa Barbara, California 93106, USA

Received 18 October 1982; accepted for publication 30 December 1982

The band offsets occurring at abrupt hetero-interfaces in heterostructure devices serve as potential steps acting on the mobile carriers, in addition to the macroscopic electrostatic forces already present in homostructure devices. Incorporation of hetero-interfaces therefore offers a powerful device design parameter to control the distribution and flow of mobile carriers, greatly improving existing kinds of devices and making new kinds of devices possible. Unusual device requirements can often be met by band lineups occurring in suitable semiconductor combinations. Excellent theoretical rules exist for the semi-quantitative ($< \pm 0.2$ eV) prediction of band offsets, even unusual ones, but no quantitatively accurate ($< \pm 1\,kT$) purely theoretical predictive rules are currently available. Poorly-understood second-order nuisance effects, such as small interface charges and small technology-dependent offset variations, act as major limitations in device design. Suitable measurements on device-type structures can provide accurate values for interface physics parameters, but the most widely used measurements are of limited reliability, with pure $I–V$ measurement being of least use. Many of the problems at interfaces between two III/V semiconductors are hugely magnified at interfaces between a compound semiconductor and an elemental one. Large interface charges, and a strong technology dependence of band offsets are to be expected, but can be reduced by deliberate use of certain unconventional crystallographic orientations. An understanding of such polar/nonpolar interfaces is emerging; it is expected to lead to a better understanding and control of III/V-only device interfaces as well.

1. Introduction

This paper takes a look at interfaces in submicron structures, from the point of view of a device physicist who is interested in incorporating semiconductor hetero-interfaces into future *high-performance* semiconductor devices.

A significant fraction of such devices will be compound semiconductor rather than silicon devices. Before long, most compound semiconductor devices will involve heterostructures [1,2]. Homostructure devices made from a single compound semiconductor will probably be relegated to the low-performance/low-cost end of compound semiconductor technology, although silicon device technology will very likely continue to be dominated by homostructure devices. Furthermore, high performance in devices usually means minimizing the non-active part of the device volume, to the point that the device turns from a collection of semiconductor regions separated by interfaces, to a collection of interfaces with a minimum of semiconductor between them.

As this development progresses, it calls for a constant interchange of ideas between the device physicist and the more fundamentally-oriented "basic" surface/interface physicist. This interchange goes both ways: On the one hand, the device physicist (even if inclined to do so) can less and less rely on "cookbook empiricism"; instead he must closely follow the basic physicist in assimilating and utilizing the new fundamental knowledge that the latter has acquired. On the other hand, device physics constantly poses new problems to the basic physicist; and experiments on device-type structures (sometimes deliberately "misdesigned" as devices) offer themselves as powerful tools for basic research. One of the purposes of this paper is to contribute to this necessary interchange of ideas between the device physicist and the basic physicist.

Throughout the paper, the term *heterostructure device* is to be understood in the sense that the hetero-interface plays an essential role in the operation of the device, rather than just serving as a passive interface between what is basically a homostructure device and a chemically different substrate as in silicon-on-sapphire structures. In many cases, the interface *is* the actual device. The emphasis must therefore be on "good" interfaces made by "good" technology. Various kinds of interface defects, although never totally absent, can then at least be assumed to be present in only such small densities that their effect can be treated as a perturbation of a defect-free interface model, rather than as dominating the physics. These assumptions are by no means unrealistic "academic" ones, made to simplify the problem in neglect of practical realities: They spell out the conditions that a heterointerface must satisfy to be of interest for incorporation into the active portion of a high-performance device. This poses stringent demands on the concentrations of these defects, to the point that they can rarely be neglected altogether.

The main device physics problems of hetero-interfaces can be roughly divided into problems of the static energy band structure, and problems of the electron transport *within that structure*. I shall concentrate here on the band structure aspects, and ignore the transport aspects. This is not because I consider transport problems less interesting or important (heaven forbid!), but simply because the transport aspects of the device physics are well covered by others at this Symposium. Instead, I will address myself at the end to an area of electronic structure that is not yet in the mainstream of heterostructure device development: The problems of achieving device-quality polar/nonpolar interfaces, involving such pairs as GaAs-on-Ge or GaP-on-Si. This is already an area of active interest to the basic physicist, but so far only from the structural point-of-view, largely neglecting the electrical properties that are the essence of device. Currently, the device physicist is disenchanted about the consistently miserable electrical properties that have resulted whenever device-type structures of this kind have been attempted. I believe that device-quality interfaces in such systems *can* be achieved, but only if both structural and electrical considerations are pursued jointly. This raises some new kinds of problems that simply do not exist in III/V-only systems, but the understanding of which is likely to have benefits far beyond these esoteric mixed systems themselves, feeding back even on such much simpler systems as the familiar GaAs/(Al, Ga)As systems.

2. Energy band diagrams of hetero-interfaces

2.1. Band offsets: the Shockley–Anderson model

From a device physics point-of-view the most important aspect of a semiconductor hetero-interface, and the point of departure for all subsequent considerations is the energy band diagram of the interface. We assume that the transition from one semiconductor takes place over at most a few lattice constants. For such abrupt interfaces the "canonical" energy band model is the *Shockley–Anderson model* [3–6], (Fig. 1). Its characteristic feature is an abrupt change in energy gap at the interface, leading to discontinuities or *offsets* in the conduction and valence band edges. The magnitudes of these offsets are assumed to be characteristic properties of the semiconductor pair involved, essentially independent of doping levels and hence of Fermi level considerations, but possibly dependent on the crystallographic orientation and on other factors influencing the exact arrangement of the atoms near the interface. Far away from the interface, the band energies are governed by the requirement that a bulk semiconductor must be electrically neutral, which fixes the band energies relative to the Fermi level. Except for certain fortuitous doping levels, the combination of specified band offsets with specified band energies at infinity calls for band bending, accommodated by space charge layers near the interface, similar to the space charge layers at p–n homojunctions. The calculation of the exact shape of this band bending is an exercise in electrostatics and Fermi statistics, not of interest here [5].

The band diagram shown in fig. 1 is for an n–n structure (often written n–N structure, to indicate the change in energy gap). As the figure shows, the conduction band offset then leads to a shallow potential notch and a Schottky-barrier-like potential spike barrier, both of which play large roles in the electrical properties of such junctions. Fig. 2 shows two other possibilities, an N–p junction and an n–P junction.

From the device physics point-of-view, the band offsets are the dominant aspect of heterostructure interfaces, and their existence is in fact the principal

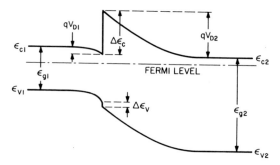

Fig. 1. Band diagram of the Shockley–Anderson model for an abrupt unbiased n–N heterojunction, showing the band edge discontinuities (or offsets) that are the characteristic feature of the model. The specific lineup shown is the "normal" lineup, for whch the narrower forbidden gap falls within the wider gap at the interface.

reason why heterostructures are incorporated into semiconductor, devices: The band offsets act as potential barriers, exerting very strong forces on electrons and holes. These quantum-mechanical "quasi-electric" forces exist in addition to those purely classical electrostatic forces that are due to space charges and applied voltages, which govern carrier flow and distribution in homostructures made from a single semiconductor. The band offset forces may be made either to assist or to counteract the classical electrostatic forces. This gives the device

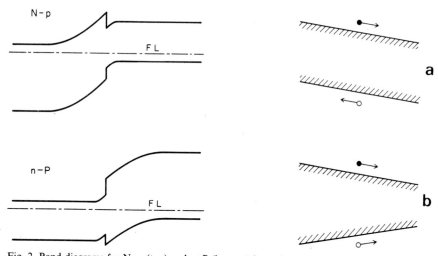

Fig. 2. Band diagrams for N–p (top) and n–P (bottom) heterojunctions.

Fig. 3. Forces on electrons and holes. In a uniform-gap semiconductor (top) the two forces are of equal magnitude but opposite direction, equal to the electrostatic forces $\pm qE$. In a graded-gap structure (bottom) the forces on electrons and holes may be in the same direction. From ref. [2].

physicist an extraordinary new degree of design freedom in controlling the distribution and flow of carriers, to improve the performance of existing devices, and to make possible new kinds of devices.

Basically, it is not the electrostatic force $\pm qE$ that acts as force on the carrier, but the slope of the band edge of the band containing the carrier, multiplied by the sign of the charge of the carrier. In a homostructure, the slopes are necessarily equal to each other and to qE (fig. 3a). But in a heterostructure, energy gap variations cause the slopes of the conduction and valence bands to differ from each other and from the electrostatic force. The case of abrupt band offsets is simply a limiting case; the underlying physics is perhaps clearer by considering the more general case of a graded energy gap, as in fig. 3b, in which only band edge slopes are visible, with no hint as to the magnitude or even the direction of the electric field.

This *general heterostructure design principle* [1,2] may be used in many different ways. A judicious combination of classical electrostatic forces and band gap variations (fig. 3b) makes it possible in a bipolar structure on control the flow of electrons and holes separately and independently. This principle is the basis of operation of the double-heterostructure laser [7,6] that serves as the heart of emerging light-wave communications technology. It also forms the basis of new kinds of improved bipolar transistors [2], and probably of other future devices.

In unipolar devices only one kind of carriers, usually electrons, are present. Here the band offset force has been used with great success in at least two different ways: (a) to confine electrons in quantum wells [8] that are much narrower and have much steeper walls than would be achievable by classical electrostatic forces (= doping) alone; (b) to spatially separate electrons from the donors, against their mutual Coulomb attraction [9]. The latter possibility forms the basis of a rapidly developing new class of field effect transistors [10]. Quantum well structures form the basis of new classes of lasers [11], and they will probably also be responsible for fundamentally new kinds of future device that would not exist at all without quantum wells.

In the energy band diagrams shown in figs. 1 and 2 the signs and magnitudes of the two band offsets were such that at the interface the narrower of the two gaps fell energetically within the wider gap. This "straddling" lineup is the most common case. The most extensively studied of all hetero-interfaces, $GaAs/Al_xGa_{1-x}As$, is of this kind, and its lineup is known to a higher accuracy than that of any other system: For $x < 0.45$, the range in which (Al, Ga)As is a direct-gap semiconductor, the conduction band offset is $85\% \pm 3\%$ of the total energy gap discontinuity ("Dingle's rule" [8]), which translates into a conduction band offset of 10.6 meV per percent of Al. For higher Al concentrations see Casey and Panish [6].

Although the "straddling" lineup, with varying ratios of $\Delta\epsilon_c : \Delta\epsilon_v$, appears to be the most common case, "staggered" lineups, as in fig. 4a, can also occur.

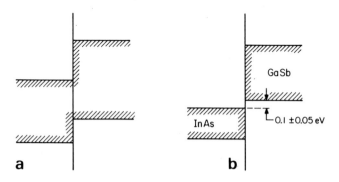

Fig. 4. (a) "Staggered" lineups are expected to occur in many semiconductor pairs. (b) The InAs/GaSb lineup has a broken gap, as shown.

One of the most extreme (and most interesting) lineups is the "broken-gap" lineup at the InAs/GaSb interfaces (fig. 4b): The conduction band edge of InAs falls below the valence band edge of GaSb, by an amount somewhere between 60 and 150 meV [12].

Such different kinds of lineups give the device physicist a powerful device design tool. One of the purposes of this paper is to give a few examples illustrating this point, another is to give some guidance about what governs the lineups in several basic heterosystems. But first we must turn to some of the nuisance effects that complicate considerably the simple Shockley–Anderson model.

2.2. Interface charges

The Shockley–Anderson model in its simplest form described above, is an oversimplification in that it neglects the possibility that there might be inter-face charges associated with the hetero-interface. Any such interface charge would deform the energy band diagram from that in figs. 1 and 2. Fig. 5 shows the results for an n–N heterostructure, for both signs of the charge. A negative interface charge raises the height of the spike barrier, a positive charge lowers it, and if the positive charge is large enough, the barrier is obliterated altogether, creating instead a potential well. Evidently, interface charges – if strong enough – can have a significant effect on the overall barrier heights seen by the carriers, and hence on the properties of any heterostructure device employing the offset barriers.

Interface charges may arise either from the accumulation of chemical impurities at the interface during growth, or from various kinds of structural defects at the interface. An additional mechanism discussed in detail in section 5 occurs at hetero-interfaces that combine two semiconductors from different columns of the periodic table (example: GaAs/Ge), in which case there will often exist a large net interface charge due to non-cancellation of the ion core charges at the interface.

Major modifications of the band diagram occur already for interface charge densities that are still small compared to monolayer densities. Hence, interface charges can play a non-negligible role even at hetero-interfaces which by any other criterion might be considered interfaces with a high degree of perfection.

Consider GaAs, with a lattice constant $a = 5.653$ Å and a dielectric constant $\epsilon_r = 13$. The density of atoms in a monolayer is $2/a^2 = 6.23 \times 10^{14}$ atoms per cm^2. Suppose the GaAs is doped to a level of 10^{17} cm^{-3}, and a region of $d = 10^{-5}$ cm thickness is depleted at a heterojunction, corresponding to $\sigma = 10^{12}$ charges per cm^2, a number certainly very small compared to a monolayer. The electric field supported by such a charge is $E = q\sigma/E\epsilon_0 \cong 1.4 \times 10^5$ V/cm. The accompanying band bending is $\Delta\epsilon_c = \frac{1}{2}qEd = 0.7$ eV, about twice the band bending occurring at a typical GaAs/(Al, Ga)As n–N heterojunction. Evidently, an interface charge density due to defects of, say, 10^{12} charges per cm^2, equivalent to 1.6×10^{-3} monolayer charges, will change the energy band diagram of such a heterojunction completely, and with it the electrical properties of any device containing this heterojunction. Even much smaller interface charge densities, of the order 10^{-4} monolayers, will still have a significant effect. Unfortunately, effects apparently attributable to interface charges of such small but non-negligible magnitude appear to occur frequently

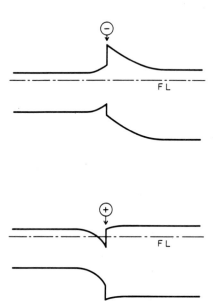

Fig. 5. Band deformation due to a negative (top) or positive (bottom) interface charge.

[13–15]. Evidently, the interface charge is an example of the high degree of sensitivity of the performance of heterojunction devices on the exact atomic structure at or near the hetero-interface, and hence an example of the interrelation between "nanostructure" and device performance.

To a basic physicist, an interface charge of, say, 10^{-3} monolayers may be all but indistinguishable from a "perfect" interface with zero interface charge. But to a device physicist such a small change is a major effect, whose neglect would be unrealistic, and which must be considered along with the band offsets. Still, the roles of the two effects are different: Whereas the band offsets are fundamental and are usually the reason for using heterostructures in devices, the interface charges are almost always a nuisance. Hence we will continue to stress the effects of offsets, raising the issue of interface charges only where necessary.

Unfortunately, interface charges are not the only nuisance: The band offsets themselves appear to be at least somewhat sensitive to exactly how the heterostructure is grown [16], on a level that is not negligible for the device properties, even though it may again be of minor concern to the basic physicist. This introduces another element of uncertainty into the device design, about which we will have to say more later.

3. Band offsets as central device design parameters

3.1. General comments

The extent to which band offsets influence device performance varies tremendously from device to device. At one extreme, the abrupt band offsets may be a nuisance. The heterojunctions in double heterostructure lasers are a good example: Although a varying energy gap is an essential ingredient of the device, a gradual variation would, for various reasons, be greatly preferable over an abrupt step [6,7]. Similar considerations apply to the p–n heterojunctions in heterostructure bipolar transistors [2]. If the semiconductors involved exhibit a continuous mutual solid solubility, the abrupt offsets are easily eliminated by gradient the transition, and this is frequently done.

Of greater interest in the context of this Symposium are devices that call for the retention of the sharp band edge discontinuities, usually with a highly specific kind of mutual band lineup. Many of the more recent heterostructure device concepts are of this kind. Such devices call for a good understanding and knowledge of the band offsets, but exactly what is needed in the way of understanding and knowledge varies greatly from case to case. It depends strongly on the nature of the device; for a given device it changes with the state of development of that device; and more often than not, the needs of the device physicist are again quite different (usually much more severe) than those of the basic physicist. Roughly, the device physicist needs three different levels of knowledge about band offsets:

(a) Semi-quantitative theoretical predictions of the band offsets for as wide a range of semiconductor pairs as possible, to assist in the selection of promising semiconductor pairs to implement new device concepts.

(b) Quantitative data about band offsets, much more accurate than ± 0.1 eV, for those semiconductor pairs that are of clear interest for practical devices, to assist in the detailed development of such devices. Ideally, this should not be restricted to accurate empirical data, but would include a theoretical understanding on a level permitting theoretical predictions with this accuracy.

(c) Data about, and a theoretical understanding of, such nuisance effects as offset variations and interface charges.

In the following three sub-sections of this paper (3.2 through 3.5), these three items are taken up, one by one. Only with respect to item (a) does a satisfactory solution exist, and only with respect to this item have the needs of the device physicist been fully met by the interests of the basic physicist. One of the hopes of this writer is that this paper might stimulate the basic physicist to take up a similar interest in the other two problem areas, to contribute to a satisfactory resolution to those problems as well.

3.2. Rough device design: semi-quantitative theoretical offset rules

New heterostructure device concepts, especially the truly novel ones, usually start out as a hypothetical energy band diagram which, if it could be realized in an actual semiconductor structure, would presumably lead to the desired device properties. The solid state photomultiplier proposed by Williams, Capasso and Tsang (= WCT) [17], and discussed by Capasso earlier at this Symposium, is an excellent example. It requires a highly unsymmetric band lineup, with a conduction band offset that is larger than the gap of the narrower-gap semiconductor, and a valence band offset as small as possible. In such cases, in which the choice of semiconductors is not obvious, the first task is to determine whether the needed energy band diagram is in fact achievable by a real semiconductor combination, and whether or not any such combination is compatible with whatever other constraints may be present (lattice matching, mobilities, overall energy gap constraints, etc.). To this end, semiquantitative predictive lineup rules are required.

The oldest and still widely used such rule is Anderson's *Electron Affinity Rule* [4–6], according to which the conduction band offset should equal the difference in electron affinities between the two semiconductors. Although the rule has been repeatedly criticized on various grounds [18–20], it is better than nothing at all. In fact, it has found vocal defenders [21,22], and it continues to be widely used despite all criticism, largely because its principal competitors, the Frensley–Kroemer theory [23] and the Harrison theory [19,24] are not so overwhelmingly superior to have caused its abandonment.

Although none of these three rules or theories are accurate enough to base a quantitative device design on their predictions, all of them are very useful as semi-quantitative guides. In fact, in simple cases, such as the WCT device [17], even rougher guides may be useful, such as the *Equal Anion Rule* [25]. It states that, for heterojunctions in which the anion atom (the column V or VI element) is the same on both sides, most of the energy gap discontinuity occurs in the conduction band, and the valence band offset is small compared to the conduction band offset. The GaAs/(Al, Ga)As pair has a common anion, and the comparatively small valence band discontinuity in that system $\Delta\epsilon_v \sim 0.15$ $\Delta\epsilon_g$ (for an Al fraction less than 0.45) demonstrates both the rule itself and its approximate nature. The rule has a theoretical foundation: For the III/V and II/VI semiconductors, the valence band wave functions are heavily concentrated around the anion atoms, with only a small part of the wave function being near the cation atom. Equal anion atoms thus naturally mean similar valence band energies [26].

Inasmuch as the WCT solid state photomultiplier calls for as small a valence band offset as possible, it naturally calls for a semiconductor pair that shares the anion species, such as a pair of phosphides, arsenides, or antimonides. Lattice matching is an additional important consideration, and because all Al and Ga compounds with the same anion tend to have very similar lattice constants [6], we can restrict the consideration further to the pairs AlP/GaP, AlAs/GaAs, and AlSb/GaSb, or related alloys. A look at the energy gaps eliminates all but the last pair, which remains as the natural candidate. With energy gaps of 1.60 eV (AlSb) and 0.72 eV (GaSb) [27], the equal anion rule predicts a conduction band offset of 0.88 eV, more than enough to exceed the gap of GaSb, and making some allowance for the approximate nature of that rule. In fact, the Harrison theory [19,24,28] predicts a valence band offset of only 0.02 eV, with the GaSb valence band edge actually the lower of the two semiconductors, that is, a very slightly staggered arrangement. Such a 20 meV prediction should not be taken seriously – the whole theory is probably not better than ± 0.2 eV – but it certainly suggests that the predictions of the equal-anion rule cannot be far off, and it makes AlSb/GaSb a natural candidate for the WCT device. This is in fact one of the two systems discussed by WCT [17] for their device; the foregoing discussion was intended to illustrate by what simple considerations one arrives at this kind of selection. Because AlSb and GaSb do not lattice-match perfectly (2.66 versus 2.65 Å), the addition of a few percent of As to the AlSb is desirable and probably necessary, but this is a refinement going beyond the semi-quantitative considerations discussed here [17].

The Frensley–Kroemer theory [23] (without the doubtful dipole corrections of that theory) predicts an only slightly different band lineup: $\Delta\epsilon_v \cong 0.05$ eV, with AlSb having the lower valence band. Evidently, this changes little. The widely-uded electron affinity rule [4–6] cannot be applied to this system, because the electron affinity of AlSb is unknown, and we do not consider the use of Van Vechten's theoretical values [29] – suggested by Shay et al. [21] and by Philips [22] – as a reliable substitute: The Harrison theory tends to give more accurate values.

The equal-anion rule can be extended into a prediction of how valence band edges vary as the anion is changed: With increasing electronegativity of the anion, the valence bands tend to move to lower energy [25], essentially because the increase in electronegativity reflects a lowering of the valence electron states within the anion atomic potential. In the case of Au Schottky barriers, a quantitative correlation was found [25] between valence band energies relative to the Fermi level, and the anion electronegativity. In the case of semiconductor heterojunctions, no *quantitative* correlation exists, but the anion electronegativity rule remains a useful *qualitative* predictor – see the broken-gap lineup in InAs/GaSb [12,30] – especially if one compares semiconductors whose energy gaps are not too dissimilar. In such cases the valence bands of the phosphides should be lower than those of the arsenides, which in turn should be lower than those of the antimonides.

This kind of prediction can be of great help if – for whatever reasons – a staggered band lineup is desired. As a good example, consider a superlattice with staggered band lineup as shown in fig 6. There has recently been a strong interest in such superlattices [31], for the following reasons. In a staggered structure, any electrons would accumulate in the low-ϵ_c layers, any holes in the high-ϵ_v layers. If both kinds of layers are thin enough (≤ 100 Å), there would be significant tunneling of both electrons and holes, and the entire superlattice would behave essentially as a homogeneous substance with an overall energy gap smaller than that of either constituent compound, slightly larger than the separation between the highest valence band and the lowest conduction band. Suppose next that the low-ϵ_c layer is n-type doped, and the high-ϵ_v layer p-tape. If selective contacts are made to the n-type and p-type layers, and a bias voltage applied, the effective energy gap is varied. But a voltage-adjustable

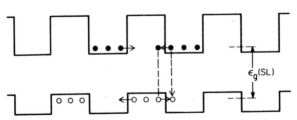

Fig. 6. Staggered-offset superlattice, in which electrons and holes (if present) accumulate in alternating layers. Because of electron tunneling, such structures can have an effective gap narrower than the gaps of both bulk semiconductors.

energy gap would of course be an extremely valueable new phenomenon.

The whole concept is simply an elaboration of the n–i–p–i superlattice concept of Döhler and Ploog [32], except that the spatial separation of the high concentrations of electron and hole from each other is now achieved very easily by the band offset forces, rather than purely electrostatically, by heavy doping.

The occurrence of a broken gap in the InAs/GaSb system suggests that less extreme cases of staggering are indeed achievable, but are they achievable in semiconductors with much larger energy gaps? The anion electronegativity rule [25] suggests that combinations of a phosphide with an antimonide form a promising point of departure. Because phosphides tend to have smaller lattice constants (and larger energy gaps) than antimonides, it is advisable to start with the phosphide that has the largest lattice constant (and the smallest gap), InP, and the combine it with the largest-gap antimonide, AlSb. For this system the Harrison theory [28] predicts indeed staggered band offsets, with a conduction band well depth $\Delta\epsilon_c = 1.20$ eV, a valence band well depth $\Delta\epsilon_v = 0.97$ eV, and a net band separation

$$\epsilon_g(SL) > \epsilon_c(InP) - \epsilon_v(AlSb) \cong 0.4 \text{ eV}.$$

The actual superlattice gap should be somewhat larger, increasing with decreasing superlattice period.

Although the estimate was rough, the message is clear: Staggered superlattices with usefully large gaps should be achievable! Whether or not the simple InP/AlSb pair is indeed a promising pair, remains to be seen, but it is certainly a useful point of departure. If anything, the staggering is larger than needed and the effective gap (≥ 0.4 eV) too small to be useful. Evidently, the conditions to obtain staggering may be relaxed somewhat. Now, one of the drawbacks of the InP/AlSb pair is a large lattice mismatch ($\cong 4.6\%$). Such a lattice mismatch, would almost certainly be fatal to device performance in a single-interface heterostructure device due to inevitable misfit dislocations. But it might be quite acceptable in a short-period superlattice, where the lattice misfit can be taken up by elastic strain, a point recently elaborated upon by Osbourn [31] in the context of strained-layer staggered superlattices based on the GaP/Ga(P, As) system. If necessary, the lattice misfit could be reduced by replacing AlSb with an Al(Sb, As) alloy. This would make the valence band well shallower and increase the net gap, but the Harrison theory predicts that even for perfect lattice match to InP, that is, for $AlAs_{0.56}Sb_{0.44}$ [33], a valence band well of 0.46 eV and a net gap of 0.91 eV should remain. Further fine-tuning could be achieved by replacing some of the Al by Ga [6].

Two other lattice-matched pairs for which staggered lineups can be safely predicted are InP/$Al_{0.50}In_{0.50}As$ ($\epsilon_g \geq 1.1$ eV) and $Ga_{0.52}In_{0.48}P$/AlAs ($\epsilon_g \geq 1.6$ eV).

There is some evidence [34] that the GaP_xAs_{1-x} system for $x > 0.5$ leads to staggered lineups with large net gaps, but for this system the theoretical predictions are not as clear-cut as for the above combination. We will return to this point later.

3.3. Quantitative device design: the absence of theoretical guidance

Although semi-quantitative lineup prediction rules are very useful in identifying promising hetero-pairs for hypothetical device applications, a detailed device design requires far more accurate values. In any device in which current flows *across* a heterostructure barrier, the current depends on the barrier height $\Delta\epsilon$ at least like a Boltzmann factor $\mathrm{Exp}(-\Delta\epsilon/kT)$, implying a factor e for every change in barrier height by $1\ kT\ (\cong 26$ meV at 300 K). If the current is tunneling rather than thermionic current, the dependence tends to be even steeper. There is no need to discuss here whether a prediction to some fraction of kT is necessary or whether $\pm 1\ kT$ or even $\pm 2\ kT$ would be sufficient: None of the predictive theories comes anywhere near even the less demanding limit. Those physicists (not involved in actual device design) who have expressed their satisfaction with either the electron affinity rule [21,22] or the Harrison theory [20], quote examples of "excellent agreement" between theory and experiment, in which the predicted offsets vary by 0.2 eV ($\cong 8\ kT$) or more from reliable experimental data. Presumably, then, this is roughly the level of reliability of existing predictive rules or theories. This degree of agreement may indeed be very satisfactory to the fundamental physicist, who wants a general understanding of band offsets; it is totally unsatisfactory as a quantitative basis for device design.

Nor is the need for an accurate prediction significantly less demanding in those devices in which current does not flow *across* a hetero-barrier, but *along* it, as in the new high electron mobility transistor (HEMT) [10] which represents one of the most active areas of heterostructure device research and development, also discussed (from a physics- rather than device-oriented point-of-view) by Störmer at this Symposium. One of the most important design parameters in these devices is their threshold voltage, that is, the gate voltage at which the conductance along the 2D conducting channel is effectively turned on or off (it may be either a positive or a negative voltage, depending on the desired design). To be useful in future high-performance IC's (their dominant area of interest), the threshold voltages of these devices must be predictable much more accurately than ±0.1 V, preferably to ±0.1 V, which calls for a knowledge of the band offsets to within a similar accuracy.

As the HEMT case shows, the absence of any purely theoretical predictive tools with the desired accuracy is not preventing the design of this particular device to go forward. The band offsets at the (Al, Ga)As-on-GaAs (100) interface *are* known to the required degree of accuracy [6]. But this accurate knowledge is the result of accurate *experimental* measurements [8], not of an accurate predictive theory. Once the evolution of a new heterostructure device has progressed beyond the initial speculative stage, to the point of practical device development, it is necessary that the band offsets be accurately *known*, but the knowledge need not come from a predictive theory; knowledge from accurate experimental data may actually be preferable to a theoretical prediction. This de-facto status of the band offsets is similar to that of energy gaps: Whenever available, we use accurate experimental values of energy gaps, rather than theoretical values. Only when experimental data are missing, will we use theoretical ones.

Does any of this mean, however, that the attempts to predict band offsets theoretically have no value beyond the crude semi-quantitative value discussed earlier? Far from it! First of all, the purpose of theories of band offsets (e.g. electron affinity rule, Harrison's theory, etc.) is only partially to provide the device physicist with quantitative design data. A more important role is to test the assumptions that go into each theory, and thereby to test our fundamental understanding of what determines the band offsets. This is similar to the way band structure calculations test our understanding of band structures more than providing accurate theoretical gap values when accurate experimental values are already available. All these are *retrodictive* theories more than *predictive* ones! By that standard, neither the electron affinity rule not the Harrison theory, with their ±0.2–0.3 eV accuracy, are doing badly (nor does the Frensley–Kroemer theory, which is of similar accuracy). Inasmuch as the present paper is to describe a device physicist's view of hetero-interfaces, it does not provide a suitable forum to discuss exactly how well these theories meet the needs of the basic physicist, and much less to discuss critically the enthusiastic support that Shay et al. [21] and Philips [22] have expressed for the electron affinity rule, and Margaritondo and his co-workers [20] for the Harrison theory.

A second reason why more accurate theoretical predictions could be useful as quantitative rather than merely semi-quantitative predictive tools occurs whenever the accuracy of the existing theories is insufficient to yield a clear-cut yes–no decision about a speculative device, but in which experimental data would require the prior development of an elaborate technology. A theoretical guidance on whether or not the development of this technology is worthwhile would be highly useful in such cases [18].

A good example is once again at hand. There has been considerable speculation [31] that a $GaP/GaP_{0.6}As_{0.4}$ superlattice would be of the interesting staggered variety shown in fig. 6. This speculation is partially based on the electron affinity rule, using the electron affinity value of 4.3 eV quoted by Milnes and Feucht [5] without giving any source. Partially it is based on a highly indirect claim by Davis et al. [35] (contradicting other data) that the conduction band offset in the GaP/GaAs system should be near zero. A very careful measurement of the electron affinity has recently been performed by Guichar et al. [36], yielding 3.70 ± 0.05 eV. Using this value, and the known electron affinity for GaAs, 4.07 eV, and making due allowance for the change from direct gap to indirect gap in going from GaAs to GaP, one predicts a conduction band offset for the superlattice of only 0.02 eV, just very slightly staggered. The Harrison theory predicts the same value [28]. With the reliability of both the electron affinity and the Harrison theory rule being no better than ±0.2 eV, this prediction is simply a draw. Inasmuch as the development of an entire superlattice technology hinges on this prediction, it is an excellent example of why more accurate predictions would indeed be desirable.

Recent experiments suggest [34] that the superlattice is indeed staggered, by about 0.2 eV. If future measurements confirm this result, this would show that both theoretical predictions are indeed incorrect by about 0.2 eV.

3.4. The nuisance effects: offset variations and interface charges

In the preceding discussion we pointed out the device physicist's need for knowing band offsets to an accuracy much better than ±0.1 eV. But this request implicitly assumed that the band offsets are in fact constants that characterize a given semiconductor pair, rather than being variables themselves. As was pointed out by Bauer [37] and by Margaritondo [20] at this Symposium, evidence is accumulating [16,38] that the offsets are process-dependent, changeable over a finite range outside of the tolerance limits of the device designer. A dependence on crystallographic orientation is almost to be expected, and while it might be a nuisance, it does not introduce any problems into device design. Nor do we need to be surprised about large offset variations in systems in which a compound semiconductor (GaAs, GaP) is grown on an elemental semiconductor (Ge, Si), or vice versa, the cases of particular interest to Bauer [37] and Margaritondo [20]. We shall argue in section 5 that in such systems technology-dependent offset variations and interface charges *are to be expected*. What *is* disturbing are offset variations and interface charges in such supposedly well-behaved lattice-matched systems as GaAs/(Al, Ga)As. It was found by Waldrop et al. [16] that for {110}-oriented MBE growth at a substrate temperature of 580°C the band lineup depends noticeably on whether AlAs is grown on GaAs ($\Delta\epsilon_v \cong 0.15$ eV), or GaAs on AlAs ($\Delta\epsilon_v \cong 0.40$ eV). By comparison, the {100}-lineup data of Dingle [8] for GaAs/(Al, Ga)As, extrapolated to Gas/AlAs, corresponds to an in-between value of $\Delta\epsilon_v \cong 0.20$ eV.

Although differences between {100} and {110} might have been expected, the strong growth sequence dependence for the {110} orientation comes as a rude shock. For a given orientation, band offsets can depend on growth sequence only through differences in the exact atomic arrangement near the interface. Evidently the atomic arrangements for {110} interfaces depend strongly on growth sequence. Put bluntly: At least for this orientation the offsets depend quite strongly on technology [39] rather than being a fundamental materials parameter! The question naturally arises whether or not this might quite

generally be the case. Might there be a similar growth sequence dependence for
⟨100⟩ growth? I find it hard to believe that any significant growth sequence
asymmetry of ⟨100⟩ band offsets would leave intact the superb fit of Dingle's
superlattice data (which automatically involve both growth sequences) to a
single-offset model, especially considering Dingle's wide range of layer thicknesses. Yet there exists strong evidence that, if not the band offsets, at least the
transport properties in the 2D electron gas along GaAs/(Al, Ga)As ⟨100⟩
heterojunctions, depend quite strongly on the growth sequence [38], with
higher mobilities occurring for (Al, Ga)As-on-GaAs than for GaAs-on-
(Al, Ga)As. In fact, it appears that in structures containing multiple interfaces,
the properties of the interfaces grown first differ from those grown later [40]!

One frequently hears the argument that effects such as these are somehow
artifacts of the growth process, reflecting "bad" interfaces. While in a practical
sense this might be true, it avoids the fundamental issue: Even a "bad"
interface must have some atomic configuration that causes these effects, and
which configuration constitutes "badness"? And can this "badness" in fact be
avoided under the numerous constraints imposed upon the growth of an actual
device?

We clearly need an understanding of these effects, and this may indeed by
one of the most urgent research topics in which the device physicist would like
to see the basic physicist take an active interest. To the basic interface
physicist, offset variations of ≡ 50 meV might be a minor nuisance, negligible
to the basic understanding of the interface physics. But the degree to which
these offset variations can be controlled, may be decisive for the role heterostructure FET's will play in future high-speed VLSI technology.

A return to the earlier example of HEMT threshold voltages will illustrate
the urgency. As we stated, these threshold voltages depend on several structural parameters, one of which is the conduction band offset. Now the most
important envisaged applications of this transistor is in future very fast
large-scale digital integrated circuits which may contain anywhere from 10^3 to
10^6 identical FET's per chip. For a variety of reasons, it is necessary that the
threshold voltages of all transistors on the same chip have essentially the same
value, *and* that this design value can be technologically maintained from chip
to chip and even from wafer to wafer. Threshold voltage variations far below
±0.1 V on a single chip are essential, or else the IC will simply not work, and
variations below 10 mV are desirable. Worse, the variations from chip to chip
should not be much larger. Evidently this calls for tight tolerances on the band
offsets and on residual interface charges.

To achieve these tolerances requires an understanding of what causes offset
variations and interface charges, not just purely empirical tight process control.
In fact, it is probably more important to develop a physical understanding of
offset *variations* on the ±5 meV level than to be able to *predict* the exact
magnitude of these offsets to better than ±0.1 eV.

4. On measuring band offsets experimentally

4.1. Introductory comments

There does not exist any experimental technique to determine band offsets that is simultaneously simple, reliable, and universally applicable.

The most careful and presumably most accurate determination of any band lineup is Dingle's well-known work [8] on the infrared absorption spectra of superlattices of weakly-coupled multiple GaAs/(Al, Ga)As quantum wells. Dingle was able to fit large numbers of data, for wells of various widths, to a single model in which the conduction band offset is 85% ± 3% of the energy gap difference.

For sufficiently narrow wells, the method is fairly insensitive to errors by small interface charges. Major distortions in the well shape would quickly destroy the excellent fit of the experimental data to the simple square-well model. Dingle's data prominently include transitions involving the higher energy levels in the wells, which would be especially sensitive to any distortions of the well shape. It is hard to believe that the large number of observed transitions, over a wide range of well widths, could be fitted just as well to a significantly different well shape. This same quality-of-fit argument also speaks against various kinds of modifications in the band offsets, such as growth sequence asymmetries, etc. Certainly, the burden of the proof for any such modifications lies with those who would propose such modifications. Note, however, that Dingle's data, being strictly {001} data, in no way rule out any offset dependence on crystallographic orientation.

A second widely used technique to determine band offsets is based on photoelectron spectroscopy [20,41], executed with various levels of sophistication. It is even less sensitive to interface charges, and is in principle capable of giving quite accurate offsets, perhaps more directly than Dingle's technique. Especially the Rockwell group of Kraut, Grant, Waldrop and Kowalczyk [41] has cultivated this technique to a high level of perfection, to the point that in favorable cases offsets with (believable) uncertainties of ±0.03 eV were obtained. Inasmuch as Margaritondo, another practitioner of this technique, has discussed it at this Symposium, we refer to his paper [20] for more information and references.

Both the superlattice absorption technique and the photoelectron spectroscopy technique are "physicist's techniques", rather than device-type techniques. Now we argued earlier in this paper that the properties of heterostructure devices depend sensitively on band offsets. It should therefore be possible to extract accurate band offsets from measurements on devices. Because of the simplicity of purely electrical measurements, such attempts have indeed often been made [5], and many band offsets found in the literature were in fact obtained from purely electrical measurements, usually on simple p–n or n–n heterojunctions. Unfortunately, such measurements are sensitive not only to band offsets; they are just as sensitive to other phenomena that deform the band diagram, especially interface charges. Most of the electrical measurements have difficulty separating these effects. More often than not the data are merely fitted to the simple Shockley–Anderson model ignoring such complications, which can lead to grossly inaccurate band offsets.

Inasmuch as this paper represents a review of hetero-interfaces from the device physicist's point-of-view, a critical review of the main techniques is in order.

4.2. Capacitance–voltage profiling

Probably the best of the purely electrical measurement techniques is based on a powerful adaptation of conventional C–V impurity profiling, recently developed by Kroemer et al. [14,42]. It can, under favorable circumstances, give reliable *separate* values for both the band offsets and any interface charges. The method requires an n–n heterojunction whose doping profile is known, a condition often satisfied for junctions grown by highly developed technologies such as MBE. A Schottky barrier is placed on the outer surface of the heterostructure, parallel to the hetero-interface, and the C–V relation of the Schottky barrier rather than of the heterojunctions itself is measured. The method works best with heterojunctions exhibiting poor rectification, which are particularly hard to evaluate by other means. An *apparent electron concentration* $\hat{n}$ is determined by the conventional interpretation of C–V profiling theory [42],

$$\frac{\mathrm{d}}{\mathrm{d}V}\frac{1}{C^2}=\frac{2}{q\epsilon}\frac{1}{\hat{n}(x)},\tag{1}$$

where C is the capacitance per unit area, and $x=\epsilon/C$. The $\hat{n}(x)$ profile will differ both from the doping profile $n_{\mathrm{d}}(x)$ and from the true electron concentration $n(x)$. But if the doping distribution $n_{\mathrm{d}}(x)$ is known, the interface charge is easily obtained by integrating the apparent difference distribution $\hat{n}(x)-n_{\mathrm{d}}(x)$, and the conduction band offset is obtained from the first moment of this difference distribution. The true electron distribution is not needed! The method is simple and powerful, and readily applicable to any technology that permits the growth of heterostructures in which the doping level can be kept accurately constant on both sides of the interface, with an abrupt switch at the interface. The two constant doping levels need not even be predetermined; they may be extracted from the C–V profile itself.

The method may be made self-checking, by using the two doping values, the interface charge, and the band offset, to simulate on a computer the C–V profile that *should* have been seen experimentally, and by comparing this reconstructed profile with the profile actually observed.

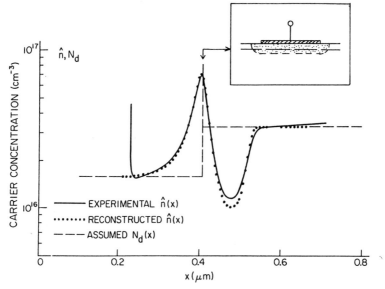

Fig. 7. *C–V* profiling through an LPE-grown GaAs/(Al, Ga)As n–N junction, after ref. [14]. From the measured apparent electron concentration $\hat{n}(x)$ (solid curve) and the assumed donor distribution $n_D(x)$ (broken curve) one can calculate a conduction band offset $\Delta\epsilon_c = 0.248$ eV and an interface charge density $\sigma_1 = +2.7 \times 10^{10}$ cm^{-3}. The inset shows the basic test arrangement.

Fig. 7 shows an example, from ref. [14], for an LPE-grown n–N heterostructure, not ideally suited for the purpose, but so far the only published result in which the method has been used for a quantitative determination of both a band offset and an interface charge, including the self-consistency check. The technique should be even better suited to MBE- or MOCVD-grown interfaces, in which an abrupt transition with flat adjacent doping levels is more easily achieved, and this writer does in fact expect that it will be widely used in the future.

4.3. The C–V intercept method

When the doping level n_d and hence the electron concentration n in an n-type semiconductor is position-independent, the *C–V* profiling theorem (1) yields a linear C^{-2}-versus-V plot. This remains true for the capacitance of a p–n junction, including a p–n heterojunction, if the carrier concentrations on both sides are constant. This has led to the *C–V intercept method*, which claims that the intercept voltage V_{int} in such a linear C^{-2}-versus-V plot is exactly equal to the total built-in voltage of the heterojunction (fig. 8), sometimes called the *diffusion voltage*,

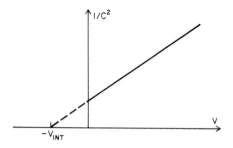

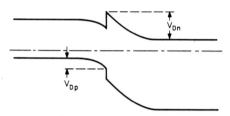

Fig. 8. The C–V intercept method of determining the band offsets at p–N heterojunctions. *If the* heterojunction is abrupt, with constant doping levels right to the interface (no grading), and without any interface charges, then the intercept voltage V_{int} in a C^{-2}-versus-V plot is related to the two diffusion voltages V_{Dn} and V_{Dp} via eq. (3). If both doping levels (and hence both Fermi energies) are known, this permits a determination of the band offsets. The method is sensitive to errors caused by grading or interface charge effects.

$$V_{int} \overset{?}{=} V_{Dn} - V_{Dp}. \tag{2}$$

For known doping levels, the energy separations between the bulk band edges and the Fermi level are known, and hence the band offsets are known if $V_{Dn} + V_{Dp}$ is known. Unfortunately, the accuracy of eq. (2) is largely a (persistent) muth. First of all, (2) neglects the so-called Gummel–Scharfetter correction [43]; it should really read

$$V_{int} = V_{Dn} + V_{Dp} + 2kT/q, \tag{3}$$

a small correction, but not a negligible one. More important: Even in the form (3), the intercept rule is strictly valid only if both doping levels are constant right to the hetero-interface, forming an abrupt transition there, and if no interface charges are present [42,44]. Interface charges tend to lower the intercept voltage, whereas impurity grading effects raise it. A small region right at the interface always remains inaccessible, even if C–V profiling is extended to forward bias values. Any space charge re-adjustments entirely inside this region will not affect the linearity of the C^{-2}-versus-V plot unless the charge inside the depletion region somehow depends on the applied voltage (which may be the case for deep levels, but not otherwise). Although these facts have been established for some 25 years [44], they remain strangely ignored except by a small fraternity of semiconductor device physicists intimately familiar with $C-V$ profiling theory. Even as astute a researcher as Phillips [22] writes in a recent paper: "The great merit of this technique is that it is self-checking, i.e., when chargeable traps are present at the interface, C^{-2} is not a linear function of V_a. The deviations from linearity automatically provide estimates of the accuracy of the determination of V_D and from it the accuracy of ΔE_C and ΔE_V." Well, they don't. To get experimental access to the charges located right near the interface, one must profile *through* the heterojunction from the outside, as described earlier, not from the interface outward.

Considering this inherent weakness of the intercept method, it is not surprising that the offset values determined by it have fluctuated widely whenever data from more than one investigator have been available, and often even for the data from the same group. Two examples are provided by the chaos in the offset data reported for GaP/GaAs and Ge/GaAs. In most of these measurements, $C-V$ intercept data were not used alone, but in conjunction with current–voltage ($I-V$) data. However, this hardly excuses the failure of the intercept method to "catch" the ever greater inadequacies of the $I-V$ techniques.

In the case of GaP/GaAs, the reported conduction band offsets vary by at least 0.65 eV: Weinstein et al. [45] claim $\Delta\epsilon_c \cong 0.22$ eV, Alferov et al. [46], $\Delta\epsilon_c \cong 0.65$ eV, and Davis et al. [35], $\Delta\epsilon_c \cong 0$. It is anybody's guess which of these values is least far away from the truth – if there is in fact a single "true" value.

The situation for Ge/GaAs is, if anything, even worse. Conduction band offsets varying from 0.09 to 0.54 eV can be found in the literature, a range corresponding to 68% of the energy gap of the narrower-gap semiconductor, Ge. The reason is probably only partially due to erratic measurements. As we shall see later, for polar/nonpolar systems such as GaAs/Ge, an erratic technology-dependence of the offsets should be expected.

Despite this history of unreliable results, the intercept method should be capable of yielding accurate offsets *if* the uncertainties inherent in it are treated with due respect, and are eliminated by suitable complementary data, especially for interfaces grown by one of the better and more tractable technologies, such as MBE or MOCVD. There is something inherently satisfactory about $C-V$ profiling measurements: They are essentially purely electrostatic measurements of equilibrium charge distributions versus position, almost completely unencumbered by transport effects.

4.4. Current–voltage measurements

Whatever criticisms one might have of band offsets based primarily on $C-V$ intercepts, most of those based on current–voltage ($I-V$) data on p–n or n–n heterojunctions are even less well-founded. Exceptions tend to occur for systems with unusual band lineups, in which the $I-V$ data on heterojunctions differ already qualitatively in drastic ways from those of ordinary p–n homojunctions. The outstanding (but not the only) example is the striking broken-gap lineup at the InAs/GaSb interface (fig. 4b), for which the first experimental evidence was obtained [12] from systematic rectification experiments with lattice-matched Ga(As, Sb)/(Ga, In)As p–n heterojunctions of varying (lattice-matched) alloy compositions. As the GaSb/InAs end was approached, all rectification effects suddenly disappeared, due to the "uncrossing" of the forbidden gaps.

But $I-V$ data on p–n heterojunctions without special lineup feature tend not to contain enough qualitatively different detail to be useful for quantitative offset determination, although they may be useful to supplement other data.

Worst, $I-V$ data on n–N rather than p–n heterojunctions, although they could in principle be quite informative, have in the past been largely worthless. For example, the claim that the conduction band offset of GaP–Si interfaces is essentially zero, is based on nothing more than the failure to observe any rectification effects in Si-on-GaP n–n junctions even at liquid nitrogen temperature [47]. More recent data on this system show [48,49] this claim to be quite false. How erroneous such absence-of-rectification data can be, is illustrated by what is now the best understood heterostructure of all, the GaAs/(Al, Ga)As structure: Most early data on this system showed a more or less complete absence of rectification in n–N junctions [50]. The explanation in terms of zero conduction band offset flatly contradicted Dingle's lineup data. The problem seems to have gone away with subsequent improvements in technology; it was almost certainly due to donor-like defects at the interface, as first proposed by Kroemer et al. [13]. Similar donor-like defects were probably responsible for the lack of rectification in Si/GaP heterojunctions [47].

5. Polar / nonpolar heterostructures

5.1. Motivation

Almost all heterostructure *device* structures currently under active investigation employ heterostructures between III/V compounds only. There are strong incentives to extend heterostructure device technology to other systems, especially to combinations of a III/V semiconductor with one of the elemental semiconductors, Ge or Si. Natural pairs, because of their close lattice match, would be GaAs/Ge and GaP/Si. The latter is particularly interesting. If device-quality interfaces between GaP and Si could be achieved, this would be a major advance towards bridging the wide gap between highly-developed Si technology and the rapidly developing technology of III/V compounds, with potentially far-reaching device applications.

A number of attempts to grow such polar/nonpolar heterostructures have led to disappointing results: These systems are clearly far more difficult than III/V-only heterosystems. However, a physical understanding of these systems is beginning to emerge that explains why many of the earlier purely empirical "cookbook" approaches *should* have failed, and which suggests that a better understanding of both the growth mechanism and the electronic structure of these interfaces might make possible substantial progress towards the elusive goal of device-quality polar/nonpolar heterostructures.

In fact, the incentives to achieve such a better understanding go far beyond the device utilization of polar/nonpolar interfaces themselves: It would also advance the understanding of more "ordinary" III/V-only interfaces. Many of the problems that occur at polar/nonpolar interfaces are simply hugely magnified versions of problems that occur already at the GaAs/(Al, Ga)As interface. Examples: Residual interface charges, offset variations, crystallographic orientation dependence, and technology dependence. The difference is purely quantitative: In the III/V-only cases these problems are second-order nuisances, in the polar/nonpolar cases they dominate. I believe this dominance is the reason why polar/nonpolar interfaces have so far proven so intractable. It is reasonable to expect that a better understanding of these effects, leading to control in the polar/nonpolar case, will also greatly benefit the III/V-only case.

5.2. Interface neutrality and crystallographic orientation

In 1978, Harrison, Kraut, Waldrop and Grant (HKWG) published a classical paper [51] that forms the point of departure for any rational understanding of the problems of polar/nonpolar interfaces. The authors studied the electrostatics of the simplest possible atomic configurations for the three lowest-index orientations of an ideal GaAs/Ge hetero-interface. They showed that for both the {100} and {111} orientations these atomic configurations correspond to a huge net electrostatic interface charge, of the order of one-half of a monolayer charge. The argument is brought out in fig. 9 for the (001) interface, viewed in the [$\bar{1}$10] direction. The black circles represent Ga atoms, the white circles As atoms, and the shaded ones, Ge. An alternate possibility has Ga and As interchanged. An important point in the HKWG argument is a point emphasized earlier by Harrison [52]: The tetrahedral bond configuration *guarantees* that each of the bonds connecting each atom to its four nearest neighbors contains exactly two electrons, just as in Ge, and regardless of whether the bonds are Ge–Ge, Ga–As, or mixed Ga–Ge or Ge–As bonds. Only the electron distribution along each bond depends on these details, not the overall bond charge. This means that the net electrical charge associated with the overall interface region can be determined by simply counting each column-V atom as having one extra proton charge relative to a neutral column-IV atom, and each column-III atom as missing one such charge. The overall interface charge is easily obtained by a fictitious process, whimsically called "theoretical alchemy", in which one pretends that the GaAs portion of the heterostructure has been obtained from a Ge single crystal by moving a proton lattice from one-half of the Ge atoms to the other half of the Ge atoms, creating Ga and As in the process. Depending on whether the fictitious proton motion is away from the interface or towards it, a negative or positive charge imbalance is thereby created at the interface. The bottom half of fig. 9 shows the electrostatic potential resulting from a proton transfer away from the interface, with the electron distribution along the bonds initially kept fixed. The potential staircase on the GaAs side is evident. The average slope of this staircase represents a net electric field, which is easily shown to be that of a charge of $-q/2$ per interface atom. With an interface atom density of $2/a^2$, this is a charge density $-q/a^2$. The important point is now that the bond charge relaxation following the proton transfer does not change the net

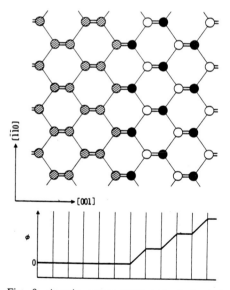

Fig. 9. Atomic arrangement and electrostatic potential at an idealized unreconstructued Ge/GaAs(001) interface, from ref. [51]. The idealized atomic arrangement exhibits a large charge imbalance at the interface, leading to a staircase potential with a large net electric field on the GaAs side. The full circles represent Ga atoms, the open circles As atoms.

interface charge, even though it is strong enough to actually reverse the sign of the net charge on the Ga and As atoms inside the GaAs side. But the total charge per bond always remains at exactly two electrons; no net charge crosses the Ga and As atomic planes inside the GaAs side, implying conservation of net interface charge during the relaxation. In terms of the potential diagram in fig. 9, the shape of the individual steps in the staircase changes, but the net *average* slope remains unchanged.

As HKWG point out, the field supported by the net interface charge is huge ($E = q/a^2\epsilon \cong 4 \times 10^7$ V/cm, assuming the dielectric constant of GaAs), sufficient to guarantee an atomic re-arrangement during the crystal growth itself, to minimize those interface charges. The authors give two specific atomic configurations which lead to zero interface charge, shown in figs. 10 and 11. The first of these contains one mixed-composition layer, but it retains a finite interface dipole. In the second configuration, containing two mixed-composition layers, the interface dipole has also been obliterated. The authors speculate that the second configuration might actually arise during epitaxial growth.

It is at this point that we must differ from HKWG. Although there can be no doubt that a drastic atomic re-arrangement will take place, and almost certainly in the general direction postulated by HKWG, it appears inconceiva-

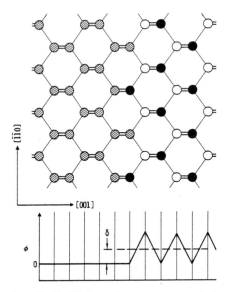

Fig. 10. Modified atomic arrangement and electrostatic potential at a Ge/GaAs(001) interface, containing one atomic plane of mixed composition, with zero net interface charge, but retaining a finite interface dipole. From ref. [51].

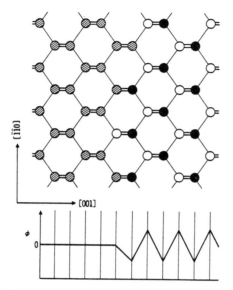

Fig. 11. Further modification of the atomic arrangement at a Ge/GaAs (001) interface, containing two atomic planes of mixed composition yielding both zero interface charge and a zero interface dipole. From ref. [51].

ble that any such re-arrangement goes sufficiently far towards completion that the remaining interface charge becomes negligible for device purposes. We recall that even a charge of only 10^{-3} monolayers is still a large interface charge for device purposes; even if the interface atomic re-arrangement goes 99% towards completion, this would still leave an intolerably large charge five times as large.

We therefore conclude that, at least for the ⟨100⟩ orientation, large residual interface charges must be expected at GaAs/Ge and similar polar/nonpolar interfaces. Worse, the exact amount of interface charge left must be expected to depend on the growth process. Hence the interface charges will not only be large, but technology-dependent. Finally, because even for zero interface charge the residual interface dipoles still depend on exactly which atomic re-arrangement was created, the band offsets must also be expected to be technology-dependent and hence poorly reproducible.

There are mitigating circumstances present if the growth sequence is non-polar-on-polar. Harrison has pointed out [53] that the electrostatic arguments of HKWG also apply, with some modification, to the free surface of a compound semiconductor. A GaAs ⟨001⟩ surface terminating in complete Ga or As planes is electrostatically just as unfavorable as an ideal GaAs/Ge interface. The actual atomic configuration present at a free GaAs ⟨100⟩ surface will already be such that the net surface charge is minimized. If all dangling surface bonds dimerize, apparently a good first-order approximation, an atomic arrangement leading to a neutral surface will also lead to a neutral Ge/GaAs interface, if the vacuum is subsequently replaced by Ge.

But this argument does not apply if GaAs is grown on Ge. Thus we are led to a second prediction: Polar/nonpolar interfaces must be expected to exhibit drastic growth sequence dependences, much stronger than those observed in the GaAs/(Al, Ga)As system. Unfortunately, the more difficult polar-on-non-polar growth sequence is demanded in the majority of device applications. In my opinion, attempts to grow GaAs/Ge or similar polar-on-nonpolar {100} heterojunctions or – worse – polar/nonpolar superlattices with this orienta-tion, in the hope that device-quality interfaces will somehow result, are likely to be little more than a waste of time. The fact that this orientation is so successful for III/V-only growth is quite irrelevant. The likely answer – if any – to the quest for successful polar-on-nonpolar growth lies in the use of one of the nonpolar orientations to be discussed presently.

The HKWG argument is by no means restricted to the {100} orientation. Qualitatively similar arguments with only minor quantitative modifications can be made for {111}-oriented interfaces, and in fact for all interface orientations except those in which the interface is parallel to one of the ⟨111⟩ bond direction.

The condition for this can be expressed as a mathematical condition on the Miller indices (hkl) of the interface [54]. Let $[hkl]$ be the direction perpendicu-lar to the interface plane. The plane is parallel to one of the ⟨111⟩ bond

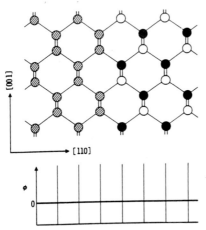

Fig. 12. Atomic arrangement and electrostatic potential at an ideal Ga/GaAs(110) interface. Each GaAs plane parallel to the interface contains an equal number of Ga and As atoms and is hence electrically neutral. From ref. [51].

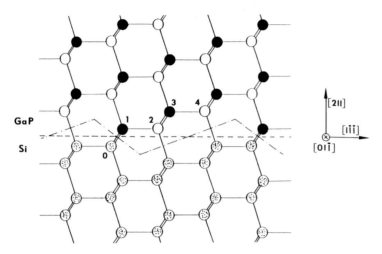

Fig. 13. Atomic arrangement at idealized GaP/Si(211) interface, from ref. [54]. As in the ⟨110⟩ case, each GaP plane parallel to the interface contains an equal number of Ga and P atoms and is hence electrically neutral. But in addition, the bonding of the "black" sublattice sites across the interface is much stronger (two bonds) than that of the "white" sublattice sites (one bond). When GaP is grown on Si, this bonding difference can be utilized to achieve growth free of antiphase disorder, with the "black" sublattice occupied by P atoms, the white by Ga atoms.

directions if $[hkl]$ is perpendicular to that direction. This implies

$$[hkl] \cdot \langle 111 \rangle = \pm h \pm k \pm l = 0,$$

for at least two of the eight possible independent sign combinations. The simplest such orientation is the {110} orientation, already recognized as such and intensively discussed by HKWG. The next-simplest orientation is {112}, followed by {123}, etc. Figs. 12 and 13 show the atomic arrangements at a (110) and at a (112)-oriented polar/nonpolar interface, both viewed again in the [$\bar{1}$10] direction.

In the absence of specific reasons to do otherwise, it is probably advisable to use the lowest-index orientation for the epitaxial growth. If only the nonpolar-on-polar growth sequence is needed for a particular device, the {110} orientation may indeed be the preferred orientation. Inasmuch as the {110} planes are the natural cleavage planes of III/V compounds, this happily coincides with the natural interest of the surface physicist in this orientation: Most of the non-device studies of the initial growth of Ge on GaAs have indeed used these planes. However, if the polar-on-nonpolar growth sequence is demanded (which automatically induces polar/nonpolar superlattices), altogether new considerations intervene.

5.3. Polar-on-nonpolar growth: the site allocation problem

When, in a polar/nonpolar heterosystem, the polar (compound) semi-conductor is to be grown on the nonpolar (elemental) one, a new problem arises [54,55]: Avoiding antiphase disorder in the growing compound semi-conductor. This problem does not exist at all in element-on-compound growth, and it is at most a minor problem in compound-on-compound growth. But for compound-on-element growth it is as severe and fundamental as the interface neutrality problem at {001} polar/nonpolar interfaces, and it totally dominates the problem of polar-on-nonpolar growth for nonpolar orientations, such as {110} and {112}.

When a binary compound with two different atoms per primitive cell (e.g. GaAs, GaP) is grown on an elementary substrate (e.g. Ge, Si) in which the two atoms are identical, there exists an inherent ambiguity in the nucleation of the compound, with two different possible atomic arrangements, distinguished by an interchange of the two sublattices of the compound. If different portions of the growth exhibit different sublattice ordering, antiphase domains result, separated by antiphase domain boundaries, a defect similar to grain and twin boundaries. For high-performance devices, antiphase domain boundaries must almost certainly be avoided, which calls for a rigorous suppression of one of

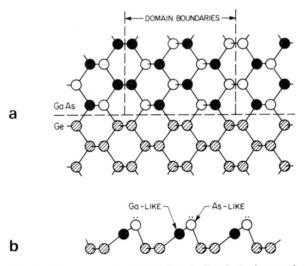

Fig. 14. (a) Occurrence of antiphase domain disorder in the growth of GaAs on an unreconstructed Ge {110} surface, due to the absence of a built-in bonding difference for the as-yet unoccupied surface sites belonging to the two sublattices. (b) Creation of Ga-like and As-like electronic configurations in the top Ge {110} atomic layer, due to reconstruction, aiding in the suppression of antiphase disorder inside the GaAs. From ref. [55].

the two nucleation modes. The problems in doing so depend very strongly on the exact atomic arrangement and on the dangling-bond configuration at the surface of the elemental semiconductor substrate. Unfortunately, they are particularly severe for the simplest nonpolar interface orientation, the {110} orientation. The situation is illustrated in fig. 14a, which shows that on an ideal and perfectly flat (= unreconstructed) Ge {110} surface the sites subsequently to be occupied by Ga and by As atoms have no built-in distinction between themselves. The relative Ga/As ordering at different nucleation sites should therefore be perfectly random, which in turn would lead to a high degree of antiphase domain disorder, with domain sizes of the order of the nucleation site separation, which is usually very small for good epitaxial growth.

The situation on the {112} surface is far more favorable. As fig. 13 shows, the unoccupied sites ahead of an ideal (112) surface are of two quite different kinds: Sites (labelled 1 in fig. 13) with two back bonds to the Si surface, and sites (Nos. 2 and 4) with only one back bond. One easily sees that the two kinds of sites belong to the two different sublattices. Now it is well known that the column-V elements P, As, and Sb, form chemical compounds with Ge and Si, whereas the column-III elements Al, Ga and In do not. One might therefore expect that the strongly-bonding column-V atoms might displace any column-III atoms from the doubly back-bonded sites (No. 1). But once site No. 1 has been occupied by a column-V atoms, site No. 2 becomes more favorable for occupancy by a column-III atom than by a column-V atom. This, in turn, favors occupancy of site No. 3 by another P atom, followed by another Ga atom on site No. 4. Apparently, this is indeed that happens: We have grown GaP on Si {112} by MBE [54], and tests show that the observed sublattice ordering is as described here, with no evidence of antiphase domains. Furthermore, although the electrical properties of these first GaP-on-Si {112} interfaces are still far from ideal, we were able to build bipolar n–p–n transistors with an n-type GaP emitter on a Si p–n base/collector structure, with emitter injection efficiencies up to 90%. This is still far below what would be desirable for practically useful devices (> 99%), but is far better than anything else ever achieved in the very difficult GaP-on-Si system. It raises the hope that device-quality polar-on-nonpolar hetero-interfaces might in fact be achievable.

Our above theoretical speculation was oversimplified in that the reconstruction of the free Ge or Si surface, which is unquestionably present, was ignored. because of the strong bonding difference present already in the unreconstructed {112} surface, any reconstruction on that surface [56] should be little more than a quantiative complication, unless the reconstruction somehow destroys the strong inherent surface site inequivalence, which is extremely unlikely. The situation on the {110} surface is entirely different. Here any reconstruction would *create* a site inequivalence (see fig. 14b), and if this inequivalence is of the right kind, it might convert a hopeless orientation into a promising one. As we have pointed out elsewhere [55], the simplest possible reconstruction, a bond rotation similar to that on GaAs {001}, and postulated by Harrison [57] to occur on Si {110}, is exactly of the most desirable kind. In fact, growth of GaAs on Ge {110} apparently free from antiphase disorder can be achieved under certain growth conditions [55], which unfortunately however do not appear to lead to device-quality electrical properties. The {112} surface, which has a built-in strong site inequivalence, is therefore preferable over the reconstructed {110} surface, which must rely on a tenuous surface reconstruction to achieve site selection. Our experimental experience [54] strongly confirms this expectation. We therefore consider our own former advocacy [55] of the reconstructed {110} surface as having been superseded by the subsequent realization of the inherently greater promise of the {112} orientation.

5.4. Small misorientations: nuisance or design parameter?

There is no such thing as a perfectly-oriented crystallographic interface. Any real interface will have deviations from perfect flatness and perfect orientation, as a result of which the ⟨111⟩ bonds are rotated out of the true hetero-interface plane by a small but non-zero angle θ. At apolar/nonpolar interface this will cause a finite built-in interface charge to appear, and even for small misorientations the resulting charge may be large by device standards. For the {112} interface, the charge density is easily shown to be

$$\sigma = \left(q\sqrt{3} / a^2 \right) \sin \theta.$$

If the tilt angle is small enough, this charge is not likely to be removed by the HKWG atomic re-arrangement, but is likely to act as a permanent *tilt doping*. A wafer orientation to within $\pm 0.5°$ ($\cong 10$ milliradian) is roughly the practical limit of current *routine* wafer orientation techniques. Assuming the lattice constant of GaAs, such a misorientation corresponds to an interface charge density of 4.7×10^{12} elementary charges per cm². This is a large charge, and much more accurate wafer orientation techniques than are in current use will be necessary. This is of course possible, but is a major nuisance. A highly (112)-selective etch would certainly help. However, one man's nuisance is often the next man's design parameter. *If* the orientation could be controlled to significantly better than 10^{-3} radian, a deliberate misorientation might become a practical means of introducing desirable interface charges into devices such as HEMT's. Because the interface charges would not be randomly distributed, but be located on quasiperiodic interface steps, they would scatter less, and even new superlattice effects might arise. Finally, by deliberately creating a controlled local variation in the interface tilt, one might even introduce lateral "doping" variations into device structures. It is a fitting notion on which to close a paper that addresses itself to the role of interfaces in submicron structures, more specifically, to the role of the interface nanostructure in determining the properties of devices containing those interfaces.

Acknowledgments

It is a pleasure to thank Dr. R.S. Bauer for inviting me to present this paper at this Symposium, and thereby providing the stimulus to order my thoughts on the topics discussed and to put them down on paper, something that otherwise would have been unlikely to occur. Many thanks are due to Drs. E.A. Kraut, J.R. Waldrop, R.W. Grant, D.L. Miller and S.P. Kowalczyk, for uncounted discussions. Last, but not least, I wish to acknowledge the profound influence that Professor W.A. Harrison has had on my thinking.

References

[1] See, for example, H. Kroemer, Japan. J. Appl. Phys. 20, Suppl. 20-1 (1981) 39.

[2] H. Kroemer, Proc. IEEE 70 (1982) 13.

[3] W. Shockley, US Patent 2,569,347, issued 25 Sept. 1951.

[4] R.L. Anderson, Solid-State Electron. 5 (1962) 341.

[5] For a general review, see A.G. Milnes and D.L. Feucht, Heterojunctions and Metal–Semiconductor Junctions (Academic Press, New York, 1972).

[6] An excellent recent review is contained in chs. 4 and 5 of H.C. Casey and M.B. Panish, Heterostructure Lasers (Academic Press, New York, 1978).

[7] H. Kroemer, Proc. IEEE 51 (1963) 1782.

[8] R. Dingle, in: Festkörperprobleme/Advances in Solid State Physics, Vol. 15, Ed. H.J. Queisser (Vieweg, Braunschweig, 1975) p. 21.

[9] R. Dingle, H.L. Störmer, A.C. Gossard and W. Wiegmann, Appl. Phys. Letters 33 (1978) 665.

[10] For a recent review, see T. Mimura, Surface Sci. 113 (1982) 454.

[11] For a review, see N. Holonyak, R.M. Kolbas, R.D. Dupuis, and D.D. Dapkus, IEEE J. Quantum Electron. 16 (1980) 170.

[12] H. Sakaki, L.l. Chang, R. Ludeke, C.-A. Chang, G.A. Sai-Halasz and L. Esaki, Appl. Phys. Letters 31 (1977) 211;
see also L.L. Chang and L. Esaki, Surface Sci. 98 (1980) 70.

[13] H. Kroemer, W.-Y. Chien, H.C. Casey and A.Y. Cho, Appl. Phys. Letters 33 (1978) 749.

[14] H. Kroemer, W.-Y. Chien, J.S. Harris, Jr. and D.D. Edwall, Appl. Phys. Letters 36 (1980) 295.

[15] Y.Z. Liu, R.J. Anderson, R.A. Milano and M.J. Cohen, Appl. Phys. Letters 40 (1982) 967.

[16] See, for example, J.R. Waldrop, S.P. Kowalczyk, R.W. Grant, E.A. Kraut and D.L. Miller, J. Vacuum Sci. Technol. 19 (1981) 573.

[17] G.F. Williams, F. Capasso and W.T. Tsang, IEEE Electron Devices Letters 3 (1982) 71;
see also F. Capasso, Surface Sci. 132 (1983) 527.

[18] H. Kroemer, Critical Rev. Solid State Sci. 5 (1975) 555.

[19] W.A. Harrison, J. Vacuum Sci. Technol. 14 (1977) 1016;
see also ref. [24] below.

[20] G. Margaritondo, A.D. Katnani, N.G. Stoffel, R.R. Daniel and T.-X. Zhao, Solid State Commun. 43 (1982) 163;
see also G. Margaritondo, Surface Sci. 132 (1983) 469.

[21] J.L. Shay, S. Wagner and J.C. Phillips, Appl. Phys. Letters 28 (1976) 31.

[22] J.C. Phillips, J. Vacuum Sci. Technol. 19 (1981) 545.

[23] W.R. Frensley and H. Kroemer, Phys. Rev. B16 (1977) 2642.

[24] W.A. Harrison, Electronic Structure and the Properties of Solids: The Physics of the Chemical Bond (Freeman, San Francisco, 1980); see especially section 10F.

[25] J.O. McCaldin, T.C. McGill and C.A. Mead, Phys. Rev. Letters 36 (1976) 56. These authors expressed the correlation between valence band lineup and anion electronegativity for Schottky barriers; the approximate applicability of their result to heterojunctions appears to have been discussed first by W.R. Frensley and H. Kroemer, J. Vacuum Sci. Technol. 13 (1976) 810; see also ref. [23].

[26] For a very "physical" discussion of this theoretical foundation, see Harrison, ref. [24], especially chs. 1–3 and ch. 6.

[27] S.J. Anderson, F. Scholl and J.S. Harris, in: Proc. 6th Intern. Symp. on GaAs and Related Compounds, Edinburgh, 1976, Inst. Phys. Conf. Ser. 33b (Inst. Phys., London and Bristol, 1977) p. 346.

[28] The numerical values are based on Harrison's table 10-1 on p. 253 of ref. [24], except that we use the values from ref. [27] for the energy gaps of GaSb and AlSb.

[29] J.A. Van Vechten, Phys. Rev. 87 (1969) 1007. Van Vechten gives an extensive table of theoretical ionization energies, from which electron affinities are easily obtained by subtracting the energy gaps.

[30] This broken-gap lineup is, in fact, predicted by all three major predictive theories: The electron affinity rule, the Frensley–Kroemer theory, and the Harrison theory.

[31] G.C. Osbourn, J. Appl. Phys. Letters 53 (1982) 1536; J. Vacuum Sci. Technol. 21 (1982) 469; see also ref. [34] below.

[32] G.H. Döhler, Phys. Status Solidi (b) 52 (1972) 79,553;
G.H. Döhler, H. Künzel and K. Ploog, Phys. Rev. B25 (1982) 2365.

[33] Our calculation is to illustrate the basic idea only. The quoted composition falls into a solid solubility gap of uncertain width the existence of which has been reported. It may therefore be difficult or impossible to prepare. For a discussion and further references on this point see ch. 5 of ref. [6].

[34] P.L. Gourley and R.M. Biefeld, J. Vacuum Sci. Technol. 21 (1982) 473;
G.C. Osbourn, R.M. Biefeld and P.L. Gourley, Appl. Phys. Letters 41 (1982) 172.

[35] M.E. Davis, G. Zeidenbergs and R.L. Anderson, Phys. Status Solidi 34 (1969) 385.

[36] G.M. Guichar, C.A. Sébenne and C.D. Thuault, Surface Sci. 86 (1979) 789.

[37] R.S. Bauer and H.W. Sang, Jr., Surface Sci. 132 (1983) 479.

[38] H. Morkoc, L.C. Witkowski, T.J. Drummond, C.M. Stanchak, A.Y. Cho and J.E. Greene, Electron. Letters 17 (1981) 126;
see also H.L. Störmer, Surface Sci. 132 (1983) 519.

[39] It has been suggested by W.I. Wang (personal communication) that the {110} sequence dependence might be related to an as yet unexplained instability of {110}-oriented (Al, Ga)As growth observed by him. For another report of a different kind of {110} growth instability see P. Petroff, A.Y. Cho, F.K. Reinhart, A.C. Gossard and W. Wiegmann, Phys. Rev. Letters 48 (1982) 190.

[40] R.C. Miller, W.T. Tsang and O. Munteanu, Appl. Phys. Letters 41 (1982) 374.

[41] E.A. Kraut, R.W. Grant, J.R. Waldrop and S.P. Kowalczyk, Phys. Rev. Letters 44 (1980) 1620.

[42] H. Kroemer and W.-Y. Chien, Solid-State Electron. 24 (1981) 655.

[43] H.K. Gummel and D.L. Scharfetter, J. Appl. Phys. 38 (1967) 2148;
see also C. Kittel and H. Kroemer, Thermal Physics, 2nd ed. (Freeman, San Francisco, 1980) ch. 13. For very unsymmetrically doped junctions, the GS correction is between $1\ kT/q$ and $2\ kT/q$.

[44] H. Kroemer, RCA Rev. 17 (1956) 515.

[45] M. Weinstein, R.O. Bell and A.A. Menna, J. Electrochem. Soc. 111 (1964) 674.

[46] Zh.I. Alferov, V.I. Korolkov and M.K. Trukan, Soviet Phys.-Solid State 8 (1967) 2813.

[47] G. Zeidenbergs and R.L. Anderson, Solid-State Electron. 10 (1967) 113.

[48] N.N. Gerasimenko, L.V. Lezheiko, E.V. Lyubopytova, L.V. Sharanova, A.Ya. Shik and V. Shmartsev, Soviet Phys.-Semicond. 15 (1981) 626.

[49] S.L. Wright, PhD Thesis, University of California, Santa Barbara, CA (1982).
[50] See, for example, C.M. Garner, C.Y. Su, Y.D. Shen, C.S. Lee, G.L. Pearson, W.E. Spicer, D.D. Edwall, D. Miller and J.S. Harris, Jr., J. Appl. Phys. 50 (1979) 3383; see also the references quoted there.
[51] W.A. Harrison, E.A. Kraut, J.R. Waldrop and R.W. Grant, Phys. Rev. B18 (1978) 4402.
[52] W.A. Harrison, in: Festkörperprobleme/Advances in Solid State Physics, Vol. 17, Ed. H.J. Queisser (Vieweg, Braunschweig, 1977) p. 135.
[53] W.A. Harrison, J. Vacuum Sci. Technol. 46 (1979) 1492.
[54] S.L. Wright, M. Inada and H. Kroemer, J. Vacuum Sci. Technol. 21 (1982) 534.
[55] H. Kroemer, K.J. Polasko and S.L. Wright, Appl. Phys. Letters 36 (1980) 763.
[56] R. Kaplan, Surface Sci. 116 (1982) 104.
[57] W.A. Harrison, Surface Sci. 55 (1976) 1.

Direct observation of effective mass filtering in InGaAs/InP superlattices

D. V. Lang, A. M. Sergent, M. B. Panish, and H. Temkin

AT&T Bell Laboratories, Murray Hill, New Jersey 07974

(Received 16 June 1986; accepted for publication 24 July 1986)

By making transport measurements on InGaAs/InP superlattices we have been able to demonstrate a regime of rapid electron tunneling perpendicular to the layers in the presence of quantum hole localization, i.e., effective mass filtering. The experiments were conducted on multiple quantum well and single quantum well samples in the form of p^+-n junctions and involved low-temperature photoinduced capacitance and ac conductivity measurements, which easily resolved the 205 Å superlattice period, and deep level transient spectroscopy observations of hole trapping in 60–80 Å quantum wells.

Recently there has been considerable interest in carrier transport phenomena in semiconductor superlattices. For the case of transport perpendicular to the layers the manner in which the carriers pass through the barriers between the quantum wells is central to the problem. At low temperatures the carriers must tunnel through the barriers whereas at higher temperatures one may observe phonon-assisted tunneling and, finally, at the highest temperatures, true thermionic emission over the barriers. In the tunneling regime in direct gap III-V semiconductors electrons are much more mobile than holes due to the very large difference in effective masses between electrons and holes. As pointed out by Capasso *et al.*,[1,2] such an "effective mass filter" could lead to large photoconductive gain, in analogy to the well-known examples in II-VI semiconductors[3] where hole traps give rise to large photoconductive gain.

The photoconductivity experiments of Capasso *et al.*[1,2] on AlInAs/GaInAs superlattices showed that the electrons were clearly being controlled by the quantum mechanical aspects of the superlattice (tunneling through the barriers); the mechanism of hole trapping was not clear. For example, in addition to a very small tunneling rate of holes through the barriers due to their large effective mass (the effective mass filter effect), it could also have been possible for defect-related bulk hole traps to be introduced near the heterostructure interfaces during crystal growth which could act as photoconductive sensitizing centers.[3]

We have studied InGaAs/InP superlattices grown by gas-source molecular beam epitaxy (GS MBE) and have seen hole trapping which we can directly relate to superlattice quantum mechanical effects alone. To verify the existence of this effect in our samples we used a combination of capacitance versus voltage (*CV*), ac conductivity, and deep level transient spectroscopy (DLTS) measurements. We were able to demonstrate a regime at low temperatures where the perpendicular transport of photoinduced electrons through the superlattice was much faster than that of the photoinduced holes, similar to the photoconductivity experiments of Capasso *et al.*[1,2] In our case, however, we were also able to show by ac conductivity and DLTS measurements that the hole trapping was clearly a quantum well effect and not related to superlattice-induced defects. In this letter we will first describe our *CV* and ac conductivity measurements showing electron tunneling in the presence of hole localization. We will then describe the DLTS results which show that the holes are trapped in the potential well formed by the small-gap (0.75 eV) InGaAs layer being placed between the larger gap (1.35 eV) InP layers, i.e., quantum well trapping.

Our samples were grown by GS MBE using standard procedures that have been reported elsewhere.[4–6] The layer sequence for the sample containing the superlattice started with a 1.0-μm buffer layer of n-InP (5×10^{17} cm^{-3}) on an n^+-InP substrate. A 50-period undoped superlattice was then grown with alternating layers of 80 Å of InGaAs and 125 Å of InP. This was followed by 3000 Å of undoped InP which was n type ($\approx 10^{16}$ cm^{-3}), a 0.7-μm layer of p-InP (5×10^{17} cm^{-1}) to form a p^+-n junction, and a contact layer of 1000 Å of p-InP (1×10^{19} cm^{-3}). The single quantum well samples were essentially the same except for an undoped InP buffer and a single 60 Å layer of InGaAs in place of the superlattice. Ohmic contacts were applied and mesa diodes were formed by photolithography with either 100 or 325 μm diameters. The samples were contacted and mounted so that they could be illuminated through the substrate. The measurements were made in a helium-flow Dewar which could be varied from 20 to 500 K. The *CV* and ac conductivity measurements were made with a digital LCR meter at frequencies between 1 kHz and 10 MHz. The *CV* measurements were numerically differentiated in the standard way[7] to obtain the *CV* profile [$N(x)$ vs x], where $N(x) = (2/q\epsilon)[d(1/C^2)/dV]^{-1}$ is the apparent local carrier concentration and x is the distance from the p-n junction.

Our most dramatic results are shown in Fig. 1. This shows two *CV* profiles of the apparent carrier concentration of the 50 period superlattice measured at 23 K. One was taken in the dark and the other while the sample was illuminated with a microscope lamp through a silicon filter (to remove the wavelengths directly absorbed by InP). Note that in both cases the apparent width x of the diode depletion region moves through the superlattice at low temperatures in spite of the fact that the electrons must pass through barriers of order of 300 meV to reach x, as shown in Fig. 2. In order to obtain such profiles the electrons must be able to tunnel through the superlattice barriers on a time scale fast compared to the 1 MHz capacitance measurement frequency. If the electrons could not move this fast through the barriers at 23 K, the superlattice region would appear in a 1-MHz capacitance measurement to be an insulator and the *CV* profile would never come closer than 1.3 μm to the junction. This point is clear when one recalls that the apparent

150

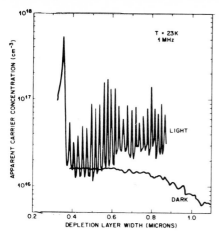

FIG. 1. Apparent carrier concentration profile obtained from differentiating 1 MHz CV data taken at 23 K under illuminated and dark conditions.

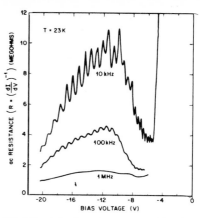

FIG. 3. Photoinduced ac resistance vs reverse bias voltage at 23 K for three measurement frequencies.

value of x in a CV profile of a p^+-n junction corresponds to the first moment of the ac charge distribution in the diode induced by the ac drive voltage of the capacitance meter.[8] Thus if the electrons cannot move through a particular region of the material near the junction on this time scale, the ac charge distribution, and hence x, can never appear within this region on a CV profile.

The presence of the superlattice is clearly seen in the illuminated profile. The superlattice period inferred from the GS MBE growth rate is 205 Å. The period in Fig. 1 is 205 ± 5 Å and the width of the peaks (full width at half-maximum) is 80 ± 10 Å. The Debye length for a doping of 10^{17} cm^{-1} at 23 K is 35 Å.[8] The peaks beyond 0.6 μm are not fully resolved due to the bias voltage supply being limited to steps larger than 0.1 V beyond 10 V. Presumably the oscillations would come down to the dark CV profile if fully resolved. Normally such narrow peaks in a CV profile indicate sharp peaks in the spatial profile of the donor doping concen-

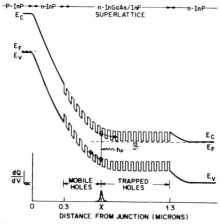

FIG. 2. Potential energy vs distance of a p-n junction with the edge of the depletion region x, within the superlattice. The location of the ac charge distribution is also shown.

tration. However, in this case where the peaks are only present under illumination, they must be due to trapped photoinduced holes. These trapped holes affect a CV profile in the same way as ionized donor impurities, which are also positively charged. When the holes are no longer trapped, e.g., at temperatures above 100 K, the superlattice peaks disappear.

The dynamics of hole transport can be directly studied by ac conductivity measurements in the low-temperature tunneling regime as a function of frequency. Figure 3 shows the ac resistance of the sample at 23 K under the same illumination conditions as Fig. 1. The oscillations in the 10 and 100 kHz resistance have the same period as the CV profile oscillations in Fig. 1 and correspond to very small steps which can be observed in the dc photocurrent versus reverse bias. But whereas the oscillations in Fig. 1 were due to mobile electrons screening trapped, photoinduced holes at frequencies of 1 MHz or greater, the resistance oscillations in Fig. 3 are a direct manifestation of the motion of holes on a slower time scale. The 1 MHz resistance in Fig. 3 shows no such oscillations and is equal to the slope of the photoinduced IV curve with the bias-dependent steps smoothed out. Thus the holes cannot tunnel fast enough to follow a voltage modulation on the MHz time scale. This is consistent with Fig. 1 where the CV profile oscillations could only occur if the electrons were mobile and the holes immobile on a MHz time scale. Indeed, CV measurements at frequencies less than 1 MHz are so dominated by photocurrent oscillations that CV profiles are meaningless.

The mechanism giving rise to the resistance oscillations can be understood with reference to Fig. 2. In the high field part of the depletion region (left of x) the holes can tunnel rapidly and give rise to the dc photocurrent. In the low field and neutral region (right of x) the holes are immobile and recombine with the electrons. At location x near the outer edge of the depletion region the tunneling rate for holes is critically dependent on the applied bias and will be modulated by a small oscillation in the voltage. The resistance oscillations in Fig. 3 correspond to the holes in the quantum well at x experiencing a rapid increase in tunneling rate with bias until the depletion region widens to include the next quan-

151

tum well and the process repeats with a new value of x.

The first quantum well in the superlattice (at $0.3\,\mu$m in Fig. 2) has a much thicker barrier than the others and hence has a much slower tunneling rate. Thus even though this well is within the dark zero-bias depletion region in Fig. 1, very little photocurrent is detected until more than 4 V is applied to the diode, as seen in Fig. 3 where $R \geqslant 10^8\ \Omega$ for $V < 4$ V. The trapped holes in this first well effectively pin the depletion layer width and screen the superlattice from the junction field over this voltage range. This gives rise to the large peak in the photoinduced CV profile at $0.35\,\mu$m in Fig. 1. Because of the wider barrier, this peak also persists to higher temperatures (200 K) than the other superlattice peaks (100 K). Only thermal emission over the entire valence-band discontinuity or an electric field induced lowering of this barrier can remove trapped photoinduced holes, and hence the profile peak, from this first well. We have verified this interpretation by making the same CV measurement on a sample with a single quantum well of 60 Å thickness in place of the superlattice. In this case the illuminated CV profile has a large peak exactly like that at $0.35\,\mu$m in Fig. 1 with none of the superlattice oscillations.

We were able to measure the thermal activation energy associated with hole emission from the single quantum well and the first quantum well in the superlattice by using DLTS.[9] For these experiments we maintained the sample at a fixed reverse bias and used a strobe lamp (1 μs pulse length) through a silicon filter as the minority-carrier injection pulse. The DLTS spectrum of the single quantum well at various bias voltages for a rate window of 200 μs is shown in Fig. 4. The disappearance of the peak at low bias voltages corresponds to the quantum well no longer being within the depletion layer, i.e., the DLTS peak in Fig. 4 is spatially correlated with the location of the quantum well. Note also the shift in the DLTS peak to lower temperatures and the eventual broadening of the peak as the voltage is increased. This corresponds to a transition from thermally activated emission (sharp DLTS peak) to a temperature-independent emission rate (broad DLTS peak) as the bias is increased.

The bias voltage (4 V) corresponding to temperature-independent hole emission in the DLTS spectrum is exactly the same as that corresponding to the onset of hole tunneling in Figs. 1 and 3, as mentioned above.

By varying the rate window and measuring the activation energy of the DLTS peak as a function of voltage (inset in Fig. 4) we can see that the shift in this peak is due to a very strong reduction in energy with increasing voltage. Our largest activation energy appears to extrapolate to a zero-field value of 260 meV. When the bound state energy of 20 meV is added to this, we have a total of 280 meV for the apparent valence-band discontinuity ΔE_v. From the CV measurements of Forrest and Kim[10] the valence-band discontinuity is $\Delta E_v = 370$ meV, although this value seems to be somewhat sample dependent.[11] Photoluminescence and absorption measurements[12] are not very conclusive, but indicate that $\Delta E_v \sim \Delta E_g/2 = 300$ meV. Our corrected DLTS energy of 280 meV is smaller than other estimates of ΔE_v. However, one should expect thermal emission in the presence of an electric field to underestimate the true band-edge discontinuity. Our result is thus a lower limit on ΔE_v and is consistent with thermal emission of holes from a valence-band quantum well in InGaAs/InP.

In summary, we believe that these results constitute a convincing demonstration of a truly quantum mechanical effective mass filtering effect. The observation of photoinduced superlattice oscillations in the low-temperature CV profile can only be interpreted as electrons being mobile and holes being immobile on the time scale of 1 MHz. Furthermore, the observation of oscillations in the photoinduced ac conductivity at low frequencies and a strongly field-dependent DLTS hole emission peak indicate that the holes are trapped in the quantum wells and not at deep level defects. Thus we have all of the elements of effective mass filtering. The electrons, by virtue of their extremely small effective mass, can tunnel easily through the 125 Å barriers in less than 0.1 μs, while the holes, with much larger effective mass, are trapped in the quantum wells for times of order 1 μs and begin to move between wells only for times longer than 10 μs.

We wish to acknowledge the technical assistance of R. A. Hamm, A. Savage, S. Sumski, and H. Wade in various aspects of this project.

[1]F. Capasso, K. Mohammed, A. Y. Cho, R. Hull, and A. L. Hutchinson, Appl. Phys. Lett. **47**, 420 (1985).
[2]F. Capasso, K. Mohammed, A. Y. Cho, R. Hull, and A. L. Hutchinson, Phys. Rev. Lett. **55**, 1152 (1985).
[3]R. H. Bube, in *Physics and Chemistry of II-VI Compounds*, edited by Aven and Prener (North-Holland, Amsterdam, 1967), p. 659.
[4]M. B. Panish, H. Temkin, and S. Sumski, J. Vac. Sci. Technol. B **3**, 657 (1985).
[5]M. B. Panish, *Progress in Crystal Growth and Characterization* (in press).
[6]H. Temkin, M. B. Panish, and S. N. G. Chu, Appl. Phys. Lett. (in press).
[7]S. M. Sze, *Physics of Semiconductor Devices* (Wiley, New York, 1969), p. 90.
[8]G. L. Miller, D. V. Lang, and L. C. Kimerling, Ann. Rev. Mater. Sci. **7**, 377 (1977).
[9]D. V. Lang, J. Appl. Phys. **45**, 3023 (1974).
[10]S. R. Forrest and O. K. Kim, J. Appl. Phys. **52**, 5838 (1981).
[11]S. R. Forrest, O. K. Kim, and R. G. Smith, Appl. Phys. Lett. **41**, 95 (1982).
[12]H. Temkin, M. B. Panish, P. M. Petroff, R. A. Hamm, J. M. Vandenberg, and S. Sumski, Appl. Phys. Lett. **47**, 394 (1985).

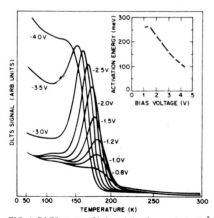

FIG. 4. DLTS spectra of hole emission from a single 60-Å InGaAs/InP quantum well for various reverse bias voltages at a rate window of 200 μs. The inset shows the measured DLTS activation energy vs voltage.

152

Angle-resolved photoemission measurements of band discontinuities in the GaAs-Ge heterojunction[a]

P. Perfetti,[b] D. Denley,[c] K. A. Mills, and D. A. Shirley

Materials and Molecular Research Division, Lawrence Berkeley Laboratory
and Department of Chemistry, University of California, Berkeley, California 94720
(Received 22 May 1978; accepted for publication 31 July 1978)

The conduction- and valence-band discontinuities for the (110) GaAs-Ge heterojunction have been measured as $\Delta E_c = 0.50$ eV and $\Delta E_v = 0.25$ eV by the angle-resolved ultraviolet photoemission (ARUPS) technique. These values are in good agreement with the theoretical predictions of Pickett *et al*.

PACS numbers: 79.60.Eq, 71.20.+c, 73.40.Lq

During the past ten years, much effort has been devoted to understanding the physical properties of Schottky barriers and heterojunctions.[1] In both systems,

[a] Work supported by the Division of Chemical Sciences, Office of Basic Energy Sciences, U. S. Department of Energy.
[b] Permanent address: PULS (CNR), Laboratori Nazionali di Frascati, 00044 Frascati, Roma, Italy.
[c] Also with Department of Physics, University of California, Berkeley, Calif. 94720.

the main effects originate at the interface; for example, it is well known that Schottky-barrier heights, as measured by capacitance voltage (C-V) or current-voltage (I-V) characteristics, are nearly independent of the metal's work function for covalent semiconductor-metal pairs. Several theoretical models have been suggested to account for this pinning of the Fermi energy (E_f).[2-4] Recent results obtained with surface-sensitive techniques such as ultraviolet photoelectron spectroscopy (UPS),[5,6] partial yield spectroscopy,[7] and electron energy-loss spectroscopy,[8] however, seem to indicate that no single theory will explain the data for all Schottky-barrier devices.

For semiconductor-semiconductor heterojunctions, as with Schottky barriers, the most serious problems arise from the misfit which necessarily occurs when materials of unequal lattice parameters are interfaced. The dangling bonds arising at the interface could be expected to provide electrically active states serving either as charge traps, in which case a modified band profile could result, or as recombination states which could affect the transport properties of the junction. The wide application of heterojunctions in electronic devices suggests that a better knowledge of their interface states would be useful for the "tailoring" of such devices. Nevertheless, up to now only one surface-sensitive experiment has been reported for such heterojunctions,[9] and that work did not establish the magnitudes of the conduction- and valence-band discontinuities at the interface, but only showed their orientational dependence, which is a consequence of the dipolelike potential produced by the interface.

These discontinuities may, however, be determined from angle-resolved ultraviolet photoelectron spectroscopy (ARUPS) data. Briefly, the method consists of detecting electrons which have been photoejected from the surface of one semiconductor single crystal which is covered with overlayers of a second semiconductor of various thicknesses; as the overlayer is produced, the top of the valence band (called here the "valence-band maximum", VBM) will shift at the surface relative to E_f. Since the escape depth of electrons at these energies may be expected to be of the order of 10 Å,[10] the data obtained for thicknesses of such magnitude reflect the electronic structure of the heterojunction at the surface and in its first few bulk layers. By increasing the overlayer thickness, it should be possible to observe a satu-

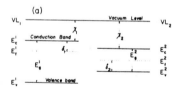

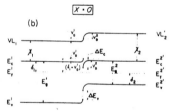

FIG. 1. Band scheme (a) before and (b) after the formation of a semiconductor-semiconductor heterojunction. The ordinate is the relative energy, and the abscissa the position in real space, with $x = 0$ at the interface. See text for explanation of symbols. The figure is drawn so that GaAs would qualitatively correspond to material 1 and Ge to material 2. Note that the spatial extent of the region where the band bending occurs is of the order of $1\,\mu$ (see Ref. 11).

668   Appl. Phys. Lett., Vol. 33, No. 7, 1 October 1978

Perfetti et al.   668

TABLE I. Values of valence- and conduction-band discontinuities.

Method	ΔE_v	ΔE_c
ARUPS[a]	0.25	0.50
EPM[b]	0.35	0.40
C-V[c]	0.6—0.19	0.15—0.56
Eq. (5)	0.69	0.06

[a] This work.
[b] Reference 15.
[c] Reference 16.

ration in the substrate band bending; for overlayers of sufficient thickness (i.e., several times the inelastic mean free path), no photoemission from the substrate should be detectable, and the overlayer band bending should become observable.

The band scheme usually employed to describe a semiconductor-semiconductor heterojunction is shown in Fig. 1. In Fig. 1(a), the energy levels are shown for two noninteracting (separated) materials; the characteristic parameters are the band gaps (E_g), electron affinities (χ), and positions of the Fermi levels relative to the bulk valence- or conduction-band energies (δ). As shown in Fig. 1(b), the formation of the heterojunction leaves these parameters invariant, but the charge flow across the junction, induced by the difference in chemical potential of the two materials, results in the equalization of the Fermi levels and a concomitant bending of the bands at the interface, given by V_D^1 and V_D^2. The discontinuities in the valence and conduction bands, ΔE_v and ΔE_c, are given by

$$\Delta E_v + \Delta E_c = \Delta E_g, \tag{1}$$

while the relation

$$E_f^2 - E_f^1 = V_D^1 + V_D^2 \equiv V_D \tag{2}$$

describes the band bending. The partitioning of the induced potential into V_D^1 and V_D^2 is determined by the doping of the semiconductors, with the bending being larger for the less conducting material. As is clear from Fig. 1, a simple equation obtains for the conduction-band discontinuity:

$$\Delta E_c = (\delta_1 + V_D^1) - (E_g^2 - \delta_2) + V_D^2. \tag{3}$$

Thus, a determination of V_D will also allow the calculation of ΔE_c. This has typically been done in the past by analyzing the C-V characteristics of the heterojunction,[1] according to the relationship

$$1/C^2 \sim (V_D - V). \tag{4}$$

Such results have frequently been explained by equating the conduction-band discontinuity at the interface to the difference in electron affinities,

$$\Delta E_c \cong \chi_2 - \chi_1, \tag{5}$$

but the validity of this approach has been questioned by several authors, who have proposed different theoretical models.[12-15] For the (110) interface of the GaAs/Ge heterojunction, the predictions of the self-consistent

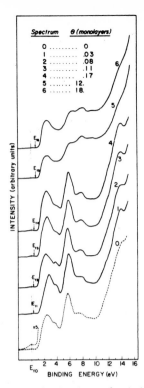

Spectrum	θ (monolayers)
0	0
1	.03
2	.08
3	.11
4	.17
5	12.
6	18.

FIG. 2. Spectra at normal emission and 21.2 eV photon energy of clean (dotted) and Ge-covered (solid) (110) GaAs. The magnified portion of spectrum 0 shows the Ga 3d peak due to 40.8 eV radiation. Approximate coverages are as given.

pseudopotential method of Ref. 15 and the electron affinity difference method are given in Table I, along with C-V results.[16] The variation in the C-V data seems to indicate that such measurements must be made in conjunction with microscopically sensitive techniques in order to fully characterize the interface.

We report here the first ARUPS determination of the GaAs/Ge (110) heterojunction discontinuities. The experimental apparatus was described earlier.[17] Briefly, electrons were photoejected by 21.2-eV (He I) photons from an in situ cleaved n-type GaAs (110) single crystal (carrier concentration of 8×10^{14} cm^{-3} by Hall effect measurements) which was subsequently covered by evaporative deposition with successively thicker overlayers of Ge. The electrons were energy analyzed using a cylindrical mirror analyzer with a resolution of 0.07 eV. During the depositions the temperature of the substrate was held at the epitaxial growth temperature[18] of 420 °C, with a base pressure of 2×10^{-10} Torr. ARUPS spectra were taken for various Ge coverages, ranging from a fraction of a monolayer to ~20 monolayers; coverages were determined by calibrating the oven with a piezoelectric thickness monitor, and one monolayer was defined as 8.85×10^{14} atoms cm^{-2}, in conformity with previous work.[5]

Selected results are shown in Fig. 2, where the spectra for clean and Ge-covered GaAs are presented for normal emission; these spectra are hereafter referred to as 0 to 6. The position of E_f was determined by measuring the photoemission edge of a tantalum strip which was in electrical contact with the GaAs.[19] This strip is in contact with the spectrometer ground, and therefore provides a stable reference level. As long as charging effects are small (as is the case here; see Ref. 20), the Fermi level of the metal and the GaAs will be equivalent. The small peak just to the left of E_f is due to emission from the Ga 3d core level, excited by 40.8 eV (He II) photons which are a subsidiary component of the spectrum of the discharge lamp used. The separation of this peak and E_f decreases with Ge coverage, indicating a change in the relative separation, near the interface, of E_f and the valence band. Accordingly, we have aligned spectra 0 to 4 by keeping fixed the Ga 3d peak position; the zero of energy is set as the Fermi level for the clean GaAs sample. Spectra 5 and 6, having thick Ge overlayers, have been aligned by keeping constant the VBM, a procedure which is sufficient for our purpose of obtaining shifts in E_f. A weak 3d signal was also observed for the spectrum 5 coverage at non-normal emission, and was used to confirm the alignment.

For the Ge-covered surfaces, E_f is displaced from the zero of energy as shown in Fig. 2. This displacement is summarized in Fig. 3 where we have plotted the Fermi level position, relative to E_f for clean GaAs, as a function of overlayer thickness. A saturation value of the band bending is reached for very low coverages; analogous behavior was observed by Gregory and Spicer for the GaAs-Cs Schottky barrier.[5] For such layer thicknesses, most of the photoelectrons observed are originating in the GaAs, and the saturation value thus corresponds to the maximal GaAs band bending. For the spectra incorporating thick Ge overlayers, we see a further movement of E_f towards the VBM. As the Ge layer thickness is much less than the characteristic distance for band bending (see Fig. 1), bending effects in the overlayer are negligible for such small coverages. This new saturation value is representative of the Ge half of the heterojunction; in fact, the photoelectron escape depth is sufficiently short as to preclude, with the coverages employed in these spectra, observation

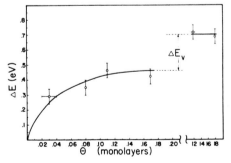

FIG. 3. Plot of the separation of the Fermi levels for clean GaAs and for Ge-covered GaAs, as a function of overlayer thickness.

155

of the substrate bands. From Fig. 3, we may immediately deduce that $\Delta E_v = 0.25$ eV and, with $\Delta E_g = 0.75$ eV, we obtain from Eq. (1) that $\Delta E_c = 0.50$ eV. These results, which qualitatively agree more closely with the values of Ref. 15 than with those obtained by the use of Eq. (5), seem to indicate that equating ΔE_c to the electron affinity difference is a poor approximation, and that a better knowledge of the interface physical properties should be obtained through the use of more sophisticated theoretical models which include the possibility of relaxation at the interface.

In conclusion, we have demonstrated the viability of ARUPS as a structure-sensitive tool, on the microscopic level, for determining band discontinuities in semiconductor-semiconductor heterojunctions, and have obtained results for the GaAs/Ge (110) couple which suggest the need for further theoretical work. A paper which describes our data in more detail is currently under preparation.

[1]A. G. Milnes and D. L. Feucht, *Heterojunctions and Metal-Semiconductor Junctions* (Academic, New York, 1972).
[2]J. Bardeen, Phys. Rev. 71, 717 (1947).
[3]V. Heine, Phys. Rev. 138, A1689 (1965).

[4]H. C. Inkson, J. Phys. C 5, 2599 (1972); 6, 1350 (1973).
[5]P. E. Gregory and W. E. Spicer, Phys. Rev. B 12, 2370 (1975).
[6]G. Margaritondo, J. E. Rowe, and S. B. Christman, Phys. Rev. B 14, 5396 (1976).
[7]W. Gudat and D. E. Eastman, J. Vac. Sci. Technol. 13, 831 (1976).
[8]J. E. Rowe, S. B. Christman, and G. Margaritondo, Phys. Rev. Lett. 35, 1471 (1975).
[9]R. W. Grant, J.R. Waldrop, and E. A. Kraut, Phys. Rev. Lett. 40, 656 (1978).
[10]W. E. Spicer, J. S. Johannessen, and Y. E. Strausser, Varian Report VR-98, p. 14 (unpublished); see also, I. Lindau and W. E. Spicer, J. Electron Spectrosc. 3, 409 (1974).
[11]S. M. Sze, *Physics of Semiconductor Devices* (Wiley-Interscience, New York, 1969).
[12]W. R. Frensley and H. Kroemer, J. Vac. Sci. Technol. 13, 810 (1976).
[13]W. A. Harrison, J. Vac. Sci. Technol. 14, 1016 (1977).
[14]G. A. Baraff, J. A. Appelbaum, and D. R. Hamann, J. Vac. Sci. Technol. 14, 999 (1977).
[15]W. E. Pickett, S. G. Louie, and M. L. Cohen, Phys. Rev. B 17, 815 (1978).
[16]R. L. Anderson, Solid-State Electron. 5, 341 (1962).
[17]J. Stöhr, G. Apai, P. S. Wehner, F. R. McFeely, R. S. Williams, and D. A. Shirley, Phys. Rev. B 14, 5144 (1976).
[18]*Handbook of Thin Film Technology*, edited by L. I. Maissel and R. Glang (McGraw-Hill, New York, 1970).
[19]P. A. Pianetta, Stanford Synchrotron Radiation Laboratory Report SSRL 77/17, 1977 (unpublished).
[20]L. Ley, R. A. Pollak, F. R. McFeely, S. P. Kowalczyk, and D. A. Shirley, Phys. Rev. B 9, 600 (1974).

156

Ge–GaAs(110) interface formation[a]

R. S. Bauer and J. C. McMenamin

Xerox, Palo Alto Research Center, Palo Alto, California 94304

(Received 18 April 1978; accepted 19 April 1978)

The heterojunction chemistry for Ge grown by molecular beam epitaxy (MBE) on *in situ* cleaved GaAs exhibits significant interdiffusion in short times at growth temperatures T_G of 430°C (significantly lower critical T_G than that reported for moderate-vacuum physical vapor deposition). This results in profound changes in the electronic properties of the interface as probed by synchrotron-radiation-excited $3d$ core electron photoemission. Even when there is significant alloying of the two lattice-matched semiconductors, there is nearly equal probability for Ge to bond to either a Ga or an As atom at the initial stage. As Ge becomes the dominant species, we find As preferentially diffusing toward the Ge side of the junction. This As is distributed throughout the overlayer in contrast to metal–semiconductor interface formation where the diffusing constituent resides only on the free, growing surface. We show that these behaviors are consistent with the kinetic and thermodynamic properties of the atomic species. The valence band discontinuity is negligible over atomic dimensions, while for an abrupt interface ($T_G = 350°C$) we measure $\Delta E_V = 0.7 \pm_{0.3}^{0.05}$ eV. The photoemission changes character rapidly with temperature, indicating an activation barrier for the diffusion below which simple expressions for attenuation of the photoelectrons by electron–electron scattering are applicable. In that case we deduce an escape depth of 7.0 ± 0.5 Å, indicating uniform growth of Ge, with composition changing abruptly from GaAs over ~1 bond length in the (110) direction. A negligible (<0.2 eV) localized interface dipole layer is formed in the process.

PACS numbers: 68.48.+f, 79.60.Eq

I. INTRODUCTION

We report the first photoelectron spectroscopy study of the initial stages of formation of a heterojunction; specifically, that produced by Ge deposited epitaxially on the nonpolar faces of n- and p-type GaAs. By probing with monochromatized soft x-ray synchrotron radiation, the core photoelectron escape depth can be set to ~5 Å, allowing microscopic band discontinuity determination, and can be tuned to probe the spatial distribution of the chemical species. Further, the dependence on Ge coverage provides a nondestructive measurement of heterojunction abruptness on an atomic scale.

We find that the interface chemistry is a critically important determinant of semiconductor–semiconductor junction properties. In particular, the interdiffusion kinetics of the atomic species is a very sharp function of temperature. There is only a narrow temperature range (<100°C) over which epitaxial growth occurs without a high mobility for Ge and GaAs intermixing. The interface electronic structure is profoundly modified when interdiffusion occurs. Only when an atomically abrupt interface is produced do we observe a significant discontinuity in the valence band edge comparable to that predicted by preservation of free-surface electron affinities between the semiconductors. For such a chemically sharp interface, the bands change over a distance of 5 Å or less with a negligible (<0.2 eV) dipole layer localized in this junction region. We emphasize the need for determining the kinetics of heterojunction formation by each and every experimental preparation technique before definitive conclusions of electronic or other properties can be made.

II. EXPERIMENT

Ge was evaporated from a molecular beam epitaxy (MBE) effusion source developed by R. Z. Bachrach *et al.*[1] or from a tungsten filament onto GaAs (110) cleaved *in situ* at 2×10^{-10} Torr. The GaAs was held at growth temperatures T_G ranging from 350 to 525°C at pressures of $\sim 2 \times 10^{-9}$ Torr during the ~10 Å per minute evaporation. The absolute temperature of the surface may be as much as 50°C lower than this T_G measured at the sample clamp ~5 mm away; the relative growth temperatures however are conservatively accurate to ±10°C. Epitaxial growth has been reported[2] at temperatures as low as 300°C under much worse vacuum conditions, where crystalline overlayer formation should be more difficult than in our preparation environment.

The interfaces were probed at room temperature using the continuum synchrotron radiation from SPEAR as shown schematically in Fig. 1; the photons were monochromatized in the soft x-ray region by the grasshopper monochromator on the 4° beam line at the Stanford Synchrotron Radiation Laboratory. The electrons thereby photoemitted were analyzed using a PHI 15-255G, double-pass CMA at 75° to the optical axis and 20° from the sample normal. By studying Ga, As, and Ge $3d$ core emission at kinetic energies $E_K \approx 60$ eV, we are sensitive to the environment of the atoms within ~3 layers of the free surface. Attenuation of the Ga and As emission by the Ge overlayer provides a sensitive measure of heteroepitaxial growth abruptness using such short photoelectron escape depths. The overlayer thickness L_{Ge} is determined by setting the Ge evaporation rate of the MBE

157

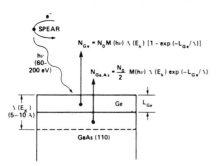

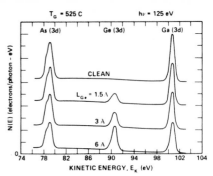

FIG.1. Schematic diagram depicting how core-level photoemission excited by synchrotron radiation can be used as a variable depth probe of interfacial characteristics. Changing photon energy $h\nu$ varies the kinetic energy E_K of the emitted electrons and thereby the distance below the surface from which they escape Λ. Variation of the overlayer coverage by a known thickness L allows the uniformity and abruptness of growth to be determined. A simple exponential attenuation model presented in Sec. IV describes the photoemitted electron intensity N (electrons/photon) by the expressions in this diagram. Note that core level optical absorption can also be measured on the same sample to characterize the unoccupied interface states by constant-final-state partial yield as described in Refs. 24 and 25.

FIG. 2. Energy distributions normalized to constant total core yield for $3d$ core electrons of As, Ge, and Ga photoemitted by 125 eV photons from GaAs (110) cleaved *in situ* and covered with equivalent thicknesses of Ge, L_{Ge}, up to 6 Å while held at a growth temperature T_G of 525°C. The curves are offset vertically by one major division from each other for clarity.

source using a Sloan 900023 quartz crystal thickness monitor placed at the deposition position with a linear motion in each experiment; the evaporation rate varied by less than 20% over periods of several hours. The chemical bonding and composition as manifest by core level changes are probed as a function of depth below the growing interface by tuning the photoelectron escape depth $\Lambda(E_K)$ through variation of the photon energy $h\nu$; while this can be simulated in XPS/ESCA measurements by tilting the sample relative to the analyzer, angular-dependent contributions of the final states can significantly alter the photoemitted electron energy distribution. Measurement of the valence band region provides emission that is unique to interface states. The valence band cutoff can be used with the core-level photoemission to provide the band edge discontinuity on a microscopic scale. Independently, Grant *et al.*[3] have used a similar method.

III. INTERDIFFUSION OF Ge AND GaAs
A. Kinetics

In reading the literature or any of a number of excellent reviews,[4,5] one is led to believe that there is a large range of temperature (some 500°C) over which abrupt heterojunctions can be formed. Variations with T_G are reported in the twinning observed in the epitaxial layer (minimum at 425–525°C)[2] or the electrical rectification depending on preferential evaporation of the substrate during growth (above 800°C).[6] We find that these processes are critically dependent on the growth environment. As we show in this section, even at moderate temperatures (~430°C), significant interdiffusion occurs during MBE growth in ultra-high vacuum, while a 2×10^{-5} Torr vacuum produces best junction behavior for an 800°C growth temperature.[6]

Typical $3d$ core photoemission is shown in Fig. 2 for a growth temperature of 525°C. The spectra are normalized to have constant yield $\mathcal{N} = N_{Ge} + N_{Ga} + N_{As}$. Then, if we

make the reasonable assumption that the matrix elements M and escape depth Λ are the same in Ge and GaAs, the integrated Ge emission N_{Ge} relative to the Ga, As, or total photoemission is a quantitative measure of the interdiffusion since the probing depth is fixed as the overlayer becomes thicker. Note then in Fig. 2 that the As $(3d)$ and Ga $(3d)$ strength decrease together, rather uniformly as L_{Ge} becomes comparable to the photoelectron escape depth. By visual inspection, the As to Ga ratio remains relatively constant to coverages of a few Ge layers. We conclude that in the initial stage there is nearly equal bonding of Ge to either a Ga or an As atom on the GaAs side of the heterojunction.

The strengths of the Ga and As $3d$ core emissions appear a bit too large for the Ge layer to be uniformly bonded over all Ga and As surface atoms to a thickness given by L_{Ge}. This discrepancy dramatically increases and exhibits interesting characteristics for very large coverages. In Fig. 3, we see the GaAs emission persisting for $L_{Ge} > 50\Lambda$. Further, the As emission remains strong for the growth conditions used in this study, while the Ga strength exhibits a continuous reduction

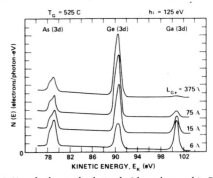

FIG. 3. Normalized energy distributions for $3d$ core electrons of As, Ge, and Ga photoemitted with 125 eV photons from GaAs (110) covered with Ge of 6 Å $\leq L_{Ge} \leq$ 375 Å for $T_G = 525$°C. The 6 Å data is the same as that shown in Fig. 2 for reference. The curves are offset by one major division from each other for clarity. The increasing emission of As relative to Ga and its persistence for overlayers much thicker than the photoelectron escape depth are indicative of interdiffusion of all the species.

J. Vac. Sci. Technol., Vol. 15, No. 4, July/Aug. 1978

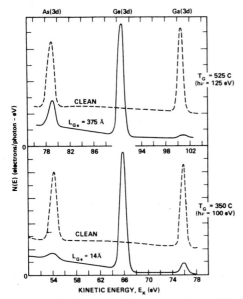

FIG. 4. Energy distributions normalized to constant total core yield for $3d$ core electrons of As, Ge, and Ga photoemitted from two GaAs (110) surfaces on which Ge was subsequently evaporated by MBE to a thickness corresponding to 375 Å when heated to 525°C and to 14 Å when heated to 350°C. The data of the upper panel is reproduced from Figs. 2 and 3, and the clean curves are offset by one major division for clarity. Note the rapid attenuation of substrate emission in the soft x-ray region and the constant ratio of Ga to As emission when an abrupt interface is formed under the conditions in the lower panel.

beginning at 15 Å. This observation indicates a signficant alloying of the two lattice-matched semiconductors. The preferential diffusion of As toward the free Ge surface region continues for coverages up to 375 Å (the extent of our measurement), where negligible Ga emission is observed.

These results are indicative of interdiffusion rather than Ge island formation on the GaAs. As summarized in Fig. 4, the relative Ga to As emission exhibits unique characteristics for high temperature growth. The lower panel shows the situation for an interface that is abrupt (as discussed below). Note the slightly lower As $(3d)$ peak height relative to the Ga $(3d)$ structure characteristic of the clean photoemission from all GaAs (110) surfaces studied at these photon energies. This characteristic is maintained in thick overlayers ($L_{Ge} \approx 2\Lambda$) for the abrupt epitaxial junction ($T_G = 350°C$) in comparison to the data for $L_{Ge} = 15$ Å at $T_G = 525°C$ in Fig. 3. In measurements we have made for room temperature growth, the GaAs emission persists for total coverages equivalent to overlayers many times the escape depth in thickness. However, in this case the As $(3d)$:Ga $(3d)$ ratio is maintained in a manner similar to the lower panel of Fig. 4. For $T_G = 25°C$, one does not expect enough mobility for appreciable diffusion of Ge and GaAs; therefore, we conclude that the 25°C data is indicative of island formation. The characteristically increased As to Ga ratio shown in the top panel of Fig. 4 must then be due to the enhanced diffusion coefficient for As in Ge compared to Ga.[7] For this high T_G data to be caused by a clus-

tering or island-type formation, one must invoke a preferential bonding of Ge to Ga rather than to As atoms. This is not observed in the low coverage data as noted with regard to Fig. 2, and further, it is likely not to persist on such an atomic scale for Ge coverages hundreds of times larger than the Ga to As interatomic spacing. Further, the success of the simple expressions of Fig. 1 in describing below the photoemission of intermediate temperature data (i.e., $T_G = 350°C$) can only occur if the overlayer is uniformly covering the substrate. The interdiffusion of Ge and GaAs is very abrupt with temperature (as discussed in Sec. IV) as is characteristic of an activated process. Thermodynamic properties[8] will also affect the interface growth kinetics and the composition of the transition region comprising the junction.

B. Bonding

When the Ge and GaAs interdiffuse under the conditions described above, the intriguing question arises as to whether the As (and Ga) are present throughout the thick Ge layer or are rather a fixed number of atoms "floating" on the surface of the growing Ge as occurs for some metal–semiconductor systems (e.g., Au on GaSb,[9] Al on GaAs[10]). For the case shown in the top of Fig. 4, one can measure the change in relative As $(3d)$ photoemission as the escape depth is varied by about a factor of 2 through a change in photon energy from 125 to 200 eV (E_K from ~80 to 155 eV).[11] We restrict ourselves to this range since the matrix element effects are small and slowly varying, and therefore they can be properly accounted for. A reduction in normalized As emission of at most 10% is measured over this E_K range. If the number of As atoms represented by the ~17% partial yield in Figs. 3 and 4 ($T_G = 525°C$, $L_{Ge} = 375$ Å) were all diffusing to and resident on the surface of the sample, then the simple photoemission expressions in Fig. 1 would predict a much more substantial variation with escape depth (on the order of 25%). The negligible measured photon energy dependence then shows that As is distributed throughout the Ge. The continuing source for the As (and Ga) in the growing Ge layer is the decomposition of the GaAs substrate itself.[12]

This uniform interdiffusion of the two semiconductors is in contrast to the surface segregation of one of the semiconductor species in metal–semiconductor interface formation. The difference in driving force is related to the relative solubilities and heats of formation of the materials as well as their diffusion.[12] For Ge or Al interfaces on GaAs, the heat of formation of AlAs is over twice that of GaAs while GeAs has less than half the enthalpy of the substrate.[8] It is then favorable for a stoichiometric compound to form at an Al interface by converting the entire surface GaAs layer to AlAs.[10] The liberated Ga can then rapidly diffuse through the depositing Al layer to reside on the room temperature Al surface; the more than three orders of magnitude smaller semiconductor interdiffusion limits the substrate dissolution to the first layers. For a Ge interface, the GaAs is stable, so a transition from pure GaAs to pure Ge can occur over a single bond length without an intermediate composition region. By extrapolating to 525°C the diffusion coefficients[7] for Ga and As atoms in Ge and in GaAs, we find that the anion diffuses rapidly enough to continually permeate the growing interface.

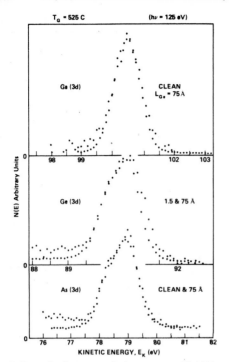

FIG. 5. Energy distributions normalized to constant partial yield for each of the 3d core photoemission peaks individually. Two sets of raw data are shown for the core level characteristics in pure GaAs and the preferentially As-diffused Ge overlayer grown at 525°C. The essential constancy in lineshape indicates a similarity in Ga–Ge and As–Ge bonding compared to Ga–As covalent bonding in GaAs.

The factor limiting the As concentration is most likely the anomalously low Ga movement out of the substrate. The diffusion rate for Ga in Ge is slightly less than the overlayer deposition rate, thereby accounting for the sharper Ga profile.

The details of the 3d core photoemission provide insight into the bonds formed by the As and Ga atoms diffused into the Ge. Notably, one would like to distinguish unique interface states caused by Ge bonding to As and Ga nearest neighbors.[13] This occurs dramatically at the Al–GaAs interface where AlAs is formed.[10] As shown in Fig. 5, we do not observe significant changes in the 3d–core electrons between clean GaAs and a 75 Å Ge film grown on GaAs at 525°C. The raw data points in this figure are two sets of normalized results from the indicated specimens of Figs. 2 and 3. One can see an increased number of scattered secondary electrons on the low kinetic energy side of each primary core peak; there is also a slight increase in Ge emission on the high E_K side but we do not consider it significant when not accompanied by other changes. The essential indistinguishability of these two sets of data suggest a negligible charge transfer for As or Ga atoms in Ge compared to the bonds existing in elemental Ge and GaAs. Further, a negligible dipole layer (as determined by 3d shifts of less than 0.2 eV) occurs in this case.

J. Vac. Sci. Technol., Vol. 15, No. 4, July/Aug. 1978

In order to determine the characteristics for 3d electrons of As atoms covalently bonded to Ge, J. C. Mikkelsen of our laboratory grew a GeAs crystal which we cleaved in situ and measured at SSRL with the same apparatus. The resulting Ge and As core level photoemission had the same width and binding energy as in either Ge or GaAs, but exhibited sharper, more pronounced spin–orbit splittings. Further, the valence band exhibited three density of states peaks common to both Ge and GaAs.[14] The differences between the layered, monoclinic structure of GeAs and the diamond/zincblende arrangement for Ge/GaAs mainly affect the broadening of the core spectra; thus, from the similarity of the features of 20% As diffused in Ge to those of the compounds, we conclude that the As is in a substitutional site with covalent bonds to near-neighbor Ge atoms characterized by negligible difference in the charge-transfer from the stoichiometric semiconductors.

IV. ABRUPT Ge–GaAs (110) INTERFACE

A. Kinetics of MBE growth

A growth temperature can be chosen which allows epitaxial formation of the Ge layer but negligible intermixing of the two species. This is seen quite clearly in the photoemission for a 15 Å overlayer. For $T_G = 525$°C in Fig. 3, the peak heights for Ga and As are roughly half the Ge 3d emission with the proportion of As now greater than in GaAs. By comparison, for $T_G = 350$°C in Fig. 4, the Ga and As emissions maintain the same proportion as the substrate and are less than 10% as strong as the yield from the Ge overlayer.

We can determine the abruptness of the transition of GaAs to Ge by using a simple model for the photoemission process. Consider the sample as the two uniform media shown in Fig. 1. Then, the first-order effect on the photoexcited electron density N_0 is an exponential attenuation due to electron-electron scattering of the number of emitted electrons N characterized by the escape depth Λ. In this case, the Ge emission from a distance x below the free surface is just given by

$$dN_{Ge} = (M_{Ge} N_0 \, dx) \, e^{-(x/\Lambda_{Ge})}.$$

The same expression describes Ga or As emission (with half the atomic density as the Ge overlayer) with the addition of a fixed attentuation caused by the Ge of thickness L_{Ge}:

$$dN_{Ga,As} = (0.5M_{Ga,As} N_0 \, dx) \, e^{-(x/\Lambda_{Ga,As})} e^{-(L_{Ge}/\Lambda_{Ge})}.$$

Integrating through the overlayer to obtain the Ge emission, and throughout the semiinfinite substrate to obtain the Ga or As emssion, the expressions shown in Fig. 1 result from assuming equal matrix elements M at each $h\nu$ and equal escape depths at each E_k for Ge, Ga, or As initial states.

The spatial dependence of the interface composition can either be determined by changing the escape depth Λ through $E_K(h\nu)$ variation for a given Ge overlayer, or by probing with a single photon energy and varying the Ge coverage L_{Ge}. Since the escape depth is so short for 3d core electrons excited by soft x rays, we can obtain a sensitive test of abruptness of the interface by the second method. The same technique would be valid for XPS measurements but thickness of 40–50

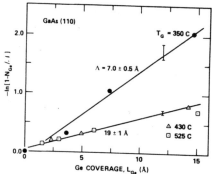

FIG. 6. Coverage dependence for Ge $(3d)$ photoemission normalized to total $3d$ core yield n when grown on GaAs (110) at temperatures ranging from 350 to 525°C. N_{Ge} is now the *area* under the $N(E)$ peaks as for the data of Figs. 2 and 3 shown here by the squares. This data analysis should result in a straight line passing through (0,0) with slope equal to the escape depth Λ if the junction is abrupt and described by the expressions in Fig. 1. The coverage scale was checked by two independent methods and the deduced Λ's are an upper limit based on the least-square straight-line fits plotted on the data. The 7.0 ± 0.5 Å value for $T_G = 350$°C compares to 5.8 ± 1.5 Å for GaAs deduced from oxidation studies in Ref. 11; it indicates uniform growth of a compositionally abrupt junction at 350°C. The sudden change in behavior which saturates quickly is expected for an activated process such as diffusion.

Å (i.e., ≈2Λ) would have to be used. The data of Figs. 2 and 3 were analyzed taking the *area* of the peaks for the number of photoemitted electrons per photon required in the expressions of Fig. 1. As shown for this data set and two others in Fig. 6, a plot of $-\ln\left[1 - N_{Ge}/(N_{Ga} + N_{As} + N_{Ge})\right]$ vs L_{Ge} should give a straight line passing through zero and having a slope equal to the escape depth. For a growth temperature of 350°C, the least squares fit of the data yields an escape depth of 7.0 ± 0.5 Å. This compares to a minimum escape depth of 5.8 ± 1.5 Å determined[11] using similar expressions for oxidized GaAs. Such good agreement could only occur if the atomic composition changes from (110) planes containing all Ga and As atoms to complete Ge layers over one bonding distance. The data for T_G of 430° and 525°C fall on the same straight line predicting a Λ of 19 ± 1 Å, clearly outside the range of uncertainty for the scattering length of moderate kinetic energy electrons. Therefore, sizable interdiffusion of the semiconductors is indicated when the changing ratio of As to Ga emission is also considered.

We then find a very abrupt transition for temperatures between 350° and 430°C, with no additional activation for the next 100°C over the short times (i.e., less than an hour) of these growths. This is quite reasonable for an activated diffusion process.[7] The temperatures at which the transition occurs and the sharpness of the change with T_G are quite dependent on the growth method. Although we obtain unsatisfactory junctions from an atomic abruptness viewpoint above 425°C, electrically acceptable Ge–GaAs heterojunctions are reported in the literature up to temperatures of 800°C.[4,6]

Foreign molecules in moderate-vacuum, physical-vapor-deposition environments inhibit the diffusion process at low temperatures; contaminants impinge on the surface and either reduce the adatom mobility or form as impurities at the interface.

terface.[12] For the MBE conditions we have used, a growth temperature of 350°C resulted in near-ideal photoelectron behavior for a single layer transition region. This agreement with the simple model of Fig. 1 further indicates homogeneous overlayers rather than island formation for even the first plane of Ge atoms.

B. Electronic structure

We measure characteristics of the Ge–GaAs (110) interface which are unique to the abrupt junction. Only for the sharp interface results ($T_G = 350$°C) discussed above do we observe a sizable valence band discontinuity. The features of the valence band photoemission change from those characteristic of GaAs to those of Ge for very small coverages. By $L_{Ge} = 7$ Å, the valence band maximum (VBM) and s–p bands measured with $h\nu = 130$ eV are fully developed into those of Ge. The band discontinuity can then be measured by the change in Ga $(3d)$ and As $(3d)$ binding energies relative to the VBM on going from GaAs to the abrupt junction formed with lower bandgap Ge. Conservatively estimating the errors in locating the $3d$ core peaks and VBM, we deduce $\Delta E_v = 0.7 \pm ^{0.05}_{0.3}$ eV for the n-type GaAs used. Further, this band structure change occurs over a distance of 5 Å or less (the limit of our depth resolution). The average[13] of other experimental values is 0.55 ± 0.15 eV, consistent with our result. It is important to note that if this microscopic probe is used to determine the valence band discontinuity, then the interfaces exhibiting interdiffusion have $\Delta E_v = 0.2 \pm 0.1$ eV.

Theoretical estimates for the band discontinuities vary by almost half the Ge bandgap. Simply preserving the free surface electron affinities of the two semiconductors at the heterojunction[15] predicts 0.69 eV for ΔE_v. More detailed microscopic theories yield 0.72,[16] 0.69,[17] 0.41,[18] and 0.35 ± 0.15 eV.[13,19] The rather large uncertainties of our present data analysis do not allow us to favor some of these theoretical methods over the others. However, we can conclude that *only* for those experimental studies where the semiconductor interfaces have been demonstrated to be compositionally abrupt can detailed theoretical calculations of interface states such as Refs. 13, 18, and 20 be applied. We caution, however, that large rearrangements of the interface atoms may occur and these are not, we believe, properly accounted for in present theories.[21] In particular, the negligible localized interfacial dipole layer and insignificant rigid GaAs band shift in the $T_G = 350$°C data may be due to sizeable reconstructions at the abrupt interface. Large dipole layers have been predicted in some theoretical treatments[22,23] of ideal, unperturbed lattices, while they are only 0.1–0.2 eV in other calculations.[19]

V. SUMMARY AND CONCLUSIONS

We studied the initial stages of formation of the interface of Ge deposited by MBE onto *in situ* cleaved GaAs (110). Synchrotron radiation excited core electron photoemission allows a variable depth probe of the interface. At temperatures of 430°C and above, both the Ga and As emission uniformly decrease starting from a fraction of a monolayer to coverages of several Ge layers; thus, in the initial stage there is equal bonding of Ge to either a Ga or an As atom on the GaAs side of the heterojunction. The GaAs emission persists for Ge thicknesses many times the photoelectron escape depth, indicating a significant alloying of the two lattice-matched semiconductors. The Ga photoemission decreases relative to that of As for Ge layers greater than ~20 Å thick. This preferential diffusion of As to the free Ge surface region is consistent with the bulk diffusion coefficients of Ga and As in Ge and GeAs it is observed to continue for coverages up to 375 Å (the extent of our measurements), where Ga emission is negligible. The As is distributed throughout the Ge rather than residing as a thin layer on the surface as occurs for metal–semiconductor interface formation. This is understood by the thermodynamics controlling compound formation at the interface. The position and width of the Ge (3d) core level is constant over the entire coverage range; further, there is negligible broadening or relative energy shift of Ga to As core levels. These measurements are inconsistent with a large interfacial dipole layer. These graded junctions exhibit a valence-band discontinuity of only 0.2 ± 0.1 eV for both n- and p-type GaAs, a value about a third that expected from most theories.

This interdiffusion is an activated process, allowing atomically abrupt composition change for a growth temperature of 350°C where epitaxy should still occur. The interface electronic structure is very different for such junctions though a negligible localized interfacial dipole layer is also measured in this case. Here the valence band discontinuity is $0.7 \pm ^{0.05}_{0.3}$ eV occurring over distances of less than 5 Å. Our indirect evidence of significant interfacial atomic rearrangement and the demonstrated crucial importance of heterojunction chemistry argue for determination of the growth kinetics for each preparation system before definitive conclusions about semiconductor–semiconductor interface properties are warranted.

ACKNOWLEDGMENTS

We gratefully acknowledge the experimental collaboration of R. Z. Bachrach and S. A. Flodstrom, and stimulating discussions with E. A. Kraut and J. C. Mikkelsen.

*Aspects of this work were performed at the Stanford Synchrotron Radiation Lab, which is supported in part by National Science Foundation Grant No. CMR 73-07692 in cooperation with the Stanford Linear Accelerator Center and U.S. Department of Energy.

[1] R. Z. Bachrach, R. D. Burnham, R. S. Bauer, and S. B. M. Hagström, Material Research Society Symposium on MBE, Cambridge, MA (1976).

[2] J. E. Davey, Appl. Phys. Lett 8, 164 (1966).

[3] R. W. Grant, J. R. Waldrop, and E. A. Kraut, J. Vac. Sci. Technol. 15, 1451 (1978).

[4] A. G. Milnes and D. L. Feucht, Heterojunctions and Metal–Semiconductor Junctions (Adademic, New York, 1972), p. 251.

[5] B. L. Sharma and R. K. Purohit, Semiconductor Heterojunctions (Pergamon, New York, 1974), p. 20.

[6] I. Ryu and K. Takahashi, Jpn. J. Appl. Phys. 4, 850 (1965).

[7] L. P. Hunter, editor Handbook of Semiconductor Electronics (McGraw-Hill, New York, 1970) 3rd edition, p. 7–22; H. S. Veloric and W. J. Greig, R.C.A. Rev. 21, 454 (1960).

[8] M. Kh. Karapet'yants and M. L. Karapet'yants, Thermodynamic Constants of Inorganic and Organic Compounds (Humphrey and Science Publishers, Ann Arbor, 1970), pp. 7, 99, 103.

[9] I. Lindau, P. W. Chye, C. M. Garner, P. Pianetta, C. Y. Su, and W. E. Spicer, J. Vac. Sci. Technol. 15, 1332 (1978).

[10] R. Z. Bachrach, J. Vac. Sci. Technol. 15, 1340 (1978); R. Z. Bachrach, R. S. Bauer, and J. C. McMenamin, Proc. 14th International Conference on Physics of Semiconductors (Edinburgh, 1978).

[11] P. Pianetta, I. Lindau, C. M. Garner, and W. E. Spicer, Phys. Rev B (to be published).

[12] J. C. Mikkelsen (private communication).

[13] W. E. Pickett, S. G. Louis, and M. L. Cohen, Phys. Rev. B 17, 815 (1978); Phys. Rev. Lett. 39, 109 (1977).

[14] R. S. Bauer, D. J. Chadi, J. C. Mikkelsen and J. C. McMenamin (to be published).

[15] R. L. Anderson, Solid State Electron. 5, 341 (1962); A. G. Milnes and D. L. Feucht, Ref. 3, p. 3.

[16] W. R. Frensley and H. Kroemer, J. Vac. Sci. Technol. 13, 810 (1976).

[17] J. L. Shay, S. Wagner, and J. C. Phillips, Appl. Phys. Lett. 28, 31 (1976).

[18] W. A. Harrison, J. Vac. Sci., Technol. 14, 1016 (1977).

[19] W. E. Pickett and M. L. Cohen, J. Vac. Sci. Technol. 15, 1437 (1978).

[20] F. Herman and R. V. Kasowski, Phys. Rev. B 17, 672 (1978).

[21] W. E. Pickett and M. L. Cohen, Solid State Commun. 25, 225 (1978).

[22] W. R. Frensley and H. Kroemer, Phys. Rev. B 16, 2642 (1977).

[23] G. A. Baraff, J. A. Applebaum, and D. R. Hamann, Phys. Rev. Lett. 38, 237 (1977); J. Vac. Sci. Technol. 14, 999 (1977).

[24] R. S. Bauer, R. Z. Bachrach, S. A. Flodstrom, and J. C. McMenamin, J. Vac. Sci. Technol. 14, 378 (1977).

[25] R. S. Bauer, D. J. Chadi, J. C. McMenamin, and R. Z. Bachrach, Proceedings of the 7th International Vacuum Congress and 3rd International Conference on Solid Surfaces (Vienna 1977), p. A-2699.

162

Observation of the Orientation Dependence of Interface Dipole Energies in Ge-GaAs

R. W. Grant, J. R. Waldrop, and E. A. Kraut

Science Center, Rockwell International, Thousand Oaks, California 91360

(Received 19 December 1977)

The interfaces between a thin (~ 20-Å) abrupt epitaxial layer of Ge grown on substrates of (111), (110), and (100) GaAs have been investigated with x-ray photoelectron spectoscopy. Observed changes in core-level binding energies have been directly related to the crystallographic orientation dependence of interface dipoles and variations of band-gap discontinuities. The orientation variation of the band-gap discontinuities is found to be a significant fraction ($\approx \frac{1}{4}$) of the total band-gap discontinuity.

There has been considerable theoretical interest in the properties of ideal abrupt interfaces between different semiconductors, stimulated in part by the recent progress in molecular beam epitaxy (MBE) whereby truly abrupt interfaces can now be achieved. A basic property of the abrupt semiconductor interface is the relative alignment of the energy bands of the two semiconductors; i.e., how the energy difference in the band gaps (ΔE_g) is distributed between the valence- and conduction-band discontinuities (ΔE_v and ΔE_c) such that $\Delta E_g = \Delta E_v + \Delta E_c$.

The first and most widely used model for estimating ΔE_c (or ΔE_v) is based on electron affinity differences.[1] Critical evaluations[2,3] have been made of this model. Alternative models for predicting ΔE_v have appeared,[4,5] and two self-consistent calculations of the Ge/GaAs-interface electronic structure have been completed.[6,7] Although it has long been recognized that interface dipoles could produce energy-band discontinuities which depend on crystallographic orientation of the interface plane, such effects have generally been ignored. Transport measurements[8] on vapor-grown Ge/GaAs heterojunctions suggested that

there could be substantial (a few tenths of an eV) changes in valence- and conduction-band discontinuities, $\delta(\Delta E_v)$ and $\delta(\Delta E_c)$, dependent on crystallographic orientation. Unfortunately, it is relatively difficult to determine these dopant-level-independent quantities from transport measurements and the scatter in these data is as large as the measured effect.

To investigate the interface dipole orientation dependence, we have developed a contactless x-ray photoemission spectroscopy (XPS) technique which allows a direct probe of interface potential variations. Herein, we report the observation of sizable and systematic variations in ΔE_v for the Ge/GaAs interface as a function of crystallographic orientation. Figure 1 is a schematic energy-band diagram of an ideal abrupt Ge/GaAs interface. The relative positions of the average bulk crystal potential within the two semiconductors determine ΔE_v and ΔE_c.[2,4-6] An orientationally dependent change in the interface dipole magnitude may shift the relative positions of the valence and conduction bands in the two semiconductors as shown schematically by dashed lines in Fig. 1. Figure 1 also shows the position of a

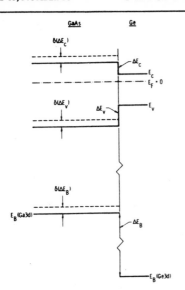

FIG. 1. Schematic energy-band diagram. The dashed lines illustrate a decreased value of ΔE_v associated with an interface dipole layer that accelerates photoelectrons from a GaAs substrate relative to Ge $3d$ photoelectrons which do not cross the interface.

core level in Ge and in GaAs. As the average bulk crystal potential changes to adjust to the dipole variation, the relative binding energies of all levels on both sides of an abrupt interface (measured relative to the common Fermi level, E_F) must also vary by the change in dipole energy with orientation; i.e., $|\delta(\Delta E_v)| = |\delta(\Delta E_c)| = |\delta(\Delta E_B)|$ also indicated by dashed lines in Fig. 1. For the Ge/GaAs interface, we will specifically consider the energy separation, ΔE_B, between the Ga $3d$ and Ge $3d$ core electron levels. A measurement of $\delta(\Delta E_B)$ by XPS thus provides a direct measure of $\delta(\Delta E_v)$. The dashed lines in Fig. 1 illustrate a change in the interface dipole which would increase the splitting between the Ga $3d$ and Ge $3d$ core levels to equal the decrease in ΔE_v.

Our experiment used Al $K\alpha$ ($h\nu = 1486.6$ eV) radiation in conjunction with an extensively modified Hewlett-Packard model 5950A ESCA (electron spectroscopy for chemical analysis) spectrometer to excite photoelectrons from Ge/GaAs interfaces for which the Ge was an ≈ 20-Å-thick layer on a thick (≈ 0.5 mm) GaAs substrate. The escape depth for the Ge $3d$ and Ga $3d$ photoelectrons is ≈ 20 Å. Thus, photoelectrons from both sides of the Ge/GaAs interface are observed simulta-

neously in the same XPS spectrum. Electrons which originate on the GaAs side of an abrupt interface pass through any dipole layer at the interface in order to be emitted from the free surface and detected, while electrons originating in the Ge do not. For example, an electron passing through a dipole layer in a direction from higher to lower electron density will experience an acceleration and, consequently, a relative increase in kinetic energy proportional to the dipole moment per unit area, τ, at the interface.[9] A kinetic-energy increase will appear as an apparent binding-energy decrease in the XPS spectrum. In terms of the average charge density $\bar{\rho}(z)$ over planes parallel to the interface, the dipole moment per unit area is

$$\tau = \int z \bar{\rho}(z) \, dz.$$

The self-consistent calculations of Baraff, Appelbaum, and Hamann[6] and Pickett, Louie, and Cohen[7] have shown that the potential variations near an interface are localized to within 1 or 2 atomic layers, a length considerably less than the Ge $3d$ and Ga $3d$ photoelectron escape depths.

Interface states and bulk doping differences which cause band bending can complicate the ability to determine ΔE_v from transport measurements. In the XPS techniques described here, however, because the photoelectron escape depth is much smaller than typical band-bending lengths $\mathcal{L}$ ($\mathcal{L} > 10^3$ Å for moderate dopant levels), the effect of interface states is to shift the potential within the sampled region on both sides of an interface by the same constant value. Therefore, since ΔE_B is the difference in core-level binding energy for photoelectrons which originate from each side of the interface, any potential shift due to interface states or other sources of band bending cancel. It is assumed that the two semiconductors are nondegenerately doped and that the dimensions perpendicular to the interface sampled by XPS are small compared to $\mathcal{L}$.

The very thin (~ 20-Å) epitaxial layers of Ge used for these interface studies were grown within the XPS apparatus on heated ($\approx 425°C$) GaAs substrates by evaporative MBE techniques similar to those previously described,[10] but at low flux rates. GaAs substrates with (100), (111), (1̄1̄1̄), and (110) faces were cut from a single boule of undoped GaAs (n-type carrier concentration 10^{16} cm³).[11] Laue back-reflection photography showed that the substrates were oriented to better than 1°. Each substrate was etched in 3:1:1 $H_2SO_4:H_2O_2:H_2O$ prior to insertion into the

657

XPS vacuum system. Substrate surfaces were
cleaned by Ar⁺-ion sputtering (750 eV) followed
by annealing at ≈575°C to remove sputter damage
(vacuum-system base pressure was low 10⁻¹⁰
Torr). Room-temperature low-energy electron-
diffraction (LEED) patterns characteristic of
(110) (1×1), (111)Ga (2×2), (1̄1̄1̄)As (1×1), and
(100)Ga c(8×2) were obtained. In addition, a
(100)As surface was also studied which was either c(2×8) or (2×4). Additional LEED measurements confirmed the epitaxy of the Ge overlayers.
Following the XPS measurements, a metal point
contact was made to the semiconductor surface
to ensure reasonable diode characteristics.

Figure 2 shows an XPS spectrum from a sample of epitaxial Ge grown on a (110) (1×1) GaAs
substrate. To determine ΔE_B, a background
function which is proportional to the integrated
photoelectron area was subtracted from the data
to correct for the effect of inelastic photoelectron
scattering. ΔE_B was measured between the centers of the peak widths at half of the peak heights.
This procedure made it unnecessary to resolve
the spin-orbit splitting of the Ge 3d and Ga 3d lev-

els (≈0.5 eV) to obtain high-precision peak positions.

Measurement results of eight different interfaces are given in Table I. In general, several
(three to five) independent determinations were
made on each interface. In all cases, measurement reproducibility was <0.01 eV and was usually <0.005 eV; calibration uncertainties increase
the error limits to 0.1 eV. The measurements
on the two samples of (110) (1×1) and (1̄1̄1̄)As
(1×1) reproduce very well. We believe the discrepancy in the two values shown for (111)Ga
(2×2) is real and represents a subtle difference
in the interface properties grown on this surface.

If we arbitrarily reference all $\delta(\Delta E_v)$ values
to the (110) charge-neutral surface such that
$\delta(\Delta E_v)_{110}\equiv0$, we obtain the values of $\delta(\Delta E_v)$ shown
in Table I. It is interesting that the (1̄1̄1̄)As and
(111)Ga and the (100)As and (100)Ga differences
are nearly symmetrically distributed around the
(110) value. However, the known complexity of
these surfaces[12] makes a simple interpretation
of the variations in valence-band discontinuity
difficult.

In summary, a technique has been developed
to observe directly variations in band-gap discontinuities at abrupt semiconductor interfaces, and
systematic changes in ΔE_v as a function of interface crystallographic orientation have been observed for Ge/GaAs. The maximum variation in
ΔE_v between the (111) and (1̄1̄1̄) interfaces is
≈0.2 eV, which is a significant fraction (≈¼) of
ΔE_g (0.75 eV). This result suggests that accurate
future models used to predict ΔE_v and ΔE_c need
to account for dipole orientation dependence.

We acknowledge helpful discussions with Profes-

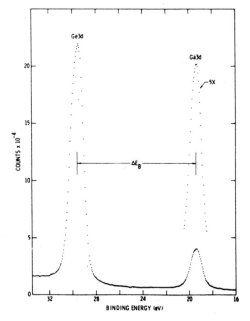

FIG. 2. XPS spectrum in the energy region of the Ga
3d and Ge 3d core levels obtained from a (110) Ge/GaAs
interface. The thickness of the epitaxial Ge overlayer
was ≈ 20 Å.

TABLE I. Ge-3d–Ga-3d binding-energy differences
and corresponding variations in valence-band discontinuity for various Ge/GaAs interfaces.

Substrate surface	ΔE_B (eV)	$\delta(\Delta E_v)$ (eV)
(111)Ga	10.27±0.01	≈−0.085
(2×2)	10.31±0.01	
(100)Ga		
c(8×2)	10.22±0.01	−0.015
(110)	10.20±0.01	0
(1×1)	10.21±0.01	
(100)As	10.17±0.01	+0.035
⋯		
(1̄1̄1̄)As	10.11±0.01	+0.10
(1×1)	10.10±0.01	

658

sor W. A. Harrison and appreciate the x-ray analysis performed by Dr. M. D. Lind. This work was supported by the U. S. Office of Naval Research, Contract No. N00014-76-C-1109.

---

[1]R. L. Anderson, Solid-State Electronic. 5, 341 (1962).

[2]H. Kroemer, CRC Crit. Rev. Solid State Sci. 5, 555 (1975).

[3]J. L. Shay, S. Wagner, and J. C. Phillips, Appl. Phys. Lett. 28, 31 (1976).

[4]W. A. Harrison, J. Vac. Sci. Technol. 14, 1016 (1977).

[5]W. R. Frensley and H. Kroemer, Phys. Rev. 16, 2642 (1977).

[6]G. A. Baraff, J. A. Appelbaum, and D. R. Hamann, Phys. Rev. Lett. 38, 237 (1977), and J. Vac. Sci. Technol. 14, 999 (1977).

[7]W. E. Pickett, S. G. Louie, and M. L. Cohen, Phys. Rev. Lett. 39, 109 (1977), and to be published.

[8]F. F. Fang and W. E. Howard, J. Appl. Phys. 35, 612 (1964).

[9]J. A. Stratton, Electromagnetic Theory (McGraw-Hill, New York, 1941), p. 190.

[10]R. F. Lever and E. J. Huminski, J. Appl. Phys. 37, 3638 (1966).

[11]Obtained from Morgan Semiconductor, Inc.

[12]See, e.g., W. Ranke and K. Jacobi, Surf. Sci. 63, 33 (1977); A. Y. Cho, J. Appl. Phys. 47, 2841 (1976); J. R. Arthur, Surf. Sci. 43, 449 (1974); L. L. Chang, L. Esaki, W. E. Howard, R. Ludeke, and G. Schul, J. Vac. Sci. Technol. 10, 655 (1973); several references to earlier work are contained in these papers.

Physica 134B (1985) 433-438
North-Holland, Amsterdam

INTERNAL PHOTOEMISSION IN GaAs/(Al$_x$Ga$_{1-x}$)As HETEROSTRUCTURES

G. ABSTREITER and U. PRECHTEL

Physik-Department, Technische Universität München, D-8046 Garching, Fed. Republic of Germany

and

G. WEIMANN and W. SCHLAPP

FTZ, D-6100 Darmstadt, Federal Republic of Germany

Band offsets in (Al$_x$Ga$_{1-x}$)As/GaAs heterostructures are determined using internal photoemission experiments. Onsets in the photocurrent are observed for photon energies exceeding the fundamental energy gaps of GaAs and (Al$_x$Ga$_{1-x}$)As. Additional onsets occur at photon energies in the infrared region due to internal photoemission from the conduction band in GaAs over the barrier into the conduction band of (Al$_x$Ga$_{1-x}$)As and in the near red region where excitations from the GaAs valence band into the (Al$_x$Ga$_{1-x}$)As conduction band are involved. From the measured energies we determine $\Delta E_c / \Delta E_g = 0.8 \pm 0.03$ for $x = 0.2$.

1. INTRODUCTION

The fascinating electrical and optical properties of semiconductor heterostructures have opened the possibility for the development of various new and future devices. For many of these new concepts an exact knowledge of the band offsets at the interface of the two semiconductors involved is important. The most widely studied heterojunction is the nearly lattice matched GaAs/(Al$_x$Ga$_{1-x}$)As system. The band offset has been studied with various techniques. Among the most prominent ones are optical absorption and excitation spectroscopy in quantum wells[1-4] and methods which use purely electrical properties which have been discussed extensively by Kroemer[5]. More recently also the comparison between calculated and measured charge transfer at single heterojunction interfaces has been used to determine values of conduction and valence band offsets[6]. X-ray or UV photoemission spectroscopy, a method which has been used extensively for other heterojunctions[7], is not very suitable because of the small energy gap differences in the GaAs/(Al$_x$Ga$_{1-x}$)As system as well as growth problems[8]. While until last

year the values of band offsets proposed in Ref. 1 had been widely accepted, most but not all of the recent work suggests considerably smaller conduction band offsets, especially since the work of Miller et al.[2] was published. The mentioned experimental techniques, however, suffer from the need of more or less complicated theoretical models for the interpretation of the experimental data or not accurately known properties of the samples. Important features which have not been taken into account properly in most of the published work are the complicated valence band structure in quantum wells[9], the nonparabolicity of the conduction band which is important especially for higher lying levels or thin quantum wells, the not well understood behaviour of impurity levels in (Al$_x$Ga$_{1-x}$)As, especially DX-centers, and excitonic effects in absorption spectroscopy. These insufficiencies in the interpretation of the experimental data and the lack of the exact knowledge of the sample parameters are probably the main reason for the large scatter of the published data[10,11]. Most of the available data were obtained in the x-region for which (Al$_x$Ga$_{1-x}$)As is still a direct band gap semiconductor. At x = 0.3 the published values range from $\Delta E_v / \Delta E_g$ = 0.52 to 0.12, where ΔE_g is the total band gap difference and ΔE_v is the valence band offset. These large uncertainties make it necessary to look for alternative experimental methods which have not to rely on sophisticated theoretical models.

2. THE EXPERIMENTAL METHOD

In this contribution we present first experimental results of band offsets determined with a simple and transparent method which we call internal photoemission. This technique has been applied already in 1965 by R. Williams[12] and shortly afterwards by A.M. Goodman[13] in a slightly different version to investigate barrier heights at Si/SiO$_2$ interfaces. The method makes use of the photoexcitation of carriers at the interface of either the valence- or the conduction band over the barrier into the conduction band of the insulator or of the wide-gap material. The internal photoemission is detected via the induced photocurrent normal to the barrier. In the following we describe the first application of this method to study the band offsets at semiconductor-semiconductor interfaces.

The samples used in the present studies were grown by molecular beam epitaxy on (100) oriented n-doped GaAs substrates. The heterostructures consist of a 3 µm thick GaAs buffer layer doped lightly with Si (n $\simeq$ 3 x 10^{16} cm^{-3}), an undoped (Al$_{0.2}$Ga$_{0.8}$)As insulating layer of nominal thickness 120 nm, and a thin ($\simeq$ 30 nm) highly doped (n $\simeq$ 2 x 10^{18} cm^{-3}) (Al$_{0.2}$Ga$_{0.8}$)As top layer. The metal-insulator-semiconductor arrangement was completed by evaporating semi-transparent NiCr Schottky gates on top of the highly doped alloy layer. The gate area was a few mm^2. The doping and thickness parameters were chosen in such a way that the total

(Al$_x$Ga$_{1-x}$)As layer was depleted at zero bias. No persistent photoconductivity was observed, in marked contrast to the situation of usual n-type modulation doped structures with higher x-values. The samples were first characterized by low temperature current-voltage and capacitance-voltage characteristics. At low temperatures, no dark current normal to the interface and Schottky barrier was detectable. From C-V measurements flat-band condition at the interface was determined at approximately zero bias. A more accurate determination of flat-band situation is obtained from the change in sign of the ac-photocurrent, as will be discussed later. The basic experimental set-up for measuring the photocurrent consisted of a single pass grating spectrometer, a continuous light source in the visible and infrared region, a chopper, and a lock-in amplifier. The light was focussed onto the sample which was mounted in a temperature-variable liquid He cryostat with optical windows at the exit slit of the spectrometer. A cut-off filter avoided higher-order radiation passing the spectrometer. The dependence of the ac-photocurrent on photon energy was measured with lock-in techniques, and in the case of infrared frequencies, which showed much weaker effects, signal averaging was used. The energy dependent photocurrent was analyzed for various applied voltages.

3. RESULTS AND DISCUSSION

Fig. 1 gives a general idea of expected and measured onsets of photocurrents in the used heterostructures. The band diagrams for applied voltages U < 0 and U > 0 are shown schematically on the top. For U < 0 both the (Al$_x$Ga$_{1-x}$)As and the GaAs layers are depleted, accompanied by electric fields which cause a separation of photoexcited electron-hole pairs in the same direction. Consequently the sign of the photocurrent is the same in the (Al$_x$Ga$_{1-x}$)As and in

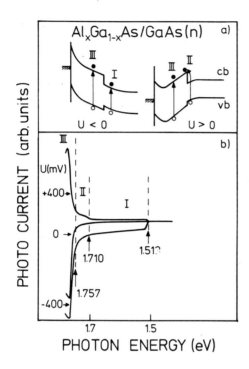

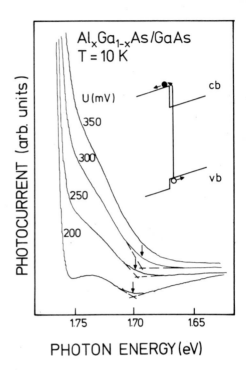

PHOTO CURRENT (arb. units)

PHOTON ENERGY (eV)

PHOTOCURRENT (arb. units)

PHOTON ENERGY (eV)

FIGURE 1
a) Schematic band diagrams of the used hetero-
   stucture under forward and reverse bias conditions
b) Dependence of the induced ac-photocurrent on
   the photon energy for different bias voltages

FIGURE 2
Detailed spectra of the induced photocurrent
due to the process shown schematically in the
insert (internal photoemission)

the GaAs layer. The onsets are expected for
photon energies larger than the energy gaps of
GaAs and (Al_xGa_{1-x})As, respectively. This can be
seen in the lower curve of Fig. 1b). For U =
- 400 mV applied bias a relatively sharp onset
of the ac-photocurrent is observed at 1.513 eV
and a much stronger one at 1.757 eV. The higher
intensity of the photocurrent for energies above
the gap of $(Al_{0.2}Ga_{0.8})$As is caused by the much
stronger electric field close to the surface and
the total separation of the electron-hole pairs.
At zero bias the induced photo-current is negligibly small for energies below the (Al_xGa_{1-x})As
fundamental gap. For higher photon energies the
same sign of the current is observed due to the

surface electric field related to the Schottky
barrier. With positive bias voltages, the sign
of the current has changed due to the inversed
directions of the internal electrical fields.
The signal size of the GaAs related photocurrent
is, however, much reduced. An electron accumulation layer is formed at the interface, accompanied by only weak electric fields in GaAs. The
change in sign of the induced photocurrent just
above the GaAs energy gap allows an exact determination of flat-band condition at the interface.
For the sample studied at present we found
$U_{FB} = (+ 30 \pm 10)$ mV. The band situation for
U > 0 is also shown in Fig. 1a) schematically.
An additional onset of photocurrent is expected
when the photon energy exceeds the separation

of the valence band in GaAs to the conduction band of (Al$_x$Ga$_{1-x}$)As marked by the arrow II in Fig. 1. A careful examination of the measured traces at forward bias indeed shows an additional onset of the photocurrent in between the GaAs and (Al$_x$Ga$_{1-x}$)As energy gap. The intensity is comparable or even larger than the GaAs related photocurrent. The onset energy determines directly the valence and conduction band offset at the interface.

A more careful examination of this additional induced photocurrent is shown in Fig. 2. Experimental traces are displayed for various applied positive gate voltages in the energy region below, but close to the (Al$_x$Ga$_{1-x}$)As gap. The insert shows schematically the process responsible for the induced photocurrent in this energy region. With increasing voltage the onset is found to shift to smaller energies. This is mainly due to the increased electric field at the interface. An exact determination of the onset energy consequently requires a back extrapolation to flat-band voltage. The simplest evaluation of the relevant energies is shown in Fig. 2. The crossing of a linear extrapolation of the background and the additional induced photocurrent is marked by arrows. In Fig. 3 these positions are plotted versus photon energy (full squares). Extrapolation to flat-band voltage leads to an onset energy of (1713 ± 5) meV. This does not change when different evaluations of the original spectra are used like marking the first measurable additional signal or plotting the spectra versus square root of the induced photocurrent. Using the measured energy gap of GaAs and (Al$_{0.2}$Ga$_{0.8}$)As as measured by the onset of the photocurrent and taking into account small shifts due to Franz-Keldysh effect by extrapolating also back to flat-band condition and bulk exciton effects, we find a total energy gap difference of ΔE_g = 244 meV and ΔE_c = 197 meV and ΔE_v = 47 meV. This leads to $\Delta E_c/\Delta E_g$ = 0.8 ± 0.03.

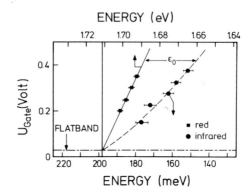

FIGURE 3
Experimentally determined onsets of the photocurrent due to internal photoemission versus gate voltage. Full squares are related to excitations from the valence band, full circles to excitations from the conduction band.

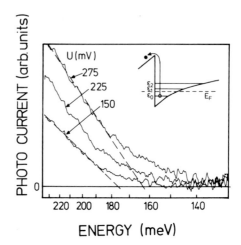

FIGURE 4
Induced photocurrent at infrared energies for different gate voltages. The insert shows the excitation process schematically.

In order to verify and support this interpretation we have also measured the induced photocurrent for photon energies in the infrared close to the energy of the conduction band offset. A current normal to the barrier is expected for energies large enough to excite electrons from the conduction band in GaAs into the conduction band of (Al$_x$Ga$_{1-x}$)As as shown in the insert of Fig. 4. This type of carrier excitation can only be observed for positive gate voltages when an electron accumulation layer is formed at the interface. Experimental traces for various bias voltages are shown in Fig. 4. With the infrared optical cryostat used for those experiments we reached sample temperatures of only 30 K. Therefore the low energy tails of the spectra are smeared out. As expected, photocurrent is only observed for U > 0. Using a similar procedure as described for the experiments in the visible we find onset energies versus gate voltage which are plotted in Fig. 3 as full circles. The error bars in this case are larger. A linear back extrapolation to U_{FB} leads to a conduction band offset of 193 ± 7 meV in excellent agreement with the earlier discussed value. The shift of the onset energies with increasing gate voltage is however much stronger than observed for the excitations of valence band electrons. This is naturally explained by the quantization of electronic states in the conduction band (see insert of Fig. 4). The energy of the lowest subband ε_0 is increasing with increasing gate voltage. At the interface the barrier in the conduction band is lowered at least by ε_0. This effect adds to the shift caused by the increased electric field which is also present in the experiments performed in the visible or near red region. In Fig. 3 the dashed line represents the sum of both shifts where the voltage dependent calculated values of ε_0 have been added to the solid line which represents the linear back extrapolation to U_{FB} of the visible experiments. The effect of the increasing Fermi energy in the conduction band has been neglected in this evaluation. The dashed line falls right on the experimentally determined onset energies of the infrared experiments. This excellent agreement gives us confidence on the reliability of this type of experiments for the determination of band offsets.

4. CONCLUDING REMARKS

We have presented first experimental results of the determination of heterostructure band offsets by internal photoemission both from the valence band and the conduction band. The derived values of ΔE_c and ΔE_v are in disagreement with most of the recent evaluations of band offsets which, however, rely on complicated theoretical models. The simplicity of the new method is evident. For x = 0.2 and (Al$_x$Ga$_{1-x}$)As grown on top of GaAs we find ΔE_c = (197 ± 5) meV and ΔE_v = (47 ± 5) meV. The new method can be applied also to p-type samples, inverted structures, and symmetric GaAs/(Al$_x$Ga$_{1-x}$)As/GaAs structures. Essential, however, is that no persistent photoconductivity is present in the samples. Further internal photoemission experiments on various structures should contribute to a better understanding of band offsets at semiconductor heterostructures and perhaps shine light on the dependence of the offsets on various parameters like composition, growth condition, and strain.

REFERENCES

1/ R. Dingle, Festkörperprobleme 15, 21 (1975), and R. Dingle, W. Wiegmann, and C.H. Henry, Phys. Rev. Lett. 33, 827 (1974)

2/ R.C. Miller, D.A. Kleinman, and A.C. Gossard, Phys. Rev. B 29, 7085 (1984), and R.C. Miller, A.C. Gossard, D.A. Kleinman, and O. Munteanu, Phys. Rev. B 29, 3470 (1984)

3/ P. Dawson, G. Duggan, H.J. Ralph, K. Woodbridge, and G.W. 't Hooft, Superlattices and Microstructures 1, 231 (1985)

4/ M.H. Meynadier, C. Dalalande, G. Bastard, M. Voos, F. Alexandre, and J.L. Liévin, Phys. Rev. B (in press)

5/ H. Kroemer, Surf. Sci. 132, 543 (1983), and J. Vac. Sci. Technol. B 2, 433 (1984)

6/ W.J. Wang, E.E. Mendez, and F. Stern, Appl. Phys. Lett. 45, 639 (1984)

7/ A.D. Katnani and G. Margaritondo,
J. Appl. Phys. <u>54</u>, 2522 (1984)

8/ J.R. Waldrop, S.P. Kowalczyk, R.W. Grant,
E.A. Kraut, and D.L. Miller, J. Vac. Sci.
Technol. <u>19</u>, 573 (1981)

9/ A. Fasolino and M. Alterelli in
"Two-Dimensional Systems", Springer Series
in Solid State Sciences <u>53</u>, 176 (1984)

10/ W. J. Wang and F. Stern, submitted for
publication

11/ G. Duggan, submitted for publication

Quantum States of Confined Carriers in Very Thin Al$_x$Ga$_{1-x}$As-GaAs-Al$_x$Ga$_{1-x}$As Heterostructures

R. Dingle, W. Wiegmann, and C. H. Henry

Bell Laboratories, Murray Hill, New Jersey 07974

(Received 24 June 1974)

Quantum levels associated with the confinement of carriers in very thin, molecular-beam–grown Al$_x$Ga$_{1-x}$As-GaAs-Al$_x$Ga$_{1-x}$As heterostructures result in pronounced structure in the GaAs optical absorption spectrum. Up to eight resolved exciton transitions, associated with different bound-electron and bound-hole states, have been observed. The heterostructure behaves as a simple rectangular potential well with a depth of $\approx 0.88\Delta E_g$ for confining electrons and $\approx 0.12\Delta E_g$ for confining holes, where ΔE_g is the difference in the semiconductor energy gaps.

One of the most elementary problems in quantum mechanics is that of a particle confined to a one-dimensional rectangular potential well.[1] In this Letter, we report the direct observation of numerous bound-electron and bound-hole states of rectangular potential wells, formed by a thin layer of GaAs sandwiched between Al$_x$Ga$_{1-x}$As slabs. The levels are observed by measuring the optical absorption of the central GaAs layer of the structure. The presence of the bound states introduces a series of resolved exciton transitions in the above–band-gap absorption spectrum of GaAs layers less than 500 Å thick. A range of heterostructures, with central GaAs layers as thin as 70 Å, has been studied. The heterostructures produce two attractive potential wells of different depths, one for electrons and one for holes. Analysis of the spectra shows that the wells are extremely rectangular and that the electron and hole well depths are approximately 88 and 12% of ΔE_g, respectively.

The investigation was made possible by two recent developments. The first is the emergence of molecular-beam epitaxy[2,3] (MBE) as a technique for the growth of layers of III-V semiconductors. Our observations demonstrate the great precision of MBE in fabricating thin and uniform layers. The second is the development of selective chemical etches[4] for the removal of the GaAs substrate without damaging the thin epitaxial layers of the heterostructure.

During the last decade there has been intense activity in the study of electrons confined to thin layers. These studies were primarily experiments on metals, superconductors, and metal-oxide-semiconductor devices.[5] Recently, Chang, Esaki, and Tsu[6] reported observing two levels in tunneling experiments involving GaAs-Al$_x$Ga$_{1-x}$As heterostructures, grown by MBE, with GaAs thicknesses of 40–50 Å. This confining layer is thinner than any we have studied and in their experiment the applied electric field distorts the rectangular well into a trapezoidal shape. Nevertheless, the energies they quote are consistent with our more detailed observations.

With the use of MBE, the precision growth of multilayer GaAs-Al$_x$Ga$_{1-x}$As heterostructures has been possible. The usual growth conditions are as follows: vacuum before growth, $\leqslant 1\times10^{-9}$ mm; vacuum during growth, $\sim 1\times10^{-7}$ mm (arsenic); As$_4$ source; {100} GaAs substrate; temperature, 600°C; and semiautomatic shuttering on the Al oven. At our growth rate, 1 μm per hour, the shutter time is equivalent to ~0.5 Å of growth. To increase the GaAs optical absorption, as many as fifty GaAs layers have been grown in a single structure. These GaAs layers are separated by Al$_x$Ga$_{1-x}$As layers which are normally > 250 Å thick. The observed bound states penetrate only about 25 Å into the Al$_x$Ga$_{1-x}$As layers. Consequently, the carriers are tightly bound to individual layers. Hence, we are studying energy levels of a single well, not energy bands of a superlattice. Although it is not possible to measure the electrical properties of the layers themselves, thicker layers, grown under identical conditions, are $p \sim 10^{14}$–10^{15} cm^{-3} (GaAs) and $p \sim 10^{16}$–10^{17} cm^{-3} (Al$_x$Ga$_{1-x}$As). As a consequence of the $\approx 0.12\Delta E_g$ discontinuity in the valence band, the Al$_x$Ga$_{1-x}$As layers will be depleted. Band bending of 1–10 meV in the Al$_x$Ga$_{1-x}$As layers, caused by this depletion, should have a negligible effect on the energy levels. This was confirmed by the fact that the spectral features did not change when the thickness of the Al$_x$Ga$_{1-x}$As layer was varied from 125 to 500 Å. Most data to be discussed here were obtained from structures with $x = 0.2 \pm 0.01$.

If a particle is completely confined to a layer of thickness L_z (by an infinite potential well) then the energies of the bound states are

$$E = E_n + (\hbar^2/2m)(k_x^2 + k_y^2),   (1)$$

where

$$E_n = (\hbar^2\pi^2/2m)(n/L_z)^2, \quad n = 1, 2, 3.   (2)$$

In reality the potential well is finite and the above solutions are inadequate for a quantitative analysis of the data. We have used a computer to obtain the eigenvalues for a well depth V_0. The behavior of energy levels relative to those obtained for $V_0 = \infty$ is shown in Fig. 1. Both the level spacing and the number of bound states decrease as V_0 is decreased, but the $n = 1$ state exists for all positive values of V_0. Thus, for all attractive potential wells, at least one bound state will exist for each type of carrier.

There will be two series of bound-hole states associated with the $\pm\frac{3}{2}$ and $\pm\frac{1}{2}$ valence bands, quantized in the z direction. We will refer to these as the states of the heavy and light hole. The appropriate masses for calculating these states are $(\gamma_1 - 2\gamma_2)^{-1}m_0 \approx 0.45m_0$ and $(\gamma_1 + 2\gamma_2)^{-1}m_0 \approx 0.08m_0$, respectively.[7] These masses deter-

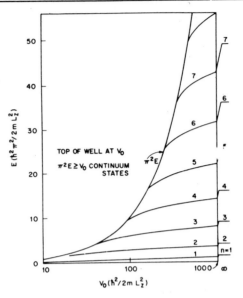

FIG. 1. Calculated energy levels of a particle in a symmetrical rectangular potential well of depth V_0.

mine the k_z (001) dispersion of these bands. Coulomb attraction correlates the motion of the carriers in x and y directions, forming exciton states with peaks in the optical absorption spectrum. States with the same quantum number n have a substantially greater electron-hole overlap. Consequently, excitons with these states will dominate the optical absorption spectrum. Therefore we expect two series of exciton peaks, one series associated with equal-n states of the electron and the heavy hole and one series associated with equal-n states of the electron and the light hole.

Figure 2 displays typical absorption spectra of our structures in the band-edge region of GaAs at 2 K. Roughening the external surfaces with an etch removed all structure due to interference effects. There is negligible absorption in the Al$_{0.2}$Ga$_{0.8}$As layers below 1.75 eV. The trace labeled $L_z = 4000$ Å is typical of high-purity bulk GaAs and it shows none of the quantum effects central to this paper. It does, however, show the dominant excitonic[8] contribution to the bulk GaAs band-edge absorption. The traces $L_z = 210$ Å and $L_z = 140$ Å show well-developed structure above the usual GaAs band gap. Moreover, the exciton peak of bulk GaAs moves smoothly to higher energy as L_z is reduced below 500 Å,

828

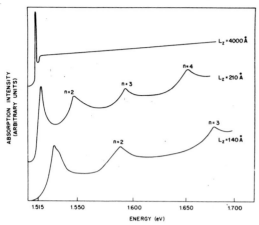

FIG. 2. Typical absorption spectra at 2 K. The traces labeled $L_z = 210$ Å and $L_z = 140$ Å show excitons associated with the electron and hole, each in the nth bound state. For $L_z = 4000$ Å, the absorption coefficient α (cm^{-1}) is about 2.5×10^4 at the exciton peak and $\approx 1 \times 10^4$ in the band-to-band region. Similar values are obtained for the thinner multilayers.

thereby becoming the lowest absorption feature in the quantum limit. The faint doubling of the lowest peak in the $L_z = 140$-Å spectrum is real. This splitting increases as L_z^{-2} and results in two resolved peaks in thinner layers. No doubling is observed for the $n = 2$ peak.

The single exciton series and the doubling of the lowest peak can be explained by assuming that the potential well for holes is weak. Then for layers with small L_z for which two exciton series could be resolved, there is only one bound state for the light holes and consequently only the lowest exciton peak will double. The well depth for holes was determined by fitting the splitting of the lowest peak in a series of samples (see Fig. 3). It was found to be about 28 meV or $\approx 12\%$ of ΔE_g. In making this fit, the known heavy- and light-hole masses and the measured L_z were used. The well depth for electrons must therefore be about $0.88\Delta E_g = 220$ meV.

Figure 3 is a plot of L_z versus the measured exciton energies. L_z was determined, within $\pm 10\%$, from the measured rate of growth of the epitaxial layer. The solid theoretical curves were constructed from the known electron mass $m_e = 0.0665 m_0$,[9] the known heavy-hole mass, and the known well depths. Absolute energies were determined by extrapolating the measured ener-

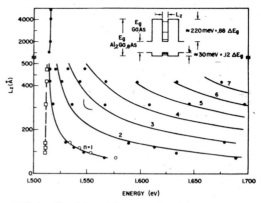

FIG. 3. The data points are a plot of the measured L_z versus the measured exciton energies. The open circles are resolved excitons associated with the light-hole $n = 1$ state. The open squares are the extrapolated energies for $n = 0$. The solid curves are the calculated energies for the excitons associated with the electron and heavy hole, each in the nth bound state.

gies to $n = 0$. These extrapolated energies are shown by the open squares. The calculated energies of the bound states for $n = 1, 2, \ldots$ were then added to these energies. The excellent agreement between theory and experiment is a confirmation that the well depths and masses are correct and that the potential wells are quite rectangular. By varying the depth of the potential well for electrons, we found that a depth of 220 ± 30 meV was required to fit the data in Fig. 3, confirming the value deduced above. Attempts to fit the energies with eigenvalues of nonrectangular wells indicated that the potential step forming the side of the well occurs in less than 5 Å.

The energy of the open squares in Fig. 3 is equal to the band-gap energy of GaAs minus the exciton binding energy. The energy of the squares decreases slowly with L_z and eventually saturates at $L_z < 200$ Å at 1.512 ± 0.001 eV, about 3 meV below the bulk exciton energy of 1.515 eV (2 K), indicating that the exciton binding energy increases from 4 meV[8] to ~ 7 meV as a consequence of carrier confinement. This increase in binding energy agrees quite well with that expected[10] for a three-dimensional exciton as it approaches the two-dimensional limit.

We have benefited from conversations with E. O. Kane, G. A. Baraff, A. C. Gossard, J. C. Hensel, and M. B. Panish. L. Kopf rendered valuable technical assistance.

829

[1]W. Kauzmann, *Quantum Chemistry* (Academic, New York, 1957), p. 183.

[2]J. R. Arthur, J. Appl. Phys. **39**, 4032 (1968).

[3]A. Y. Cho, J. Vac. Sci. Technol. **8**, 531 (1971).

[4]R. A. Logan and F. K. Reinhart, J. Appl. Phys. **44**, 4172 (1973).

[5]Some recent publications are J. M. Rowell, Phys. Rev. Lett. **30**, 167 (1973), and J. Vac. Sci. Technol. **10**, 702 (1973); R. C. Jaklevic, J. Lambe, M. Mikkar, and W. C. Vassell, Phys. Rev. Lett. **26**, 88 (1971), and Solid State Commun. **10**, 199 (1972); R. A. Wheeler and R. W. Ralston, Phys. Rev. Lett. **27**, 925 (1971); D. C. Tsui, in *Proceedings of the Eleventh International Conference on the Physics of Semiconductors, Warsaw, Poland, 1972,* edited by The Polish Academy of Sciences (PWN— Polish Scientific Publishers, Warsaw,

1972), p. 109; A. Kamgar, P. Kneschawek, G. Dorga, and J. F. Koch, Phys. Rev. Lett. **32**, 125 (1974).

[6]L. L. Chang, L. Esaki, and R. Tsu, Appl. Phys. Lett. **24**, 593 (1974).

[7]The dispersion of the bands is discussed by J. C. Hensel and G. Feher, Phys. Rev. **129**, 1 (1963). The band parameters were taken from R. L. Aggarwal, in *Semiconductors and Semimetals,* edited by R. K. Willardson and A. C. Beer (Academic, New York, 1972), Vol. 9, p. 239.

[8]D. D. Sell, R. Dingle, S. E. Stokowski, and J. V. DiLorenzo, Phys. Rev. Lett. **27**, 1644 (1971).

[9]G. E. Stillman, C. M. Wolfe, and J. O. Dimmock, Solid State Commun. **7**, 921 (1969).

[10]E. O. Kane, Phys. Rev. **180**, 852 (1969), and private communication.

Schottky barriers: An effective work function model

J. L. Freeouf and J. M. Woodall

IBM Thomas J. Watson Research Center, Yorktown Heights, New York 10598

(Received 2 July 1981; accepted for publication 18 August 1981)

The experimental observations of metallurgical interactions between compound semiconductor substrates and metallic or oxide overlayers have stimulated a new model of Fermi level "pinning" at these interfaces. This model assumes the standard Schottky picture of interface band alignment, but that the interface phases involved are not the pure metal or oxide normally assumed by other models. For both III-V and II-VI compounds, the barrier height to gold is found to correlate well with the anion work function, suggesting the interface phases are often anion rich. This correlation holds even for cases in which the "common anion rule" fails, and explains both successes and failures of this earlier model.

PACS numbers: 73.30. + y, 85.30.Hi, 68.48. + f

Metal-semiconductor contacts, while crucial to semiconductor devices and studies, are still not well understood. Models relating Schottky barrier heights to metal workfunction, electronegativity, and heats of condensation and reaction with substrate constituents, as well as semiconductor properties such as surface and interface states, heats of formation, polarizability, ionicity, band gap, and defect energy levels can all be found in the recent literature. Some of these models assume the interface to occur abruptly between the two desired phases, while other models require the occurrence of the metallurgical interactions recently observed. The wealth of models available, and the diversity of assumptions they invoke, imply that the fundamentally important aspects of Schottky barrier formation have not yet been established.

In spite of the rich array of various models there are some notable experimental results which remain unexplained. One is the fact that liquid gallium will make a temporary ohmic contact to lightly doped n-type GaAs under the conditions in which the native oxide to GaAs is disrupted exposing clean gallium to an oxide free GaAs surface.[1] With time and exposure to air the contact will become rectifying as predicted by previous models. The second and more convincing result is the Okamoto et al. study[2] of Schottky barrier heights for the Al-(GaAs-AlAs) interface prepared by molecular beam epitaxy. They find barrier heights, particularly to AlAs, which are significantly different from those predicted by previous models and which are significantly different from those for Au-AlAs.[3] We have reexamined earlier models in light of the recent observations of interface intermixing and propose that the simple Schottky picture of work-function matching—if coupled with mixed phases at the interface—appears to account for a large amount of experimental data and suggests directions for research in controlling Fermi level pinning.

Our model begins with that of Schottky,[4] which assumes an ideal metal-semiconductor interface, i.e., one in which the interface is inert and there are no appreciable surface or induced interface states in the semiconductor. The Schottky barrier height is given by[4]

$$\phi_{bn} = \Phi_M - \chi,$$
$$\phi_{bp} = (E_G/q) + \chi - \Phi_M,$$

where $\phi_{bn} (\phi_{bp})$ is the Schottky barrier height to an n-type (p-type) semiconductor, Φ_M is the metal work function, q is the electron's charge, and χ is the electron affinity of the semiconductor. Thus, for the ideal case and for a given semiconductor, ϕ_b should be determined by the metal work function. Unfortunately, this is not the case for GaAs and many other semiconductors.[3]

Our model, called the effective work function model (EWF), suggests that the Fermi level at the surface (or interface) is not fixed by surface states but rather is related to the work functions of microclusters of the one or more interface phases resulting from either oxygen contamination or metal-semiconductor reactions which occur during metalization. The theory requires that when a metal is deposited, or an oxide is formed, there is a region at the interface which contains a mixture of microclusters of different phases, each having its own work function. We should therefore modify the "ideal" surface discussion as follows:

$$\phi_{bn} = \Phi_{\text{eff}} - \chi,$$

where Φ_{eff} is an appropriately weighted average of the work functions of the different interface phases. Thus the measured ϕ_{bn} can depend somewhat on the measurement technique, i.e., C-V or I-V.

For most of the compounds under discussion, metalization and/or oxidation results in a condition in which Φ_{eff} is due mainly to Φ_{Anion}, the work function of the anion; we suggest that this occurs as a result of one or both of the following reactions:

Anion oxide + Compound→Anion + Cation oxide,

M + compound→(Anion or Metal-Anion complex) + (M-Cation).

The condition for driving this reaction to the right and hence generating excess Anion at the interface is that the Gibbs free energy ΔF is negative. Such oxide reactions have been examined,[5] and excess group V anions have been experimentally observed when ΔF is negative, i.e., for GaAs, InAs, and InSb.[6,7] This has not been observed when ΔF is positive, i.e., for GaP.[7] It is interesting to note that for InP, $\Delta F \approx 0$; it has been possible to form metal-oxide semiconductor field-effect transistor (MOSFET) structures using SiO_2, which exhibit a low interface state density[8] on this

0003-6951/81/210727-03$00.50

TABLE I.[a] Au Schottky barriers.

Compound	$E_G/q + \chi$	ϕ_{bp}	$\Phi_{Au} = 5.1-5.5$ eV[b] $E_G/q + \chi - \phi_{bp}$	Φ_{Anion}
GaP	5.86[c]	0.96[j]	4.9	5.0[r]
InP	5.75[c]	0.85[k]	4.9	5.0[r]
AlAs*	5.6–6.0[d,e]	0.9[l] (1.4)**[m]	4.7–5.1 (4.2–4.6)**	5.0[r](4.8)[s] ($\Phi_{Al} = 4.0$–4.3[b,r])
GaAs	5.5[f]	0.5[l]	5.0	5.0[r](4.8)[s]
InAs	5.3[f]	0.3–0.5[n,o]	4.8–5.0	5.0[r](4.8)[s]
AlSb*[g]	5.2[g]	0.54[l]	4.7	4.8[r](4.7)[t]
GaSb	4.76[f]	0.1[l]	4.7	4.8[r](4.7)[t]
InSb	4.77[f]	≈0.1[l]	4.8(77 K)	4.8[r](4.7)[t]
ZnO	7.92[h]	2.7[l]	5.2	7.3[r]
ZnS	7.5[h]	1.6[l]	5.9	5.74[r]
CdS	7.21[h]	1.63[l]	5.58	5.74[r]
GaS*	6.5[l]	0.75[p]	5.75	5.74[r]
ZnSe	6.76[h]	1.31[l]	5.45	5.7[r]
CdSe	6.65[h]	1.21[l]	5.44	5.7[r]
GaSe*	5.4[l]	0.5[p]	4.9	5.7[r]
ZnTe	5.79[h]	0.65[q]	5.14	4.88[r]
CdTe	5.72[h]	0.78[l]	4.94	4.88[r]
GaTe*	4.95[l]	0.45[p]	4.5	4.88[r]

*Does not obey common anion rule.
**Al-AlAs barriers.
[a]Band gaps were taken from A. G. Milnes and D. L. Feucht, *Heterojunctions and Metal-Semiconductor Junctions* (Academic, New York, 1972), p. 8.
[b]Reference 14.
[c]J. Van Laar, A. Huijser, and T. L. Van Rooy, J. Vac. Sci. Technol. **14**, 894 (1977).
[d]R. Dingle, A. C. Gossard, and W. Wiegmann, Phys. Rev. Lett. **4**, 1327 (1975).
[e]A. H. Nethercot, Jr., Phys. Rev. Lett. **33**, 1088 (1974).
[f]G. W. Gobeli and F. G. Allen, Phys. Rev. **137**, A245 (1965).
[g]T. E. Fischer, Phys. Rev. **139**, A1228 (1965).
[h]R. K. Swank, Phys. Rev. **153**, 844 (1967).
[i]R. H. Williams and A. J. McEvoy, Phys. Status Solidi A **12**, 277 (1972).
[j]B. L. Smith and M. Abbott, Solid-State Electron. **15**, 361 (1972).
[k]B. L. Smith, J. Phys. D **6**, 1358 (1973).
[l]C. A. Mead, Solid State Electron. **9**, 1023 (1966).
[m]Reference 2.
[n]K. Kajiyama, Y. Mizushima, and S. Sakata, Appl. Phys. Lett. **23**, 458 (1973).
[o]J. N. Walpole and K. W. Nill, J. Appl. Phys. **42**, 5609 (1971).
[p]S. Kurtin and C. A. Mead, J. Phys. Chem. Solids **30**, 2007 (1969).
[q]W. D. Baker and A. G. Milnes, J. Appl. Phys. **43**, 5152 (1972).
[r]K. W. Frese, Jr., J. Vac. Sci. Technol. **16**, 1042 (1979).
[s]As on native oxide of GaAs, J. L. Freeouf and J. M. Woodall (unpublished).
[t]J. L. Freeouf, M. Aono, F. J. Himpsel, and D. E. Eastman, J. Vac. Sci. Technol. (to be published).

material. This is consistent with our model that would predict either no or very little excess free phosphorus at the interface. A GaP MOSFET structure with low interface-state densities would be predicted, since no free P is expected at this interface. It should also be noted that for GaAs it is well known that MOSFET structures have notoriously high "interface-state densities" (10^{13}–10^{14} cm^{-2}) and that excess arsenic is usually observed at the interface.[9] Again this is consistent with the model, since the ϕ_{bn} expected for the As-GaAs interface is about 0.8 eV (the usually observed barrier height for most metal depositions as well). Since workers have reported a large density of mid-gap states for MOSFET GaAs structures, the model would ascribe these "states" to arsenic clusters at the interface which act as Schottky barrier contacts with $\phi_{bn} \approx 0.8$ eV embedded in an oxide matrix.

Excess anions can also be generated by reaction of metals with the substrate. For example, it is known that Au deposited on GaAs and GaP results in excess Ga in the Au film.[10] Also preliminary phase diagram data[11] show that an arsenic phase is expected at equilibrium for Au-GaAs and Au-InSb. Thus a knowledge of both oxide and reactive metal chemistry should enable accurate predictions of the transport properties of metal-semiconductor devices (including Schottky barrier heights).

The current status of this model[12] is shown in Table I, which lists the experimentally derived values of ϕ_{bp} and $E_G/q + \chi - \phi_{bp}$ for Au/III-V and Au/II-V contacts. There are three points to note in this table. First, the Schottky model ($E_G/q + \chi - \phi_{bp} = \Phi_{Au} \equiv 5.1-5.5$ eV) is not obeyed. Second, the EWF model agrees well, as expected, for these data by assuming Φ_{eff} to be dominated by Φ_{Anion}, i.e., $\Phi_{Anion} = E_G/q + \chi - \phi_{bp}$. Third, the common anion rule[13] is not obeyed for AlAs and AlSb. We believe that the common anion rule followed more directly from the anion than initially suggested; in fact, we believe that this rule followed from the formation of microclusters of anions at the interface which dominated the Fermi level position determinations cited . The common anion rule asserts that ϕ_{bp} depends only upon the semiconductor anion. Since, in our model, $\phi_{bp} = ((E_G/q) + \chi) - \Phi_{\text{Anion}}$, a common anion would lead to a constant ϕ_{bp} only for a constant $E_G/q + \chi$; Table I shows that those cases following the common anion rule also obey that constraint.

The EWF model also explains such departures from "normal" behavior as the Al–AlAs result,[2] also shown in Table I. For the Al-AlAs case, the metalization was performed in an ultrahigh vacuum molecular beam epitaxy system, where the AlAs surface was very clean, and subsequently annealed. Under these conditions, excess As should react with Al rather than forming microclusters of As. Thus, it is expected that Φ_{eff} should be dominated by $\Phi_{\text{Al}} = 4.0\text{–}4.3$ eV. We believe that this explanation is correct, since $\chi + \phi_{bn} \approx 4.2\text{–}4.6$ for this case, which is much closer to Φ_{Al} than to Φ_{As}. Similarly, the Ga-GaAs ohmic contact mentioned earlier can be explained since $\Phi_{\text{Ga}} = 4.36 \text{ eV}$ (Ref. 14) and $\phi_{bn} = 0\text{–}0.3$ (for ohmic behavior); $\chi_{\text{GaAs}} + \phi_{bn} = 4.1\text{–}4.4 \approx \Phi_{\text{Ga}}$.

The electrical behavior of most covalent semiconductor interfaces is dominated by the apparent pinning of the Fermi energy level at the interface. We are proposing a model of this behavior which assumes work function matching and (typically) mixed phase behavior at the interface; "pinning"

normally observed is shown to follow naturally from microclusters of anions at the interface, which are expected from chemical arguments and observed in some recent experiments.

[1] J. M. Woodall, C. Lanza, and J. L. Freeouf, J. Vac. Sci. Technol. **15**, 1436 (1978).

[2] K. Okamoto, C. E. C. Wood, and L. F. Eastman, Appl. Phys. Lett. **15**, 636 (1981).

[3] C. A. Mead and W. G. Spitzer, Phys. REv. **134**, A713 (1964).

[4] W. Schottky, Zeitschrift fur Physik **118**, 539 (1942).

[5] J. L. Freeouf, J. M. Woodall, and L. M. Foster, Bull. Am. Phys. Soc. **26**, 285 (1981).

[6] R. L. Farrow, R. K. Chang, S. Mroczkowski, and F. H. Pollak, Appl. Phys. Lett. **31**, 768 (1977).

[7] G. P. Schwartz, G. L. Gualtieri, J. E. Griffiths, C. D. Thurmond, and B. Schwartz, J. Electrochem. Soc. **127**, 2488 (1980).

[8] L. G. Meiners, D. L. Lile, and D. A. Collins, J. Vac. Sci. Technol. **16**, 1458 (1979).

[9] H. H. Wieder, J. Vac. Sci. Technol. **15**, 1498 (1978).

[10] A. Hiraki, K. Shuto, S. Kim, W. Kammura, and M. Iwami, Appl. Phys. Lett. **31**, 611 (1977).

[11] R. S. Williams (private communication).

[12] A similar model has been applied to the silicide-silicon interface; see J. L. Freeouf, Sol. State Commun. **33**, 1059 (1980).

[13] J. O. McCaldin, T. C. McGill, and C. A. Mead, Phys. Rev. Lett. **36**, (1976).

[14] H. B. Michaelson, J. Appl. Phys. **48**, 4729 (1977).

179

Surface Science 168 (1986) 518–530
North-Holland, Amsterdam

DEFECTIVE HETEROJUNCTION MODELS

J.L. FREEOUF and J.M. WOODALL

IBM T.J. Watson Research Center, P.O. Box 218, Yorktown Heights, New York 10598, USA

Received 10 June 1985; accepted for publication 14 June 1985

Fermi-level pinning behavior has been observed at the free surface, oxide interface, metal interface, MBE grown surface, stop-regrown homojunction, and misfit-dislocation pinned heterojunction of GaAs. Theories of such behavior are numerous and disparate. Theories of ideal heterojunction band offsets are less diverse, but have still not converged to a single mechanism. Recent studies of heterojunctions suggest that the conduction-band offsets are relatively independent of interface Fermi-level position, including situations in which the interface Fermi-level appears to be strongly "pinned". In "ideal" heterojunctions, the conduction-band offsets and bulk doping determine interface Fermi-level location; among other results, this mechanism allows the two-dimensional electron gas at modulation-doped AlGaAs–GaAs heterojunctions. If "pinned" heterojunctions involve charge densities comparable to those inferred for Schottky barriers, then the pinning interface states should set up a dipole sufficient to alter the band offsets; the interfacial band alignment should then be dominated by the alignment of the pinning states, rather than that of the bulk bands. The experimentally suggested lack of sensitivity of band offsets to changes in pinning at heterojunction interfaces suggests that the mechanisms involved in band line-ups at "ideal" heterojunctions may be related to those mechanisms involved in Fermi-level "pinned" systems. A simple mechanism is that of work function matching, in which the transition to "pinned" behavior involves the generation of a new material at the interface; since the work function difference between heterojunction materials is unaffected, the band offset would likewise be unaffected. The effective work function model explains the pinning phenomenon on the basis of anion clusters, which have been observed at all classes of pinned interfaces involving III–V compounds. The application of other models to both pinned and unpinned interfaces is less clear; more information is required. Pinning models which involve interface state densities within each semiconductor must address the lack of sensitivity of band offset to different interface Fermi-level locations.

1. Introduction

Heterojunction-based device structures offer very intriguing possibilities [1,2]. The possibility of separately "biasing" electrons versus holes or of confining one or both classes of carriers without applied bias and separately from the source ions permits new device structures; implementing such structures in III–V materials to exploit greater mobilities and possible ballistic transport could permit the increased fabrication costs to be justified by improved device performance. Furthermore, heterojunctions provide an ordered single-crystal interface between two well-characterized semiconductors; their fabrication is becoming increasingly controlled and routine via several techniques, thereby permitting extensive experimental characterization of the formation and electronic properties of these interfaces. Theoretical analysis of such interfaces faces few of the difficulties encountered in similar treatments of Schottky barriers; in at least some cases, the interfaces are known to be ordered and abrupt on an atomic scale, and the crystal structure and band structures of the two sides of the interface are quite similar. We regard these interfaces as the likely first candidates for "complete" understanding, at least at the level currently achieved for the cleaved free surface of III–V compound semiconductors.

However, the hoped-for understanding has not yet been achieved; in fact, there is little agreement on so fundamental an issue as the "correct" division of the total change in band gap (between the two semiconductors) between a valence-band offset and a conduction-band offset. The literature contains many studies asserting that the band offsets depend upon growth conditions, growth sequence [3–5], and crystallographic orientation. Such studies typically conclude that, since more than one band offset is possible, a simple prediction ignoring such complications is demonstrably false and unwarranted [3]; furthermore, authors still willing to rashly ascribe some validity to a simple "zeroth-order" approximation [1] often assert that a work function matching approach [6] is demonstrably less accurate than a "first-principles" bulk approximation.

In this paper, we wish to suggest that simple models are both useful and valid – both as initial guides in device design and in terms of addressing theoretical aspects of interfaces. Furthermore, we assert that at present a work function matching scheme of heterojunction band offsets is the "simple model" of choice; it is at least as consistent with current acceptable data as any competitor, and it describes a wider range of experimental conditions than do alternatives. In support of this suggestion, we shall first briefly describe our modified work function matching scheme for "pinned" compound semiconductor surfaces and interfaces – and explain why we feel these modifications are justified. Next, we shall discuss why this model is also consistent with band offsets as currently measured. We shall exclude some interfaces from current attack by our model; we add that such an exclusion should also be the current approach of theorists wishing to achieve a first-principles full understanding of these interfaces. Finally, we shall point out that the interfaces in which a range of band offsets have been observed may well not be of any interest to device designers, since the electrical properties of such interfaces are unknown. In fact, the alteration of band offsets (at least under most current models) would appear to require a charged dipole and/or graded, mixed, disordered and/or roughened interface; all of these possibilities offer reasons for concern to a device designer, at least for specific classes of devices.

2. Work function matching

Conceptually the most straightforward mechanism for understanding of the bands of two dissimilar materials is to seek some means of aligning them both with respect to the same reference. If we make this reference the vacuum level, then we are invoking a work function matching scheme. Advantages of such a choice include thermodynamic validity and observability. A detailed treatment of the thermodynamic basis for work functions has been available for many years, as have many caveats which we shall discuss shortly [7]. The observability of work functions by many techniques has been exploited for an even longer time; the fruits of such labor are the availability of data concerning the work functions of many materials under many conditions as measured by many techniques. For many materials, the work function of specific, ordered surfaces are known to two or three decimal places, without debate. Objections to the application of such data to schemes matching two work functions may be divided into three classes:
(1) They don't work [3].
(2) The difference of two large numbers to derive a small number is risky [1].
(3) The surface/interface dipole invalidates the thermodynamic treatment [8].
We shall present data suggesting that such schemes do work fairly well, if one properly allows for interface metallurgical effects. We agree that the limited precision in defining the difference in two large (≈ 5 V) numbers to infer a small (≤ 1 V) number limits the utility of the method; however, this is a practical question, not one of basic concepts [9]. The primary conceptual objection to this scheme is one of surface and interface dipoles and their effect upon both measured work functions and actual interface band alignments. This question is difficult to address experimentally, since separation of "bulk" work functions from surface dipole effects is basically not possible [7,10]; the magnitude of variation of work functions from one crystal surface to another suggest possible dipole-induced errors of ≈ 1 V [11] suggesting that this effect could dominate the desired answer in a fashion that cannot be addressed by simple techniques. Theoretically, the answer is "maybe": some recent models describing band line-ups of either metals or semiconductors on semiconductors have ascribed the *entire* observed band line-ups to interface dipole-driven effects [12,13]; another recent calculation suggests that interface dipoles deviate from surface dipoles by ≤ 0.1 V [14], suggesting that the *intrinsic* error to the concept may be only comparable to the present measurement precision.

3. Effective work function model

The question of why the model "doesn't work" despite these conclusions must now be addressed; in point of fact, we believe that the model [15] does work! We first discuss metal contacts on compound semiconductors, in which purported disagreement is larger. We note that in-situ surface-sensitive studies of the formation of such interfaces have often observed metallurgical disruption of the semiconductor [16], giving rise to several "explanations" of Fermi-level pinning based upon such disruption [15,17]. The effective work function (EWF) model assumes such disruption occurs, leading to mixed phase interfaces. On chemical grounds, the interface between most III–V compounds and oxygen or Au should lead to cation compounds and/or alloys with the overlayer, along with free excess anion. The work function of this postulated anion corresponds to the work function observed for the barrier height.

To summarize, the EWF model assumes that, at pinned interfaces, the interface Fermi-level location is determined by the work function of the anion released from the substrate by the processes used to generate the interface. A major implication of this model is that the Fermi-level position is relatively independent of the bulk metallurgy, since the anion at the interface is the determing factor. Another implication is that the observed pinning position should have a work function ($\phi = \chi + \phi_{bn}$) equal to that of the anion. This prediction is compared with experiment in fig. 1 for some materials; note that

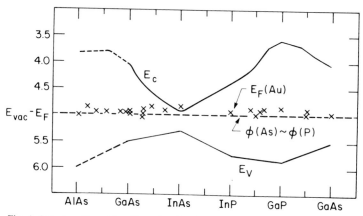

Fig. 1. Interface Fermi-level location for (mostly) Au contacts on various compounds and alloys. Except AlAs affinity, the electron affinities, band gaps, and barrier heights for compounds from references in ref. [15]; AlAs electron affinity derived by *assuming* the EA model for heterojunction band alignments. All alloys *assume* that the valence bands vary linearly with composition between the end-point compounds; the barrier heights for alloys are from ref. [39] (GaAsP), ref. [40] (InGaAs), ref. [41] (InGaP) and ref. [42] (AlGaAs).

$\phi_P \approx \phi_{As}$, so the model predicts a constant work function in this series, as is observed. In ref. [15] we compare the predictions with experiment for other materials. Pinning behavior has been observed on free surfaces and at III–V interfaces with oxides, metals, and misfit dislocations [17]; anion clusters or excesses have been observed on free surfaces [18], metal interfaces [19], oxide interfaces [20], and at dislocations [21]. We believe the correlation of measured barrier heights to this simple model, which is based upon effects that have actually been observed to occur in at least some instances, is highly suggestive. We further note that the mixed phase assumption embedded in our model provides a natural explanation for the discrepancies between different measurement "definitions" for a single barrier height, since different techniques will average via *different weighting functions* over the mixed phase interface. However, we add that valid questions concerning the model include:

(1) The work function of small anion clusters, and in fact the size distribution of such clusters.

(2) The persistence of these "pinning" positions under conditions where surface probes suggest depletion of the anion.

(3) The manifold parameters available to vary alloy and/or compound distribution to "explain" deviations from our modified "anion rule".

Further, all models based upon metallurgical interactions must assume a "universality" of such interactions; given the small degree of disruption necessary for such a model to "explain" experimental results ($\approx 10^{12}$ cm^{-2} for chemisorbed surfaces and $\approx 10^{14}$ cm^{-2} for metal–semiconductor interfaces), the exclusion of such effects at any interface may not be possible with current techniques. While we are uncomfortable invoking a "Maxwell's demon", we advocate use of the model for its practical utility, its chemical intuition, and the *observation* of our proposed demon in the same classes of interfaces that are pinned.

We note that the defect model [17] also must assume "universality", has a large number of defect parameters, assumes "demons" never observed at these interfaces, and must further assume that interface defects are not subject to the metal screening demonstrated by Heine [22] to apply to surface states at metal–semiconductor interfaces. Finally, the defects must be stable under very substantial fields – of order $\approx 2 \times 10^7$ V cm^{-1} [23]. Furthermore, the calculations for such levels [24] demonstrate that the energies are a strong function of whether the defects are at the surface or in the bulk; this suggests that pinning positions should differ between chemisorbed surfaces and stop-regrow interfaces.

4. Electron affinity rule

The original "rule" for ascertaining heterojunction band alignments was due to Anderson [6]; this rule is simply work function matching, and is typically called the Anderson rule or the electron affinity rule (EA). Despite the compelling nature of the thermodynamic argument in favor of such a description, the electron affinity rule has come under strong attack for the same reasons as the work function matching scheme for metal–semiconductor contacts. Further objections specific to the heterojunction literature refer to various observed "non-linear" results [3–5]; this term is used to refer to aspects of interfaces not addressable by any model that defers treatment of interface dipoles, grading, etc. The magnitude of such effects seems to be large; in favorable cases heterojunction band offsets have been observed to vary by ≈ 0.3 V depending upon which semiconductor is deposited upon the other [3]. It is clear that the electron affinity rule cannot predict such behavior; it is less clear that predicting such behavior is a desirable aspect of even a first-order theory.

At present, one cannot exclude the possibility that such effects are extrinsic, and not related to abrupt ideal heterojunctions at all. These effects have been reported for heterojunctions involving Ge with some compound semiconductors; such interfaces are experimentally attractive because germanium is easy to deposit "stoichiometrically". Problems with comparing such experiments with any theory include:

(1) The theoretical "certainty" of atomic rearrangement/intermixing at most such interfaces [25].

(2) The uncertain growth morphology on the (110) growth plane less susceptible to the above effect [26].

(3) The absence of electrical characterizations of such heterojunctions grown in the same systems under the same conditions as were the interfaces for which a band offset was measured.

The experimental situation becomes even more uncertain given the current status of the device world's favorite heterojunction, AlGaAs/GaAs. The "best guess" value for that band offset has recently been revised by of order 0.3 V [27]. The revision of the band offset in this case is largely due to the application of new techniques for determining the offset, coupled with re-evaluation of the theoretical basis for the original technique. Although such revisions should not have impacted the surface studies, the only previous surface study of that interface was consistent with the old "right" answer [4]. We feel that this result reflects the quality of those samples, which unfortunately were not of state-of-the-art caliber even for that time. Those results were obtained for growth in the (110) orientation, an orientation notoriously difficult to grow well [26], at an unusually low growth temperature; a year after those measurements, segregation or spinodal decomposition of AlGaAs alloys grown in that orientation at that growth temperature was reported by another group [28].

However, we must work with what we have; in fig. 2 we compare the electron affinity model to the selected heterojunction band offset values presented by Kroemer in a recent critical review [29], with the AlAs value altered to reflect the current understanding. In fig. 3 we present the same comparison after exclusion of the homopolar/heteropolar interfaces for which we have strong questions. For comparison, we include the Harrison atomic orbital model (HAO) [30], and an empirical model of Katnani and Margaritondo (KM) [31]. In fig. 3 we have added a point with question marks; AlSb/GaSb appeared a possibly useful device interface [2], and the HAO model disagrees with the EA model. Current data are intermediate, but place only a lower bound on the discrepancy [32] with the HAO model; the data are not yet sufficiently complete to warrant strong conclusions, however. Based on the new AlAs/GaAs, and possibly the AlSb/GaSb, values, we suggest that the EA model fits the data at least as well as the HAO model; we suggest that the HAO model incorporate modifications to deal with the potential of aluminum compounds differently. We further note that the empirical model of Katnani and Margaritondo seems in error for InAs/GaSb, CdS/InP, and (possibly) GaAs/InAs. The EA model shows errors only for Si/Ge and for ZnSe/GaAs; while the former interface was found to be inde-

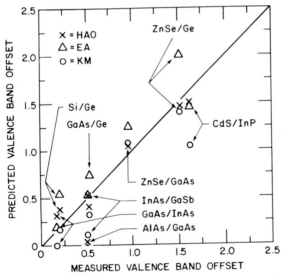

Fig. 2. Heterojunction band offsets as predicted by refs. [6, 31, 30] compared with "critically selected" values of ref. [8]. Electron affinities for compounds from references cited in ref. [15]; those for sillicon and germanium from ref. [43].

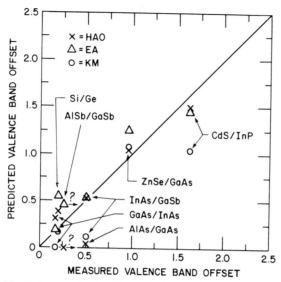

Fig. 3. Same as fig. 2, without the homopolar/heteropolar heterojunctions.

pendent of deposition order (the latter was not studied in the inverse order), there are insufficient data to establish the electrical perfection and abruptness of these interfaces. We note that the experimental [31] offset plotted implies that the band gap of Ge is entirely contained within that of Si ("normal offset") and that $\Delta_v \leq \Delta_c$. Some strained layer superlattice studies suggest [33] that $\Delta_v \gg \Delta_c$. Further, preliminary data on strained layer SiGe/Si super-lattices suggest a "staggered" configuration [34], as the EA model predicts, but of unknown magnitude; further, we do not know whether this staggering would be observed in the absence of strain, which is the configuration being predicted by all the models. From fig. 2 we could infer a general failure of the EA model to explain heterojunction offsets involving germanium; should this failure persist in "perfect" heterojunctions, perhaps a calculation of ex-pected "dipole changes" for this indirect small band-gap semiconductor might be appropriate.

In general, however, we believe that current "best data" on "best ordered" and understood heterojunctions permit the EA rule to be applied with at least the confidence due other first-order theories. We feel that a large discrep-ancy between the EA rule and experiment or theory should stimulate study to ascertain the cause of the discrepancy. We do not feel the EA rule will always be the state of the art heterojunction band offset theory; in fact, we encourage theorists to attack this problem on systems currently under experimental con-trol (AlAs/GaAs, GaSb/InAs) where interfaces of near ideal abruptness and order appear feasible. We note, however, that since the band-gap discontinuity and its allocation between the valence and conduction bands are a crucial aspect of the issue, theories which properly obtain the band gaps (including subsidiary minima at L and X) would be highly desirable.

5. Pinned heterojunctions

A work function matching picture appropriate to both pinned and unpinned heterojunctions is clearly possible; the pinned heterojunction is assumed to have some metallurgical interactions giving rise to the excess anion postulated by the effective work function model. Such an interface will have band bending in both semiconductors permitting the interface Fermi level to coincide with the work function of the postulated anion. For an interface between materials using different anions, such as InAs/GaSb, one would presumably have some form of alloy of the two anions, with a possibly different work function. However, unless there are two *layers* of anions each contacting only the other and the semiconductor (i.e. InAs/As/Sb/GaSb) there should be a single interface work function; both semiconductors should line up with respect to this single work function by standard work function matching arguments, and the *difference* between the two semiconductor band alignments should again correspond to the difference in their electron affinities, so the band offset should be independent of pinning or interface Fermi level location. For this argument to apply, we need not even assume the pinning is due to excess anion; *any* single work function interface layer leads to this conclusion. This does assume that the interface disruption does not lead to graded or intermixed interfaces leading to unusual dipoles.

Such an argument leads to the conclusion that heterojunction band offsets should be independent of interface Fermi level location; this argument applies to both pinned and unpinned heterojunctions. Many heterojunction band offset models assume that interface Fermi level position has no effect upon band alignment, but it is not clear to us that this is true for all such models. The model of Tersoff discusses band alignments in terms of dipole formation at the interface; this dipole involves states tunneling from one material into the other. The "neutrality level" of this theory is somewhere in mid gap, and the imaginary states (and their decay length) used to derive a dipole are related to this point; we would expect their occupation to be dependent upon Fermi level position, especially as the Fermi level moves across the band gap, as it can do in an ideal heterojunction interface such as the AlAs/GaAs case, where both electron and hole two-dimensional gases have been observed. Further, we do not understand why the band offset for Ge/GaAs appears independent of Fermi level location when the interface Fermi level is varied from well within the fundamental band gaps to the case where the Ge is apparently degenerate n-type even at the interface [34]. A final problem with the model of Tersoff [13] is that it apparently assumes that dielectric screening affects both electrostatic *and chemical* potential changes [10]; we are unable to reconcile band offsets depending upon, e.g., growth sequence, by ≈ 0.3 V [3] with such screening affecting all possible sources of chemical potential shifts.

We believe that these arguments also have some relevance to other models of Fermi-level pinning; a standard current model for Fermi-level pinning is that of metallurgical interaction-induced defects [17]. Within this model, both the pinning position [17] and the sign of the charge induced in these defects is a function of the bulk doping [35]. Defect densities adequate to pin Schottky barriers would also strongly alter band offsets [35] and align pinned Fermi levels rather than bulk bands [36]; we would predict that a pinned n–p heterojunction should exhibit a different band offset than would a pinned p–n heterojunction grown under the same conditions and the same order. Both values should be different from that of an unpinned heterojunction. With some effort, one should be able to perform the same band offset tuning for pinned homo-junctions! We suggest that some effort in realizing this intriguing possibility is warranted; aside from intrinsic interest in interface state densities arising solely from band offsets and totally independent of material and band structure changes (since there are none!), such studies should be directly relevant to an important device issue: namely, that of stop-regrow epitaxial growth techniques. Performing such studies on GaAs(100) surfaces and interfaces should strongly impact both device technology and our understanding of pinning mechanisms. We note that this raises the possibility of experimentally performing the gedanken experiment discussed by Tersoff of fabricating a heterojunction between two semiconductors of identical band structure but containing a band offset [13]. However, we must note that this inferred offset arises from an *electrostatic* potential shift; unlike chemical potential shifts [10], the screening discussed by Tersoff [13] should apply to such abrupt potential shifts, possibly reducing the magnitude by a factor of the dielectric constant of the semiconductor. Furthermore, the postulated existence of such strong fields at the interface could well destabilize the pinning centers; this suggests that stop-regrow interfaces without Fermi-level pinning may prove simpler to achieve at n–p junctions than at n–n or p–p junctions!

We have performed some preliminary measurement on misfit-dislocation pinned heterojunctions [37]. These measurements are fully consistent with the band offset being independent of the existence of interface pinning; the observed rectification can be modelled semiquantitatively (including its temperature dependence) in terms of pinning at the dislocation and screening of this potential along the interface away from the dislocation. The point is that there is no indication of a dependence of band offsets upon Fermi-level position within the interface. We have not yet attempted to repeat this study for p–p, or the p–n and n–p possibilities, so we cannot rule out the occurrence of these intriguing possibilities of band-offset tuning.

A further relationship between Fermi-level pinning and heterojunction band offsets has been noted by Katnani and Margaritondo [31]; this is that, if one "lines up" the pinned Fermi-level positions on two semiconductors, the resultant band offsets are in reasonably good agreement with experiment. This relationship between Schottky barrier heights and heterojunction band offsets follows naturally from the work of Tersoff [13], leading to answers presumably similar to those of ref. [31]. We note that the effective work function model, coupled to the EA model, requires such behavior for cases in which the pinning material (usually the anion) for the two materials has similar work functions. For anions having different work functions, our model would predict a discrepancy between Schottky barrier heights and heterojunction band offsets; this error is ≈ 0.1–0.3 V for GaSb/InAs, and nearly 0.75 V for CdS/InP, the two discrepancies between the KM model and experiment noted earlier in this paper.

6. Conclusion

In this paper we have demonstrated the ability of work function matching schemes to organize and predict experimental band alignments at Schottky barriers and at both pinned and unpinned heterojunctions. We have raised some questions as to the ability of alternative first-order models of such interfaces to account for some general aspects of these interfaces. Further, we have suggested several classes of experiments that appear likely to elucidate the role of at least some classes of pinning theories.

We note that no current model, including the work function models espoused herein, is truly sufficient for device design needs. A convincing, first-principles treatment of the underlying band structure is likely required to fully understand the manifold details of carrier confinement and transport both parallel and perpendicular to these very ordered interfaces. We strongly suggest that theorists address some specific interfaces where interface structure appears simple and well defined, and data appear likely to be both reproducible and available. However, we also point out that theories without an adequate description of band gaps are unlikely to provide convincing first-principles allocation of band offsets, let alone a detailed description of transport, tunneling, and carrier confinement.

Acknowledgement

We gratefully acknowledge helpful discussions with S. Iyer, T. Jackson, W. Wang, S. Wright, and J. Van Vechten.

References

[1] See, e.g., H. Kroemer, Surface Sci. 132 (1983) 132, for an excellent overview.

[2] F. Capasso, J. Vacuum Sci. Technol. B1 (1983) 457.

[3] R.S. Bauer, P. Zurcher and W. Sang, Jr., Appl. Phys. Letters 43 (1983) 663.

[4] J.R. Waldrop, S.P. Kowalczyk, R.W. Grant, E.A. Kraut and D.L. Miller, J. Vacuum Sci. Technol. 19 (1981) 573.

[5] S.P. Kowalczyk, E.A. Kraut, J.R. Waldrop and R.W. Grant, J. Vacuum Sci Technol. 21 (1982) 482.

[6] R.L. Anderson, Solid State Electron. 5 (1962) 341.

[7] C. Herring and M.H. Nichols, Rev. Mod. Phys. 21 (1949) 185.

[8] H. Kroemer, CRC Critical Rev. Solid State Sci. 5 (1975) 555.

[9] Note that one of the "best" current measurements of band offsets, used in refs. [3–5], obtains the band offset from a difference between two numbers typically 5 times larger than the work function.

[10] J.A. Van Vechten, presented at the 12th Ann. Conf. on the Physics and Chemistry of Semiconductor Interfaces (PCSI-12), Tempe, AZ, January 29–31, 1985, J. Vacuum Sci. Technol., to be published.

[11] H.B. Michaelson, J. Appl. Phys. 48 (1977) 4729.

[12] C. Tejedor, F. Flores and E. Louis, J. Phys. C10 (1977) 2163; C. Tejedor and F. Flores, J. Phys. C11 (1978) L19.

[13] J. Tersoff, Phys. Rev. Letters 52 (1984) 465; Phys. Rev. B30 (1984) 4874.

[14] C. Mailhot and C.B. Duke, presented at the 12th Ann. Conf. on the Physics and Chemistry of Semiconductor Interfaces (PCSI-12), Tempe, AZ, January 29–31, 1985, J. Vacuum Sci. Technol. to be published.

[15] J.L Freeouf and J.M. Woodall, Appl. Phys. Letters 39 (1981) 727.

[16] I. Lindau, P.W. Chye, C.M. Garner, P. Pianetta, C.Y. Su and W.E. Spicer, J. Vacuum Sci. Technol. 15 (1978) 1332.

[17] W.E. Spicer, I. Lindau, P. Skeath and C.Y. Su, J. Vacuum Sci. Technol. 17 (1980) 1019.

[18] F. Bartels, H.J. Clemens and W. Mönch, Physica 117/118B (1983) 801.

[19] F.A Ponce and S.J. Eglash, Thin Solid Films 104 (1983) 317.

[20] R.L. Farrow, R.K. Chang, S. Mroczkowski and F.H. Pollak, Appl. Phys. Letters 31 (1977) 768.

[21] A.G. Cullis, P.D. Augustus and D.J. Stirland, J. Appl. Phys. 51 (1981) 2556.

[22] V. Heine, Phys. Rev. 138 (1965) A1689.

[23] A. Zur, T.C. McGill and D.L. Smith, Surface Sci. 132 (1983) 456.

[24] R.E. Allen and J.D. Dow, Solid State Commun. 45 (1983) 379.

[25] W.A. Harrison, E.A. Kraut, J.R. Waldrop and R.W. Grant, Phys. Rev. B18 (1978) 4402.

[26] W.I. Wang, J. Vacuum Sci. Technol. B1 (1983) 630.

[27] See, e.g., W.I. Wang and F. Stern, presented at the 12th Ann. Conf. on the Physics and Chemistry of Semiconductor Interfaces (PCSI-12), Tempe, AZ, January 29–31, 1985, J. Vacuum Sci. Technol., to be published.

[28] P.M. Petroff, A.Y. Cho, F.K. Reinhart, A.C. Gossard and W. Wiegemann, Phys. Rev. Letters 48 (1982) 170.

[30] H. Kroemer, J. Vacuum Sci. Technol. B2 (1984) 433.

[30] W.A. Harrison, J. Vacuum Sci. Technol. 14 (1977) 1016.

[31] A.D. Katnani and G. Margaritondo, Phys. Rev. B28 (1983) 1944.

[32] C. Tejedor, J.M. Calleja, F. Meseguer, E.L. Mendez, C.A. Chang and L. Esaki, presented at the 17th Intern. Conf. on the Physics of Semiconductors, San Francisco, Sept. 1984, to be published.

[33] R. People, J.C. Bean, D.V. Lang, A.M. Sergent, H.L. Störmer, K.W. Wecht, R.T. Lynch and K. Baldwin, Appl. Phys. Letters 45 (1984) 1231.

[34] H. Jorke and H.-J. Herzog, presented at the 167th Meeting of the Electrochemical Society, Toronto, Ontario, May 12–17, 1985, extended abstracts, p. 211.

[35] A.D. Katnani, P. Chiaradia, H.W. Sang, Jr. and R.S. Bauer, J. Electron. Mater. 14 (1985) 25.

[36] A. Zur and T.C. McGill, J. Vacuum Sci. Technol. B2 (1984) 440.

[37] Please note that the difference between p–n and n–p junctions derives from the dependence of pinning position upon doping type. For a pinning position independent of doping type, there should be no effect in a homojunction; the effect in a pinned heterojunction would depend upon the difference between pinned Fermi-level alignment and unpinned band alignments, as discussed later in this manuscript in relating band offsets and Schottky barrier heights, and would be independent of doping types.

[38] J.M. Woodall, G.D. Pettit, T.N. Jackson, C. Lanza, K.L. Kavanagh and J.W. Mayer, Phys. Rev. Letters 51 (1983) 1783.

[39] D.A. Neamen and W.W. Grannemann, Solid State Electron. 14 (1971) 1319.

[40] H.H. Wieder, Appl. Phys. Letters 38 (1981) 170;
K. Kajiyama, Y. Mizushima and S. Sakata, Appl. Phys. Letters 23 (1973) 458.

[41] T.F. Kuech and J.O. McCaldin, J. Vacuum Sci Technol. 17 (1980) 891.

[42] J.S. Best, Appl. Phys. Letters 34 (1979) 522.

[43] C.A. Sébenne, Il Nuovo Cimento 39B (1977) 768.

PHYSICAL REVIEW B   VOLUME 34, NUMBER 4   15 AUGUST 1986

Heterojunctions: Definite breakdown of the electron affinity rule

D. W. Niles and G. Margaritondo

Department of Physics and Synchrotron Radiation Center, University of Wisconsin, Madison, Wisconsin 53706
(Received 20 March 1986)

We performed a simple and straightforward synchrotron-radiation photoemission test of the electron affinity rule, the oldest and most widely used model to predict semiconductor-semiconductor band lineups. The results show, beyond any experimental uncertainty, that the rule is incorrect. The elimination of the rule and of all models related to it considerably simplifies the theoretical situation of this fundamental area of solid-state physics.

When two different semiconductors are brought together to form a heterojunction interface, the mismatch in forbidden gaps must be accommodated by discontinuities in their band edges.[1] The resulting conduction-band valence-band discontinuities, ΔE_c and ΔE_v, are the most important parameters in determining the behavior and performance of heterojunction systems. The strong fundamental and practical interest of such systems has stimulated much research to understand and predict the band discontinuities.[2] In fact, it is not clear *a priori* how the forbidden-gap difference is shared between ΔE_v and ΔE_c.

Many models have been developed to solve this problem.[3-19] Several of these models [3,4,6] are related to the so-called electron affinity rule,[3] originally proposed in 1962. This rule simply states that the conduction-band discontinuity equals the difference between the electron affinities of the two semiconductors.

For 24 years the electron affinity rule has been very popular and widely used in fundamental research and in technology.[1,2] Recently, it came under strong theoretical criticism, which prompted the development of alternate approaches.[4-19] Experimental tests of the rule have been made difficult by the chronic unreliability of the electron affinity data. The uncertainty has left this fundamental area of solid-state physics in a state of underlying confusion, which has certainly contributed to some notorious problems such as the errors in estimating the $Ga_{1-x}Al_xAs$-GaAs band lineup.

We present here a simple, straightforward, and unambiguous test of the electron affinity rule for the prototypical interface ZnSe-Ge. The test is based on synchrotron-radiation photoemission measurements of all the physical quantities involved in the rule. The results clearly demonstrate that the rule is not correct.

The experimental approach is somewhat related to that used by Zurcher and Bauer[20] to test the rule in the case of the GaAs-Ge interface. ZnSe-Ge, however, has clear advantages which eliminate the uncertainties affecting the test of Ref. 20. In particular, ΔE_v is very large for ZnSe-Ge, and therefore can be directly derived from the double-edge structure of valence-band photoemission data without relying on an indirect derivation from core-level peak data.[2] Furthermore, the large magnitude of the discontinuity enhances the discrepancy between the predictions of the rule and the experimental findings, to values well beyond any reasonable experimental uncertainty.

The simple philosophy of the test is explained by Fig. 1. Here DOS labels the density of states of a semiconductor in the energy region close to its forbidden gap, E_g. E_v and E_c are the band edges. The distance between E_c and the vacuum level (VL) is by definition the electron affinity, χ. EDC labels the energy distribution curve of photoelectrons emitted by the semiconductor under bombardment by photons of energy $h\nu$. The shaded area corresponds to the secondary electrons created by inelastic scattering processes. The low-energy cutoff of the distribution corresponds to the vacuum level. The upper edge corresponds to $E_v + h\nu$.

The distance in energy between the two EDC edges D equals $h\nu - (E_g + \chi)$. Thus, the electron affinity can be directly derived from the EDC spectra. Calling D_1 and D_2 the values of D for two different semiconductors, the electron affinity rule for their interface trivially predicts that

$$\Delta E_v = D_1 - D_2 . \tag{1}$$

Equation (1) can be used to directly test the rule with photoemission methods. This is done by comparing the value of ΔE_v predicted by Eq. (1) with the measured discontinuity. In turn, the discontinuity is measured[2,21-23] by taking EDC's on thin overlayers of one semiconductor deposited on the other. This approach has been discussed in detail in several recent reviews.[2]

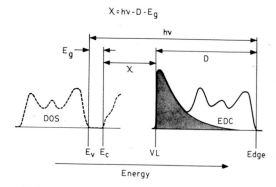

FIG. 1. Schematic explanation of the test. The distance in energy D between the upper and lower edge of a photoemission spectrum is related to the electron affinity. For a detailed explanation, see text.

The above approach is not immune from experimental difficulties. The low-energy EDC cutoff can be due to the electron analyzer rather than to the sample vacuum level. This problem is solved by electrostatically biasing the sample to move the low-energy cutoff of its spectra to higher energies. The sample can become charged when exposed to the electron beam, and this affects the EDC's. However, charging problems are easily corrected by illuminating the sample with an intense visible light which generates photoconductivity. Of course, the results are crucially dependent on the cleanliness of the system and the test must be performed *in situ* under ultrahigh-vacuum conditions.

ZnSe-Ge offers the additional advantage of being a very extensively studied interface.[21-24] Several photoemission experiments measured ΔE_v,[21-23] with results between 1.29 and 1.52 eV. In the present case, from double-edge spectra like the two top EDC's of Fig. 2, we derive $\Delta E_v = 1.44$ eV. These curves were taken on as-grown Ge overlayers on cleaved ZnSe. Extensive experiments have demonstrated that the discontinuity measured at these Ge thicknesses

coincides with the final ΔE_v for very thick overlayers.[22] Furthermore, they also demonstrated that ΔE_v does not change (within 0.1 eV at most) in going from disordered to ordered overlayers.[21]

Measurements of the distances D for ZnSe substrates and very thick (> 50 Å) Ge overlayers were performed on six different systems, with three different photon energies, $h\nu = 17$, 20, and 23 eV, and with a variety of bias voltages and intensities of the discharging light. The combined results of all these measurements give $D_1 - D_2 = 2.21$ eV. This value is 0.77 eV larger than the experimental valence-band discontinuity.

This complete breakdown of the electron affinity rule is directly visualized in Fig. 2. The two bottom curves show the EDC's of ZnSe and Ge, aligned with respect to each other so that their low-energy cutoffs coincide (as shown in the inset). The two top curves have been aligned with respect to the clean-ZnSe EDC so that the ZnSe-related features coincide (e.g., the Zn 3d peak). Thus, if the electron affinity rule was valid, the upper edges of the two top curves would coincide with the upper edge of the Ge EDC. The dashed vertical line shows that they do not, and dramatically so.

Of course, the validity of this test depends on its combined accuracy. Contributing to this accuracy are the uncertainty in deriving the edge positions from the experimental curves, and the uncertainty in measuring ΔE_v.[2] From the extensive experiments performed by different authors on this interface,[21-23] we can derive a conservative uncertainty for our present ΔE_v value, $1.44^{+0.08}_{-0.15}$ eV. The combined uncertainties in deriving the four required edge positions give an uncertainty of the order of 0.4 eV for $D_1 - D_2$. This is consistent with the standard deviation of our $D_1 - D_2$ data, 0.46 eV. Thus, in the worst case there is still a large difference of 0.24 eV between the minimum possible value of $D_1 - D_2$, 1.76 eV and the maximum possible value of ΔE_v, 1.52 eV.

We emphasize that our test has several self-consistency features which increase its reliability. For example, one could argue that we are not really measuring the electron affinity of Ge, but that of whatever species we obtain by depositing Ge on ZnSe. However, the electron affinities which must be used for the electron affinity rule are specifically those of the interface species.[6] Thus, we are measuring exactly the quantities which are relevant for the rule. As a limit case, the test would be valid even if our system was heavily contaminated—which it was not. The EDC's taken at low Ge coverage are affected by signal related to localized states.[24] However, this is irrelevant to the huge discrepancy between the upper edge of the two top curves of Fig. 2 and of the bottom curve.

This result should put an end to a long and bitter controversy, and simplify the theoretical situation by eliminating one class of models. Obviously, it does not *per se* endorse a specific alternate model. For example, in the framework of Mailhiot-Duke approach,[4] it can be interpreted as evidence that there is substantial relaxation of the interface atomic positions with respect to their bulk values.

We emphasize, however, that recent evidence was provided for a correlation between Schottky barrier heights and heterojunction valence-band discontinuities.[2,25] This

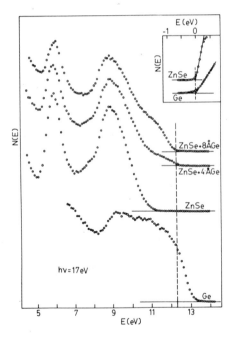

FIG. 2. A direct illustration of the breakdown of the electron affinity rule. The two upper curves are photoemission spectra taken on ZnSe covered by 4 and 8 Å of Ge. They exhibit the characteristic double edge due to the valence-band discontinuity. The two other curves refer to clean ZnSe and Ge. These last two curves were aligned to each other so that the low-energy cutoffs coincide (see inset). The two upper curves were aligned with respect to the ZnSe curve so that the bulk-ZnSe features coincide. Thus, the electron affinity rule would predict that their upper edges coincide with the upper edge of the lower Ge curve. The dashed line emphasizes that this prediction is wrong.

result could either be explained[25] by the midgap-energy approach proposed by Tersoff,[16,26] or by a combination of the electron affinity rule and of the Schottky model for metal-semiconductor interfaces. The breakdown of the electron affinity rule leaves Tersoff's approach[16] as the only heterojunction model consistent with all present experimental data.

This work was supported by the NSF, Grant No. DMR-84-21292. The Synchrotron Radiation Center (SRC) of the University of Wisconsin–Madison is supported by the NSF and by the Wisconsin Alumni Research Foundation. We are grateful to the SRC technical staff for their dedication and perseverance in supporting the users' research during a very critical time.

[1]A. G. Milnes and D. L. Feucht, *Heterojunctions and Metal-Semiconductor Junctions* (Academic, New York, 1972).

[2]For a recent review, see G. Margaritondo, Solid State Electron. **29**, 123 (1986); Z. Phys. **61**, 447 (1985); G. Margaritondo and P. Perfetti, in *Heterojunctions: Band Discontinuities and Device Applications,* edited by F. Capasso and G. Margaritondo (North-Holland, Amsterdam, in press).

[3]R. L. Anderson, Solid State Electron. **5**, 341 (1962).

[4]C. Mailhiot and C. B. Duke, Phys. Rev. B **33**, 1118 (1986).

[5]C. Tejedor and F. Flores, J. Phys. C **11**, L19 (1978).

[6] J. L. Freeouf and J. M. Woodall, Appl. Phys. Lett. **39**, 727 (1981); Surf. Sci. (to be published).

[7]J. M. Langer and H. Heinrich, Phys. Rev. Lett. **55**, 1414 (1985); in *Proceedings of the International Conference on Hot Electrons in Semiconductors, Innsbruck, July 1985* [Physica (to be published)].

[8]M. J. Caldas, A. Fazzio, and Alex Zunger, Appl. Phys. Lett. **45**, 671 (1984); Alex Zunger, Annu. Rev. Mater. Sci. **15**, 411 (1985); in Second Brazilian School on Semiconductors, 1985 (unpublished); Solid State Phys. (to be published).

[9]W. A. Harrison, J. Vac. Sci. Technol. **14**. 1016 (1977).

[10]W. A. Harrison, J. Vac. Sci. Technol. B **3**, 1231 (1985).

[11]P. Vogl, H. P. Hjalmarson, and J. D. Dow, J. Phys. Chem. Solids **44**, 365 (1983).

[12]Z. H. Chen, S. Margalit, and A. Yariv, J. Appl. Phys. **57**, 2970 (1985).

[13]F. Bechstedt, R. Enderlein, and O. Heinrich, Phys. Status Solidi (b) **126**, 575 (1984).

[14]W. R. Frensley and H. Kroemer, Phys. Rev. B **16**, 2642 (1977).

[15]J. A. Van Vechten, J. Vac. Sci. Technol. B **3**, 1240 (1985); Phys. Rev. **182**, 891 (1969).

[16]J. Tersoff, Phys. Rev. B **30**, 4875 (1984).

[17]M. L. Cohen, Adv. Electron. Electron Phys. **51**, 1 (1980), and references therein.

[18]A. Nussbaum, in *The Theory of Semiconducting Junctions,* edited by R. K. Willardson and A. C. Beer, Semiconductors and Semimetals, Vol. 15 (Academic, New York, 1981), and references therein.

[19]O. Von Ross, Sol. State Electron. **23**, 1069 (1980).

[20]P. Zurcher and R. S. Bauer, J. Vac. Sci. Technol. A **1**, 695 (1983).

[21]G. Margaritondo, C. Quaresima, F. Patella, F. Sette, C. Capasso, A. Savoia, and P. Perfetti, J. Vac. Sci. Technol. A **2**, 508 (1984).

[22]A. D. Katnani and G. Margaritondo, Phys. Rev. B **28**, 1944 (1983).

[23]S. P. Kowalczyk, E. A. Kraut, J. R. Waldrop, and R. W. Grant, J. Vac. Sci. Technol. **21**, 482 (1982).

[24]G. Margaritondo, F. Cerrina, C. Capasso, F. Patella, P. Perfetti, C. Quaresima, and F. J. Grunthaner, Solid State Commun. **52**, 495 (1984).

[25]D. W. Niles, G. Margaritondo, P. Perfetti, C. Quaresima, and M. Capozi, J. Vac. Sci. Technol. (to be published).

[26]See Ref. 16. Also, see Ref. 5.

RAPID COMMUNICATIONS

PHYSICAL REVIEW B   VOLUME 29, NUMBER 6   15 MARCH 1984

Parabolic quantum wells with the GaAs-Al$_x$Ga$_{1-x}$As system

R. C. Miller, A. C. Gossard, D. A. Kleinman, and O. Munteanu

AT&T Bell Laboratories, Murray Hill, New Jersey 07974

(Received 3 November 1983)

Photoluminescence measurements at 5 K on wafers containing parabolic quantum wells fabricated by molecular-beam epitaxy with the GaAs-Al$_{0.3}$Ga$_{0.7}$As system reflect harmonic oscillator-like electron and hole levels. The many observed heavy-hole transitions can be fitted accurately with a model that divides the energy-gap discontinuity ΔE_g equally between the conduction and valence-band wells. This is in marked contrast to the usual $\Delta E_c = 0.85 \Delta E_g$ and $\Delta E_v = 0.15 \Delta E_g$ generally assumed for square wells. Experiment and theory show that parabolic wells can lead to parity-allowed $\Delta n = 2$ ("forbidden") transitions with strengths greater than that of nearby $\Delta n = 0$ ("allowed") transitions.

INTRODUCTION

It is well known that molecular-beam epitaxy (MBE) readily lends itself to the growth of structures requiring smooth and abrupt GaAs-Al$_x$Ga$_{1-x}$As heterointerfaces.[1,2] In addition, the MBE growth method is well suited to the fabrication of structures with various potential profiles, e.g., triangular quantum wells have been grown by MBE.[3] Recently, a pulsed Al source has been used with MBE to allow the growth of Al$_x$Ga$_{1-x}$As with an arbitrary Al profile.[4] This Rapid Communication describes the MBE growth and some of the characteristics of multiquantum well GaAs-Al$_x$Ga$_{1-x}$As structures with parabolic potential wells. As expected, these structures result in exciton transitions in the excitation spectra that reflect a uniformly spaced density-of-states function for the electrons and holes. The photoluminescence data also show enhanced "forbidden transitions,"[5,6] transitions with $\Delta n = 2$ but parity allowed. Analyses of the energies of the various exciton transitions suggest that the partitioning of the energy-gap discontinuities between the electron and valence-band wells may not be the same as that utilized for square wells.[7-9]

For square GaAs wells of width L_z and infinite height, the energy levels of a particle of mass m_i^* depend on L_z according to

$$E_{ni} = \frac{1}{2 m_i^*} \left(\frac{n \pi \hbar}{L_z} \right)^2 , \qquad (1)$$

where $n = 1, 2, 3$, etc. With parabolic wells

$$E_{ni} = (n - 1/2) \hbar \omega_{0i} , \qquad (2)$$

where again $n = 1, 2, 3$, etc, and

$$\omega_{0i} = \sqrt{K_i / m_i^*} , \qquad (3)$$

with K_i equal to the curvature of the parabolic well. Defining the curvature K_i by the potential height of the finite parabolic well at $z = \pm L_z/2$, namely, $Q_i \Delta E_g$, where ΔE_g is the total energy-gap discontinuity between the GaAs at the bottom of the wells and the Al$_x$Ga$_{1-x}$As at the top of the wells and Q_i is the fraction of ΔE_g for the ith particle well, Eq. (2) becomes

$$E_{ni} = 2 (n - \tfrac{1}{2}) \frac{\hbar}{L_z} \left[\frac{2 Q_i \Delta E_g}{m_i^*} \right]^{1/2} \qquad (4)$$

It is interesting to note that the partitioning of the energy-gap discontinuity Q_i comes in directly in Eq. (4) but not in

Eq. (1). Equations (1) and (4) are, of course, only approximations since the finite well heights should be taken into account as well as the dependence of the effective mass on the Al$_x$Ga$_{1-x}$As alloy composition x.

RESULTS

Parabolic compositional profiles were generated by alternate deposition of thin undoped layers of GaAs and Al$_x$Ga$_{1-x}$As of varying thickness. Computer control was employed in the deposition. The relative thicknesses of the Al$_x$Ga$_{1-x}$As layers increased quadratically with distance from the well centers while that of the GaAs layers decreased. Average layer thickness of approximately 10 Å were employed in order to permit the GaAs layers to be sufficiently thick to produce surface smoothing and cleaning,[10] while still allowing ample electron and hole tunneling to average the effective potentials to parabolic profiles. Each well contained 20 layers of Al$_x$Ga$_{1-x}$As and 21 layers of GaAs. The thickness of the Nth layer of Al$_x$Ga$_{1-x}$As from the center of the well was $[(N - 0.5)/10]^2 \times L_z/20$. The Al$_xGa_{1-x}$As layers are centered at distances $(N - 0.5) L_z/20$ from the well center, and the remaining material is GaAs.

Figure 1 shows the photoluminescence and excitation spectra at 5 K from a parabolic well sample with ten periods where each period consists of a parabolic well estimated from the growth parameters to have $L_z = 510 \pm 35$ Å and barriers of width $L_B = 237 \pm 16$ Å composed of $x = 0.30 \pm 0.06$ alloy. The photoluminescence spectra were obtained with an excitation intensity $I_p = 0.14$ W/cm^2. The photoluminescence is relatively sharp, 2.2 meV full width at half maximum, and sufficiently intense to demonstrate that the Al-containing layers do not seriously degrade the recombination efficiency. The excitation spectrum with detection set at the photoluminescence peak exhibits much structure and shows essentially no Stokes shift between the $n = 1$ heavy-hole exciton E_{1h} and the emission peak. Thus, the main recombination from this sample is intrinsic and due to E_{1h} exciton emission as in the better quality square potential well samples.[11] Any electron density in the wells cannot exceed 5×10^{10} cm^{-2}.

Assignments of the various exciton transitions are also indicated in Fig. 1. Circular polarization excitation and detection techniques aided in the identification of some of the lower energy peaks.[5,11] The allowed transitions, $\Delta n = 0$, are identified by E_{nm}, where n is defined by Eq. (4) and m signi-

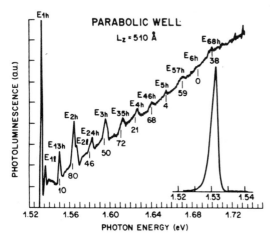

FIG. 1. The photoluminescence spectrum obtained at 5 K with 0.14 W/cm^2 excitation at 1.6 eV is shown in the insert. The excitation spectrum was taken with the same intensity as above and with the detection set at the peak of the photoluminescence, 1.531 eV. Various exciton transition peaks are labeled in the figure. Exciton transition energies for the heavy-hole excitons calculated using parabolic wells of equal height for the electrons and holes are shown as short vertical bars below the peaks. Their calculated strengths normalized to 100 for E_{1h} (without the resonant enhancement) are given as integers below the peaks. For $\Delta n \neq 0$, the sum of strengths of overlapping transitions, e.g., E_{24h} and E_{31h}, are included in the strength given.

fies whether the exciton transition involves a light or heavy hole, l or h, respectively. For the parity allowed "forbidden transitions,"[5,6] $\Delta n \neq 0$ but even, the designation $E_{nn'_m}$ is used, where n refers to the electron level as above and n' the quantum number for the hole designated by m as above. Differences of the energies of the various transitions were then used to determine the energy-level ladders for electrons, heavy holes, and light holes, ΔE_e, ΔE_h, and ΔE_l, respectively. For these estimates, the binding energies of all the excitons were assumed equal.[12] The experimental values of ΔE_i are given in Table I along with estimates from Eq. (4) using the commonly accepted values for m_i^* and Q_i, namely, $m_e^*/m_0 = 0.0665$,[13] $m_h^*/m_0 = 0.45$,[13] $m_l^*/m_0 = 0.088$,[14] $Q_e = 0.85$, and $Q_h = Q_l = 1 - Q_e = 0.15$.[7] Data from two other parabolic well samples, $L_z = 325 \pm 25$ Å and $L_z = 336 \pm 25$ Å, are also given in Table I.

DISCUSSION

The agreement between the measured energy-ladder spacings and that calculated via Eq. (4) from the known growth parameters as given in Table I is poor. The L_z dependence of the calculated results can be removed by taking ratios of these energy ladders which then points up wherein the major problem lies. The average of the measured ratios are $\Delta E_e/\Delta E_h = 2.6$, $\Delta E_e/\Delta E_l = 1.4$, and $\Delta E_l/\Delta E_h = 1.9$. These ratios are to be compared to calculated values; $\Delta E_e/\Delta E_h = 6.0$, $\Delta E_e/\Delta E_l = 2.7$, and $\Delta E_l/\Delta E_h = 2.3$. The agreement between these two sets of numbers is also very poor

TABLE I. Experimental and calculated energy-level spacings for parabolic quantum wells.

	Expt.	Calc. Eq. (4)	"Exact" calc.
$L_z = 510 \pm 35$ Å, $x = 0.30 \pm 0.06$			
ΔE_e (meV)	22.3	33.5	31.3
ΔE_h (meV)	8.4	5.4	5.2
ΔE_l (meV)	16.9	12.2	11.8
$\dfrac{\Delta E_e}{\Delta E_h}$	2.65	6.19	6.02
$\dfrac{\Delta E_e}{\Delta E_l}$	1.32	2.73	2.65
$\dfrac{\Delta E_l}{\Delta E_h}$	2.01	2.26	2.27
$L_z = 325 \pm 25$ Å, $x = 0.29 \pm 0.06$			
ΔE_e (meV)	40.1	51.6	48.9
ΔE_h (meV)	15.6	8.33	8.1
ΔE_l (meV)	27.9	18.9	17.9
$\dfrac{\Delta E_e}{\Delta E_h}$	2.57	6.19	6.04
$\dfrac{\Delta E_e}{\Delta E_l}$	1.44	2.73	2.73
$\dfrac{\Delta E_l}{\Delta E_h}$	1.79	2.27	2.21
$L_z = 336 \pm 25$ Å, $x = 0.30 \pm 0.06$			
ΔE_e (meV)	33.1	50.8	48.2
ΔE_h (meV)	12.4	8.20	8.0
ΔE_l (meV)	23.7	18.5	17.7
$\dfrac{\Delta E_e}{\Delta E_h}$	2.67	6.19	6.03
$\dfrac{\Delta E_e}{\Delta E_l}$	1.40	2.73	2.72
$\dfrac{\Delta E_l}{\Delta E_h}$	1.91	2.27	2.21

except for $\Delta E_l/\Delta E_h$ and hence raises questions about the validity of Eq. (4) which assumes one effective mass throughout and parabolic wells of infinite height. With this in mind Eq. (4) was modified to include by perturbation theory the variation of the effective masses with z which results in a correction to the energy levels determined from Eq. (4) of

$$\delta E_i = - \frac{3.81 \times 10^3}{[L_z \,(\text{Å})]^2 m_i^*/m_0} f(m_i^*, x)(3 - 2n + 2n^2) \text{ meV},$$

$$(5)$$

where for $x = 0.3$, $f(m_i^*, x) = 0.27$ for electrons and 0.17 for heavy holes. For the sample with $L_z = 510$ Å this correction reduces the ladder spacings given in Table I by from 1.5% to 3.0% and hence for this L_z has little effect on the calculated ratios derived from Eq. (4). However, the correction to Eq. (4) given by Eq. (5) does result in calculated energy-level spacings that decrease slightly with increasing n as is usually observed and predicted by the more exact calculation given below.

A better calculation of the energy levels has also been made using a program that determines the transmission of an arbitrary sequence of square-shaped wells and barriers as a function of energy. This computation includes any standing wave effects due to the discontinuous growth profile, the variation of the effective masses with x, the finite well height, and the boundary conditions for GaAs-Al$_x$Ga$_{1-x}$As interfaces proposed by one of us (D.A.K.) and independently by Bastard.[15] The results of these calculations are also given in Table I and they are found to differ by only a few percent from those determined from Eq. (4). The relatively good agreement between the experimental and calculated values of $\Delta E_l / \Delta E_h$ (15% $\pm$4%) suggests that the main difficulty involves the partitioning of the energy-gap discontinuity and not the hole masses. Since Eq. (4) gives results on ΔE_l that are only a few percent smaller than the values given by the better computation, Eq. (4) will be used to illustrate the problem with the partitioning of the energy-gap discontinuity. Equation (4) leads to

$$\frac{\Delta E_e}{\Delta E_h} = \left[\frac{Q_e}{1-Q_e} \frac{m_h^*}{m_e^*} \right]^{1/2} = 2.6 \ , \qquad (6)$$

which with the conventional masses m_i^* (Refs. 13 and 14) yields $Q_e \approx 0.50$. Thus there is a discrepancy when compared with the generally accepted value of $Q_e = 0.85$ (Ref. 7) based on square-well spectra. However, there is some evidence that Q_e is sensitive to certain growth parameters.[16]

At present we have no explanation for the discrepancy between our value for Q_e and the accepted value. The parabolic wells we require to explain the observed ladder of levels could be produced by a combination of the accepted value $Q_e \approx 0.85$ and a negative space charge due to a density $n_{2D} \approx 1 \times 10^{12}$ cm^{-2} of either electrons or negatively charged acceptors. The absence of a Stokes shift between the emission peak and the $1h$ excitation peak rules out such a density of electrons in these samples. Also, with this density of electrons one would not see the $1h$ exciton peak in excitation at all. We believe the presence of such a density of acceptor or donor impurities is also ruled out by the fact that the same MBE apparatus produces quantum wells in modulation-doped samples exhibiting very high carrier mobilities. Therefore we believe space-charge effects in these samples are negligible.

Short vertical bars under the various peaks in Fig. 1 indicate energies of the heavy-hole transitions determined via the exact program using $Q_e = 0.51$, $L_z = 507$ Å, and $x = 0.25$. Values of L_z and x employed are within the estimated uncertainties of these quantities given earlier. The calculation gives $\Delta E_e = 22.8$ meV, $\Delta E_h = 8.0$ meV, and $\Delta E_l = 19.4$ meV. The calculated and experimental energies

of the E_{1h} transition were set equal. The agreement between these calculated and experimental heavy-hole exciton transitions is considered excellent, but the L_z and x used are not unique. On the other hand, the calculated light-hole transitions (not shown) are too high in energy as expected since the calculated $\Delta E_l / \Delta E_h$ is too large.

One of the more striking characteristics of the data in Fig. 1 is the large strength of the "forbidden transitions" (parity allowed, $\Delta n = 2$), especially those for large n. Strong forbidden transitions E_{13h} with resonant-type line shapes like that shown in Fig. 1 have been seen previously in multi-quantum square-well structures.[17] For the undoped square-well case, theoretical estimates of the strengths of the forbidden transitions using finite square-well eigenfunctions which take into account different effective masses for the wells and barriers give values that are many orders of magnitude too small. These estimates have now been repeated using infinite parabolic-well eigenfunctions that include only GaAs masses. The calculated strengths (matrix elements squared) for E_{ijh} and E_{jih} are equal.[5,6] Also, since the spacing of the energy level ladder for the $\Delta n = 0$ heavy-hole transitions is almost four times that of the heavy-hole ladder, transitions E_{24h} and E_{31h}, E_{35h} and E_{42h}, etc. are at nearly the same energy and hence are not expected to be resolved. Therefore to compare the calculated strengths with the excitation spectra, the strengths of overlapping transitions have been added together. The numbers under the various heavy-hole exciton transitions in Fig. 1 represent the integer values of the calculated strengths normalized to 100 for the calculated strength of E_{1h}. (The resonant enhancement of E_{1h} in Fig. 1 due to resonant Rayleigh scattering renders direct comparisons with this experimental peak meaningless.[18]) These results explain the large strengths of the $\Delta n \neq 0$ transitions and the decreasing strength of the $\Delta n = 0$ transitions as n increases. The strengths of these parity-allowed transitions arise from the fact that, in contrast to the square-well case, the hole and electron wave functions for parabolic wells have different spatial ranges for the same n, and for different n are not even approximately orthogonal. Thus, with parabolic wells, the $\Delta n \neq 0$ parity-allowed transitions are not really "forbidden."

CONCLUSIONS

Photoluminescence spectra of GaAs-Al$_x$Ga$_{1-x}$As parabolic quantum-well samples reflect the expected harmonic oscillator levels. The observed level intervals suggest that the energy-gap discontinuity between the GaAs and Al$_x$Ga$_{1-x}$As layers is evenly split between the electron and valence-band wells. Theory and experiment show that the $\Delta n = 2$ parity-allowed transitions are enhanced relative to the $\Delta n = 0$ allowed transitions as n becomes large.

[1]C. Weisbuch, R. Dingle, A. C. Gossard, and W. Wiegmann, Solid State Commun. 38, 709 (1981).

[2]P. Petroff, A. C. Gossard, W. Wiegmann, and A. Savage, J. Cryst. Growth 44, 5 (1978).

[3]A. C. Gossard, W. Brown, C. L. Allyn, and W. Wiegmann, J. Vac. Sci. Technol. 20, 694 (1982).

[4]M. Kawabe, M. Kondo, N. Matsuura, and Kenya Yamamoto, Jpn. J. Appl. Phys. 22, L64 (1983).

[5]R. C. Miller, D. A. Kleinman, W. A. Nordland, Jr., and A. C. Gossard, Phys. Rev. B 22, 863 (1980).

[6]R. C. Miller, D. A. Kleinman, O. Munteanu, and W. T. Tsang, Appl. Phys. Lett. 39, 1 (1981).

[7]R. Dingle, Festkorperprobleme 15, 21 (1975).

[8]R. People, K. W. Wecht, K. Alavi, and A. Y. Cho, Appl. Phys. Lett. 43, 118 (1983).

[9]J. Sanchez-Dehesa and C. Tejedor, Phys. Rev. B 26, 5824 (1982).

[10]A. C. Gossard, W. Wiegmann, R. C. Miller, P. M. Petroff, and W. T. Tsang, in Proceedings of the 2nd International Conference on Molecular-Beam Epitaxy, Tokyo, 1982, p. 39 (unpublished). Also R. C. Miller, A. C. Gossard, and W. T. Tsang, Physica 117&118 B+C, Part II, 714 (1983).

[11]C. Weisbuch, R. C. Miller, R. Dingle, A. C. Gossard, and W. Wiegmann, Solid State Commun. 37, 219 (1981).

[12]R. C. Miller, D. A. Kleinman, W. T. Tsang, and A. C. Gossard, Phys. Rev. B 24, 1134 (1981).

[13]Q. H. F. Vrehen, J. Phys. Chem. Solids 29, 129 (1968).

[14]A. L. Mears and R. A. Stradling, J. Phys. C 4, L22 (1971).

[15]G. Bastard, Phys. Rev. B 24, 5693 (1981).

[16]See for example, R. S. Bauer and H. W. Sang, Jr., in *Surfaces and Interfaces: Physics and Electronics,* edited by R. S. Bauer (North-Holland, Amsterdam, 1983), p. 479.

[17]R. C. Miller (unpublished).

[18]J. Hegarty, M. D. Sturge, C. Weisbuch, A. C. Gossard, and W. Wiegmann, Phys. Rev. Lett. 49, 930 (1982).

Role of d Orbitals in Valence-Band Offsets of Common-Anion Semiconductors

Su-Huai Wei and Alex Zunger

Solar Energy Research Institute, Golden, Colorado 80401
(Received 13 February 1987)

We show through all-electron first-principles electronic structure calculations of core levels that, contrary to previous expectations, the valence-band offsets in the common-anion semiconductors AlAs-GaAs and CdTe-HgTe are decided primarily by intrinsic bulk effects and that interface charge transfer has but a small effect on these quantities. The failure of previous models is shown to result primarily from their decision to omit cation d orbitals.

PACS numbers: 73.40.Lq, 71.20.Fi, 73.30.+y

Measurements[1-3] and theoretical modeling[4-7] of the lineup between the top of the valence bands of two semiconductors forming a heterojunction have recently been revived[8] in light of new results which cast doubt on both previous measurements[9] and theories.[4,5] Textbook descriptions[10] of the zone-center valence-band maximum (VBM) in a binary zinc-blende semiconductor (the Γ_{15v} state) suggest that it consists almost exclusively of anion valence p orbitals. It was therefore initially expected[4,5,8-11] that the VBM energies of two common-anion semiconductors which share the same crystal structure and lattice constant (e.g., the AlAs-GaAs or CdTe-HgTe pairs), would be nearly equal. These expectations were formulated in terms of the hitherto successful "common-anion rule"[11] (stating that the offset ΔE_{VBM} between the VBM energies of two covalent semiconductors reflects primarily different anion energies, and hence would nearly vanish for semiconductors sharing a common anion), and simple tight-binding[4] and dielectric[5] models, all predicting nearly vanishing (< 0.1 eV) band offsets for such common-anion systems. While these predictions were in agreement with the then-available experimental data on AlAs-GaAs (Ref. 9) and HgTe-CdTe,[12] more recent measurements on AlAs-GaAs [$\Delta E_{VBM} = 0.45 \pm 0.05$ eV (Ref. 2)] and CdTe-HgTe ($\Delta E_{VBM} = 0.35 \pm 0.06$ eV (Ref. 1)] have shown previous expectations and models to be substantially in error. It has been recognized,[8,13] however, that the band offset ΔE_{VBM} could be thought to consist of an intrinsic "bulk" (b) contribution ΔE_{VBM}^b, reflecting the disparity between the VBM energies of two *isolated* semiconductors (when their energies are compared on the same, absolute scale), and an "interface specific" (IS) contribution ΔE_{VMB}^{IS} reflecting chemical events at the *interface* (hence, depending on interfacial charge transfer, orientation, dipole layer, interdiffusion, defect structure, etc.):

$$\Delta E_{VBM} = \Delta E_{VBM}^b + \Delta E_{VBM}^{IS}. \qquad (1)$$

The failure of previous models[4,5] in the crucial common-anion test was recently interpreted[4c,6] as reflecting the neglect of ΔE_{VBM}^{IS}—in particular the omission of interfacial charge-transfer (screening) effects. This inter-

pretation granted a decisive physical role to interfacial dipoles in establishing ΔE_{VBM} for these systems.

In this Letter we contest this basic physical interpretation. We first calculate the valence-band offsets of the four basic common-anion semiconductors AlAs-GaAs, CdTe-HgTe, CdTe-ZnTe, and HgTe-ZnTe in a way that parallels their measurement in photoemission core-level spectroscopy[1,3]: from the core levels. We find our calculated ΔE_{VBM} values to be in good agreement with experiment. We then use a simple electrostatic model for core shifts to show that interface-specific dipole contributions to ΔE_{VBM} are small in these systems. We show furthermore that the failure of earlier models[4,5] does not result primarily from neglect of ΔE_{VBM}^{IS}, but is predominantly a consequence of imperfect representation in simple tight-binding models[4,10] of ΔE_{VBM}^b. In particular, the omission of the outermost *cation d* orbitals explains most of the incorrect magnitudes.[4] This approach hence provides a fundamentally different interpretation of the physical mechanism governing band lineups in common-anion systems, and provides a simple correction which fixes previous models.[4] Predictions for band lineups for two hitherto unreported systems (CdTe-ZnTe and ZnTe-HgTe) are given.

We begin by reviewing the tight-binding viewpoint on the problem. In this approach[4,10] the energy of the Γ_{15v} VBM of a zinc-blende semiconductor AC is expressed solely in terms of nonmetal (C) and metal (A) p-orbital atomic energies (ϵ_p^C and ϵ_p^A, respectively) and their interaction (V_{pp}) as

$$\epsilon_{VBM}^{AC} = (\epsilon_p^A + \epsilon_p^C)/2 - [(\epsilon_p^A - \epsilon_p^C)^2/4 + V_{pp}^2]^{1/2}. \qquad (2)$$

The bulk-intrinsic ("natural"[4]) valence-band offset between two semiconductors AC and BC is then simply given as the difference between the respective VBM energies as

$$\Delta E_{VBM}^b = \epsilon_{VBM}^{BC} - \epsilon_{VBM}^{AC}. \qquad (3)$$

The charge-transfer term is approximated as the difference[4c]

$$\Delta E_{VBM}^{IS} = \epsilon_h^{AC} - \epsilon_h^{BC}. \qquad (4)$$

TABLE I. Tight-binding, all-electron, and experimental valence-band offsets (in electronvolts) in common-anion pairs, calculated from different core levels (nl). SO is the spin-orbit splitting.

Systems	Tight-binding[a]				Average (with SO)	Average (no SO)	All-electron (Present results)			
	$\Delta E\ell_{BM}$	ΔE^{IS}_{VMB}	ΔE^{tot}_{VMB}	ΔE^{expt}_{VMB}			Using 1s	Using 2s	Using $3p_{1/2}$	δ_{pd}
CdTe-HgTe	0.00	0.09	0.09	0.35 ± 0.06[b]	0.37	0.39	0.377	0.388	0.400	0.34
CdTe-ZnTe	−0.07	0.00	−0.07	···	0.13	0.12	0.125	0.122	0.108	0.04
ZnTe-HgTe	0.07	0.09	0.16	···	0.26	0.29	0.277	0.286	0.289	0.30
AlAs-GaAs	0.01	0.15	0.16	0.45 ± 0.05[c]	0.42	0.41	0.41	0.40	···	0.31

[a]Using data from Ref. 14.
[b]Reference 1.
[c]Reference 2.

between the average s-p hybrid energies ϵ_h of the semiconductor AC,

$$\epsilon_h^{AC} = (\epsilon_s^A + 3\epsilon_p^A + \epsilon_s^C + 3\epsilon_p^C)/8, \qquad (5)$$

and that of BC. From Eqs. (1)–(5), Table I[14] exhibits the following features of the tight-binding band offsets: (i) Relative to experiment, the calculated $\Delta E\ell_{BM}$ is far too small [reflecting the fact that the differences of p-orbital energies[4,10,14] for (Zn,Cd,Hg) and (Al,Ga) are small too], and may even have the wrong sign (CdTe/ZnTe), and (ii) the charge-transfer correction of Eq. (4) improves the results but still falls short (by a factor of 2–4) of experiment. For InAs-GaAs,[4c] not depicted here,[14] dipole effects alone incorrectly reverse the sign of $\Delta E\ell_{BM}$, yielding $\Delta E_{VBM} \approx -0.13$ eV, whereas the experimental result (quoted in Ref. 4c) is positive (+0.17 eV).

The reader should note at this point that it has long been customary, both in tight-binding[4] [Eqs. (2) to (5)] and in empirical[15] or first-principles[7] pseudopotential calculations for semiconductors, to neglect cation d bands, despite the fact that they reside inside[16] the valence band (for II-VI's) or close to its minimum (for III-V's). These cation d bands may, however, selectively alter the VBM energies of such compounds, and hence contribute to the band offset between two materials. Observe that in tetrahedral symmetry the cation d and anion p states share the same symmetry representation (Γ_{15}), and hence can interact through the potential matrix element $\langle \phi_A^d | V | \phi_C^p \rangle \equiv V_{dp}$. This interaction repels the VBM by $V_{dp}^2/(\epsilon_p^C - \epsilon_d^A)$. This repulsion varies significantly from one compound to another since both the variations in spatial distributions of the cation d orbitals (hence, V_{dp}) and the variations in the energy denominator ($\epsilon_p^C - \epsilon_d^A$) along the II-VI series are substantial (for ZnTe, CdTe, and HgTe the atomic-orbital energy difference[14] $\epsilon_p^C - \epsilon_d^A$ is 4.3, 5.8, and 3.9 eV, respectively; the energies of the Γ_{15d} d band relative to the VBM are 7.3, 8.4, and 7.4 eV, respectively). Such p-d repulsion effects have been previously shown to reduce significantly the band gaps of II-VI's,[16] to explain the

"band-gap anomaly" in chalcopyrites,[17] and to clarify the reason why Cu impurity acceptor states (exhibiting p-d repulsion) are abnormally deep in II-VI's relative to the isovalent Na impurity[17] (which lacks p-d repulsion). We will show below that this effect also controls much of the band offsets in common-anion semiconductors, and that alternative contributions (e.g., charge transfer) are negligibly small.

We have calculated self-consistently the band structures of ZnTe, CdTe, HgTe, AlAs, and GaAs, treating core states relativistically and valence orbitals semirelativistically, using the general potential linear augmented plane-wave method[18] in the local-density formalism. For each common-anion pair, we also calculated the band structure of the 50%-50% ordered compounds CdHgTe$_2$, CdZnTe$_2$, HgZnTe$_2$, and GaAlAs$_2$ in the ordered CuAu-I–like structure[19] [space group D_{2d}^5, identical to an alternating monolayer superlattice in the (001) orientation] to find the cation core-level difference [see Eq. (6) below]. All structural parameters of the ternary compounds[17,19] are relaxed to attain the minimum total energy.

To parallel the measurement of ΔE_{VBM} in photoemission core-level spectroscopy[1,3] ΔE_{VBM} is expressed as[1] (see Fig. 1)

$$\Delta E_{VBM} \cong (\epsilon\ell_{BM}^C - \epsilon_{nl,A}^{AC}) - (\epsilon\ell_{BM}^{BC} - \epsilon_{nl,B}^{BC}) \\ + (\epsilon_{nl,A}^{ABC_2} - \epsilon_{nl,B}^{ABC_2}). \qquad (6)$$

Here, for example, $\epsilon_{nl,A}^{AC}$ is the core level nl of atom A in AC and $\epsilon\ell_{BM}^{AC}$ is the VBM energy of AC. The first two bracketed terms in Eq. (6) are calculated from core levels nl obtained from the band structures of AC and BC, respectively, whereas the third term is calculated from the band structure of ABC_2. We assume (and demonstrate below) that the core-level difference ΔE_{CL} [last bracketed term in Eq. (6)] in common-anion superlattices (which include information on $\Delta E\ell_{BM}$) has but a negligible dependence on the superlattice thickness. We hence calculated ΔE_{CL} from a simple (1,1) superlattice ABC_2. This assumption reflects the fact that for

145

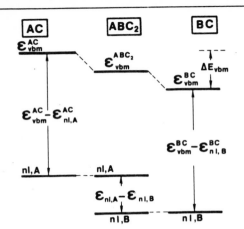

FIG. 1. Schematic energy-level diagram used to deduce the valence-band offset between AC and BC.

common-anion systems interface-induced effects are both small (see discussion below) and localized near the interface. The band offsets calculated from Eq. (6) for different choices of the core levels nl are shown in columns 7–10 of Table I. They exhibit a near independence of ΔE_{VBM} on the core level chosen. Table I also shows that the ΔE_{VBM} values are transitive, i.e., $\Delta E_{\text{VBM}}(AC\text{-}BC) = \Delta E_{\text{VBM}}(AC\text{-}DC) + \Delta E_{\text{VBM}}(DC\text{-}BC)$ to within a precision of 0.02 eV. We test independently our assumption that ΔE_{CL} in common-anion systems is insensitive to the details of the superlattice structure by comparing ΔE_{CL} calculated from the ABC_2 system to that calculated from the A_3BC_4 or AB_3C_4 systems, where the ternary phases are represented in the T_d "luzonite" structure.[19] We find these values to agree within 0.02 eV. The fifth and sixth columns of Table I compare the calculated ΔE_{VBM} values [including spin-orbit interactions][20] with experiment,[1,2] showing nearly perfect agreement for the two cases where data are currently available, and offering two predictions where they are not.

We have suggested (see also Ref. 13) that the deep core levels of *cations* in common-anion pairs are nearly unchanged relative to a *common reference energy* (e.g., vacuum) in going from a binary to a ternary (including alloy[13]) system, i.e.,

$$\epsilon_{nl,A}^{AC} \cong \epsilon_{nl,A}^{ABC_2}, \quad \text{and} \quad \epsilon_{nl,B}^{BC} \cong \epsilon_{nl,B}^{ABC_2}.$$

This "new common-anion rule" can be deduced by calculation of the change $\Delta \bar{V}_A$ in the electrostatic potential at the cation site A upon replacement of one cation (A) in the binary A_2C system by another (B), producing thereby the (1,1) superlattice $AC\text{-}BC$ (i.e., ABC_2). This electrostatic potential involves two (competing) contributions: (i) the intersite Madelung potential produced at A

by all other charges and (ii) the on-site Coulomb repulsion due to the altered charge at A. Denoting by $\Delta q = (q_B^{BC} - q_A^{AC})/2$ and $\Delta Q = (q_B^{ABC_2} - q_A^{ABC_2})/2$ the cation charge disparities in the binary compound (BC,AC) and in ABC_2, respectively, and by d the nearest-neighbor anion-cation distance (assumed equal in the lattice-matched AC and BC semiconductors), the change of the electrostatic potential at A is calculated to be

$$\Delta \bar{V}_A = (\Delta q/d)[\alpha^*\lambda - \alpha_{ZB} + (d/R_A)(1-\lambda)], \quad (7)$$

where $\alpha_{ZB} = 1.638$ and $\alpha^* = 0.976$ are the Madelung constants for the zinc blende and for the cation lattice[19] in ABC_2 structures, respectively, $\lambda = \Delta Q/\Delta q$, and R_A is the effective radius for atom A where the charge transfer (in forming ABC_2) occurs.[21] Our self-consistent calculations show that the charge differences inside the muffin-tin spheres are $\Delta q = 0.024e$, $\lambda = 0.86$ for CdTe-HgTe and $\Delta q = 0.043e$, $\lambda = 0.82$ for AlAs-GaAs. Using the experimental bond lengths ($d = 2.80$ Å for CdTe and HgTe; $d = 2.45$ Å for AlAs and GaAs) and estimating[22] R_A as $\approx 0.3d$ we find $\Delta \bar{V}_A$ to be as small as 0.04 eV. This small value (comparable to the uncertainty of the calculation and experimental error bars[1-3] for ΔE_{VBM}) suggests that interface dipole contributions to ΔE_{VBM} are equivalently small, and justify the use of a thin (1,1) superlattice. It is further supported by recent experimental observations[23] [note that if $\Delta \bar{V}_A = 0$, *any* size of the superlattice will give the same ΔE_{CL}].

Since the difference between the two cations is the only factor distinguishing any pair of lattice-matched common-anion binary semiconductors, the substantial ΔE_{VBM} values obtained here for AlAs-GaAs and HgTe-CeTe necessarily reflect participation of *cation* orbitals in the VBM. We find that the cation d *orbitals*, omitted in previous studies[4,7] are the major contributors. *First*, our self-consistent band calculations show directly substantial hybridization of cation d character in the Γ_{15v} VBM state: Within the cation muffin-tin sphere we find 7.5%, 6.9%, and 12.2% d character for Γ_{15v} of ZnTe, CdTe, and HgTe, respectively. [For comparison, note that the cation p *character* (4.3%, 4.0%, and 4.0%, respectively) is actually lower than the d character in the II-VI systems!] *Second*, one can independently model the amount ΔE_{pd}^{AC} by which the VBM of AC is repelled upwards by the cation d band, and hence find the pd correction $\delta_{pd} = \Delta E_{pd}^{BC} - \Delta E_{pd}^{AC}$ to the band offset between BC and AC. ΔE_{pd}^{AC} can be obtained as the amount by which the Γ_{15d} cation-d-band energy shifts upwards as anion p orbitals are removed from the linear augmented plane-wave basis set. Alternatively, one can calculate ΔE_{pd}^{AC} by subtracting from the total Γ_{15v}-Γ_{1v} valence-band width (calculated with cation d bands) the corresponding tight-binding value (calculated without cation d bands). Both models yield to within ± 0.1 eV the δ_{pd} corrections to the band offsets 0.04, 0.34, 0.30, 0.31, 0.04, and 0.35 eV for the CdTe-ZnTe, CdTe-HgTe,

146

ZnTe-HgTe, AlAs-GaAs, GaAs-InAs, and AlAs-InAs pairs, respectively. Note that δ_{pd} is large for Hg-containing semiconductor pairs (since the Hg 5d orbitals are shallower than other column-II cation d orbitals, and hence repel the VBM more effectively) and for Al-containing compounds (since the *empty* Al 3d orbitals are higher in energy than the anion p orbitals, and hence ΔE_{pd} is *negative*). For all other common-anion pairs ΔE_{pd}^{AC} is similar, and hence the tight-binding model is expected to work well for these systems.

In conclusion, we find that the principal error in previous tight-binding models for band lineup in lattice-matched compounds (and to a lesser extent also in plane-wave pseudopotential models[7] which also neglect occupied cation d bands) is omission of cation d orbitals, and that the assertion that interface dipole effects are needed to obtain the correct lineup is not tenable.

After the results of this work were circulated privately, Duc, Hsu, and Faurie informed us of their new photoemission measurements[24] of the band offsets in CdTe-ZnTe (0.10 ± 0.06 eV) and ZnTe-HgTe (0.25 ± 0.05 eV), in excellent agreement with our independent predictions of Table I.

This work was supported by the Office of Energy Research, Materials Science Division, U.S. Department of Energy, Grant No. DE-AC02-77-CH00178.

[1]S. P. Kowalczyk, J. T. Cheung, E. A. Kraut, and R. W. Grant, Phys. Rev. Lett. **56**, 1605 (1986).

[2]W. I. Wang and F. Stern, J. Vac. Sci. Technol. B **3**, 1280 (1985).

[3]A. D. Katnani and G. Margaritondo, Phys. Rev. B **20**, 1944 (1985).

[4a]W. Harrison, J. Vac. Sci. Technol. **14**, 1016 (1977).

[4b]W. Harrison, J. Vac. Sci. Technol. B **3**, 1231 (1985).

[4c]W. A. Harrison and J. Tersoff, J. Vac. Sci. Technol. B **4**, 1068 (1986).

[5]J. A. Van Vechten, Phys. Rev. **187**, 1007 (1964), and J. Vac. Sci. Technol. B **3**, 1240 (1985).

[6]J. Tersoff, Phys. Rev. Lett. **56**, 2755 (1986).

[7]C. G. Van de Walle and R. M. Martin, Phys. Rev. B **34**, 5621 (1986).

[8]See recent review by R. Bauer and G. Margaritondo, Phys. Today **41**, No. 1, 26 (1987).

[9]See review by G. Duggan, J. Vac. Sci. Technol. B **3**, 1224 (1985).

[10]W. A. Harrison, *Electronic Structure and Properties of Solids* (Freeman, San Francisco, 1980), pp. 74–80; A. A. Levin, *Solid State Quantum Chemistry* (McGraw Hill, New York, 1977).

[11]J. O. McCaldin, T. C. McGill, and C. A. Mead, Phys. Rev. Lett. **36**, 56 (1976).

[12]D. J. Olego, J. P. Faurie, and P. M. Raccah, Phys. Rev. Lett. **55**, 328 (1985).

[13]See, for example, A. Zunger, in *Solid State Physics*, edited by H. Ehrenreich and D. Turnbull (Academic, New York, 1986), Vol. 39, p. 275 (see Sec. VI.29); A. Zunger, Phys. Rev. Lett. **54**, 849 (1985).

[14]We evaluate the tight-binding expressions of Eqs. (2)–(5) using semirelativistic local-density orbital energies calculated with the Hedin-Lundqvist exchange correlation. The input data are s energies -15.43, -6.31, -6.04, -7.21, -14.77, -7.91, -9.25, and -8.56 eV; and p energies -6.19, -1.31, -1.41, -1.26, -5.42, -2.86, -2.82, and 2.78 eV for Te, Zn, Cd, Hg, As, Al, Ga, and In, respectively. V_{pp} is taken from Ref. 4c. Combination of relativistic with nonrelativistic (Hartree-Fock) orbital energies (Ref. 4c) has previously produced erroneous results for the CdTe-HgTe pair. *Relativistic* orbital energies produce better results for GaAs-InAs than those given in Ref. 4c.

[15]M. L. Cohen and V. Heine, in *Solid State Physics*, edited by H. Ehrenreich, F. Seitz, and D. Turnbull (Academic, New York, 1970), Vol. 24, p. 38.

[16]J. Bernard and A. Zunger, Phys. Rev. B (to be published).

[17]J. E. Jaffe and A. Zunger, Phys. Rev. B **29**, 1882 (1984).

[18]S.-H. Wei, H. Krakauer, and M. Weinert, Phys. Rev. B **32**, 7792 (1985).

[19]D. M. Wood, S.-H. Wei, and A. Zunger, Phys. Rev. Lett. **58**, 1123 (1987).

[20]The spin-orbit–corrected ΔE_{VBM} is obtained by our adding $\Delta_0/3$ to each compound, where Δ_0 is the calculated spin-orbit splitting at VBM. The calculated Δ_0 are 0.89, 0.87, 0.79, 0.34, and 0.30 eV for ZnTe, CdTe, HgTe, GaAs, and AlAs, respectively.

[21]N. J. Shevchik, J. Tejeda, and M. Cardona, Phys. Rev. B **9**, 2627 (1974).

[22]In estimating the charge transfer radius R_A we note that, in the formation of ABC_2 from $AC + BC$, the charge *transfer* $(\Delta q - \Delta Q)$ occurs on the *cation* sublattice. Since cation radii are smaller than anion radii in these systems, $R_A < d/2$.

[23]C. K. Shih, J. A. Silberman, A. K. Wahi, G. P. Carey, I. Lindau, W. E. Spicer, and A. Sher [Proceedings of the U.S. Workshop on Mercury Cadmium Telluride, Dallas, TX, October 1986, edited by G. Lucovsky, J. Vac. Sci. Technol. A (to be published)] have indeed observed that if cation core levels are assumed to be unshifted in the alloying process ($\Delta \bar{V}_A \cong 0$), the composition variation of the energy difference between VBM and core level in the $Hg_{1-x}Cd_xTe$ system gives the same band offset as that measured directly (Ref. 1).

[24]T. M. Duc, C. Hus, and J. P. Faurie, Phys. Rev. Lett. **58**, 1127, 2153(E) (1987). We thank these authors for communicating their results before they were submitted for publication.

147

RAPID COMMUNICATIONS

PHYSICAL REVIEW B   VOLUME 36, NUMBER 17   15 DECEMBER 1987-I

Common-anion rule and its limits: Photoemission studies of CuIn$_x$Ga$_{1-x}$Se$_2$-Ge and Cu$_x$Ag$_{1-x}$InSe$_2$-Ge interfaces

D. G. Kilday and G. Margaritondo

Department of Physics and Synchrotron Radiation Center, University of Wisconsin, Madison, Wisconsin 53706

T. F. Ciszek, S. K. Deb, S.-H. Wei, and Alex Zunger

Solar Energy Research Institute, Golden, Colorado 80401

(Received 16 September 1987)

Synchrotron-radiation photoemission data show that the valence-band discontinuities of CuIn$_x$Ga$_{1-x}$Se$_2$-Ge and Cu$_x$Ag$_{1-x}$InSe$_2$-Ge interfaces are independent of x within the experimental accuracy of 0.1 eV. We argue that this result is consistent with the Wei-Zunger explanation of the breakdown of the common-anion rule, which is based on a substantial role of the cation d states in determining the valence-band-edge position.

The breakdown of the common-anion rule,[1] demonstrated by a number of recent experiments,[2-9] has a major impact in semiconductor physics. The thought underlying the common-anion rule was that the valence-band maximum of covalent semiconductors is composed mostly of anion p orbitals, hence a pair of semiconductors sharing the same structure, lattice constant, and a common anion (e.g., GaAs-AlAs, CdTe-HgTe) would exhibit very similar valence-band-edge energies E_v, or a vanishing valence-band discontinuity, ΔE_v. This viewpoint is anchored in simple tight-binding approaches[10] which describe covalent semiconductors solely in terms of anion and cation *valence s* and p orbitals (note that[11] isovalent cations have similar p orbital energies, hence contribute little to ΔE_v). Wei and Zunger[12] have shown that the fallacy in this approach is that valence-band states of common-anion semiconductors manifest *cation* components in addition to anion states. Whereas such *extravalence* states (e.g., Al $3d$) or *subvalence* states (e.g., Zn, Cd, and Hg outer d orbitals), were previously thought[10,13] to be unimportant near E_v, Wei and Zunger have demonstrated through all-electron first-principle calculations that these cation orbitals provide *the main discriminating factor* between a pair of lattice-matched common-anion semiconductors, and hence the band offset.

Our understanding of band discontinuities is based primarily on binary (elemental or compound) materials. There exist, however, a large number of ternary semiconductors (e.g., $T^I B^{III} X_2^{VI}$ chalcopyrites) which allow us to test our current understanding of this problem by providing a choice of *two* cations (T^I =Cu,Ag and B^{III} =Al,Ga,In). Furthermore, since they include a noble metal T^I with relatively shallow d states, these materials offer a natural test to theories which emphasize the role of d states. We present here photoemission measurements of clean and Ge-covered CuInSe$_2$, CuGaSe$_2$, and AgInSe$_2$, and of two series of alloys of these ternary chalcopyrites. The results are consistent with the predictions derived from the Wei-Zunger model.

The experiments consisted of measuring ΔE_v for interfaces obtained by depositing Ge on freshly cleaned chalcopyrite surfaces. This approach has been extensively used to estimate the empirical position in energy of the

valence-band edges of semiconductors on an absolute scale whose zero is arbitrarily set to coincide with the top of the Ge valence band.[14,15] The value of ΔE_v for any pair of semiconductors can be estimated, in first approximation, by taking the difference between the corresponding empirical valence-band-edge positions.[14,15] There is ample evidence[14] that the accuracy of this empirical method is, on the average, 0.1–0.15 eV.

Before discussing the results of our experiments, we will extract from the detailed model of Wei and Zunger the essential physics, as it pertains to both binary and ternary semiconductors. This will establish the expectations to be tested experimentally. Consider, first, two binary common-anion semiconductors AX and BX and neglect the effect of d states on the band discontinuity between them [Fig. 1(a)]. The cation p orbitals (A^{II},p) and (B^{II},p) can couple with the anion p orbital (X,p), as all have the same symmetry (t_2, or Γ_{15}) in the zinc-blende lattice. This coupling results in the two bonding states $\Gamma_{15v}(A-X)$ and $\Gamma_{15v}(B-X)$, whose energy difference provides in this model (i.e., neglecting charge-transfer effects[10] the valence-band discontinuity ΔE_v for the AX-BX interface.

We see that each of these bonding states is repelled to *deeper* energies relative to (X,p) since the cation p orbital energy is *above* the anion p orbital energy. This repulsion, $|V(A,p;X,p)|^2/[\varepsilon(A,p)-\varepsilon(X,p)]$, is proportional to the square of the coupling matrix element, $V(A,p;X,p)$, and hence it increases as the $A-X$ bond length becomes shorter. If AX and BX have the same bond length and similar cation p energies (as is the case[11] in CdTe-HgTe or AlAs-GaAs), this model predicts a vanishing valence-band offset, in contrast with the experimental results.[2-9] This model correctly predicts, however, that the valence-band maximum (VBM) energy of the material with shorter bond length is deeper than that of the material with the longer bond length, e.g., $E_v(GaX)<E_v(InX)$. It hence gives an essentially correct value[16] for the valence-band discontinuity of the GaAs-InAs heterojunction, 0.17 eV. Corrections due to Ga and In d states were found to be small (0.04 eV in Ref. 12).

In many binary materials, e.g., AX and BX, the cation d states cannot be neglected. If these orbitals are *below* the anion p state (e.g., Cu $3d$, Zn $3d$, Ag $4d$, Cd $4d$, Au $5d$, or Hg $5d$), they will repel *upwards* the valence-band maximum, as shown in Fig. 1(b). Since this repulsion is proportional to $|V(A,d;X,p)|^2/[\varepsilon(A,d)-\varepsilon(X,p)]$, it becomes larger when the $A-X$ bond becomes shorter and when the d orbital energy of the cation A becomes more shallow. Hence, since the Hg $5d$ state is shallower than the Cd $4d$ state, HgTe would have a *higher* VBM energy than CdTe.[12] This model leads to a finite band effect between common-anion pairs with the same bond length and similar cation p orbital energies, in agreement with experiment. (Note that Al has empty, high-energy $3d$ orbitals which *lower* the VBM and AlAs relative to GaAs. This leads[12] to $\Delta E_v = 0.4-0.45$ eV for the AlAs-GaAs heterojunction).

This model can be extended to ternary materials in a simple way. We will consider two common-anion cases: (i) In Fig. 2, we show the model for ternary materials with a common noble metal, TAX_2-TBX_2 (e.g., CuInSe$_2$ and CuGaSe$_2$) and (ii) in Fig. 3, we show ternary materials with a common group-III element, $T^{(1)}BX_2$ and $T^{(2)}BX_2$ (e.g., CuInSe$_2$ and AgInSe$_2$).

In the case of TAX_2 and TBX_2, if the effect of the cation d orbitals is neglected, our simple model predicts [Fig.

2(a)] that the material with the *shorter* bonds (TAX_2 here, or CuGaSe$_2$) will have a *deeper* E_v than the material with the longer bond (TBX_2 here, or CuInSe$_2$). In contrast, if the d bands of T are taken into account [Fig. 2(b)], they repel the VBM *upwards*, i.e., on an opposite direction to the effect of the cation p states [Fig. 2(a)]. This repulsion is naturally larger for the material with shorter bonds. Hence, *the noble metal T has an opposite effect on the band offset than the main group cations A and B.* This partial cancellation suggests

$$E_v(TAX_2) > E_v(AX) ,$$
$$E_v(TBX_2) > E_v(BX) ,$$
$$\Delta E_v(TAX_2-TBX_2) < \Delta E_v(AX-BX) ,$$

i.e., that the common-anion rule is better obeyed in ternary systems than in binary systems. Since the valence-band discontinuity of the GaX-InX interface, ΔE_v (GaX-InX), is already small (0.1–0.2 eV for A =Ga, B =In, and X =As or Sb and nearly X independent in binaries,[14] this model predicts an essentially vanishing valence-band discontinuity for the interface between the two ternary compounds, i.e., ΔE_v (CuInSe$_2$-CuGaSe$_2$) $\simeq$0. Note also that this model predicts a *deeper* d-band energy [$\Gamma_{15}(pd)$

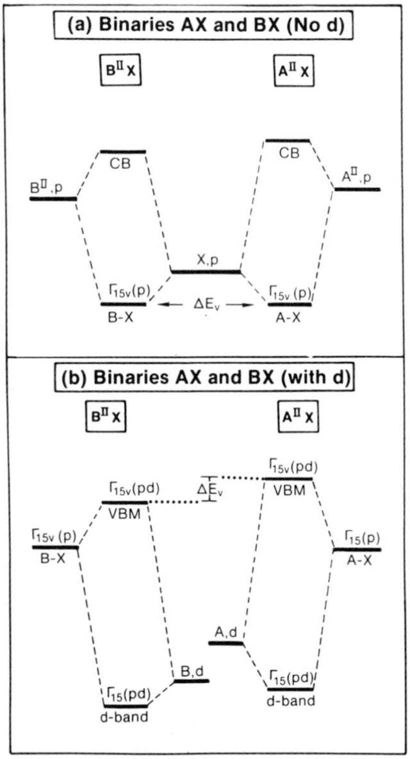

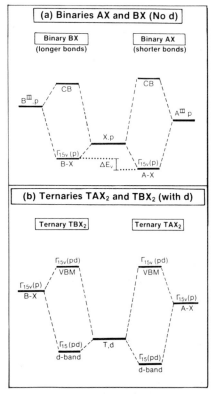

FIG. 1. Energy-level diagram for the band offset of binary semiconductor heterojunctions, (a) neglecting or (b) considering the cation d states. CB denotes conduction band; VBM denotes valence-band maximum.

FIG. 2. Energy-level diagram for the band offset of a heterojunction formed by a pair of ternary semiconductors with a common noble metal, (a) neglecting or (b) considering the role of the cation d state.

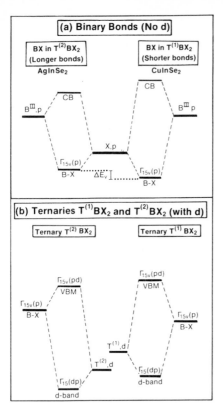

FIG. 3. Energy-level diagram for the band offset of a hetero-junction formed by a pair of ternary semiconductors with a common group-III element (a) neglecting or (b) considering the role of the cation d states.

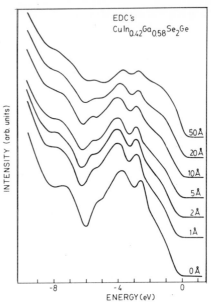

FIG. 4. Valence-band discontinuities for $CuIn_xGa_{1-x}Se_2$-Ge (top) and $Cu_xAg_{1-x}InSe_2$-Ge (bottom) interfaces, derived from our spectra and plotted as a function of x. The values of x were 0, 0.17, 0.42, 0.72, 0.89, 1.0 for $CuIn_xGa_{1-x}Se_2$, and 0, 0.35, 0.67, 0.85, 0.96, 1.0 for $Cu_xAg_{1-x}InSe_2$.

in Fig. 2(b)] for the material with the shorter bonds (Cu-GaSe$_2$), in agreement with band-structure calculations.[17]

In case (ii), T^1BX_2 and T^2BX_2, if the noble metal d bands are ignored [Fig. 3(a)], this model predicts again that the material with the shorter bonds (T^1BX_2 or CuInSe$_2$) has a deeper E_v relative to the material with the longer bonds (T^2BX_2 or AgInSe$_2$). Inclusion of d bands [Fig. 3(b)] shows again an *upward* repulsion of the VBM. This repulsion is larger the shorter the bonds are and the shallower the noble metal d orbitals are. Since in our example, T^1BX_2 (i.e., CuInSe$_2$) has both shorter bonds and shallower (Cu $3d$) orbital energies relative to T^2BX_2 (i.e., AgInSe$_2$ with Ag $4d$), these factors will make the upward repulsion larger in CuInSe$_2$. However, the Ag $4d$ orbitals are more extended than the Cu $3d$ orbitals, contributing therefore to a larger pd matrix element for Ag containing systems. The effects of upward displacements of the VBM partially offsets the effects of the main group cation [Fig. 3(a)], leading to a reduction of ΔE_v in the ternary compounds. Whereas detailed calculations may be needed to determine ΔE_v accurately, this model suggests again a small band offset.

In summary, the Wei-Zunger model predicts that the

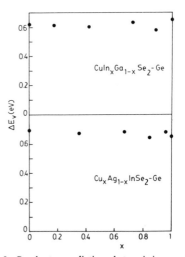

FIG. 5. Synchrotron-radiation photoemission spectra taken on a clean $CuIn_{0.42}Ga_{0.58}Se_2$ surface, and on the same surface covered with *in situ* deposited Ge overlayers of increasing thickness (shown in Å for each curve). The spectra were taken with a photon energy of 20 eV, obtained by filtering the emission of the 1-GeV storage ring at the Wisconsin Synchrotron Radiation Center with an Al Seya-Namioka monochromator. The horizontal scale is referred to the leading edge of the spectrum corresponding to a thick Ge overlayer.

position of the VBM is only weakly modified in going from $CuInSe_2$ to $CuGaSe_2$ and from $CuInSe_2$ to $AgInSe_2$. Our experimental data are consistent with these predictions. They show, in fact, that the empirical position of the valence-band edge is constant, within an accuracy limit of the order of 0.1 eV, for all compounds in the two alloy series $CuIn_xGa_{1-x}Se_2$ and $Cu_xAg_{1-x}InSe_2$. The magnitude of the accuracy is much smaller than the typical ΔE_v's observed when the common-anion rule fails.[2–9]

Crystals of $CuIn_xGa_{1-x}Se$ and $Cu_xAg_{1-x}InSe_2$ were grown from the melt by a liquid-encapsulated Bridgman-Stockbauer method after *in situ* synthesis from the elements.[18] Values of x equal to 0, 0.25, 0.5, 0.75, 0.9, and 1.0 were chosen for the starting melt composition. Electron microprobe analysis was used to determine the crystal compositions at the 15%-fraction-solidified point, the region used for the investigation. These compositions are indicated in Fig. 4.

Figure 5 shows a typical set of data, obtained for the compound $CuIn_{0.42}Ga_{0.58}Se_2$. The figure includes synchrotron-radiation photoemission spectra taken on the clean surface of this compound (obtained by operating a diamond grinder under ultrahigh vacuum) and on the same surface covered by Ge overlayers of increasing thickness. The method for extracting the valence-band discontinuity of the $CuIn_{0.42}Ga_{0.58}Se_2$-Ge interface from these data is well established, and described in detail in Refs. 14 and 15. In essence, ΔE_v is given by the distance in energy between the leading edges of the clean-surface spectrum and of the spectrum corresponding to a thick Ge overlayer, corrected for the Ge-induced changes in the substrate band bending. In turn, the latter can be derived

from the rigid shift in energy of the clean-surface spectral features, which occurs upon Ge deposition and is clearly visible in Fig. 5. From the data of this figure, we deduce a ΔE_v of 0.60 eV. Thus, the position of the $CuIn_{0.42}$-$Ga_{0.58}Se_2$ valence-band edge on the empirical scale referred to germanium is -0.60 eV.

Previous measurements on the end compositions of the first alloy series were inconclusive as far as the possible compositional dependence of the empirical valence-band-edge position is concerned.[19] The measured values of ΔE_v with respect to Ge were closer to each other than their combined experimental uncertainties. The present study, extended to a number of intermediate compositions, offers much better evidence for the substantial constancy of the valence-band-edge position. This point is emphasized in Fig. 4, which shows ΔE_v for the two alloy series as a function of composition.

In conclusion, our data are entirely consistent with the predictions of the Wei-Zunger model, extended to ternary semiconductor heterojunctions. In essence, we find that the common-anion rule, which is strongly violated in numerous binary semiconductors, works instead reasonably well for ternary materials and for their alloys. This is due to the interplay of several different factors, and among these the shallow d states play an important role.

This research was supported in part by the National Science Foundation, Grant No. DMR-84-21212 (University of Wisconsin) and by the U.S. Department of Energy, Contract No. De-ACO-2-83CH10093 (SERI). The Wisconsin Synchrotron Radiation Center is a National Science Foundation supported national facility.

[1]J. O. McCaldin, T. C. McGill, and C. A. Mead, Phys. Rev. Lett. **36**, 56 (1976).

[2]W. I. Wang and F. Stern, J. Vac. Sci. Technol. B **3**, 1280 (1985).

[3]J. Menéndez, A. Pinczuk, D. J. Werder, A. C. Gossard, and J. H. English, Phys. Rev. B **33**, 8863 (1986); D. Arnold, A. Ketterson, T. Henderson, J. Klem, and H. Morkos, J. Appl. Phys. **57**, 2880 (1985); J. R. Waldrop, S. P. Kowalczyk, R. W. Grant, E. A. Kraut, and D. L. Miller, J. Vac. Sci. Technol. **19**, 573 (1981); B. A. Wilson, P. Dawson, C. W. Tu, and R. C. Miller, Appl. Phys. Lett. **48**, 341 (1986); J. Batey and S. L. Wright, J. Appl. Phys. **59**, 200 (1985).

[4]A. D. Katnani and R. S. Bauer, Phys. Rev. B **33**, 1106 (1986).

[5]G. Duggan, J. Vac. Sci. Technol. B **3**, 1224 (1985), and references therein.

[6]S. P. Kowalczyk, J. T. Cheung, E. A. Kraut, and R. W. Grant, Phys. Rev. Lett. **56**, 1605 (1986).

[7]C. Tejedor, J. M. Calleja, F. Mesenguer, E. E. Mendez, C. A. Chang, and L. Esaki, in *Proceedings of the International Conference on the Physics of Semiconductors, San Francisco, 1984,* edited by D. J. Chadi and W. A. Harrison (Springer, Berlin, 1985), p. 559.

[8]G. J. Gualtieri, G. P. Schwartz, R. G. Nuzzo, and W. A. Sunder (unpublished).

[9]D. W. Niles, B. Lai, J. T. McKinley, G. Margaritondo, G. M. Wells, F. Cerrina, G. J. Gualtieri, and G. P. Schwartz, J. Vac. Sci. Technol. (to be published).

[10](a) W. A. Harrison, *Electronic Structure and Properties of Solids* (Freeman, San Francisco, 1980); (b) J. Vac. Sci.

Technol. **14**, 1016 (1977).

[11]For example, according to Ref. 10(a) the cation p orbital energies for (Al,Ga,In) are (-4.86 4.90, -4.69 eV) and for (Zn,Cd,Te) they are (-3.38, -3.38, -3.48 eV).

[12]S.-H. Wei and A. Zunger, Phys. Rev. Lett. **59**, 144 (1987), and references therein.

[13]W. R. Frensley and H. Kroemer, Phys. Rev. B **16**, 2642 (1977); W. A. Harrison, J. Vac. Sci. Technol. B **3**, 1231 (1985); O. F. Sankey, R. E. Allen, S. F. Ren, and J. D. Dow, *ibid.* **3**, 1162 (1985); A. B. Chen and A. Sher, Phys. Rev. B **32**, 3695 (1985).

[14]G. Margaritondo and P. Perfetti, in *Heterojunctions Band Discontinuities: Physics and Device Applications,* edited by F. Capasso and G. Margaritondo (North-Holland, Amsterdam, 1987), Chap. II, and references therein.

[15]R. S. Bauer and G. Margaritondo, Phys. Today **41** (No. 1), 26 (1987); A. D. Katnani and G. Margaritondo, Phys. Rev. **28**, 1944 (1983); A. D. Katnani and G. Margaritondo, J. Appl. Phys. **54**, 2522 (1983).

[16]S. P. Kowalczyk, W. J. Schaffer, E. A. Kraut, and R. W. Grant, J. Vac. Technol. **20**, 705 (1982).

[17]J. Jaffe and A. Zunger, Phys. Rev. B **28**, 5822 (1983).

[18]T. F. Ciszek, J. Cryst. Growth **79**, 689 (1986).

[19]M. Turowski, G. Margaritondo, M. K. Kelly, and R. D. Tomlinson, Phys. Rev. B **31**, 1022 (1985). The present data imply a slightly different value of the empirical valence-band-edge position for $CuInSe_2$. However, such a correction does not modify the conclusion that the conducting-band discontinuity is small for $CuInSe_2$-CdS heterojunctions.

Elementary theory of heterojunctions

Walter A. Harrison[a]

*Department of Applied Physics, Stanford University, Stanford, California 94305
and Science Center, Rockwell International, Thousand Oaks, California 91360*

(Received 10 February 1977; accepted 18 March 1977)

An LCAO theory of heterojunction band-edge discontinuities is formulated and tested for approximate self-consistency. It leads to a table of valence-band maxima for all tetrahedral semiconductors; discontinuities can be obtained from the table directly by subtraction. The discrepancies with the current scattered data do not appear significantly larger than the uncertainty in those data, a few tenths of electron volts. A pseudopotential theory of such discontinuities is also formulated, based upon self-consistent atomic pseudopotentials. This leads to valence-band maxima reasonably consistent with the LCAO theory, except for junctions between materials of significantly different bond length. It also suggests that the Frensley–Kroemer scheme does produce self-consistency for systems of matching lattice constant, but produces incorrect trends with mismatch in lattice constant. The goal in any case is taken to be a table of valence-band maxima. LCAO values seem a better standard than photoelectric thresholds, though a comparison of the two indicates them to be roughly consistent for treating junctions if both sides are homopolar or if both sides are polar.

PACS numbers: 73.40.Lq, 71.10.+x

I. INTRODUCTION

The junction between two materials in which the crystal structure is continuous is the simplest kind of interface. The most apparent, and perhaps the most important, property of such a junction is the shift in the band edges across the junction. One might at first think that the loss of the translational symmetry of single-crystal materials would require extensive calculation in order to solve the electronic structure from which the question could be answered. Such calculations can be done be replacing the system by a periodic set of slabs.[1] However, in this particular case the central problem is the potential to be used. Once the potential is selected, the shift in the band edges can be obtained immediately without solving for the electronic states. This can be done, for example, by examining a part of the system well removed from the junction and asking what is the maximum energy for which a propagating valence-band state can be constructed; this becomes the valence-band maximum in that region. The calculation of that maximum energy could be done by setting periodic boundary conditions on a limited region of material and thus carrying out a local band calculation. In fact that will not even be necessary. We will be able to extract the result of carrying out these calculations without going through the numerical details.

A plausible semiempirical theory of the discontinuities has been given by Anderson[2,3] and used by Shay *et al.*[4] In this approach the shift in the conduction-band minimum is set equal to the difference in electron affinities for the two materials. In some sense, however, this replaces one simple problem by two very difficult problems since the junction between the material and vacuum is intrinsically much more uncertain.[5] It is not clear that the effects of surface dipole layers and surface reconstruction would cancel out to the extent required though we will see that they appear to on

equivalent faces. In addition, the wide variation in experimental values for the electron affinities makes the model very difficult to test. We therefore seek a microscopic theory of the heterojunction itself.

II. LCAO THEORY

In the LCAO theory of perfect crystals the individual electron states are written as linear combinations of atomic orbitals

$$|\psi_k\rangle = \sum_\alpha u_\alpha |\alpha\rangle. \tag{1}$$

In covalent solids (polar as well as nonpolar) a rather good description can be given using only a single atomic s state and three atomic p states on each atom,[6,7] and these are all taken to be orthogonal to each other. This works particularly well for the valence-band states and we will concentrate on them. Translational periodicity allows one to relate the coefficients u_α on translationally equivalent orbitals using a phase factor $e^{ik \cdot (r_\alpha - r_\beta)}$ for a state of wave number k and reduces the problem to the diagonalization of an eight-by-eight matrix, corresponding to the four orbitals on each of the two atoms per primitive cell. At special wave numbers it is possible to diagonalize the matrix exactly. In particular, the valence band maximum at $k = 0$ has been found to be given by[6]

$$E_v = \frac{\epsilon_p^c + \epsilon_p^a}{2} - \left[\left(\frac{\epsilon_p^c - \epsilon_p^a}{2} \right)^2 + V_{xx}^2 \right]^{1/2}, \tag{2}$$

where ϵ_p^c is the p-state energy on the metallic atom (cation) and ϵ_p^a is the p-state energy on the nonmetallic atom (anion). The matrix element V_{xx} is an appropriate interatomic matrix element between atomic p states on adjacent atoms; matrix elements between states on more distant atoms are neglected. A triply degenerate conduction-band state at $k = 0$ is given

208

TABLE I. Atomic p-state energies, taken from Herman and Skillman (Ref. 11) or extrapolated from their values, all in eV.

Be	4.2	Cu	1.83
B	6.64	Zn	3.38
C	8.97	Ga	4.90
N	11.47	Ge	6.36
O	14.13	As	7.91
F	16.99	Se	9.53
		Br	11.20
Mg	2.99	Ag	2.05
Al	4.86	Cd	3.38
Si	6.52	In	4.69
P	8.33	Sn	5.94
S	10.27	Sb	7.24
Cl	12.31	Te	8.59
		I	9.97

by the same expression with the sign before the square root changed.

It has further been found that a reasonably good description of the bands in the perfect crystal is obtained if the p-state and s-state energies in the crystal are taken as atomic term values[6-8] and that interatomic matrix elements such as V_{xx} tend to vary with bond length d from material to material as d^{-2}.[9,10] This particular matrix element may be estimated as[8]

$$V_{xx} = 2.16\hbar^2/md^2, \tag{3}$$

where the coefficient 2.16 was chosen to accord with the values fit to the true bands for silicon and germanium by Chadi and Cohen.[6] The term values may be taken from the Herman–Skillman tables[11] and are listed in Table I for convenience. Such parameters are all that are needed for an approximate description of the energy bands for any of the tetrahedral solids and those few we have given suffice to give the splitting between the triply degenerate levels at Γ. It may be of interest that in a recent reformulation[7] of a bond-orbital theory of tetrahedral solids' V_{xx} and $(\epsilon_p^c - \epsilon_p^a)/2$ are directly identified with the covalent and polar energies V_2 and V_3 (see Ref. 9 for a definition of these parameters). We do not, however, make use of that fact here.

Use of differences in atomic-term values has given a good description of the electronic structure of the polar semiconductors and we can hope that they may be used also in the heterojunction where the different atom types are separated to different parts of the crystal; the only question is the extent to which this corresponds to a self-consistent potential. We may see that it should be reasonably self-consistent by imagining construction of the electron states in the system by a series of unitary transformations.[7,9] We begin with atomic orbitals on each atom, form sp^3 hybrids, and finally bond orbitals in each bond; two electrons are associated with each bond. All of this is done independently for each bond and should not entail approximations other than that made for the perfect crystal. If we were then to make the bond-orbital approximation[10] of forming the valence-band states for the heterojunction from bond orbitals alone (no antibonding orbitals), the corresponding unitary transformation would not entail any charge redistribution and the parameters of the theory (the ϵ_p in particular) would remain rigorously un-

changed; Eq. (2) could be used to obtain the valence-band maximum in each region.

The only error in this solution is the neglect of the admixture of antibonding orbitals as the final electron states are obtained. We may estimate the error by calculating the admixture of antibonding states in perturbation theory. (Here it is matrix elements for hybrids rather than for p states which enter.) These effects tend to be small; each silicon bond orbital has admixed contributions from each of the six nearest-neighbor antibonding orbitals corresponding to about 0.005 electrons.[7] In the bulk material such effects are included in the potential when the energy bands are fit, but at a silicon–germanium junction, for example, the germanium antibonding orbitals added to the silicon–germanium bond at the junction produces a dipole which is not exactly cancelled by the silicon–germanium antibonding orbital added to the germanium bond orbital. An additive effect occurs on the silicon side. The corresponding dipole layer, reduced by a factor of the dielectric constant (dividing by the dielectric constant should approximate a self-consistent solution), gives a shift of the band edges on the two sides of about 0.01 eV. (The largest effect was the stronger coupling with germanium, due to the larger sp-splitting, giving a dipole which lowers the electron energy in silicon relative to that in germanium.) These effects appear to be negligible, and are omitted in this first treatment. Effects appear to be similarly small in the compound semiconductors. (See Appendix)

Neglecting these shifts due to charge redistribution, we may immediately evaluate the energy of the valence-band maximum using Eqs. (2) and (3) and Table I. The results are given in Table II. Note that these energies are not meaningful as photoelectric thresholds (energies relative to vacuum) because of surface effects, as we will see in Sec. IV. However, they can be subtracted to obtain the difference in valence-band maximum on the two sides of a heterojunction. The results of this calculation for four systems where there is experimental information are given in Table III. The value for GaAs–Ga$_{0.8}$Al$_{0.2}$As is, of course, one-fifth the GaAs–AlAs value. The agreement with experiment can be considered very good.

The theory is directed at the valence-band edges, and these give the clearest specification of parameters since that edge occurs always at Γ. However, we may also obtain the conduction-band discontinuities

$$E_c - E_c' = E_v - E_v' + E_g - E_g' \tag{4}$$

from the experimental band gaps, E_g or E_0. A number of these are included in Table II. These also show experimental uncertainties of up to 0.2 eV, so the uncertainties here are even greater. The values of $E_c - E_c'$ for the four junctions of Table III are included there. Ga$_{0.8}$Al$_{0.2}$As is a direct-gap semiconductor so the relevant gap is E_0 and $E_g - E_g'$ in Eq. (4) is taken as one-fifth $E_0(\text{GaAs}) - E_0(\text{AlAs})$. Comparable agreement to that for the valence-band edge is essentially guaranteed by the use of Eq. (4). Similar agreement was obtained by Shay et al.,[4] using electron affinities for Ge–Si, Ge–GaAs, and InP–CdS. (The extrapolations and assumptions used for GaAs–AlAs make it difficult to assess.) Because of the problems involved in semiconductor-vacuum interfaces and because of the uncertainties in electron affinity we prefer the

TABLE II. Valence-band edge E_v [Eq. (2)], direct gap E_O, and indirect gap E_g, all in eV, and bond length d.

	d(Å)	$-E_v$	E_O	E_g
C	1.54	15.91		5.5[a]
BN	1.57	16.16		
BeO[d]	1.65	16.27		
Si	2.35	9.50	4.18[a]	1.13[a]
AlP	2.36	10.03		
Ge	2.44	9.12	0.89[a]	0.76[a]
GaAs	2.45	9.53	1.52[b]	
ZnSe	2.45	10.58	2.82[b]	
CuBr	2.49	11.90		
Sn	2.80	8.04		
InSb	2.81	8.41	0.24[b]	
CdTe	2.81	9.32	1.60[b]	
AgI	2.80	10.49		
SiC	1.88	12.56	7.75[a]	2.3[a]
BP	1.97	11.81		
AlN[d]	1.89	13.84		
BeS	2.10	12.05		
BAs	2.07	11.17		
GaN[d]	1.94	13.66		
BeSe	2.20	11.19		
ZnO[d]	1.98	15.58	3.40[b]	
CuF	1.84	18.41		
InN[d]	2.15	13.00		
BeTe	2.40	10.00		
AlAs	2.43	9.57	2.77[c]	
GaP	2.36	10.00	2.77[a]	2.38[a]
ZnS	2.34	11.40	3.80[a]	
CuCl	2.34	13.11		
AlSb	2.66	8.67	2.5[a]	1.87[b]
InP	2.54	9.64	1.37[a]	
MgTe[d]	2.76	9.33		
CdS	2.53	11.12	2.56[b]	
GaSb	2.65	8.69	0.81[b]	
InAs	2.61	9.21	0.42[b]	
ZnTe	2.64	9.50	2.39[b]	
CdSe[d]	2.63	10.35	1.84[b]	
CuI	2.62	10.62		

[a] J. C. Phillips, *Bonds and Bands in Semiconductors* (Academic, New York, 1973), p. 169.
[b] P. Lawaetz, Phys. Rev. B **4**, 3460 (1971).
[c] Extrapolated from Ref. b of Table III.
[d] Wurtzite structure. The three bands are split at Γ. This gives the center of gravity.

theoretical values obtained by subtracting the E_v's of Table II.

III. PSEUDOPOTENTIAL THEORY

A natural alternative to an LCAO theory is a pseudopotential-based theory, such as that of Frensley and Kroemer.[12]

Again the principal uncertainty is in the choice of a self-consistent potential, and indeed the matching of potential values at interstitial positions used by Frensley and Kroemer was viewed by them as a speculation. We will instead directly superimpose self-consistent atomic pseudopotentials and compare with LCAO results and with the Frensley–Kroemer results. Atomic pseudopotentials, such as those of Animalu and Heine,[13] are individually screened (metallic screening). Direct superposition should give approximate self-consistency, particularly in view of our finding in the LCAO theory that the effects of charge redistribution are small, and in view of the familiar transferability of pseudopotentials from one situation to another.[14] Selection of the pseudopotential again makes the problem unambiguous, and again we can extract the answer without detailed calculation.

The pseudopotential is written as a sum of atomic potentials, w_i,

$$W(\mathbf{r}) = \sum_i w_i(\mathbf{r} - \mathbf{r}_i). \qquad (5)$$

The pseudopotentials for $x < 0$ are to correspond to one compound, those for $x > 0$ are to correspond to another. The first question is the difference $\delta \overline{W}$ in average pseudopotential on the two sides, which may be obtained from

$$\delta \overline{W} = \int d^3 r \theta(x) W(\mathbf{r}) \Big/ \int d^3 r \qquad (6)$$

where $\theta(x) = 1$ for $x > 0$ and $\theta(x) = -1$ for $x < 0$. We may evaluate this by making a Fourier expansion of $\theta(x)$ and of W to obtain the result entirely in terms of the pseudopotential form factors w_q for the individual atomic pseudopotentials. Note that w_0 does not enter, but the result for a large system is found to depend only upon the $\lim_{q \to 0} W_q = -2E_F/3$ with E_F the free-electron Fermi energy. In fact the result, once obtained, is obvious. Since the self-consistency was obtained as for a metal, the Fermi–Thomas result of

$$\delta \overline{W} = -(2E_F/3N)\delta N \qquad (7)$$

must be obtained, where δN is the difference in electron density on the two sides and the lower $\overline{W}$ is on the side with higher electron density. We may in fact write $\overline{W}$ for each material (measured from a common metallic Fermi level) as

$$\overline{W} = -\left(\frac{9\sqrt{3}\pi^2}{4\sqrt{2}}\right)^{2/3} \frac{\hbar^2}{md^2} = -\frac{68.9 \text{ eV} - \text{Å}^2}{d^2} \qquad (8)$$

and then perform band calculations (adding W_q's for the

TABLE III. Valence-band and conduction-band discontinuities for S-S' heterojunctions, in eV.

S	S'	Theory $E_v - E_v'$	Experiment	Theory $E_c - E_c'$	Experiment
Ge	Si	0.38	(0.24 to 0.17)[a]	0.01	−0.12 to −0.19[d]
Ge	GaAs	0.41	(0.36 → 0.76)[a]	−0.35	−0.40 to 0[e]
GaAs	$Ga_{0.8}Al_{0.2}As$	0.01	0.03[b]	−0.24	−0.22[b]
InP	CdS	1.48	1.63[c]	0.29	0.56[c]

[a] Obtained from the experimental values of $E_c - E_c'$ using the difference in band gaps, Eq. (4).
[b] R. Dingle, W. Wiegmann, and C. H. Henry, Phys. Rev. Lett. **33**, 827 (1974).
[c] Reference 4.
[d] Reference 3, pp. 52 and 105.
[e] Reference 3, p. 110.

TABLE IV. Pseudopotential calculation of valence-band maxima and comparison with other values. All values in eV.

	$\overline{W}$ [Eq. (8)]	W_q[a]	Bandwidth[a]	$-E_v$ [Eq. (10)]	$-E_v$ (FK)[b]	$-E_v$ (LCAO)[c]
			Variations with metallicity			
Si	−12.48	−2.47	12.5	10.77	8.68	9.50
Ge	−11.57	−2.13	12.6	9.12[d]	9.12[d]	9.12
Sn	−8.79	−1.83	11.3	7.57		8.04
			Variations with polarity			
Ge	−11.57	−2.13	12.6	9.12[d]	9.12[d]	9.12
GaAs	−11.48	−2.12	12.55	9.08	9.72	9.53
ZnSe	−11.48	−2.49	12.25	10.35	10.65	10.58

[a] Reference 12.
[b] Reference 12.
[c] From Table II.
[d] These values were fit by the choice of the zero of energy.

lattice wave numbers) in each region separately in order to obtain the relative band edges.

The valence-band minimum Γ_1 will lie in the neighborhood of $\overline{W}$ since $\overline{W}$ is the expectation value of the Hamiltonian with respect to a constant pseudowave function. We immediately see a difficulty in the pseudopotential approach in that the above calculation is giving the valence-band minimum and we seek energy discontinuities at much different energies. Nonetheless, the results of the calculation will be informative. We will use the $\overline{W}$ calculated above and the pseudopotential of Chelikowski and Cohen[15] to calculate the position of the valence-band maximum. The valence-band minimum could be obtained by diagonalizing a Hamiltonian matrix based upon the Chelikowsky–Cohen pseudopotential, but perturbation theory will suffice. The $k = 0$ plane wave, of energy-expectation value $\overline{W}$, is coupled by pseudopotential matrix elements W_q to plane waves differing in energy by $\hbar^2 q^2/2m$. We will see that in silicon, for example, the coupling for $q = [111]2\pi/a = 3\pi/2d$ and the seven other equivalent q will lower the energy more than 3 eV, while those for wave numbers of the $[220]2\pi/a$ type, and others, give shifts less than 0.1 eV; we may neglect the variation of these small terms from system to system and drop them. We have then

$$E_{v\,\text{min}} = \overline{W} - \frac{8W_q^2}{\hbar^2(3\pi/2d)^2/2m} \qquad (9)$$

and the valence-band maximum is obtained from

$$E_v = E_{v\,\text{min}} + \text{bandwidth} \qquad (10)$$

with bandwidths given by Chelikowsky and Cohen.

We carry out this estimate for an interesting sample of materials in Table IV. The values of E_v are all measured from the same artificial zero of energy and only the differences between them are meaningful. It is easier to compare with other calculations if we alter the zero of energy to bring all values for germanium in register with that in Table II: This entailed adding −7.60 eV to the values from Eq. (10).

The LCAO values are listed in the final column for comparison. The first three entries reflect the variation in values with metallicity (or with bond length), while the second three reflect variations with polarity. If we take the LCAO values as a standard, we see that the pseudopotential calculation overestimates the effects of metallicity but gives a fair account of the variation with polarity; the latter trend came principally

from the final term in Eq. (9). The degree of consistency is in fact somewhat remarkable for both sets in view of the addition and subtraction of two large terms, reflected in the first and third columns. In addition, we have treated the Chelikowsky–Cohen pseudopotential as if it was local. The nonlocal terms in the pseudopotential will not influence the discontinuity across the heterojunction which we associate with the change in $\overline{W}$ since the nonlocal terms do not affect the $\lim_{q\to 0} W_q$; this aspect of the nonlocality is screened out. (For this reason it would be absolutely essential to any nonlocal calculation to carry it out self-consistently; otherwise a spurious contribution to the discontinuity would be introduced.) It would, however, affect the second term in Eq. (9). Because of the use of perturbation theory, however, a careful treatment of the nonlocality did not seem warranted.

A second interesting comparison is with the calculation by Frensley and Kroemer, also listed in Table IV. This provides a test of their matching procedure. We see that in fact their procedure may be justified for systems of varying polarity, and it is not difficult to see why this might be so. The matching is of course exact for a Ge–Ge junction (no junction at all). As we increase polarity on one side, forming GaAs, the interstitial potential might well be expected to remain constant. Similarly, the transfer of charge across the junction might well be negligible; the increased electronegativity of the As is compensated by the decreased electronegativity of the Ga.

On the other hand, Table IV suggests that the trend with metallicity is not even of the correct sign. In a junction with higher electron density on one side a self-consistent potential difference will arise stabilizing the differences in electron density. This shift in average potential level on the two sides is missed by the Frensley–Kroemer scheme, but is included, and perhaps overestimated, in our pseudopotential approach. For studies of heterojunctions this effect may not be of such major importance because junctions are difficult to make if there is appreciable lattice mismatch where the effects would be largest.

IV. SEMIEMPIRICAL APPROACH AND PHOTOTHRESHOLDS

We return finally to the semiempirical approach based upon electron affinities. The equivalent approach based upon photothresholds (the photothreshold equals the electron affinity plus the band gap) compares more directly with the tables here. The magnitudes of the E_v's given in Table II are not directly meaningful as photoelectric thresholds because of surface dipoles, because of the neglect of nonorthogonality of the orbitals making up the LCAO states,[7,9] and because of image-potential effects.[7] We test the extent to which these shifts are the same in all materials by plotting the observed photothreshold Φ against the values of E_v from Table II in Fig. 1. It is in fact seen that to a remarkable extent the data are described by the line drawn there,

$$\Phi = |E_v| - 3.8 \text{ eV} \cdot \qquad (11)$$

To this extent these photothresholds can be subtracted directly to obtain heterojunction discontinuities.

One reason this comes as a surprise is that photothresholds on different faces of the same crystal can differ appreciably due to the differences in dipole layers appearing on different surfaces. (Of course, the energy to remove the electron to infinity cannot differ and electric fields must arise outside the specimen to make this true; however, these are bypassed in the experiment.) All of the measurements on polar semiconductors were made on (110) surfaces and the effects appear to have been similar on all of them. The measurements on Si and Ge were made on (111) surfaces and indeed they are displaced from the line; they would be fit better with a 4.4 eV rather than 3.8 eV in Eq. (11). This difference in surface dipole would not enter the heterojunction step so significant errors could arise in this method as applied to a junction between a homopolar and a heteropolar semiconductor.

V. CONCLUSIONS

The energies of the valence-band maxima, as calculated in an approximately self-consistent LCAO method and as calculated in an approximately self-consistent pseudopotential method, are in approximate agreement with each other. We prefer the LCAO result, partly because in this approach approximate self-consistency comes almost automatically, and partly because it is possible to treat the levels of interest directly rather than calculating the band minimum and adding a large bandwidth to obtain the maximum. The uncertainties due to close cancellation in the latter approach may be signaled by the lack of smooth variation from material to material in the bandwidth column of Table IV.

The discontinuities in the valence-band maximum in a heterojunction, as deduced from the LCAO theory, are obtained simply by subtracting the corresponding E_v values in Table II. These can be interpolated for mixed compounds. The conduction-band minima are obtained by adding and subtracting appropriate band gaps. These also, for a given direct or indirect gap, can be interpolated for mixed compounds. The values obtained may only be reliable on the scale of a few tenths of electron volts because of the uncertainties in the potentials discussed. However, it does seem likely that a table of E_v values *could* be constructed from experiment and that it would be accurate on a much finer scale for all combinations of compounds in the junctions. This would not only be useful for study of heterojunctions, but would also be useful in many other parts of the theory of covalent solids. In the

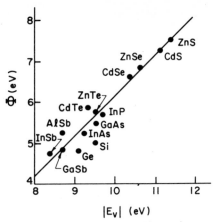

FIG. 1. A plot of the experimental photothreshold against the energy of the valence-band maximum from Table II. The line corresponds to the empirical relation, Eq. 10. Experimental values were taken from G. W. Gobeli and F. G. Allen, Phys. Rev. **127**, 141 (1962); **127**, 150 (1962); **137**, A245 (1965); R. Swank, Phys. Rev. **153**, 884 (1967); and T. E. Fisher, Phys. Rev. **139**, A1228 (1965); **142**, 519 (1966).

mean time, values from Table II may provide the most reliable and most general guide.

ACKNOWLEDGMENT

The author has benefited from discussion of this problem with Dr. Edgar Kraut.

APPENDIX: NOTE ADDED AFTER THE CONFERENCE

W. R. Frensley (private communication) has correctly pointed out that for oppositely skewed compounds, such as GaSb–InAs, Table II, can lead to the valence band on one side of a junction overlapping the conduction band on the other. In such a case the estimate of the dipole made for Ge–Si in Sec. II is quite inappropriate. Clearly, when the valence band on one side approaches in energy the conduction band on the other the tails of these valence-band states will extend far into the opposite material producing a larger dipole. A similar effect occurs at a metal–semiconductor junction when the Fermi energy in the metal approaches a band edge.

R. Sokel (private communication) has treated the latter case using a WKB approximation in the semiconductor and found that the dipole does not diverge as one might guess. For a symmetric treatment of conduction and valence bands in the semiconductor and a free-electron metal he finds that the dipole contributes a step δE in the energy given by

$$\frac{\delta E}{E_c - E_v} = \frac{e^2}{4\pi\epsilon_1}\left(\frac{2m^*}{\hbar^2(E_c - E_v)}\right)^{1/2}$$

$$\times \left[\left(\frac{E_c - E_F}{E_c - E_v}\right)^{1/2} - \left(\frac{E_F - E_v}{E_c - E_v}\right)^{1/2}\right] \qquad (12)$$

where E_F is the metal Fermi energy (Its value includes the step; it is the final self-consistent value), and E_c and E_v are the conduction-band and valence-band edges, and m^* is the effective mass in the semiconductor. ϵ_1 is the dielectric constant in the semiconductor. δE in all cases contributes with a sign such that it reduces the tendency for the bands to overlap.

This result is most applicable to the light hole and conduction bands in direct-gap semiconductors. The expression in square brackets varies between -1 and 1 with E_F in the gap. $m^*/(E_c - E_v)$ tends to be independent of material, as suggested by the $k \cdot p$ formula for the effective mass. Evaluation of Eq. (12) for GaAs gives a maximum shift of 0.02 eV. Part of the reason this is so small is because of the dielectric constant of 10.9.

This contribution, Eq. (12), arises from electrons which in the metal propagate nearly perpendicular to the interface. There are contributions from other electrons associated with different bands in the semiconductor but we may expect them to be similarly small. This metal–semiconductor problem also is different from the heterojunction problem discussed here. However, Sokel's result suggests that indeed the dipoles are not strong enough to prevent overlapping of valence bands on one side of a heterojunction with conduction bands on the other. Our estimate of small dipole effects may not have been misleading.

When we have such an overlapping we will have band bending on both sides with a degenerate hole gas on one side and a degenerate electron gas on the other and the classical band-bending analysis becomes appropriate, though with band curvatures dependent upon the electron- and hole-gas densities.

[a]Supported in part by NSF Grant DMR73-02351.

[1]See, for example, J. R. Chelikowsky and M. L. Cohen, Phys. Rev. B **13**, 826 (1976).

[2]R. L. Anderson, *Proceedings of the International Conference on Semiconductors*, Prague, 1960 (Czech. Acad. Sci., Prague, 1960), p. 563.

[3]A. G. Milnes and D. L. Feucht, *Heterojunctions and Metal-Semiconductor Junctions* (Academic, New York, 1972).

[4]J. L. Shay, Sigurd Wagner, and J. C. Phillips, Appl. Phys. Lett. **28**, 31 (1976).

[5]H. Kroemer, Crit. Rev. Solid State Sci. **5**, 555 (1975).

[6]D. J. Chadi and M. L. Cohen, Phys. Status Solidi B **68**, 405 (1975).

[7]W. A. Harrison, *The Physics of the Chemical Bond* (Freeman, San Francisco, to be published).

[8]W. A. Harrison, Bull. Am. Phys. Soc. **21**, 1315 (1976).

[9]W. A. Harrison and S. Ciraci, Phys. Rev. B **10**, 1516 (1974).

[10]S. T. Pantelides and W. A. Harrison, Phys. Rev. B **11**, 3006 (1975).

[11]F. Herman and S. Skillman, *Atomic Structure Calculations* (Prentice Hall, New Jersey, 1963).

[12]W. R. Frensley and H. Kroemer, J. Vac. Sci. Technol. **13**, 810 (1976).

[13]A. O. E. Animalu and V. Heine, listed for example in W. A. Harrison, *Pseudopotentials in the Theory of Metals* (Benjamin, New York, 1966), 309ff.

[14]M. L. Cohen and V. Heine, *Solid State Physics* (Academic, New York, 1970), Vol. 24, p. 37.

[15]J. R. Chelikowsky and M. L. Cohen, Phys. Rev. B **14**, 556 (1976).

213

VOLUME 52, NUMBER 6   PHYSICAL REVIEW LETTERS   6 FEBRUARY 1984

Schottky Barrier Heights and the Continuum of Gap States

J. Tersoff

AT&T Bell Laboratories, Murray Hill, New Jersey 07974

(Received 7 November 1983)

Simple physical considerations of local charge neutrality suggest that near a metal-semiconductor interface, the Fermi level in the semiconductor is pinned near an effective gap center, which is simply related to the bulk semiconductor band structure. In this way "canonical" Schottky barrier heights are calculated for several semiconductors. These are in excellent agreement with experiment for interfaces with a variety of metals.

PACS numbers: 73.30.+y, 73.40.Ns

Despite decades of intense study, there exists no quantitative, predictive theory of Schottky barrier heights. Simple models[1-3] and phenomenological theories[4-7] have had some success in explaining barrier formation and chemical trends in barrier heights. Also, a few calculations for model systems[8,9] have yielded reasonable agreement with experiment. In general though, the complexity of real interfaces and the subtlety of the effects involved have frustrated attempts at a truly predictive theory. In view of the intrinsic interest and technological importance of metal-semiconductor (M-S) interfaces, this inability to understand their most crucial electronic characteristic is quite disappointing.

For the more covalent semiconductors, the barrier height is independent of the metal used to within ± 0.1 eV for metals of practical interest.[4] Thus, one hopes for a rather simple explanation of the roughly "universal" barrier heights for M-S interfaces with these semiconductors. Explanations so far have focused on the possible pinning of the Fermi level (E_F) by states associated with defects in the semiconductor.[10-12] I argue below that these explanations, while successful in describing surfaces with submonolayer metal coverages, are inappropriate for bulk interfaces.

Here I show that a simple parameter-free model for Fermi-level pinning by metal-induced gap states (MIGS) can predict quantitatively the observed values of the "universal" barrier heights, as well as explaining why more-ionic semiconductors do not exhibit such universality. Such MIGS pinning has been found in numerical calculations by Louie, Chelikowsky, and Cohen[8,9] but the simplicity and generality of the mechanism has not been recognized. In particular, the behavior seen is by no means peculiar to the ideal planar interface as has been suggested.[10] While following Heine[3] in spirit, I stress here the continuum nature of gap states, and the resulting locally metallic character of the semiconductor near the interface.

The various models of Schottky barrier formation are discussed in several excellent reviews.[13,14] The crucial point is simply that the barrier height is determined by the position of E_F within the semiconductor gap. The barrier is the energy needed to excite an electron from E_F to the conduction minimum (for n-type semiconductors). Band bending due to doping can be neglected in the region of interest, which extends only ~ 10 Å from the interface.

At a M-S interface there is a continuum of states around E_F because of the metal. As first discussed by Heine,[3] those states within the gap decay exponentially inside the semiconductor, but still have significant amplitude a few layers from the interface. Any deviation from local charge neutrality in this region results in "metallic" screening by the MIGS.

A very small density of MIGS is sufficient to pin E_F. With use of Thomas-Fermi screening, a local density of only 0.02 state/atom eV in the gap gives a screening length of about 3 Å. Numerical results of Louie and Cohen[8] show a density much greater than this for the first 6 Å or so at a metal-Si interface.

It is therefore convenient to consider the barrier height as having two contributions—a short-range part, which may be related to surface dipoles,[1] the M-S electronegativity difference,[4] or more subtle details of bonding; and an additional dipole from metallic screening by MIGS, which tends to pin E_F so as to maintain local charge neutrality. Here I argue that whether short-range or screening effects dominate (E_F unpinned or pinned) depends simply on bulk semiconductor properties, as does the barrier height in the pinned limit. Moreover in the strongly pinned limit appropriate to Si, Ge, and GaAs, the pinning occurs relatively deep in the semiconductor.

It is important to remember that the MIGS are actually Bloch states of the bulk semiconductor

465

with complex wave vector.[15-17] The formal properties of gap states have been studied extensively.[15-17] There is a sum rule[18,19] on the density of states (DOS) whereby any weight in gap states must come from the valence and conduction bands. Gap states take their weight primarily from those bands that are nearest in energy (allowing for wave-vector and symmetry selection rules). Charge neutrality thus requires occupation of those MIGS which come primarily from the valence band, while leaving those of mainly conduction-band character empty.

I therefore propose that E_F must fall at or near the energy where the gap states cross over from valence- to conduction-band character. In one dimension this energy corresponds[16] to the branch point E_B of the complex band structure, as discussed by Kohn and Rehr.[15,16] The generalization to three dimensions is discussed below. (Of course, there is no discontinuous change in the character of the wave functions at E_B.[19] Rather, states at E_B derive their weight equally from valence and conduction bands. The net effect is still to pin E_F at or near E_B.) For covalent semiconductors, E_B is closely related to the surface-state and vacancy levels, explaining why different theoretical approaches[3,12] yield similar results.

By finding the branch point in the complex energy bands, we immediately have a "canonical" barrier height for the given semiconductor. The barrier heights which are determined in this way from the bulk band structure of several semiconductors are in excellent agreement with experimental values for interfaces with a variety of metals.

The expected behavior can be seen in the self-consistent calculation of Louie and Cohen[8] for a "jellium"-Si(111) interface. At Si atoms a few layers from the metal, states "spill over" into the gap from the conduction and valence bands above and below. In between there is a minimum in the calculated local DOS, presumably at E_B (where the MIGS decay length is shortest). The Fermi level is pinned precisely at this minimum; this is viewed as a natural consequence of the principle of local charge neutrality.

One begins by defining the cell-averaged real-space Green's function,

$$G(\vec{R},E) = \int d^3r \sum_{nk} \frac{\psi_{nk}{}^*(\vec{r})\psi_{nk}(\vec{r}+\vec{R})}{E-E_{nk}}$$

$$= \sum_{nk} \frac{e^{i\vec{k}\cdot\vec{R}}}{E-E_{nk}}, \qquad (1)$$

where $\vec{k}$ is the Bloch wave vector, n is the band index, and ψ and E are the corresponding wave function and energy.

In one dimension, for sufficiently large R, $G(R,E)$ changes sign at the energy of the branch point.[17] In higher dimensions we pick a direction by specifying $\vec{R}$; then for each $k_\parallel$ there is a branch point as the longitudinal wave vector is varied. By integrating over the entire Brillouin zone, for large $\vec{R}$ we automatically pick out the contribution to $G(\vec{R},E)$ with longest range.

For an ideal interface, $\mathrm{Im}(\vec{k})$ must be normal to the interface. It would therefore appear that we should pick $\vec{R}$ in that direction. However, for a disordered interface it seems preferable to assume that all directions are permitted, as in an impurity problem. Then the direction which gives the most slowly decaying MIGS is the important one. Experimentally there appears to be no dependence of barrier height on orientation for interfaces prepared in the usual manner. On the other hand, there is evidence of strong orientation dependence for ideal epitaxial interfaces,[20] consistent with the model here but not with defect models.

Fortunately, the appropriate choice of direction is obvious. The fcc (cell) nearest-neighbor lattice vector is $(a/2)\langle 110 \rangle$. This is also the direction along the chains of bonded atoms in the diamond and zinc-blende structures. Numerical calculations have shown that charge disturbances propagate farthest along these (110) chains.[21] We therefore consider $\vec{R}_m = m(a/2)\langle 110 \rangle$. For various $\vec{R}_m$ $(m=1,\ldots,10)$ we calculate $G(\vec{R},E)$ and locate the energy in the gap where this changes sign. This energy approaches a constant value for large $\vec{R}_m$ $(m>3)$. The direction dependence is discussed below.

In general, E_F must depend on the details of the interface, since the density of MIGS is determined by the boundary condition. For example, if the metal continuum were replaced by a few discrete levels in the gap (e.g., defect levels), then E_F would be pinned at one of these levels, possibly quite far from E_B. Such an effect has been seen at surfaces.[10] If, however, the metal DOS is relatively featureless throughout the gap, and the MIGS penetrate deep enough to screen the interface, then the position of E_F in the semiconductor gap will be determined primarily by the complex band structure. We ignore conservation of $k_\parallel$ across the interface, since microscopically M-S interfaces are disordered.[20]

The calculation was carried out using energy

466

bands obtained with the linearized augmented plane-wave method, summing 152 points in the irreducible wedge of the bulk Brillouin zone. Because the local density approximation used for correlation and exchange gives poor values for the absolute gap but good dispersions, I rigidly shift the conduction bands to give the correct room-temperature energy gap, following Baraff and Schluter.[22] I stress that it is unimportant how the band structure used in (1) is obtained, as long as it is reasonably accurate.

The results are tabulated in Table I, which gives the calculated barrier height for n-type semiconductors, i.e., the difference between the inferred Fermi level and the conduction-band minimum; and the asymptotic charge decay length λ for states at this energy. Results are numerically accurate to better than 0.05 eV (0.4 for ZnS) and 0.5 Å, respectively.

Only results for (110) lattice vectors are given in Table I. The (100) and (111) directions require much larger wave-vector samples for good convergence. Preliminary results with a 1186-point sample indicate a smaller barrier for Si(100), suggesting a possible explanation for new experimental results on orientation dependence for epitaxial films.[20] Results also confirm that the (110) direction gives the longest decay length, justifying the choice of (110) to determine the barrier height for disordered interfaces.

Also shown are experimental Schottky barrier heights.[23-25] I give both the range for a number of "ordinary" metals, and specific values for Au and Al, which have been extensively studied and represent the normal range in metal electronegativity. [C/V measurements for intimate contacts were chosen where available (for Si and GaAs).] In all cases the theoretical value falls within the scatter of barrier heights for typical metals. The decay lengths found here are consistent with those calculated by Louie, Chelikowsky, and Cohen for semiconductor-jellium interfaces.[9] For the elemental semiconductors the measured barriers are smaller (larger) than given by the theory for metals with lower (higher) electronegativity than the semiconductor. Thus, the deviations from the canonical barrier heights calculated here may be attributed, at least in part, to the M-S electronegativity difference (cf. Ref. 4).

In fact, the predicted barrier heights are in as good agreement with experiment as any calculations reported to date, despite the fact that no allowance has been made for the properties of the metal or the geometry of the interface. I believe that this strongly supports the correctness of the underlying physical idea, that is, the necessity of occupation of the MIGS according to their degree of valence or conduction character in order to maintain layer-by-layer charge neutrality.

Large-gap ionic semiconductors have barrier heights which vary considerably depending upon the metal used.[4] Results in Table I suggest an explanation consistent with that of Louie, Chelikowsky, and Cohen.[9] The short decay length of MIGS in midgap for ZnS results in a negligible DOS in the gap except very near the interface. The MIGS are therefore unable to screen the effect of the metal electronegativity.

If, however, the MIGS decay length is large, the pinning is metallic (Thomas-Fermi-like), and any deviation of E_F from its canonical position is screened exponentially with distance from the interface. The screening is cut off effectively at the MIGS decay length. Reexamining results of Louie and Cohen[8] for jellium-Si(111) in this light clarifies the mechanism at work there. The first Si layer sees a self-consistent potential significantly lowered by proximity to the metal, and the corresponding local DOS is shifted downward in energy (i.e., E_F is near the local conduction minimum). However, band bending (screening by MIGS) between the first and second double layer restores E_F to its canonical midgap position in the Si by moving the conduction minimum up 0.3–0.4 eV. Thus even for the ideal planar interface, pinning takes place far inside the Si, explaining the relative insensitivity to interface details.

Other models of barrier formation have been proposed based on pinning by defect levels.[10-12] At the free surface a very small number of states in the gap (defect or intrinsic surface states) can pin the Fermi level.[2] Since the screening length is hundreds or thousands of angstroms (depend-

TABLE I. Schottky barrier heights.

	Gap (eV)	Barrier heights (eV)				λ (Å)
		Au	Al	Other	Theory	
Si [a]	1.12	0.83	0.70	0.70–0.82	0.76	3.0
Ge [b]	0.66	0.59	0.48	0.38–0.64	0.48	4.0
GaAs [c]	1.42	0.94	0.78	0.71–0.94	0.74	3.0
ZnS [b]	3.60	2.00	0.80	0.80–2.00	1.40	1.5

[a] Ref. 23.   [b] Ref. 24.   [c] Ref. 25.

ing on doping), any charging gives rise to an enormous dipole, which shifts E_F so as to maintain neutrality. At a M-S interface, however, the metal will screen any defects nearby. Since the screening charge is only a few angstroms away, each defect contributes at most a modest local dipole. Defects are therefore orders of magnitude less effective at pinning E_F at a M-S interface than at a surface.

Experimentally there is evidence for defect pinning of E_F on surfaces with submonolayer metal coverage.[10] However, results of these experiments are inconsistent with bulk barrier-height measurements[24,25] and must reflect a different mechanism than the true bulk interface. In particular, submonolayer metal coverages neither screen the defect charge effectively nor provide a continuum of states in the gap. The metal atoms themselves are in effect merely local defects.

Defect models of Schottky barrier formation, while appropriate for bare surfaces, have not established any direct relevance to bulk M-S interfaces. In contrast, the continuum model described here is specifically appropriate to bulk M-S interfaces; it has immediate predictive value, and is in excellent agreement with experiment.

It is a pleasure to acknowledge many stimulating discussions and helpful suggestions by D. R. Hamann. Conversations with E. O. Kane, J. C. Phillips, M. Schluter, and R. T. Tung are also gratefully acknowledged.

[1]W. Schottky, Phys. Z. **41**, 570 (1940).
[2]J. Bardeen, Phys. Rev. **71**, 717 (1947).
[3]V. Heine, Phys. Rev. A **138**, 1689 (1965).
[4]S. Kurtin, T. C. McGill, and C. A. Mead, Phys. Rev. Lett. **22**, 1433 (1969); also M. Schluter, Phys. Rev. B **17**, 5044 (1978).

[5]J. C. Phillips, J. Vac. Sci. Technol. **11**, 947 (1974).
[6]J. M. Andrews and J. C. Phillips, Phys. Rev. Lett. **35**, 56 (1975).
[7]L. J. Brillson, Phys. Rev. Lett. **40**, 260 (1978).
[8]S. G. Louie and M. L. Cohen, Phys. Rev. B **13**, 2461 (1976).
[9]S. G. Louie, J. R. Chelikowsky, and M. L. Cohen, Phys. Rev. B **15**, 2154 (1977).
[10]W. E. Spicer, I. Lindau, P. R. Skeath, C. Y. Su, and P. W. Chye, Phys. Rev. Lett. **44**, 420 (1980); W. E. Spicer, P. W. Chye, P. R. Skeath, C. Y. Su, and I. Lindau, J. Vac. Sci. Technol. **16**, 1422 (1979).
[11]R. E. Allen and J. D. Dow, Phys. Rev. B **25**, 1423 (1982).
[12]O. F. Sankey, R. E. Allen, and J. D. Dow, to be published.
[13]M. Schulter, Thin Solid Films **93**, 3 (1982).
[14]E. H. Rhoderick, *Metal-Semiconductor Contacts* (Clarendon, Oxford, 1978).
[15]W. Kohn, Phys. Rev. **115**, 809 (1959).
[16]J. J. Rehr and W. Kohn, Phys. Rev. B **9**, 1981 (1974), and **10**, 448 (1974).
[17]R. E. Allen, Phys. Rev. B **20**, 1454 (1979).
[18]J. A. Appelbaum and D. R. Hamann, Phys. Rev. B **10**, 4973 (1974).
[19]F. Claro, Phys. Rev. B **17**, 699 (1978).
[20]R. T. Tung, preceding Letter [Phys. Rev. Lett. **52**, 461 (1984)]. Perfectly epitaxial interfaces, prepared by novel techniques, may exhibit barrier heights which are quite different from normal interfaces, and which depend strongly on the crystal orientation. For an ideal interface those approximations made here, which were justified by interfacial disorder, break down; it remains to be seen what light the present results may shed on epitaxial M-S interfaces.
[21]E. O. Kane, private communication, and unpublished.
[22]G. A. Baraff and M. Schluter, in *Defects and Radiation Effects in Semiconductors—1980*, edited by R. R. Hasiguti, The Institute of Physics Conference Series No. 59 (The Institute of Physics, London, 1981), p. 287.
[23]A. Thanailakis and A. Rasul, J. Phys. C **9**, 337 (1976), and references therein.
[24]S. M. Sze, *Physics of Semiconductor Devices* (Wiley, New York, 1969).
[25]W. G. Spitzer and C. A. Mead, J. Appl. Phys. **34**, 3061 (1963); B. L. Smith, Ph.D. thesis, Manchester University, 1969 (unpublished) (from Ref. 14 above).

RAPID COMMUNICATIO

PHYSICAL REVIEW B   VOLUME 30, NUMBER 8   15 OCTOBER 1984

Theory of semiconductor heterojunctions: The role of quantum dipoles

J. Tersoff*

AT&T Bell Laboratories, Murray Hill, New Jersey 07974

(Received 30 May 1984)

At any semiconductor heterojunction there is an interface dipole associated with quantum-mechanical tunneling, which depends on the band "lineup" between the two semiconductors. When the interface dipolar response dominates, the actual band discontinuity must be close to that unique value which would give a zero interface dipole. A simple criterion is proposed for this zero-dipole lineup, which gives excellent agreement with experimental band lineups. The close connection between heterojunction band lineups and Schottky barrier formation is emphasized.

Semiconductor heterojunction interfaces exhibit interesting and useful electronic properties associated with the discontinuity in the local band structure at the interface. As a result, such heterostructures have become important as a basis for novel devices. However, the fundamental understanding of their electronic structure is still far from satisfactory.[1,2]

The most important single property of a semiconductor heterojunction is the band "lineup," i.e., the relative position in energy of the band gaps in the two semiconductors. This lineup determines the conduction- and valence-band discontinuities, and hence the effective barrier for electron or hole transport across the interface.

This paper presents a theory for the band lineup at ideal semiconductor interfaces. The central idea is that there is in general a dipole at the interface, associated with gap states induced by the band discontinuities. This dipole depends on the band offset, and tends to drive the band lineup toward that value which would give zero dipole. A simple estimate of this zero-dipole lineup gives excellent agreement with experimental band lineups for a number of heterojunction interfaces.

Figure 1 illustrates schematically the local band discontinuities at a heterojunction, for two possible cases (which are discussed in detail below). If the semiconductors are doped, there is also band bending on a length scale ~ 1000 Å. However, such band bending can be treated semiclassically, and is not of interest here.

Experimentally, the interface properties often depend on growth conditions, so that relatively reliable experimental values for band lineups are available only for a few systems.[1] Such effects may be due to imperfections (e.g., high densities of misfit dislocations[3]) which extend beyond the interface region, and so are outside the scope of microscopic theories of interface electronic structure. Theoretical attempts to calculate band lineups for ideal interfaces have had mixed success,[1,2] with the most realistic calculations being typically less successful than some model approaches.

The simplest theories of band lineup have supposed that the problem consists simply of relating the bands of the two semiconductors to a common absolute energy scale.[1,4-6] Such an approach assumes that no significant additional dipole is formed at the interface. Harrison[4] in particular has argued that this is the case, and has obtained reasonably good predictions of band offsets on this basis.

In general, however, the interface induces states in the gap of one or both semiconductors, analogous to the so-

called "metal-induced gap states" (MIGS) at a metal-semiconductor interface.[7,8] As with MIGS, bulk electronic states in one semiconductor which fall energetically in the band gap of the other tunnel a few angstroms into the latter. The mere presence of these gap states is enough to generate an interface dipole.

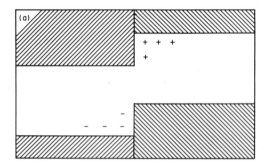

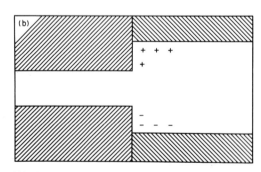

FIG. 1. Two simple examples of the relation between the band lineup and interface dipole. The band gap is shown schematically vs position normal to interface. Crosshatching shows projected bulk bands. Net charge associated with gap states is shown schematically as (+) (electron deficit) and (-) (electron excess), where states at the bottom and top of the gap are occupied or unoccupied, respectively (see text). (a) A single semiconductor in which a band discontinuity is artificially induced, e.g., by an external step potential. (b) An interface between two semiconductors, both with "symmetric" valence and conduction bands (i.e., same electron and hole effective mass, etc.), but with unequal band gaps.

For a given system, there exists a unique band lineup which gives a zero interface dipole. The actual lineup will not, in general, coincide with this "canonical" lineup; however, any deviation from this position gives rise to an interface dipole, which acts to drive the lineup towards the canonical value. If, as is argued below, a small displacement from the canonical lineup gives rise to a large restoring dipole, the actual lineup will be forced very close to the canonical position. Then the lineup in the *absence* of interface dipoles, which plays the central role in most previous theories, becomes relatively unimportant here.

In understanding interface dipoles, the conceptual starting point is the case of a single gap state in one dimension. Such a state may be associated with a surface, interface, or defect. The properties of gap states have been studied extensively.[9-13] Any state in the band gap is necessarily a mixture of valence- and conduction-band character; moreover, there is a sum rule on the local density of states, so that the gap state takes its spectral weight from the local valence and conduction bands, in proportion to its wavefunction character. Charge neutrality occurs if the valence band is completely filled, and the conduction band completely empty. Therefore, occupying a state in the gap leads to excess net charge locally, in proportion to its degree of conduction character. Leaving the gap state empty gives a local charge deficit, in proportion to the state's valence character.

If the state lies near the bottom of the gap, filling it corresponds to only a slight excess charge, since it typically has only a little conduction character. Leaving that state empty, however, results in a charge deficit of almost one electron (i.e., almost one hole in the valence band). Conversely, filling a state high in the gap gives a large excess charge, while leaving it empty gives a small local charge deficit.

It is not hard to see how, even when there are no states at the Fermi level, changing the band lineup can give a net dipole. Two particularly simple (though artificial) cases are illustrated in Fig. 1. Consider first an interface between two semiconductors which are identical, except for an overall shift in energy. In other words, the band structures and wave functions are the same, but the zero Fourier components of the two potentials (and, hence, the electron affinities) differ by an amount V. This is equivalent to a single homogeneous semiconductor in the presence of an external step potential of height V.

According to theories which ignore the interface dipole, the band discontinuity should be precisely V. In other words, the potential step is treated as unscreened. However, in reality, the band discontinuity induces gap states and associated charges on both sides of the "interface," as shown schematically in Fig. 1(a). The resulting dipole acts to cancel the potential step. From electrostatics, one knows that the induced local dipole reduces the step by a factor of ϵ, the bulk dielectric constant. The lineup is then within $\epsilon^{-1}V$ of the canonical lineup which would give zero induced dipole (in this case the trivial homogeneous lineup). This is not to say that the induced dipole is nearly zero, but only that a very small deviation from the canonical lineup is needed to provide the screening, since ϵ is large.

At a real heterojunction between two different semiconductors, the analogy to the response of a homogeneous semiconductor to a step potential is still qualitatively correct; the effect of gap states at the interface will be to screen any deviation from the canonical lineup by a characteristic dielectric constant. For covalent and III-V semiconductors, this represents an order of magnitude reduction in the deviation. Thus, the dipole response indeed dominates the difference in electron affinity. Since dipole-free theories[4-6] give lineups typically within ~ 0.5 eV of the canonical lineups tabulated below, the screened deviations from canonical lineups should be only ~ 0.05 eV, comparable to the numerical accuracy of the calculations here.

Another simple example is the case of "symmetric" valence and conduction bands, where the bands are mirror reflections (with respect to energy) across the center of the gap. In that case the condition of zero dipole requires that the bands of the two semiconductors be aligned symmetrically, i.e., that the centers of the gaps coincide. In that way the charges induced by gap states cancel, as illustrated in Fig. 1(b). Again, any deviation from this lineup results in a restoring dipole. Numerical calculations for model one-dimensional interfaces[14] confirm that this effect can be comparable in magnitude to the Fermi-level pinning by MIGS at a metal-semiconductor interface, and that both mechanisms drive the lineup towards the canonical position.

Both these examples illustrate the remarkable fact that the relative band positions are "pinned" by the interface electrostatics, even though there are no states at the Fermi level. Dielectric screening plays a role here similar to that attributed to metallic screening in the treatment of Schottky barriers.[8]

Real semiconductors have complicated band structures, so the lineup condition for zero dipole is not obvious. Clearly one must occupy the primarily valencelike states on both sides of the interface, while leaving the conductionlike states empty, so as to achieve local charge neutrality throughout. At some effective midgap energy E_B, the states in the gap are on the average nonbonding in character. States higher or lower in the gap have, respectively, more conduction or valence character on the average. The energy E_B thus plays a role analogous to that of the Fermi level in metals, as discussed in Ref. 8. A reasonable estimate of the zero-dipole lineup is, therefore, to align E_B for the respective semiconductors. This reduces to the symmetric lineup in the case of symmetric valence and conduction bands discussed above.

If one of the semiconductors is replaced with a metal, the heterojunction becomes a Schottky barrier. Then the band lineup suggested above reduces to aligning E_B in the semiconductor with the metal Fermi level, as in Ref. 8. Thus heterojunction band lineups and Schottky barrier heights are here treated within a single unified approach. For both types of systems, the agreement with experiment obtained below is at least as good as any other theoretical treatment to date.

The effective midgap point E_B is calculated exactly as in Ref. 8. One begins by defining the cell-averaged real-space Green's function (restricted to propagation by a lattice vector).

$$G(\vec{R},E) = \int d^3r \sum_{nk} \frac{\psi_{nk}^*(\vec{r})\psi_{nk}(\vec{r}+\vec{R})}{E-E_{nk}} = \sum_{nk} \frac{e^{i\vec{k}\cdot\vec{R}}}{E-E_{nk}} ,$$

(1)

where $\vec{k}$ is the Bloch wave vector, n the band index, and ψ_{nk} and E_{nk} the corresponding wave function and energy. Then E_B is the energy where valence and conduction bands contribute equally to $G(R,E)$ in (1) (typically with opposite

TABLE I. Semiconductor "midgap" energy E_B, and Fermi-level positions at metal-semiconductor interfaces, relative to valence maxima (eV).

	E_B	$E_F(\text{Au})^a$	$E_F(\text{Al})^a$
Si	0.36	0.32	0.40
Ge	0.18	0.07	0.18
AlAs	1.05	0.96	
GaAs	0.70	0.52	0.62
InAs	0.50	0.47	
GaSb	0.07	0.07	
GaP	0.81	0.94	1.17
InP	0.76	0.77	

[a]Reference 18.

TABLE II. Valence-band discontinuities at selected[a] heterojunctions (eV).

	Experiments	Theory	Difference
AlAs/GaAs	0.19[b]	0.35	0.16
InAs/GaSb	0.51	0.43	−0.08
GaAs/InAs	0.17	0.20	0.03
Si/Ge	0.20	0.18	−0.02
GaAs/Ge	0.53	0.52	−0.01

[a]Reference 1.
[b]However, see text and Refs. 1, 19, and 20.

sign). In one dimension this corresponds to the branch point in the (complex) energy bands.[9,10] In three dimensions E_B has no such precise meaning, but provides a convenient criterion for the energy at which the gap states, on the average, cross over from primarily valence to conduction character.

Equation (1) requires the band structure E_{nk} as input. This is calculated as in Ref. 8, with a linearized augmented-plane-wave method.[15] The conduction bands are rigidly shifted to give the correct band gap, following Baraff and Schlüter.[16]

The calculated position of E_B with respect to the valence maximum is given in Table I for a number of covalent and III-V semiconductors. (For GaSb the effect of spin-orbit splitting is included in an approximate way.) Also given for each semiconductor is the Fermi level at interfaces with Au and (where available) Al, based on Schottky barrier measurements. According to Ref. 8, as well as the arguments above, the Fermi level at a metal-semiconductor interface should be pinned at E_B, to within the ~ 0.1-eV variation of barrier height with metal. (This variation can be understood as deriving from the electronegativity difference between different metals.[17]) Experimental values in Table I are from the classic tabulation of Sze.[18] While more recent measurements are available, a critical evaluation of barrier-height data is outside the scope of this paper.

Table II gives the most reliably known band lineup results for semiconductor heterojunctions, according to a recent review by Kroemer.[1] (Calculations of E_B have not yet been carried out for II-VI semiconductors, and so those are excluded.) Theoretical valence-band discontinuities inferred directly from results of Table I are also given. The excellent agreement between the experimental results and the theory described here, shows at the very least that available data are consistent with the assumption that quantum-mechanical dipoles are the dominant factor determining heterojunction band lineups (as well as Schottky barrier heights).

The quantitative comparison of theory and experiment in Table II must be made with some caution. The band lineup even for the extensively studied AlAs/GaAs system remains controversial[19,20] (see especially Ref. 19). Also, calculated band structures are only reliable to 0.1–0.2 eV in general. Any agreement between theory and experiment better than that in Table II would probably be fortuitous.

Note that in the present approach, the band discontinuities could be estimated by taking the difference in E_F rather than E_B in Table I. The resulting predictions are only slightly less accurate than the theoretical values given in Table II, though obtained without any calculation.

The suggestion that heterojunction band lineups and Schottky barrier heights are correlated has been made previously, but on the basis of radically different arguments. Spicer et al.[21] had suggested that at metal-semiconductor interfaces, the Fermi level in the semiconductor is pinned by intrinsic defects. Katnani and Margaritondo[22] pointed out that were this the case, then such defect pinning at heterojunction interfaces might also account for the band lineups. This would imply Fermi-level pinning, however, which is not observed. More recent experiments,[23,24] demonstrate that the band lineup at Ge-GaAs interfaces is not determined by such defect pinning. These studies also suggest that native defects do not play a crucial role in Schottky barrier formation.

In contradiction to previous assertions,[4] simple estimates based on dielectric screening suggest that the interface dipole is the dominant factor determining band lineup. In conjunction with a simple criterion for the zero-dipole band lineup, this view leads to quantitative predictions of both heterojunction band offsets and Schottky barrier heights. These predictions are typically accurate to ~ 0.1 eV; they require only one number (E_B) for each semiconductor, which depends only on the bulk band structure; and they involve no auxiliary hypotheses about interface structure, or the presence or absence of native or extrinsic defects.

Stimulating discussions with D. R. Hamann, M. Schlüter, D. E. Aspnes, and F. Capasso are gratefully acknowledged.

*Present address: IBM Thomas J. Watson Research Center, Yorktown Heights, N.Y. 10598.

[1]H. Kroemer, in *Proceedings NATO Advances Study Institute on Molecular Beam Epitaxy and Heterostructures, Erice, Sicily, 1983*; edited by L. L. Chang and K. Ploog (Martinus Nijhoff, The Netherlands, in press).

[2]J. Pollmann and A. Mazur, Thin Solid Films **104**, 257 (1983).

[3]J. M. Woodall, G. D. Pettit, T. N. Jackson, C. Lanza, K. L. Ka-

vanagh, and J. W. Mayer, Phys. Rev. Lett. **51**, 1783 (1983).

[4]W. A. Harrison, J. Vac. Sci. Technol. **14**, 1016 (1977).

[5]W. R. Frensley and H. Kroemer, Phys. Rev. B **16**, 2642 (1977); J. Vac. Sci. Technol. **13**, 810 (1976).

[6]R. L. Anderson, Solid State Electron. **5**, 341 (1962).

[7]V. Heine, Phys. Rev. A **138**, 1689 (1965).

[8]J. Tersoff, Phys. Rev. Lett. **52**, 465 (1984).

[9]W. Kohn, Phys. Rev. **115**, 809 (1959).

[10]J. J. Rehr and W. Kohn, Phys. Rev. B **9**, 1981 (1974); **10**, 448 (1974).

[11]R. E. Allen, Phys. Rev. B **20**, 1454 (1979).

[12]J. A. Appelbaum and D. R. Hamann, Phys. Rev. B **10**, 4973 (1974).

[13]F. Claro, Phys. Rev. B **17**, 699 (1978).

[14]J. Tersoff (unpublished).

[15]D. R. Hamann, Phys. Rev. Lett. **42**, 662 (1979).

[16]G. A. Baraff and M. Schlüter, Inst. Phys. Conf. Ser. No. **59**, 287 (1981).

[17]S. G. Louie, J. R. Chelikowsky, and M. L. Cohen, Phys. Rev. B **15**, 2154 (1977), and references cited therein.

[18]S. M. Sze, *Physics of Semiconductor Devices* (Wiley, New York, 1969).

[19]R. C. Miller, A. C. Gossard, D. A. Kleinman, and O. Munteanu, Phys. Rev. B **29**, 3740 (1984). These results for alloy heterojunc-
tions, if extrapolated to the pure AlAs/GaAs interface, imply a valence-band discontinuity somewhat larger than given by the theory here, whereas the experimental value given in Table II, also based on extrapolation, is smaller than the theory.

[20]J. R. Waldrop, S. P. Kowalczyk, R. W. Grant, E. A. Kraut, and D. L. Miller, J. Vac. Sci. Technol. **19**, 573 (1981).

[21]W. E. Spicer, I. Lindau, P. R. Skeath, C. Y. Su, and P. W. Chye, Phys. Rev. Lett. **44**, 420 (1980); W. E. Spicer, P. W. Chye, P. R. Skeath, C. Y. Su, and I. Lindau, J. Vac. Sci. Technol. **16**, 1422 (1979).

[22]A. D. Katnani and G. Margaritondo, Phys. Rev. B **28**, 1944 (1983).

[23]P. Chiaradia, A. D. Katnani, H. W. Sang, Jr., and R. S. Bauer, Phys. Rev. Lett. **52**, 1246 (1984).

[24]H. Brugger, F. Schaffler, and G. Abstreiter, Phys. Rev. Lett. **52**, 141 (1984).

PHYSICAL REVIEW B   VOLUME 31, NUMBER 4   15 FEBRUARY 1985

Comments

*Comments are short papers which comment on papers of other authors previously published in the **Physical Review**. Each Comment should state clearly to which paper it refers and must be accompanied by a brief abstract. The same publication schedule as for regular articles is followed, and page proofs are sent to authors.*

Comment on "Theory of semiconductor heterojunctions: The role of quantum dipoles"

G. Margaritondo

Department of Physics and Synchrotron Radiation Center, University of Wisconsin, Madison, Wisconsin 53706
(Received 12 October 1984)

Tersoff's heterojunction model is critically analyzed using extensive experimental data obtained from photoemission measurements of the valence-band discontinuity.

A recent article by Tersoff proposed[1] a new theoretical approach to the crucial problem of understanding and estimating semiconductor heterojunction band discontinuities. The approach assumes that the dominant factor in the relative alignment of the band edges is the minimization of the interface dipoles. The zero-dipole condition is estimated by assuming that the dipoles are due to the occupancy of gap states induced by the band discontinuities. This condition corresponds to the alignment of the midgap energies E_B of the two semiconductors—where E_B is the energy at which the prevailing character of the gap states changes from valencelike to conductionlike.[2] Reference 1 strongly criticizes the discontinuity models that neglect the role of interface dipoles,[3,4] and is likely to become controversial among semiconductor theorists. This Comment, however, does not deal with the theoretical aspects of that work. I discuss instead some interesting tests not included in Ref. 1 that compare its theoretical predictions with photoemission measurements of the valence-band discontinuity ΔE_v at interfaces involving Ge or Si.[5] These tests are interesting since they show that (i) Tersoff's model does reach better accuracy than any previous discontinuity theory, and (ii) it is the first model able to reach the absolute accuracy limits underlying *all* linear models.[5]

Tersoff's model belongs to a general class of theories that estimate ΔE_v (and the conduction-band discontinuity) as the difference between two terms determined by the bulk properties of the two component semiconductors.[1] Other examples of these linear models are discussed in Refs. 3, 4, and 6–8. In 1983, Katnani and I discussed[5] the general accuracy limitations of this class of theories. The experimental data basis for this discussion was provided by extensive photoemission measurements of ΔE_v on heterojunction interfaces involving Ge or Si overlayers deposited on different semiconducting substrates under ultrahigh vacuum conditions. We concluded that the underlying accuracy limit of all linear model is ~ 0.15 eV in estimating ΔE_v. We also found that none of the models existing at that time reached the above accuracy limit—and we proposed an empirical tabulation of valence-band-edge terms deduced from our experimental ΔE_v's to bypass this problem.[9]

Instead of using our approach, Ref. 1 tested its model with a limited number of experimental ΔE_v's measured for

"selected" heterojunctions. This was inspired by the possible role of misfit dislocations in heterojunctions not included in the "selected" lists.[1,10] The importance and magnitude of these and other spurious effects is indeed extremely difficult to estimate from a theoretical point of view. I argue, however, that without a better knowledge of these factors *any* assumption about them involves some risk. Specifically, rejecting a large portion of the existing data basis on these grounds is a most risky assumption. I emphasize that the empirical accuracy limit deduced[5] for linear models implies that possible "nonlinear" factors—including interface imperfections—have limited the average effect on ΔE_v, and should not be used as a justification for the use of "selected" experimental results. The few experiments that *directly* explored the effects of imperfections strongly support this conclusion.[11-14]

Paradoxically, the unbiased use of all the existing experimental data to test Tersoff's theory considerably strengthens his case. Table I shows the comparison between experimen-

TABLE I. Comparison between ΔE_v's for heterojunctions involving Ge and Si and theoretical predictions.

Heterojunction	ΔE_v, experiment[a,b]	ΔE_v, theory[a,c]
Si-Ge	0.17	0.18
AlAs-Ge	0.95	0.87
GaAs-Ge	0.35	0.52
InAs-Ge	0.33	0.32
GaSb-Ge	0.20	−0.11
GaP-Ge	0.80	0.63
InP-Ge	0.64	0.58
GaAs-Ge	0.35	0.52
Ge-Si	−0.17	−0.18
GaAs-Si	0.05	0.34
InAs-Si	0.15	0.14
GaSb-Si	0.05	−0.29
GaP-Si	0.80	0.45
InP-Si	0.57	0.40

[a]Values in eV.
[b]From Ref. 5, except for GaP-Si (Ref. 11) and AlAs-Ge (Ref. 15).
[c]From Ref. 1.

tal ΔE_v's for heterojunctions involving Ge or Si (Refs. 5, 11, and 15) and theoretical predictions.[1] These data were analyzed coherently with the approach used in Ref. 5. The average magnitude of the discrepancy between theory and experiment is 0.15 eV. For comparison, the average magnitude of the accuracy achieved by other linear models[3,4,6-8] for the same set of interfaces ranges between 0.20 and 0.51 eV. In particular, the average accuracy of Harrison's approach,[3] specifically discussed in Ref. 1, is 0.39 eV. Thus, the accuracy of Tersoff's model is apparently better than that of other linear models. This conclusion is not affected by the experimental uncertainty due to the large data base used to reach it. Furthermore, it does not change if one extends the data base of Table I to include the interfaces used as a test in Ref. 1. Particularly interesting is the fact that the model here discussed is the only one reaching the underlying accuracy limits of its class of theories: 0.15 eV. In the framework of this theory, the empirical terms we introduced in Ref. 9 are an empirical optimization of the midgap energy E_B.[1]

The agreement found in Table I is further evidence of the small magnitude of nonlinear factors, and strengthens the above arguments against the use of "selected" experimental data. A conclusive test of Tersoff's model should be based on its predicted correlation between heterojunction discontinuities and Schottky barriers.[1,2] The existing data are somewhat contradictory and inconclusive.[1] Most Schottky barriers measurements have been performed on heavily contaminated interfaces. Photoemission studies of ultraclean heterojunction interfaces[11,16] merely demonstrate that there is no correlation between ΔE_v and the initial pinning position of the Fermi level, which Refs. 1 and 2 do *not* identify with the Schottky barrier. Thus, more experiments are required to investigate Tersoff's hypotheses, which are nevertheless strengthened by our tests.

This research was supported by the National Science Foundation under Grant No. DMR-82-00528, and by the University of Wisconsin Graduate School through a Romnes Faculty fellowship.

[1]J. Tersoff, Phys. Rev. B **30**, 4874 (1984).
[2]J. Tersoff, Phys. Rev. Lett. **52**, 465 (1984).
[3]W. A. Harrison, J. Vac. Sci. Technol. **14**, 1016 (1977).
[4]W. R. Frensley and H. Kroemer, Phys. Rev. B **15**, 2642 (1977).
[5]A. D. Katnani and G. Margaritondo, Phys. Rev. B **28**, 1944 (1983).
[6]R. L. Anderson, Solid State Electron. **5**, 341 (1934).
[7]O. Von Ross, Solid State Electron. **23**, 1069 (1980).
[8]M. J. Adam and A. Nussbaum, Solid State Electron. **22**, 783 (1969).
[9]A. D. Katnani and G. Margaritondo, J. Appl. Phys. **54**, 2522 (1983).
[10]H. Kroemer, *Proceedings NATO Advanced Study Institute on Molecular Beam Epitaxy and Heterostructures, Erice 1983*, edited by L. L.

Chang and K. Ploog (Martinus Nijhoff, The Netherlands, in press).
[11]P. Perfetti, F. Patella, F. Sette, C. Quaresima, C. Capasso, A. Savoia, and G. Margaritondo, Phys. Rev. B **30**, 4533 (1984).
[12]G. Margaritondo, C. Quaresima, F. Patella, F. Sette, C. Capasso, A. Savoia, and P. Perfetti, J. Vac. Sci. Technol. A **2**, 508 (1984).
[13]P. Chiaradia (private communication).
[14]D. W. Niles, P. Perfetti, C. Quaresima, and G. Margaritondo (unpublished).
[15]M. K. Kelly, D. W. Niles, E. Colavita, G. Margaritondo, and M. Henzler (unpublished).
[16]P. Chiaradia, A. D. Katnani, H. W. Sang, Jr., and R. S. Bauer, Phys. Rev. Lett. **52**, 1246 (1984).

Tight-binding theory of heterojunction band lineups and interface dipoles

W. A. Harrison[a)] and J. Tersoff

IBM Thomas J. Watson Research Center, Yorktown Heights, New York 10598

(Received 28 January 1986; accepted 18 March 1986)

A tight-binding theory of semiconductor heterojunction band lineups is presented. Interface dipoles are shown to play a crucial role in determining lineups, so that lineups obtained by using the vacuum level as a reference (e.g., the electron affinity rule) are not reliable. Instead, the self-consistent lineup can be obtained approximately by aligning the average sp^3 hybrid energies in the respective semiconductors. Numerical results are provided and compared with experiment, and the approximations and accuracy in this approach are discussed. The application of these ideas to Schottky barriers is also considered.

I. INTRODUCTION

In tight-binding theory the energy bands are obtained in terms of the energy of atomic-like states and interatomic couplings. For many purposes it has proven adequate to use free-atom term values for the atomic-like states.[1,2] Then the energy bands for every semiconductor are placed upon the same energy scale, i.e., they are given in effect relative to the vacuum level. In particular, in the scheme outlined by Harrison,[1,2] with nearest-neighbor interactions, the valence band maximum ϵ_v is given by

$$\epsilon_v = \frac{\epsilon_p^c + \epsilon_p^a}{2} - \left[\left(\frac{\epsilon_p^c - \epsilon_p^a}{2} \right)^2 + \left(\frac{1.28\hbar^2}{md^2} \right)^2 \right]^{1/2}. \quad (1)$$

Here ϵ_p^a and ϵ_p^c are the p-state eigenvalues of the anion and cation atoms, d is the bond length, and the last term derives from a universal form for the interatomic coupling. Then the relative position of the valence-band maxima for two semiconductors A and B is obtained directly by comparing the absolute energies from Eq. (1). These may be called "natural lineups," and are appropriate estimates of heterojunction valence-band discontinuities in this simplest tight-binding context, *if and only if* interface dipoles may be neglected. In that case the valence-band discontinuity $\Delta\epsilon_v$ is[3]

$$\Delta\epsilon_v \text{ ("natural")} = \epsilon_v(A) - \epsilon_v(B). \quad (2)$$

Tersoff[4] has argued, on the other hand, that at semiconductor heterojunctions, interface dipoles arise which can be quite large. These would shift the free-atom energies on the two sides and modify the natural band lineups. He has calculated an energy E_B, ordinarily in the gap, at which evanescent states may be thought of as equally conduction-band-like and valence-band-like, and argued that the dipoles will shift the bands of two semiconductors such that these energies line up across the interface, in analogy to Fermi-level pinning in Schottky barriers. Here an analysis is developed in the context of a tight-binding approach, and the role of interface dipoles in "pinning" the relative band lineup is explicitly verified. The average hybrid energy in the semiconductor is identified as playing a role like that of E_B in the previous theory.[4]

The critical dipole shifts may be directly estimated using self-consistent tight-binding theory. Harrison[5] in fact made estimates of the interface dipole by careful treatment of the bonds crossing the interface, and found the dipole arising from these alone to be quite small. We now consider the question of interface dipoles in tight-binding theory more generally, and show that when the entire interface region is incorporated in the calculation the total dipole must turn out to be large in the sense of Ref. 4.

II. TIGHT-BINDING THEORY OF INTERFACE DIPOLES

A. Elements of self-consistent tight-binding theory

Tight-binding theory, as indicated above, is based upon a Hamiltonian matrix with diagonal elements given here by "atomic" energy levels, and off-diagonal matrix elements representing interatomic couplings, here taken to depend only upon s or p character and internuclear distance.[1,2] The theory is made self-consistent by allowing the diagonal atomic energies to be shifted by coulomb interactions; such a self-consistent scheme has recently been codified[6] in terms of intra-atomic coulomb interactions U, interatomic interactions given by e^2/d, and bond polarizations that give rise to the dielectric constant. Within this context we can make some rigorous statements about the interface dipoles.

We begin with a semiconductor which is identical on both sides of an "interface." This is a simple bulk semiconductor, and by symmetry there is obviously no dipole at the interface. We then proceed to make modifications on the two sides, and calculate any dipole shifts which arise. We can thus construct the self-consistent tight-binding representation of a real heterojunction step by step.

B. The simplest model

As the simplest nontrivial model for a heterojunction, we first shift all of the term values on the right by $-\Delta/2$ and those on the left by $\Delta/2$. We imagine then solving the self-consistent tight-binding problem exactly to find the net self-consistent shift of the term values from the interface on the two sides, relative to each other. That shift will contain the starting shift Δ, and the effects of any dipole δ arising at or near the interface. If Δ is small, the dipole may be written $\delta = -\chi(\Delta + \delta)$, where χ is the appropriate susceptibility, and $\Delta + \delta$ is the net (total) shift. From linear response theory, $\Delta + \delta = \alpha\Delta$, with $\alpha = (1 + \chi)^{-1}$. We regard this as an exact solution of the linear response in the tight-binding context.

224

Theories which neglect interface dipoles correspond to the limit $\chi = 0$, where $\alpha = 1$, and the net shift $\Delta + \delta$ is just the starting shift Δ. In contrast to this, for a metal–metal interface $\chi \to \infty$ and $\alpha = 0$, and so the starting shift does not affect the final "lineup." We can estimate α most accurately by noting the connection between heterojunction dipoles and bulk dielectric response, as in Ref. 4, and thus show that α is simply the reciprocal of the long wavelength dielectric constant ϵ_∞. We do this by inserting a shift $\Delta = - eE_0 l_0$ between every plane of atoms, with l_0 being the interplanar spacing. This is in fact just the application of a uniform electric field E_0 in the tight-binding context. This shifts the relative energy of two points separated by a large distance L by $- eE_0 L$. Using the result above we may say that the net relative shift in the term values separated by a large distance L will be given by $- e\alpha E_0 L$. We know that in the real system, and in its tight-binding representation, that energy difference, arising from the applied field, will be reduced by the dielectric constant ϵ_∞ and hence $\alpha = 1/\epsilon_\infty$. (The optical dielectric constant ϵ_∞ is used because shifting the atomic term values differs from a true electric field, in that it does not apply an electrostatic force to the nucleus. Although an additional dipole may arise from displacements of the nuclei, these are not treated here.)

A consequence is that if two semiconductors differ simply by a small relative shift Δ in the atomic energies, giving a band discontinuity from the point of view of natural band lineups equal to Δ, the net discontinuity—calculated self-consistently—would be reduced in comparison to the natural band lineups by a factor of ϵ_∞, to Δ/ϵ_∞, as argued elsewhere.[4] Thus, when all of the bonds are treated on the same footing, it is seen to be the band discontinuity which is screened, rather than the charge transfer from the interface dipole, as was once thought.[5] The interface dipole is in fact itself simply the associated screening charge. The essential conclusion is that for this particular system, with simple shifts in the term values, there is a very major correction to the predictions based upon natural band lineups.

C. A real heterojunction

We now proceed to the more complete description of the electronic structure, but in this case we shall need to proceed approximately. For tetrahedral semiconductors, it is convenient to first transform to a basis of bonds and antibonds, by first forming the usual sp^3 directed hybrid orbitals on each site, and then forming bonds and antibonds between collinear hybrids sharing the same "bond site," i.e., overlapping strongly. In the scheme of Harrison,[1,2] the semiconductor may be characterized moreover simply in terms of the average sp^3 hybrid energy for the semiconductor, $\bar{\epsilon}_h = (\epsilon_h^a + \epsilon_h^c)/2$, where $\epsilon_h = (\epsilon_s + 3\epsilon_p)/4$, ϵ_s and ϵ_p are the atomic s and p eigenvalues, and the superscripts a and c refer to the anion and cation; the intersite coupling V_2; the polar energy $V_3 = (\epsilon_h^c - \epsilon_h^a)/2$; and the "metallic" energy $(s-p$ splitting$)$ $V_1 = (\epsilon_s - \epsilon_p)/4$ for each atom.

We have already analyzed the case where the semiconductors A and B differ only by $\bar{\epsilon}_h(A) = \bar{\epsilon}_h(B) + \Delta$. For a more general analysis, we begin with the simplest model for a tetrahedral semiconductor, the bond orbital approximation of

neglecting all couplings except the term V_2 coupling two hybrids in the same bond site.[1] For the simple system analyzed above, with hybrid levels shifted by Δ with respect to each other across the interface, the bonds on the two sides remain simple and nonpolar, but for the bonds crossing the interface there is an interface polar energy, $V_p = \Delta/2$, and a charge $\alpha_p = V_p/\sqrt{V_2^2 + V_p^2}$ transferred to the atom with lower-energy hybrid.[1] These charges will shift the energy of the hybrids on those atoms and on neighboring atoms, modifying the V_p's which enter each bond. Thus a self-consistent solution of each bond is required in principal to obtain the final charge distribution. We showed above that the result will be a shift, far from the interface, of Δ/ϵ_∞, so the detailed local solution is not required in practice. (Within this bond orbital approximation,[1] $\epsilon_\infty = 1 + \pi\sqrt{3}e^2/2V_2 d$. A more accurate solution simply gives a more accurate value for the dielectric constant.)

Still within the bond-orbital approximation, we note that V_2 may differ on the two sides, because the bond length differs on the two sides. (We neglect strain effects, which are easily included perturbatively.) We start with an average V_2 everywhere, and then increase it on one side and decrease it an equal amount on the other side; the average value is used for the bond crossing the interface. Although we have changed the value of V_2 in each bond, every bond remains nonpolar and no charge is shifted between atoms. This is an important conclusion, and one not obvious immediately since we have lowered the bond energies on one side in comparison to those on the other and one might have expected charge redistribution of the type we found when we shifted the term values on the two sides. In fact no dipole arises even if we go beyond the bond orbital approximation and include the principal coupling,[1] $V_1/2$, between bonds and antibonds sharing the same atom. It is true, as seen in perturbation theory, that the charge transferred from a given bond to a given neighboring antibond is modified when the bond energies are modified, but the transfer from the bond in the neighboring site to the antibond in the original site is changed by exactly the same amount so that there is no charge transfer. This result is more general than perturbation theory; it follows from the symmetry of the model with respect to interchange of electrons and holes.

We conclude that a difference in V_2 on the two sides does not in itself produce a dipole shift. It can, of course, modify the dipole shift arising from a nonzero Δ. For example, it modifies the dielectric constant on the two sides (and not exactly symmetrically) and this will modify the self-consistent solution. That, however, would seem to be beyond the level of calculation appropriate if we are to linearize the problem by using a dielectric constant in the first place. We therefore simply use the average dielectric constant, when it differs on the two sides, in calculating the effect of a nonzero Δ.

We may now let one of the sides of the heterojunction be a polar semiconductor. We begin with two nonpolar semiconductors with a fixed Δ. We next shift the hybrid energies of alternate atoms on one side by $\pm V_3$. Let the geometry be such that we obtain a nonpolar interface[7] such as a (110) interface so that there will be equal numbers of $+ V_3$ atoms

225

and $-V_3$ atoms at the interface. Then it is obvious that there is no change in the dipole linear in V_3, since the physical system is entirely equivalent if the two atom types are interchanged, corresponding to a change in sign of V_3. If instead we had constructed a polar interface there must be reconstruction at the interface and dipoles may arise; they will depend upon that reconstruction.[7] This is a separate issue, treated in Ref. 7, and we shall not consider it here.

There can be changes in dipole to higher order, just as there was from the introduction of a different V_2, simply because the dielectric constant is modified. Indeed, the situations are closely analogous and again it is appropriate to proceed with an average dielectric constant.

Finally, we may allow a different sp-splitting, corresponding to a different V_1 on the two sides. We saw that the coupling V_1 of bonds and antibonds through a given atom transferred equal charges both ways, even if the bond and antibond arose from different V_2's. Thus again the difference in V_1 itself, or in conjunction with a difference in V_2, does not to first order give rise to dipole shifts, but it will affect the dipole shift when there is a nonzero Δ, and we again would take this into account by using an average dielectric constant, which in principle includes the effects of V_1, in the evaluation of α.

III. BAND LINEUPS

A. Theory and experiment

We conclude that for purposes of calculating the band lineup between semiconductors A and B, we may simply assume that a dipole will arise sufficiently to reduce the difference in average hybrids, $\bar{\epsilon}_h(A) - \bar{\epsilon}_h(B)$, by factor of ϵ_∞. This will then shift the relative band lineups which would have been obtained from Eq. (2).

We begin by evaluating the average hybrid energies for each compound using the Hartree–Fock term values obtained by Mann.[8] These are listed in Table I. We also list the difference between this average hybrid energy and the valence-band maximum from Eq. (1). Given also the dielectric constant, we have everything needed to predict the relative positions of the valence-band maxima in any heterojunction, including the effect of the interface dipole. For a heterojunction A/B between compounds A and B, we first divide the quantity $\bar{\epsilon}_h(A) - \bar{\epsilon}_h(B)$ by the average dielectric constant. Then for the heterojunction, $\bar{\epsilon}_h(A)$, (in the bulk of A) will lie above $\bar{\epsilon}_h(B)$ (in the bulk of B) by $[\bar{\epsilon}_h(A) - \bar{\epsilon}_h(B)]/\epsilon_\infty$. In fact for the heterojunctions we consider here, the difference in hybrid energies is small enough and the dielectric constants large enough that it suffices to take the average hybrid energies as equal on the two sides and obtain the energy of the valence-band maximum in B, relative to that in A, by subtracting $\epsilon_h - \epsilon_v$ for B from the corresponding entry for A, i.e.,

$$\Delta\epsilon_v (\text{"pinned"}) = [\bar{\epsilon}_h(A) - \epsilon_v(A)]$$
$$- [\bar{\epsilon}_h(B) - \epsilon_v(B)]. \qquad (3)$$

In Table II, results of Eq. (3) are compared with those experimental data chosen by Kroemer[9] as most reliable. These predictions correspond to the limit of strong screening

TABLE I. Properties of tetrahedral semiconductors: bond length d (Å); average hybrid energy $\bar{\epsilon}_h$ (eV); and average hybrid energy relative to valence maximum, $\bar{\epsilon} - \epsilon_v$ (eV). All semiconductors are treated as zinc blende structure for calculation.

	d	$-\bar{\epsilon}_h$	$\bar{\epsilon}_h - \epsilon_v$
C	1.54	13.15	2.03
Si	2.35	9.38	−0.03
Ge	2.44	9.29	−0.32
Sn	2.80	8.33	−0.33
BN	1.57	13.32	2.62
BP	1.97	10.83	0.74
BAs	2.07	10.58	0.43
AlN	1.89	11.95	2.72
AlP	2.36	9.46	0.76
AlAs	2.43	9.21	0.46
AlSb	2.66	8.54	0.23
GaN	1.94	12.04	2.55
GaP	2.36	9.55	0.66
GaAs	2.45	9.30	0.34
GaSb	2.65	8.63	0.14
InN	2.15	11.75	2.59
InP	2.54	9.26	0.77
InAs	2.61	9.01	0.47
InSb	2.81	8.34	0.28
BeO	1.65	13.75	4.04
BeS	2.10	10.58	1.78
BeSe	2.20	10.09	1.31
BeTe	2.40	9.20	1.00
MgTe	2.76	8.27	1.54
ZnO	1.98	13.03	4.16
ZnS	2.34	9.86	2.15
ZnSe	2.45	9.37	1.69
ZnTe	2.64	8.48	1.40
CdS	2.53	9.75	2.14
CdSe	2.63	9.27	1.71
CdTe	2.81	8.37	1.43

of the hybrid energy difference. Also listed are the natural band lineups obtained by omitting screening of the hybrid-energy difference. (These are also directly obtainable from Table I by combining the second and third columns to obtain ϵ_v on the same scale for each compound.) It would appear that the agreement with experiment is generally improved by including the screening, although with the striking exception

TABLE II. Valence-band discontinuities for compounds A/B: $\Delta\epsilon_v$ ("natural") [Eq. (1)]; $\Delta\epsilon_v$ ("pinned") [Eq. (3)]; and experimental values (Ref. 9).

	"natural"	"pinned"	experiment
AlAs/GaAs	0.03	0.12	0.50
InAs/GaSb	0.72	0.33	0.46
GaAs/InAs	0.16	−0.13	0.17
Si/Ge	0.38	0.29	0.20
ZnSe/GaAs	1.42	1.35	0.96
ZnSe/Ge	2.09	2.01	1.52
GaAs/Ge	0.67	0.66	0.56
CdS/InP	1.86	1.37	1.63

tion of GaAs/InAs. However, we have shown that the screening should be included in any case, and comparison with experiment should not be the basis for deciding this point. As discussed below, the numerical discrepancies between theory and experiment are of just the general magnitude expected from the inherent numerical limitations of nonrelativistic tight-binding theory.

We note that the natural band lineups by themselves do a reasonably good job of predicting band lineups. This is of course because the average hybrid energies do not differ greatly between tetrahedral semiconductors of comparable lattice constants, as can be seen directly from Table I. This strong correlation between average hybrid energy and bond length is illustrated in Fig. 1. This correlation is also the reason why many descriptions of heterojunction lineups which refer levels to the vacuum level appear to work rather well experimentally. Of course when the average hybrid energies are quite different, the dipoles will be large and the reference to the vacuum level will fail. However such large differences are expected primarily in cases of severe lattice mismatch, for which reliable data are scarce and are greatly affected by strain, which is neglected here.

B. Accuracy and limitations of method

In comparing theory and experiment, it is crucial to have some *a priori* estimate of the inherent numerical accuracy of the theory as implemented, in order to assess whether quantitative discrepancies between theory and experiment reflect a fundamental problem or merely an expected accuracy limitation. Here we consider the accuracy limits inherent in the tight-binding approach to band lineups and interface dipoles.

The results in Table II are based on the limit of large dielectric constant. However, as outlined above, the calculation can easily be carried through for arbitrary dielectric constant, and the resulting changes are well under 0.1 eV for pairs of semiconductors with reasonably small lattice mismatch. Bear in mind however that some of the semiconductors included in Table I have much smaller dielectric con-

stants ϵ_∞, and care should be taken in applying Eq. (3) to such extreme cases.

The two crucial quantities here are ϵ_v and $\bar{\epsilon}_h$. The valence maximum ϵ_v calculated with Eq. (1) neglects spin-orbit splitting, and including this correction raises ϵ_v by 1/3 of the splitting. This correction to Eq. (1) is about 0.1 eV or less for many semiconductors, but is about 0.3 eV for antimonides and tellurides. If warranted, this correction could be included by subtracting 1/3 of the experimental splitting from $(\bar{\epsilon}_h - \epsilon_v)$ in Table I.

The hybrid energy $\bar{\epsilon}_h$ depends only upon the atomic eigenvalues. Unfortunately, these are somewhat sensitive to the method (Hartree–Fock, Herman–Skillman, etc.) used in the atomic calculations, and the dependence on method only partly cancels out in the subtraction (3). The differences relate to the manner in which electron–electron interactions (i.e., correlation and exchange) are approximated, and would be present in even the complete analysis of the heterojunction. The term values also depend upon the atomic configuration (sp^3, s^2p^2, etc.) assumed in the calculation. For consistency and uniqueness we use Hartree–Fock calculations in the atomic ground state. For column II atoms, which have no p electrons in the ground state, ϵ_p is obtained by extrapolation.[1,2] However shortcomings in Hartree–Fock, and the limitations of using results for free-atom configuration and environment to describe the solid, could introduce errors which we can only roughly estimate to be of order 0.2 eV, based on comparisons of different possible approaches.

The Hartree–Fock term values used also neglect "scalar-relativistic" effects, which tend especially to lower the s eigenvalue for heavy atoms. In the extreme case of HgTe–CdTe discussed below, this correction is about 0.5 eV. However the correction is probably more typically 0.1 or 0.2 eV.

Finally, the analysis of Sec. II C is correct only to first order in V_1 and V_3. An analysis for a related problem (unpublished) showed that terms beyond first order in V_1 are not negligible. In the present context, this means that treating GaAs (and *a fortiori* ZnTe) as if both atoms were in an sp^3 configuration, while surprisingly accurate, might still introduce significant errors, which remain to be carefully estimated.

In conclusion, we may expect an accuracy of a few tenths of an eV in general. The errors may be expected to be largest for II–VI's and for highly relativistic materials, because of spin-orbit splitting, scalar-relativistic shifts, and possibly because of terms beyond first order in V_1. We believe that these quantitative limitations of the present approach are more than compensated by the advantage of tremendous simplicity, which gives direct insight into the physical mechanisms and chemical trends.

C. Relativistic term values: HgTe–CdTe

We did not include any elements from the lead row of the Periodic Table in the compounds in Table I. For these, relativistic effects are so large that use of nonrelativistic term values can be misleading. As discussed above, spin-orbit splitting is easily included. We have neglected it here, but the difference in splitting for HgTe and CdTe is small enough that no serious problem arises. The relativistic shift in the

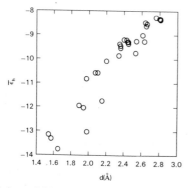

FIG. 1. Average hybrid energy $\bar{\epsilon}_h$ vs bond length d, illustrating that $\bar{\epsilon}_h$ usually does not differ greatly among tetrahedral semiconductors with comparable bond lengths.

J. Vac. Sci. Technol. B, Vol. 4, No. 4, Jul/Aug 1986

eigenvalues however turns out to be quite interesting for this system. That shift acts primarily to lower the atom s energy ϵ_s. This does not affect the p-like valence maximum, so in theories which neglect interface dipoles, this effect is unimportant. However the average hybrid energy $\bar{\epsilon}_h$ does include ϵ_s, so the relativistic shift affects the interface dipole.

CdTe and HgTe are found to have almost identical valence-band maxima. (The Cd and Hg p-state energies differ by only a few hundredths of an eV, and the bond length d is nearly the same for the two compounds.) This is true both for the relativistic and for the nonrelativistic values, and in both cases the use of natural band lineups (neglect of interface dipoles) predicts a negligible valence-band discontinuity. Moreover, in the nonrelativistic theory the s-state energies of Cd and Hg are also almost identical, so the average hybrid energies for the two compounds are almost identical, generating no dipole, and a negligible valence-band discontinuity is again predicted. However, relativistic effects lower the s-state energy in Hg drastically,[10] leading to a large dipole and a valence-band discontinuity of 0.49 eV. A very similar discontinuity for this system was recently predicted by Tersoff[11] using the approach of Ref. 4.

Experimentally the band lineup for HgTe–CdTe is quite uncertain. While Guldner et al.[12] have reported a $\Delta\epsilon_v$ of 0.04 eV, their results in fact do not rule out a large discontinuity such as found here.[13] The experimental situation for this system has been discussed elsewhere.[11]

We also give values for MnTe. Manganese is a transition metal; in the compound it contributes two s electrons to the valence bands and the remaining five d electrons may be thought of as localized in a Hund's rule atomic state. We may obtain the s-state energy from Ref. 10, and estimate the p-state energy as above. Then we have all parameters to include this system. We obtain an average hybrid energy for MnTe 1.67 eV above the valence-band maximum, in comparison to the 1.22 and 0.73 eV for CdTe and HgTe, respectively. This allows immediate prediction of band discontinuities for HgTe–MnTe (0.94 eV) and CdTe–MnTe (0.45 eV). Of course pure tetrahedral MnTe does not exist, but the alloys do and the band positions are obtained for the alloys by interpolation.

IV. SCHOTTKY BARRIERS

The same kind of tight-binding theory should be applicable to a semiconductor metal interface, though we have not yet carried out a rigorous analysis. (The detailed analysis for this case, and the relationship to the submonolayer regime, will be presented elsewhere.) It seems clear that the metallic Fermi energy should play the role of the half-filled hybrid (or average hybrid energy $\bar{\epsilon}_h$ in a compound) in the semiconductor. To the extent that the metallic density of states is structureless, and the dielectric constant of the semiconductor is large (assuring strong pinning[14]), this is the only energy describing the metal. Indeed Tersoff[15] has argued that the Fermi energy of the metal should be pinned at E_B, which plays exactly the role in that theory that the average hybrid energy does here. Then the Schottky barrier for p-type materials, $\phi_{bp} \equiv E_F - \epsilon_v$, is predicted to be

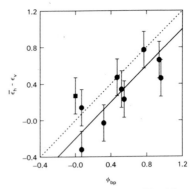

J. Vac. Sci. Technol. B, Vol. 4, No. 4, Jul/Aug 1986

FIG. 2. Calculated p-type barrier height $\bar{\epsilon}_h - \epsilon_v$ [Eq. (4)] vs experimental barrier ϕ_{bp} (Ref. 16). Error bars on calculations are a rough estimate ± 0.2 eV (see text). Dotted line represents perfect agreement. Solid line is same shifted 0.2 eV to allow for systematic errors. Square is InSb (see text).

$$\phi_{bp} = \bar{\epsilon}_h - \epsilon_v, \tag{4}$$

which is listed in Table I. We have plotted in Fig. 2 the experimental values of ϕ_{bp} (with Au contacts) for those elemental and III–V systems for which Sze[16] gives data, against the corresponding entry $\bar{\epsilon}_h - \epsilon_v$ from Table I. The accord is quite good overall, albeit with a systematic shift of about 0.2 eV between theory and experiment. There is however a serious discrepancy, even after this shift, in the case of InSb. In retrospect this is not surprising, since InSb is by far the most relativistic of these semiconductors, and the relativistic corrections to both $\bar{\epsilon}_h$ and $\bar{\epsilon}_v$ are of the right sign and magnitude to bring InSb into overall agreement. The second worst case is GaSb, which is also the second most relativistic.

This line of reasoning leads directly to the correlation between Schottky barrier heights and band lineups, which has been previously discussed.[4,17] That is, combining Eqs. (3) and (4) gives

$$\Delta\epsilon_v = \phi_{bp}(A) - \phi_{bp}(B). \tag{5}$$

This relationship has been well verified experimentally.[17,18]

Recently Sankey et al.[19] noted a correlation between Fermi level position at a Schottky barrier, and the average of the anion and cation dangling bond energies. Such an average is obviously closely related (if not identical) to $\bar{\epsilon}_h$, so that correlation is expected here. However this correlation was attributed[19] by those authors to the role of actual dangling bonds in pinning E_F. Such an interpretation, besides requiring the unnecessary step of postulating point defects for which there is no direct evidence, does not explain the compelling evidence for Eq. (5).

V. CONCLUSIONS

We have shown that interface dipoles play a dominant role in determining heterojunction band lineups, and have described a tight-binding theory which includes these dipoles self-consistently. The results are in good overall agreement with experiment. Moreover, they explain why theories which neglect interface dipoles may often give reasonably good results: The average hybrid energy, which determines the dipole, is nearly the same for nearly lattice-matched semiconductors, which are the ones for which data is available. As a result, interface dipoles are in many cases fortuitously small, although they play a critical role in certain cases of interest (e.g., HgTe–CdTe).

The same approach has been applied to Schottky barriers, although without a detailed analysis. Again the overall agreement with experiment is quite satisfying. We conclude that the tight-binding approach, with the self-consistent inclusion of interface dipoles, successfully gives a unified quantitative picture of both heterojunction band lineups and Schottky barrier heights, based on simple calculations of bulk semiconductor properties.

ACKNOWLEDGMENT

This work was supported in part by ONR Contract No. N00014-84-C-0396.

[a] Permanent address: Applied Physics Department, Stanford University, Stanford, California 94305.

[1] W. A. Harrison, *Electronic Structure and the Properties of Solids* (Freeman, New York, 1980).

[2] W. A. Harrison, Phys. Rev. B **24**, 5835 (1981), provides revised parameters, based upon Hartree–Fock, rather than the Herman–Skillman term values used in Ref. 1.

[3] W. A. Harrison, J. Vac. Sci. Technol. **14**, 1016 (1977).

[4] J. Tersoff, Phys. Rev. B **30**, 4874 (1984).

[5] W. A. Harrison, J. Vac. Sci. Technol. B **3**, 1231 (1985).

[6] W. A. Harrison, Phys. Rev. B **31**, 2121 (1985).

[7] W. A. Harrison, R. Grant, E. Kraut, and D. J. Waldrup, Phys. Rev. B **18**, 4402 (1978).

[8] J. B. Mann, "Atomic Structure Calculations, 1: Hartree–Fock Energy Results for Elements Hydrogen to Lawrencium." Distributed by Clearinghouse for Technical Information, Springfield, Virginia, 22151 (1967).

[9] H. Kroemer, in *Proceedings of the NATO Advanced Study Institute on Molecular Beam Epitaxy and Heterostructures*, Erice, Sicily, 1983, edited by L. L. Chang and K. Ploog (Martinus Nijhoff, The Netherlands, 1984). For a recent reevaluation of AlAs/GaAs, see G. Duggan, J. Vac. Sci. Technol. B **3**, 1224 (1985).

[10] C. C. Lu, T. A. Carlson, F. B. Malik, T. C. Tucker, and C. W. Nestor, Jr., At. Data **3**, 1 (1971); values were extracted for HgTe and CdTe by W. A. Harrison, J. Vac. Sci. Technol. A **1**, 1672 (1983).

[11] J. Tersoff (to be published).

[12] Y. Guldner, G. Bastard, J. P. Vieren, M. Voos, J. P. Faurie, and A. Million, Phys. Rev. Lett. **51**, 907 (1983).

[13] G. Bastard (private communication).

[14] J. Tersoff, Phys. Rev. B **32**, 6968 (1985).

[15] J. Tersoff, Phys. Rev. Lett. **52**, 465 (1984).

[16] S. M. Sze, *Physics of Semiconductor Devices* (Wiley, New York, 1969). The relationship $\phi_{bp} = E_g - \phi_{bn}$ is used where appropriate in extracting ϕ_{bp} from this tabulation.

[17] J. Tersoff, J. Vac. Sci. Technol. B **3**, 1157 (1985).

[18] G. Margaritondo, Surf. Sci. (to be published).

[19] O. F. Sankey, R. E. Allen, S.-F. Ren, and J. D. Dow, J. Vac. Sci. Technol. B **3**, 1162 (1985); O. F. Sankey (private communication).

J. Phys. C: Solid State Phys., Vol. 12, 1979. Printed in Great Britain. © 1979

Energy barriers and interface states at heterojunctions

F Flores and C Tejedor

Departamento de Física del Estado Sólido (UAM) and
Instituto de Fisica del Estado Sólido (CSIC),
Universidad Autónoma, Cantoblanco, Madrid, Spain

Abstract. In this paper homopolar and heteropolar heterojunctions have been analysed. The energy band discontinuities at heteropolar heterojunctions have been obtained self-consistently by analysing the charge induced at the interface. The crucial point of the analysis is the flow of charge between both semiconductors as a function of the difference in energy between their charge neutrality levels. For homopolar heterojunctions interface relaxation is discussed and it appears to be a function of the state of occupation of the interface. The analysis gives a relaxation of 4% for the 111-GeGaAs interface.

1. Introduction

Recently, a great deal of work has been directed towards the understanding of semi-conductor heterojunctions. Theoretical attempts at studying these junctions can be classified into two main groups: (i) by the use of self-consistent calculations (Baraff *et al* 1977, Pickett *et al* 1977, 1978); (ii) by the use of simple arguments to deduce parameters relevant to the electrical properties (Anderson 1960, Dobrzynski *et al* 1976, Frensley and Kroemer 1976, 1977, Louis 1977). In these theoretical analyses, the semiconductors are assumed to have a small lattice mismatch with a negligible number of unpaired dangling bonds. This seems to be a reasonable assumption for many cases (Milnes and Feutch 1972) and it will be also assumed in this work.

On the other hand, we can also classify heterojunctions into two groups, those with heteropolar and those with homopolar interfaces. In this description, a heterojunction built up with covalent semiconductors is considered as a heteropolar junction. The essential difference between the two classes comes from the different occupations of the inter-face states; in heteropolar interfaces these states are fully occupied whereas in homopolar interfaces the states are partially occupied. Let us remark that this classification is related to the different faces of the ionic crystal forming the junction. In heterojunctions built up with semiconductors with the zinc blende structure, a homopolar junction corre-sponds to the (111) or (100) faces, whereas the (110) face gives a heteropolar junction. From a theoretical point of view, these two types of heterojunction are very different. In homopolar heterojunctions the partial occupation of the interface states modifies the self-consistency of the interface in a way that can only be achieved by means of an ionic relaxation (Baraff *et al* 1977).

The aim of this paper is to provide a simple approach to heterojunctions. The analysis is divided into two parts. Firstly a simple self-consistent treatment is applied to hetero-polar heterojunctions, let us say the (110) faces, in line with recent work on metal–semiconductor junctions (Tejedor *et al* 1977, to be referred to hereafter as I). A brief description of this approach has already been published (Tejedor and Flores 1978).

731

Secondly, we discuss relaxation and interface states at homopolar heterojunctions. In the analysis given previously for heteropolar heterojunctions, we concluded that roughly the charge neutrality levels of both semiconductors are lined up. Based on this result, we neglect the problem of determining the energy levels at the homopolar heterojunctions assuming that the neutrality levels of both semiconductors are lined up, and concentrate on analysing the most distinctive feature of these interfaces, namely their relaxation.

In §2 we discuss heteropolar heterojunctions, introducing the charge neutrality level of each semiconductor and obtaining self-consistently an expression for the valence band discontinuities between both semiconductors. In §3 we analyse the interface relaxation of homopolar heterojunctions. Finally, in §4, we discuss the results for both interfaces.

2. Energy barriers at heteropolar heterojunctions

The fundamental magnitudes defining the electrical behaviour of heterojunctions are discontinuities in the valence and conduction bands between both semiconductors, namely ΔE_v and ΔE_c. Figure 1 shows the electron energy levels for the two semiconductors

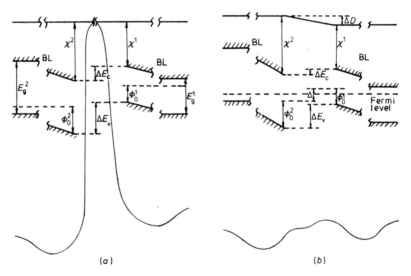

Figure 1. Electron energy level diagrams for a heterojunction contact. (*a*) The two media are isolated from each other; (*b*) intimate contact is allowed, charge overflow takes place and an induced dipole D appears at the junction. The space charge boundary layer BL on each semiconductor is symbolically indicated by the tilted straight lines of small slope to suggest a small potential change on the scale of length of the interface. $V(z)$ represents the interface potential.

before and after the intimate contact. When the contact is made, there is a flow of charge from the semiconductor on the left to the one on the right and a potential difference ΔD is created across the interface. As we are interested in relating ΔE_v (or ΔE_c) to the electron affinities of both semiconductors, we start by discussing the free surfaces of the semiconductors before the intimate junction is made; later, we shall study the heterojunction.

2.1. Semiconductor surfaces

It has been stressed elsewhere (I) that because of the cancellation between the charge placed at the surface states and the rearrangement of charge near the top of the valence band in narrow-gap semiconductors, the surface charge and surface potential should closely resemble that of a metal with the same electron density. The main changes will arise from the crystal pseudopotential but not from the appearance of surface states in the semiconductor gap (Flores and Tejedor 1977). On this basis, the dipole barrier at the semiconductor surface can be written as

$$D^s = D^s_j + D^s_L$$

(1)

where D^s_j is the surface dipole for the jellium model of a metal with the same electron density as the semiconductor and D^s_L is the correction coming from the lattice effects.

For a (110) face, similar arguments can be given to support the same conclusion, either for covalent or III–V ionic semiconductors (García-Moliner and Flores 1978). The surface dipole layer of the (110)-face of a semiconductor can be written as in equation (1). Notice that this surface dipole is contributing to the electron affinity of the semiconductor SC and therefore it must be taken into account when the energy levels before and after the intimate contact are compared.

2.2. Interface states and the induced dipole at the heterojunctions

The main problem with heterojunctions is obtaining the charge density in the interface self-consistently. Knowing this charge it is fairly easy to calculate the dipole barrier and the energy levels of the heterojunctions. Let us start by analysing the charge density of the interface in a one-dimensional mode. Figure 2 shows the model we are going to discuss. The heterojunction is formed with two covalent narrow-gap semiconductors joined at $z = 0$, with a small difference ΔV between the mean potential of both bulk crystals. Because of the good matching between the two three-dimensional lattices, we take the same reciprocal lattice vector g for both semiconductors in our one-dimensional model.

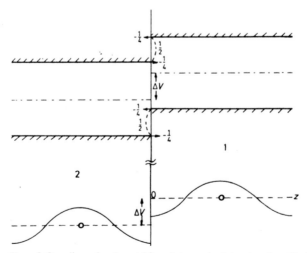

Figure 2. One-dimensional model for a heteropolar heterojunction. The integrated density of states with the band edge defects of $\frac{1}{4}$ is also shown, together with the excess of $\frac{1}{2}$ spread between the band edges.

2.2.1. Interface states. Firstly, let us look for interface states in this model. For $z > 0$ the wavefunction in the gap of the semiconductor can be written as

$$\psi^{(1)} \sim \exp(-q_1 z) \cos(\tfrac{1}{2}gz + \tfrac{1}{2}\phi_1) \tag{2}$$

where $q_1 = 2(V_1^2 - \epsilon_1^2)^{1/2}/g$ and $\exp(i\phi_1) = [\epsilon_1 + i(V_1^2 - \epsilon_1^2)^{1/2}]/V$, where V_1 is the pseudopotential component and ϵ_1 is the energy measured from the mid-gap. (Atomic units are used everywhere in this paper.) On the other hand, for $z < 0$

$$\psi^{(2)} \sim \exp(q_2 z) \cos(\tfrac{1}{2}gz - \tfrac{1}{2}\phi_2) \tag{3}$$

with an obvious notation. Matching conditions between the wavefunctions (2) and (3) gives, assuming q_1 and $q_2 \ll g$, the following equation:

$$\phi_1 + \phi_2 = 2\pi n. \tag{4}$$

This equation has no solution apart from when $V_1 = V_2$ and $\Delta V = 0$. In this particular case two interface states appear in the band edges. This is a trivial case, as it corresponds to an infinite crystal with no interface. These interface states represent the bulk states of the crystal at the edges of the valence and conduction bands. In no other case can an interface state to be found from equation (4), but the previous limit shows by an argument of continuity that there is no defect of charge in the valence band when $\Delta V \neq 0$ and that all the interface states emerging from the valence band by relaxation (see below) are fully occupied, in order to comply with the charge neutrality condition. This can also be proved by studying the density of interface states in the conduction and valence bands. This is the subject of the following section.

2.2.2. Density of interface states. The local density of states has been calculated by means of the surface Green function method (García-Moliner and Rubio 1971). From the Green function for a narrow-gap model, we obtain the following result for the change in the bulk density of states:

$$N(E, z) = \frac{1}{\pi g} \operatorname{Im}\left[\exp\left(-2\frac{(V_1^2 - \epsilon_1^2)^{1/2}}{g}\right) \left(\frac{V_1 + \epsilon_1}{(V_1^2 - \epsilon_1^2)^{1/2}} \cos\frac{g}{2}z + \sin\frac{g}{2}z\right)^2 \right.$$

$$\left. \times \left(\frac{[(V_1^2 - \epsilon_1^2)^{1/2}/(V_1 + \epsilon_1)]\{[(V_1^2 - \epsilon_1^2)^{1/2}/(V_1 + \epsilon_1)] - [(V_2^2 - \epsilon_2^2)^{1/2}/(V_2 + \epsilon_2)]\}}{[(V_1^2 - \epsilon_1^2)^{1/2}/(V_1 + \epsilon_1)] + [(V_2^2 - \epsilon_2^2)^{1/2}/(V_2 + \epsilon_2)]}\right) \right]$$

$$z > 0 \tag{5}$$

with a similar equation for $z < 0$. The whole density of states is given by

$$N(E, z) = N_0(E, z) + \delta N(E, z) \tag{6}$$

where $N_0(E, z)$ is the local density of states for an infinite crystal.

From equation (5) we can obtain the integrated density of states for the heterojunction, $N_H(E)$, defined by

$$N_H(E) = \int_{-\infty}^{\infty} \delta N(E, z)\,dz \tag{7}$$

and the interface charge, $\delta\rho(z)$, defined by

$$\delta\rho(z) = 2 \int_{-\infty}^{E_v} \delta N(E, z)\,dz \tag{8}$$

where spin degeneracy has been accounted for by the factor 2.

The integrated density of states shows band edge defects of $\frac{1}{4}$ compensated by an excess of $\frac{1}{2}$ of state spread between the band edges. Figure 2 shows these results schematically. This is in agreement with the comment of the previous section about the neutrality of the charge of the whole valence band.

On the other hand, as regards $\delta\rho(z)$ when $\Delta V \neq 0$, there is a flow of charge from the semiconductor of higher potential to the semiconductor of lower potential. This is illustrated in figure 3, where the charge $\delta\rho(z)$ is given in adimensional units by the full curve, for the particular case $V_1 = V_2$. The broken curve is an average of the oscillations and its behaviour is given by an exponential decrease which decays as $\exp[-(2|V|/g)z]$ for large z.

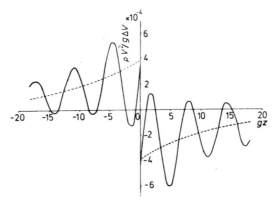

Figure 3. Charge overflow between the semiconductors shown in figure 2 for a given value of ΔV.

The essential point which emerges from the analysis of our one-dimensional model is the existence of this flow of charge between both semiconductors with an energy close to the valence band edges and creating a dipole which tends to equalise both mid-gaps. It is interesting to remark that this charge decreases exponentially as $\exp[-(2|V|/g)z]$ for large z, which is similar to the decrease found in a one-dimensional metal–semiconductor junction for the charge flowing from the metal to the semiconductor when the Fermi level of the metal is higher than the mid-gap. Let us stress that both charges play the same role in each junction and create the surface barrier which tends to equalise either both mid-gaps or the metal Fermi level and the semiconductor mid-gap (I).

Once we have discussed the one-dimensional model, let us consider the three-dimensional interface. As it has been stressed elsewhere (I), we can simplify this case if we only consider points of high symmetry in the two-dimensional Brillouin zone of the semiconductor. At these points, matching equations are split into one-dimensional-type problems (Flores and Tejedor 1977, Louis *et al* 1976). This allows us to generalise our one-dimensional results to the three-dimensional interface, taking an average of the different one-dimensional loops contributing to the restoring dipole. This average defines the charge neutrality point ϕ_0 for each semiconductor, rather like in I, in such a way that the induced interface dipole depends linearly on the difference in energy, Δ, between both charge neutrality points. In other words, as long as $\Delta = 0$, we can look at the heterojunction as a metal–metal interface, with the interface dipole barrier D^{1-2} given by

$$D^{1-2} = D_0^{1-2} + D_L^{1-2} \qquad (\Delta = 0) \qquad (9)$$

where D_0^{1-2} is the interface barrier for the jellium model and D_L^{1-2} includes all lattice effects.

In general, however, we have $\Delta \neq 0$. In this more general case, besides the metal-like interface dipole, we have a restoring dipole δD caused by a flow of charge similar to the one discussed in the one-dimensional model. We can then write

$$D^{1-2} = D_0^{1-2} + D_L^{s-s} + \delta D \qquad (\Delta \neq 0). \tag{10}$$

Now, combining equations (10) and (1) and applying these to each semiconductor, we can write the band edge discontinuity ΔE_v in a straightforward manner as

$$E_v = (\chi^{(2)} + Eg^{(2)}) - (\chi^{(1)} + Eg^{(1)}) - D_J - \delta D, \tag{11}$$

where $\chi^{(1)}$, $\chi^{(2)}$, $Eg^{(1)}$ and $Eg^{(2)}$ are defined in figure 1(b) and D_J is given by

$$D_J = D_0^{1-2} + D_L^{1-2} - (D_0^1 + D_L^1) + (D_0^2 + D_L^2). \tag{12}$$

We can now linearise for the induced dipole δD by writing $\delta D = \alpha \Delta$, where α defines the linear relation between this induced dipole and the energy difference between both charge neutrality levels Δ. Moreover, Δ can be easily related to $\phi_0^{(1)}$, $\phi_0^{(2)}$ and ΔE_v (see figure 1b) in such a way that we obtain from equation (10)

$$\Delta E_v = [1/(1 + \alpha)][(\chi^{(2)} + Eg^{(2)}) - (\chi^{(1)} + Eg^{(1)}) - D_J + \alpha(\phi_0^{(2)} - \phi_0^{(1)})]. \tag{13}$$

In this equation χ is the electron affinity associated with the crystallographic face of the junction. On the other hand, as D_J and ϕ_0 can be obtained in a similar manner to that described in (I), we can determine ΔE_v from equation (13) if we can calculate the coefficient α. This is the subject of the following discussion.

The calculation of $\alpha\Delta$ requires a model for the screening of the charge flowing between both semiconductors, where many-body interactions have to be taken into account. We are going to follow the method developed in I rather closely.

First of all, let us assume that $\Delta = 0$. Then we can describe the interface charge by

$$n_1(z) = \begin{cases} n_1 - \frac{1}{2}(n_1 - n_2) \exp(\beta_0 z) & z < 0 \\ n_2 + \frac{1}{2}(n_1 - n_2) \exp(-\beta_0 z) & z > 0 \end{cases} \tag{14}$$

where β_0 is determined by a variational method similar to the one proposed by Smith (1969).

When $\Delta \neq 0$, we have a flow of charge between both semiconductors. In a first approximation, this charge can be described by

$$\delta N_1(z) = \begin{cases} \delta N_1 \exp(Q_1 z) & z < 0 \\ \delta N_2 \exp(-Q_2 z) & z > 0 \end{cases} \tag{15}$$

where δN_1, δN_2, Q_1 and Q_2 are parameters determined to adjust equation (15) to the charge flowing between both semiconductors. This charge has been obtained for different heterojunctions by taking an average over two points of the two-dimensional Brillouin zone (Louis et al 1976). Nevertheless, this charge δN interacts with the whole system. As in I we can split this interaction into two parts, one of long range, which screens δN_1 through a dielectric constant ϵ_1 or ϵ_2 and another of short range, at the interface, which is calculated from the following model of interface charge. The point is to substitute, for $z > 0$, the charge $\delta N_1 \exp(Q_1 z)$ by

$$(\delta N_1/\epsilon_1) \exp(Q_1 z) + \delta N_1 \gamma_1 z \exp(\beta z) \tag{16}$$

where $(\delta N_1/\epsilon_1) \exp(Q_1 z)$ gives the long-range screening while the short-range second term $\delta N_1 \gamma_1 z \exp(\beta z)$ is obtained by imposing charge conservation between $\delta N_1 \exp(Q_1 z)$ and equation (16). This gives

$$\gamma_1 = (\beta^2/Q_1)[1 - (1/\epsilon_1)] \tag{17}$$

which leaves β as a parameter to be determined by the minimisation of the many-electron energy associated with the total charge of the interface.

Having obtained the charge given by equation (16), we can now proceed to write the model for the total charge at the interface. This is given by

$$N(z) = \begin{cases} n_1 + \mu_1 \exp(\beta z) + \delta N_1[(1/\epsilon_1) \exp(Q_1 z) + \gamma_1 z \exp(\beta z)] & z < a \\ n_2 + \mu_2 \exp(-\beta z) + \delta N_2[(1/\epsilon_2) \exp(Q_2 z) + \gamma_2 z \exp(-\beta z)] & z > a \end{cases} \tag{18}$$

where the parameters μ_1 and μ_2 are determined by demanding continuity of the charge and its derivative and a is calculated by imposing global charge neutrality.

In equation (18), β is a variational parameter which can be obtained by minimising the interface energy by means of the method introduced by Smith (1969). In our particular case, β has been obtained by linearising the many-electron energy in δN_1 and δN_2 and taking the limit $n_1 \rightarrow n_2$, because of the good matching between both crystals. Once β is known, it is straightforward to determine the electrostatic dipole $\alpha\Delta$ associated with the flow of charge between the semiconductors. Here, α is a complicated function of the different properties of the heterojunction components. However, for heterojunctions with covalent and/or III–V ionic semiconductors, α is almost independent of the particular components and has a value close to 2·5.

2.3. Electron energy levels for the heterojunction

Once we know α, ΔE_v can be obtained from equation (13). Let us remark that equation (13) gives Anderson's (1960) result for ΔE_v if we assume $D_J = \alpha = 0$. This implies that both the surface lattice effects D_J and the restoring dipole $\alpha\Delta$ could be neglected even though both charge neutrality levels do not coincide. Let us stress that the first approximation $D_J = 0$ is a reasonable one for the (110) faces, since for those interfaces the induced charge (220) is small (Bertoni *et al* 1973) and the effects of both crystalline pseudopotentials $V(220)$ tend to cancel out (see I). However, α cannot be neglected in equation (13), since, as has been discussed above, $\alpha \simeq 2·5$.

We have applied our model to the following heterojunctions: Ge–GaAs, Si–GaP, AlSb–GaSb and GaSb–InAs. In table 1 we give the charge neutrality level of each semiconductor as calculated by taking an average over two points of the two-dimensional

Table 1. Theoretical values of ϕ_0, ΔE_v, ΔE_c and Δ (see text) for different (110) heterojunctions.

	Ge–GaAs	Si–GaP	AlSb–GaSb	InAs–GaSb	Ge–ZnSe	GaAs–ZnSe
ϕ_0 (eV)	0·17 0·55	0·00 0·77	0·74 0·61	0·45 0·61	0·17 2·01	0·55 2·01
ΔE_v (eV)	−0·35	(i) −1·00 (ii) −1·37	0·29	0·04	1·82	1·39
ΔE_c (eV)	0·4	(i) 0·20 (ii) −0·17	−0·81	0·26	−0·12	−0·09
Δ (eV)	0·03	(i) −0·23 (ii) −0·6	0·16	0·20	0·02	0·07

Brillouin zone (Louis *et al* 1976), using the bands given by Cohen and Bergstresser (1966). In order to obtain ΔE_v from equation (13), we have neglected D_J and used the experimental values of $Eg^{(1)}$, $Eg^{(2)}$, $\chi^{(1)}$ and $\chi^{(2)}$ which were available. This was so for all III–V semiconductors. However, since we do not know the affinity of (110) covalent faces, we have proceeded for Ga–GaAs and Si–GaP as follows. For Ge–GaAs we have taken for $(\chi^{(2)} + Eg^{(2)}) - (\chi^{(1)} + Eg^{(1)})$ the theoretical value which can be deduced from the work of Picket *et al* (1977), where ΔE_v was obtained as well as the electrostatic dipole at the interface. From our equation (11) and assuming here that $D_J \simeq 0$ and that the self-consistent interface dipole coincides with $\alpha\Delta$, we can easily deduce $(\chi^{(2)} + Eg^{(2)}) - (\chi^{(1)} + Eg^{(1)})$. Thus $(\chi + Eg)\,(GaAs) - (\chi + Eg)\,(Ge) = 0.30$ eV. For Si–GaP we have taken two extreme cases. Firstly, we have assumed $\chi_{Si} = 4.05$ eV, the same value as for the (111) face. Secondly, we have taken $\chi_{Si} = 2.07$ eV which is the value that we have inferred from the UPS data given by Sakurai *et al* (1977) for Si (110) 5×1.

In table 1 we give the values deduced from equation (13) for ΔE_v and also for ΔE_c, the conduction band discontinuity for different heterojunctions. The coincidence that we have obtained with the results of Pickett *et al* (1977) for Ge–GaAs is remarkably good; in principle, our data can only be compared with this theoretical calculation and not with any empirical data, because of the way the parameters of equation (13) were estimated above. However, these results can be modified at most by 0.1 eV if the error in the estimation of $(Eg^{(2)} + \chi^{(2)}) - (Eg^{(1)} + \chi^{(1)})$ is less than 0.35 eV. As such inaccuracy in this quantity seems to be out of question, we think that the values predicted in table 1 for Ge–GaAs have an error less than 0.1 eV. The experimental evidence suggests a value of 0.2 ± 0.15 eV for ΔE_c (Pickett *et al* 1977), which seems to be in reasonable agreement with our prediction of 0.40 ± 0.1 eV.

The results for Si–GaP can be compared with the empirical data of Zeindenberg and Anderson (1967), who have deduced a value of $\Delta E_c \simeq 0$ for this heterojunction. Our estimates for the two extreme assumptions of χ_{110} (Si) are just below and above this experimental result. We believe that this is also a good confirmation of our model.

Although our analysis can only be applied to covalent or III–V ionic semiconductors with confidence, it could be of interest to extrapolate our previous analysis to II–VI ionic semiconductors. Thus, in table 1 we have also included the results obtained for Ge–ZnSe and GaAs–ZnSe. Let us notice that in these cases, the self-consistent parameter α which appears in $\alpha\Delta$ has to be changed from 2.5 to 2.1. Let us stress that the conduction band discontinuity calculated for GaAs–ZnSe seems to be in reasonable agreement with some experimental evidence (Mach *et al* 1970), which gives further support to our model.

Finally it is of interest to note that in two recent theoretical calculations for Ge–GaAs, either no surface states were found in the fundamental gap (Hermann and Kasowski 1978) or only surface states very close to the top of the valence band have been detected. This seems to be in good agreement with the analysis given above for a one-dimensional model and further extrapolated to a three-dimensional crystal.

3. Relaxation and interface states at homopolar heterojunctions

As has been mentioned above, there are some differences between homopolar and heteropolar interfaces. The crucial one comes from the partial occupation of interface states at homopolar heterojunctions. As we shall discuss presently, this can be related to the interface relaxation needed to locate the interface band above the valence band. In this section we discuss this interface relaxation, neglecting the self-consistency which gives the valence band discontinuity. We shall assume that both semiconductors have their charge neutrality level lined up, i.e. $\Delta = 0$. After the results of the last section (see table 1), this can be taken as a reasonable approximation.

3.1. One-dimensional models: sum rules

Let us start by discussing a one-dimensional model. Figure 4 shows the model under consideration: two covalent semiconductors are treated within the narrow-gap approximation and are in contact through an interface whose width is d. Notice that the essential

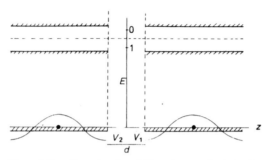

Figure 4. One-dimensional model for a heteropolar heterojunction, with a relaxation defined by the distance d between both jellium edges.

difference between figures 4 and 2 is given by this relaxation; moreover, in figure 4 we have also assumed that both mid-gaps are lined up. Let us note that this model is only appropriate to analyse heteropolar heterojunctions, but here this analysis is important as a first step to the discussion of homopolar heterojunctions further.

Firstly we are going to prove that the potential and charge distribution in the interface of the model shown in figure 4 behave rather like these parameters in an interface between two metals with the same densities as the semiconductors. Initially it is convenient to analyse the density of states at the metallic interface, i.e. let us assume $V_1 = V_2 = 0$ (see figure 4). Here, because of the symmetry around $z = 0$, the wavefunctions can be classified as symmetric or antisymmetric in such a way that, defining the phase-shifts η^S and η^A, we can write

$$\psi^S(z) \sim \sin(q|z| + \eta^S) \tag{19a}$$

$$\psi^A(z) \sim \sin(qz + \eta^A \, \text{sgn}(z)). \tag{19b}$$

Now for an infinite barrier at $z = 0$ and assuming that the crystalline potentials extend up to this barrier, we have $\eta^S = \eta^A = 0$ and the interface density of states is given by

$$N^I_\infty = -\tfrac{1}{2}\delta(E) \tag{20}$$

since a defect of $\tfrac{1}{4}$ state comes from each side of the infinite barrier (Flores and Tejedor 1977, hereafter referred as II). Having obtained the density of states for an infinite barrier, we can easily write down the density of states associated with a general interface barrier creating the phase-shift η^S and η^A (II) as

$$N^S(E) = -\tfrac{1}{2}\delta(E) + (1/\pi)(dn^S/dE) + (1/\pi)(d\eta^A/dE). \tag{21}$$

From this equation a type of Friedel sum rule can be obtained by integrating $N^S(E)$ up to the Fermi level:

$$Z = (2/\pi)(\eta^S + \eta^A) - 1. \tag{22}$$

Here Z is the interface charge defined by the distance d between both jellium edges (see figure 4). In our one-dimensional model

$$Z = -(gd/\pi), \tag{23}$$

whence

$$(\eta^S = \tfrac{\pi}{2}[1 - gd/\pi] - \eta^A. \tag{24}$$

Let us return to the semiconductor interface, switching on the pseudopotentials V_1 and V_2 (figure 4). We shall assume that the potential at the interface behaves like that for a metallic interface. Then the interface states can be studied by matching the wave-function

$$\exp(q_2 z)\sin|\tfrac{1}{2}g(z + \tfrac{1}{2}d) - \tfrac{1}{2}\Phi_2| \qquad z \ll -\tfrac{1}{2}d$$

in the semiconductor gap to

$$A\sin(q|z| + \eta^S) + B\sin(qz + \eta^A\,\mathrm{sgn}(z))$$

at the interface and further on, matching this same interface wavefunction to

$$\exp(-q_1 z)\sin[\tfrac{1}{2}g(z - \tfrac{1}{2}d) + \tfrac{1}{2}\Phi_1] \qquad z \gg \tfrac{1}{2}d.$$

The matching conditions determine the different interface states. For simplicity, let us consider $V_1 = V_2$, for which $\Phi_1 = \Phi_2 = \Phi$. Then, assuming $q_1 = q_2 \ll \tfrac{1}{2}g$, we obtain two independent equations for the existence of interface states, namely:

$$\Phi(\epsilon_1) = 2\eta^A + \tfrac{1}{2}gd \tag{25a}$$

$$\Phi(\epsilon_2) = 2\eta^S + \tfrac{1}{2}gd. \tag{25b}$$

From the relation (24), we can write instead of equation (25b)

$$\Phi(\epsilon_2) = \pi - (2\eta^A + \tfrac{1}{2}gd). \tag{26}$$

Comparing equations (25a) and (26), we see that the interface state levels ϵ_1 and ϵ_2 must be symmetrically located around the lined up mid-gaps of both semiconductors. This is shown schematically in figure 4.

On the other hand, the lowest interface state must be fully occupied while the highest one is empty. This can be proved easily by considering initially an infinite barrier which will be lowered later. With the infinite barrier at $z = 0$, we find an interface state with a double degeneracy which is half-occupied (II). When this barrier is lowered, the degeneracy is broken and the lowest state is fully occupied.

The previous arguments have shown that if it is assumed that the interface potential behaves like that in a metal–metal junction, we have two interface states at the hetero-junction, symmetrically located around the mid-gap, with occupations of 0 and 1. Now because of the symmetry around the mid-gap, the charge of the fully occupied interface state is cancelled out by the associated defect of charge in the interface density of states of the valence band. This cancellation is local (II) and finally, this shows that the assumption about the metal-like behaviour of the interface potential is self-consistent.

Although the metal-like behaviour of the one-dimensional interface of figure 4 has just been proved, let us note that the interface states have not yet been determined, as they depend on the phase shift $\eta^{\scriptscriptstyle \wedge}$ (equations 25 and 26). A reasonable assumption can be made, however, if the relaxation d is small. For this case, the interface potential must be localised near $z = 0$, which allows us to assume that $\eta^{\scriptscriptstyle \wedge} \simeq 0$, since the antisymmetric wavefunction is zero at $z = 0$. Then equations (25a) and (26) give

$$\Phi(\epsilon_1) = \tfrac{1}{2}gd \tag{27a}$$

$$\Phi(\epsilon_2) = -\tfrac{1}{2}gd. \tag{27b}$$

Before turning our attention to homopolar heterojunctions, it is of interest to discuss the implication of the previous results for heteropolar or covalent heterojunctions. Firstly let us stress that the previous arguments about one-dimensional models can be generalised to three-dimensional surfaces. Then, by comparison with the results obtained in §2 for no relaxation, we find two main differences: (i) for an outwards relaxation an interface state band emerges from the valence band; (ii) meanwhile, the restoring dipole $\alpha\Delta$ decreases depending on the value of this outwards relaxation. However, for a small relaxation, i.e. a few percent of the bond length, the coefficient α does not change very much. The implication of this is that for a small relaxation, the energy levels of the heteropolar heterojunctions are almost independent of this relaxation. This also seems to be the result obtained theoretically by Pickett and Cohen (1978) in a self-consistent calculation for a relaxed (110) Ge–GaAs heterojunction.

Let us consider now a one-dimensional model for an homopolar heterojunction. In this case the main difference, compared with the model given in figure 4, is that the minimum of the ionic crystalline potential does not coincide with the jellium edge which defines the surface charge (Flores *et al* 1978). These displacements can be defined by two phase shifts, α_1 and α_2, such that the wavefunctions inside the semiconductor gaps for $|z| > \tfrac{1}{2}d$ are given by

$$\exp(q_2 z) \cos\left[\tfrac{1}{2}g(z + \tfrac{1}{2}d) - \tfrac{1}{2}\Phi_2 - \alpha_2\right] \qquad z < -\tfrac{1}{2}d \tag{28a}$$

and

$$\exp(-q_1 z) \cos\left[\tfrac{1}{2}g(z - \tfrac{1}{2}d) + \tfrac{1}{2}\Phi_1 + \alpha_1\right] \qquad z > \tfrac{1}{2}d. \tag{28b}$$

Now, the surface charge located between the jellium edges and the minimum of each crystalline potential defines the occupation of the interface states. It is then an easy matter to see that this new occupation Q is given by

$$Q = 1 - \left[(\alpha_1 + \alpha_2)/\pi\right] \tag{29}$$

if $\alpha_1 + \alpha_2$ is positive. For $\alpha_1 + \alpha_2$ negative, $|\alpha_1 + \alpha_2|/\pi$ gives the occupation of the highest interface state which is no longer unoccupied.

The interface potential of this model can be analysed like the covalent case. Firstly, let us assume that this potential behaves like that at the interface of a metal. Then, from equations (28), we get the following equations for the interface state levels:

$$\Phi(\epsilon_1) = 2\eta_1 \equiv -\tfrac{1}{2}gd + (\alpha_1 + \alpha_2) + 2\cos^2 \tfrac{1}{2}(\alpha_1 - \alpha_2)\left(\tfrac{1}{2}\pi - \eta^{\mathrm{S}}\right) - 2\eta^{\scriptscriptstyle \wedge} \sin^2 \tfrac{1}{2}(\alpha_1 - \alpha_2) \tag{30a}$$

$$\Phi(\epsilon_2) = 2\eta_2 \equiv \pi - \tfrac{1}{2}gd + (\alpha_1 + \alpha_2) + 2 \sin^2 \tfrac{1}{2}(\alpha_1 - \alpha_2)\left(\tfrac{1}{2}\pi - \eta^{\mathrm{S}}\right)$$
$$-2\eta^{\scriptscriptstyle \wedge} \cos^2 \tfrac{1}{2}(\alpha_1 - \alpha_2) \tag{30b}$$

where the phase shifts η_1 and η_2 are defined by the last equalities. From these equations and equation (24) it can easily be shown that

$$2\eta_1 + 2\eta_2 = \pi + 2(\alpha_1 + \alpha_2). \tag{31}$$

This equation shows that for $\alpha_1 + \alpha_2 \neq 0$, the interface states are no longer symmetrically located around the midgap. For $(\alpha_1 + \alpha_2) \gtrless 0$ either interface state moves upwards/downwards in energy.

Morover, as we are going to prove, for ionic semiconductors $(\alpha_1 + \alpha_2 \neq 0)$ there is no local cancellation between the charge of the interface state and the associated charge in the valence band. We can analyse this case by treating the charge in one semiconductor as the superposition of two charges for a free semiconductor surface. As has been shown elsewhere (II), the surface electronic properties of a one-dimensional free semiconductor are determined by the electronic phase shift near the gap. In the present work, equations (30) show that we can look at the charge in one semiconductor of the heterojunction as the superposition of two charges for a free surface. From equations (30) the first surface has the phase shift η_1, while the second surface has the phase shift η_2. Now the cancellation of charge in each one-dimensional free surface can be analysed by calculating the dipole

$$D = 4\pi \int_0^\infty z\delta\rho\,(z)\,dz \tag{32}$$

where $\delta\rho(z)$ is the sum of the charges in the surface state and the associated density of states in the valence band. For a narrow-gap semiconductor

$$D(\eta,Q) = \frac{\pi g}{2V} \frac{\sin 2\eta}{1 + \cos 2\eta} + \frac{g}{V} \frac{2\eta}{\sin 2\eta} + \frac{2\pi g}{|V|} \frac{Q}{\sin 2\eta} \tag{33}$$

where Q is the occupation of the surface state (Flores et al 1978). For our present case, we can obtain the dipole layer created at the heterojunction between $z = 0$ and ∞, only across one semiconductor, by the superposition of two surfaces with $\eta = \eta_1$ and η_2 and from the occupancy given by equation (29):

$$D = \tfrac{1}{2}D\{\eta_1, 0\} + \tfrac{1}{2}D\{\eta_2, 1 - [(\alpha_1 + \alpha_2)/\pi]\}. \tag{34}$$

Here the factor $\tfrac{1}{2}$ is related to the existence of two media for $z > 0$ and $z < 0$. It is an easy matter to see that this dipole D is not zero for an ionic semiconductor. This is due to to the lack of cancellation between the charge in the interface state and the charge in the valence band. This implies that our assumption about the metal-like behaviour of the interface potential is not valid and that a perturbation of this potential must be induced at the interface. For a metal-like interface we have seen that $\eta^A \simeq 0$ and $\eta^S \simeq \tfrac{1}{2}(\pi - gd)$, after assuming that the interface potential is localised near $z \simeq 0$. Now the perturbation of this potential coming from the interface states must be extended over distances of the order of $g/|V|$, which is much greater than the relaxation d which is assumed to be a few percent of the bond length. Then, the interface potential can be split into two regions: one is localised close to the origin $z \simeq 0$ and creates the metal-like phase shifts $\eta^A \simeq 0$ and $\eta^S = \tfrac{1}{2}(\pi - gd)$; the other is much more extended and changes those phase shifts by some given amount. If we assume that this perturbative potential can be analysed by a WKB approximation, we conclude that the phase shifts η^A and η^S of the metal-like interface are both modified by the same amount. Therefore we change η^A and η^S into $\eta^A + \delta\alpha$ and $\eta^S + \delta\alpha$ for the ionic semiconductor. In order to determine $\delta\alpha$, we also assume that the induced dipole must be zero for the new phase shifts, namely

$$D\{\eta'_1, 0\} + D\{\eta'_2, 1 - [(\alpha_1 + \alpha_2)/\pi]\} = 0 \tag{35}$$

where η'_1 and η'_2 are given by equation (30), substituting η^A and η^S by $\eta^A + \delta\alpha$ and $\eta^S + \delta\alpha$. This assumption is supported by the fact that the induced potential tends to shift the interface states towards an energy level for which the cancellation of the interface charge is better. Since the dipole layer given by equation (34) is important even for a small uncompensated charge, it is reasonable to assume that the interface state is practically determined by the cancellation of this dipole.

Once $\delta\alpha$ has been obtained from equation (35), the interface states can be calculated from the equations

$$\Phi(\epsilon_1) = 2\eta'_1 \tag{36a}$$

$$\Phi(\epsilon_2) = 2\eta'_2 \tag{36b}$$

which must be used instead of equations (30). Let us note that the two dipoles appearing in equation (35) depend essentially on $(\alpha_1 + \alpha_2)$ and $\delta\alpha$. Since $(\alpha_1 + \alpha_2)$ is directly related to the interface occupation through equation (29), $\delta\alpha$ is also a function of this occupation through equation (35). The interface state can therefore be determined through equation (36) as a function of its occupation.

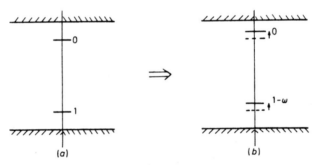

Figure 5. Scheme showing the interface states for a one-dimensional relaxed interface. (a) Heteropolar heterojunction; (b) homopolar heterojunction.

Figure 5 shows in a schematic way the conclusions that can be obtained from the above argument. While in a one-dimensional heteropolar heterojunction, for a given outwards relaxation, there are two interface states symmetrically located around the mid-gap, for a one-dimensional homopolar heterojunction, with an occupancy less/greater than one, the two interface states move upwards/downwards in energy. It is interesting to note that these results are similar to those obtained for free semiconductor surfaces (Flores *et al* 1978). In a one-dimensional covalent semiconductor we find a half-occupied surface state located at the mid-gap. However, in a one-dimensional ionic semiconductor, this surface state moves upwards/downwards in energy for an occupation less/greater than 0·5.

3.2. Three-dimensional surfaces; relaxation

The arguments used above for a one-dimensional model can be generalised to three-dimensional surfaces with some qualifications. Firstly let us point out that the 'metallic' sum rule given by equation (22) can now be applied to three-dimensional surfaces if the phase shifts are substituted by their average over the Fermi surface:

$$Z = (2/\pi)(\langle \eta^s \rangle + \langle \eta^\wedge \rangle) - 1. \tag{37}$$

On the other hand, in order to discuss the three-dimensional semiconductor surfaces, we follow the general approach developed in (II) and apply a simple surface self-consistency to a (111) covalent semiconductor. Let us summarise the main points:

(i) The corner of the two-dimensional Brillouin zone is taken as a representative of the whole zone. In the above reference reasons have been given substantiating this choice.

(ii) By analysing this point, the three-dimensional Schrodinger equation can be split into three different one-dimensional equations. Surface self-consistency is then reduced to study those three different one-dimensional models.

This method has been also applied to ionic surfaces (Flores *et al* 1978) and from this previous work we can discuss the main points for a heterojunction. In general, we have to solve the three-dimensional equation

$$-\tfrac{1}{2}\nabla^2 \psi_\kappa(\rho, z) + V(\rho, z)\psi_\kappa(\rho, z) = E\psi_\kappa(\rho, z) \tag{38}$$

where

$$V(\rho, z) = V_0(z) + \sum_G V_G(z)\exp(iG \cdot \rho) \tag{39}$$

and G is a two-dimensional reciprocal lattice vector. At the corner of the two-dimensional Brillouin zone, the three-dimensional equation (38) can be reduced to the following three one-dimensional equations:

$$[-\tfrac{1}{2}(d^2/dz^2) + \tfrac{1}{2}\kappa^2 + V^{(i)}(z)]f^i(z) = E^{(i)}f^{(i)}(z) \qquad i = 1, 2, 3 \tag{40}$$

where

$$V^{(1)}(z) = V_0(z) + V_1(z) + V_1^*(z) \tag{41a}$$

$$V^{(2)}(z) = V_0(z) + V_1(z)\omega^2 + V_1^*(z)\omega \tag{41b}$$

$$V^{(3)}(z) = V_0(z) + V_1(z)\omega + V_1^*(z)\omega^2. \tag{41c}$$

$V_1(z)$ is defined by the following approximation to $V(\rho, z)$ (the Jones zone approximation, II):

$$V(\rho, z) = V_0(z) + V_1(z)[\exp(iG \cdot \rho) + \exp(iG_2 \cdot \rho) + \exp(iG_3 \cdot \rho)] + cc \tag{42}$$

where G_1, G_2 and G_3 are the three minor vectors of the two-dimensional Brilluoin zone ($\tilde{r} \equiv (\rho, z)$ and $\omega = \exp[i(2\pi/3)]$). Inside each crystal, the potentials $V_0(z)$ and $V_1(z)$ become

$$V_0(z) = V_3 \exp(ihz) + cc \tag{43a}$$

$$V_1(z) = -V_3 \exp(ihz/3) + V_8 \exp(i4hz/3) \tag{43b}$$

where $V_3 = V(11\bar{1})$, $V_8 = V(220)$ and h is the modulus of the reciprocal vector (111). In the narrow-gap approximation, we neglect V_3 and substitute V_8 by an effective real pseudopotential V_8^{eff}, in such a way that $V_0(z)$ becomes a constant for $|z| \gg d$ and $V_1(z)$ is given by $V_8^{eff} \exp(i4hz/3)$. Then the effective pseudopotentials in equations (40) become

$$V^{(1)}(z) = V_0 + [V_8^{eff} \exp(i4hz/3) + cc] \tag{44a}$$

$$V^{(2)}(z) = V_0 + \{V_8^{\text{eff}} \exp[i(4hz/3 + 2\pi/3)] + \text{cc}\} \qquad (44b)$$

$$V^{(3)}(z) = V_0 + \{V_8^{\text{eff}} \exp[i(4hz/3 - 2\pi/3)] + \text{cc}\} \qquad (44c)$$

inside each crystal.

This analysis shows that as far as an effective two-band approximation is good, the the electronic structure of the fundamental gap is the same for both covalent and ionic semiconductors. However, there remain some important differences in relation to the position of the edge of the jellium defining the surface charge in each case. As has been shown elsewhere (Flores *et al* 1978), in an ionic crystal this edge does not coincide with the plane dividing the crystal by a mid-bond. There, it was shown how this plane edge for a III–V crystal is removed a distance *l*/24 away from the mid-bond, *l* being the bond length.

Putting together all these arguments, we can see that the three-dimensional surface can be discussed as the superposition of three different one-dimensional surfaces, each one similar to the model discussed in the last section. For a covalent–ionic heterojunction (with a III–V ionic crystal), it is easy to see that for each one of the three one-dimensional surfaces we have

$$i = 1 \qquad \alpha_2 = \pi/24 \qquad \alpha_1 = 0 \qquad (45a)$$

$$i = 2 \qquad \alpha_2 = (\pi/24) + (2\pi/3) \qquad \alpha_1 = -2\pi/3 \qquad (45b)$$

$$i = 3 \qquad \alpha_2 = (\pi/24) - (2\pi/3) \qquad \alpha_1 = 2\pi/3 \qquad (45c)$$

These are the parameters defining each surface, as in equations (28). Moreover, for each one-dimensional surface with a cation-like face there is a charge defect of $(\alpha_1 + \alpha_2)/\pi = 1/24$ units. This means that for the three one-dimensional problems we have a defect of 1/8 units, in such a way that the band of interface states has an occupation of 7/8. Notice that for an anion-like face this band is fully occupied and that the next upper band must have an occupation of 1/8.

Now we can analyse each one-dimensional problem as in §2. However, one further approximation must be made. This consists of using equation (37), an averaged equation for a three-dimensional metal-like surface as a valid equation for the phase shifts of the three one-dimensional surfaces which are taken as the average of the whole two-dimensional Brillouin zone (II). Then for each one-dimensional surface defined by equations (45) we can analyse the interface states and the local density of interface states associated with the valence band. From this analysis one finds that only interface states appear for the surface defined by equation (45a); for the other two surfaces there is no interface state. Moreover, it can be proved that the interface potential is not in this case similar to that of a metal-like interface. The phase shifts associated with this interface potential can be found, as in the one-dimensional case, by imposing the condition that the induced interface dipole vanishes. This gives the following equation:

$$\sum_{i=1}^{3} [D(\eta_1^{\prime i}, 0) + D(\eta_2^{\prime i}, Q^i)] = 0 \qquad (46)$$

which generalises equation (35) to our three-dimensional interface. In equation (46) $\eta_1^{\prime i}$ or $\eta_2^{\prime i}$ are the phase shifts given by equations (39) but writing $\eta^A + \delta\alpha$ and $\eta^S + \delta\alpha$ instead of η^A and η^S, and Q^i are the occupancies of the interface states ($Q^1 = 7/8$; in the two other cases we must take $Q^2 = Q^3 = 0$, since there is no interface state). Notice that in writing equation (46) we have assumed the effective pseudopotentials to be the same for each semiconductor. This is a reasonable approximation for the Ge–GaAs heterojunction to which we are going to apply our results.

Equation (46) determines the interface phase shifts, $\eta^A + \delta\alpha$ and $\eta^S + \delta\alpha$, from which interface state appearing for equation (45a) can be easily obtained. Knowing these phase shifts, not only the interface state at the corner of the two-dimensional Brillouin zone, but also the interface states at any other two-dimensional point can be calculated. In order to explain how we have proceeded, let us write down the wavefunctions across the interface for the one-dimensional models given above. Under these conditions, instead of equations (19) we have

$$\psi^S(z) \simeq \cos(\tfrac{2}{3}h|z| + \eta'^S - \tfrac{1}{2}\pi) \tag{47a}$$

$$\psi^A(z) \sim \sin(\tfrac{2}{3}hz + \eta'^A \, \mathrm{sgn}(z)) \tag{47b}$$

where g has been substituted by $4h/3$ (see equations 44) and we have written η'^A and η'^S instead of η^A and η^S. For a different point of the two-dimensional Brillouin zone, the wavefunctions across the interface have a dependence on z different from the one shown in equations (47). In general, other reciprocal wavevectors appear in them and also other phase shifts must be included. As the wavevector $4h/3$ appearing at the corner of the Brillouin zone is an average of the different wavevectors of the whole Brillouin zone, we have approximated the wavefunctions across the interface by

$$\psi^S(z) \sim \cos\left[\tfrac{1}{2}g|z| + \left(\frac{g}{4h/3}\right)(\eta'^S - \tfrac{1}{2}\pi)\right] \tag{48a}$$

$$\psi^A(z) \sim \sin\left[\tfrac{1}{2}gz + \left(\frac{g}{4h/3}\right)\eta'^A \, \mathrm{sgn}(z)\right]. \tag{48b}$$

Thus we define the phase shifts in such a way that they have the appropriate mean values η'^S and η'^A, since $\langle g \rangle = 4h/3$ for the (111) surface within the Jones zone approximation (II; Elices *et al* 1974).

Knowing the wavefunctions (48) across the interface, the matching equations for the interface states can be obtained at any point of the two-dimensional Brillouin zone. As an example, let us consider the point $\overline{X}$ of the two-dimensional Brillouin zone (figure 6).

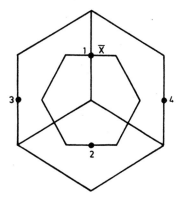

Figure 6. The two-dimensional Brillouin zone for a (111) face (inner hexagon) and the projected Jones zone (outer hexagon). Points 1–4 are equivalent to the point X of the first Brillouin zone.

In this figure the projected Jones zone for the (111) direction has also been drawn. Elices *et al* (1974) explained how to calculate the band structure of a covalent crystal in the Jones zone approximation. In the lowest approximation, the equivalent to a narrow-gap approximation, the wavefunctions along the fundamental gap can be calculated readily. For the symmetry Δ_2, this wavefunction is given by

$$\psi \sim \{\cos(\tfrac{1}{2}hz) + [(E'_1 - E_2)/(E - E'_2)]^{\ddagger} \sin(\tfrac{1}{2}hz)\} (\exp(i\kappa_3 \cdot \rho) - \exp(i\kappa_4 \cdot \rho)), \tag{49}$$

where E'_1 and E'_3 are the levels around the fundamental gap associated with the symmetry Δ_2 at the point L in the three-dimensional band structure (see figure 2 of Elices *et al* 1974) and κ_3 and κ_4 are the parallel components of the points 3 and 4 in figure 6. In equation (49) the origin $z = 0$ is taken at the minimum of the crystalline pseudopotential and we have neglected a decaying exponential factor.

A similar function can be obtained for an ionic semiconductor by using the same type of approximation. Then, for a Ge–GaAs junction we have the following wavefunctions:

$$\psi^1 \sim \cos(\tfrac{1}{2}h)[z - \tfrac{1}{2}d - (l/24)] + [(E'_1(\text{GaAs}) - E)/(E - E'_3(\text{GaAs}))]^{1/2}$$

$$\times \sin(\tfrac{1}{2}h)[z - \tfrac{1}{2}d - (l/24)] \qquad z \gg \tfrac{1}{2}d$$

$$\psi_2 \sim \cos(\tfrac{1}{2}h)(z + \tfrac{1}{2}d) + [(E'_1(\text{Ge}) - E)/(E - E'_3(\text{Ge}))]^{\ddagger} \sin(\tfrac{1}{2}h)(z + \tfrac{1}{2}d)$$

$$z \ll \tfrac{1}{2}d \tag{50}$$

where $1/24$ is the distance between the edge of the jellium and the minimum of the pseudopotential for a III–V ionic compound. In equation (50) we have neglected the component parallel to the surface given by $|\exp(i\kappa_3 \cdot \rho) - \exp(i\kappa_4 \cdot \rho)|$.

On the other hand, at the interface, the wavefunction is given by a combination of the functions (48a) and (48b), taking $g = h$. Therefore

$$\psi^1 \sim A \cos[\tfrac{1}{2}h|z| + \tfrac{3}{4}(\eta'^S - \tfrac{1}{2}\pi)] + B \sin(\tfrac{1}{2}hz + \tfrac{3}{4}\eta'^A \text{ sgn}(z)) \qquad |z| < \tfrac{1}{2}d. \tag{51}$$

Matching the wavefunctions (50) and (51), we find the energy levels of the interface state we are looking for at $\bar{X}$.

The same type of argument can be used to analyse other points of the Brillouin zone with other symmetries. We omit further details and concentrate on giving the main results obtained for the (111) Ge–GaAs interface.

(i) In this heterojunction one band of interface states has been found, with its maximum at the Brillouin zone centre and following closely the top of the valence band projection in the (111) direction.

(ii) When an outwards relaxation is considered, this band moves upwards in energy.

(iii) In order to achieve self-consistency, an outward relaxation must exist at the interface such that 1/8 of the whole band is raised above the top of the valence band. This condition comes from the partial occupation of this band (7/8) (Baraff *et al* 1977).

(iv) An outwards relaxation of 4% of the bond length is needed to comply with the last condition.

4. Concluding remarks

In this paper we have analysed homopolar and heteropolar heterojunctions. The main difference between these junctions is associated with the occupancy of the interface band. In a homopolar heterojunction the interface states are partially occupied, so that an interface relaxation is needed in order to raise part of the interface band above the top of the valence band. Heteropolar heterojunctions do not have this type of restriction.

In §2 we have given a simple approach to obtain the electronic energy levels at these heteropolar junctions. The crucial point in this approach is the introduction of a restoring surface dipole, which tends to equalise the charge neutrality points of both semiconductors. With this simple idea we have obtained the energy levels at the junction and the results are in very good agreement with other experimental and theoretical information. We then discussed the effect of relaxation on these heterojunctions. We have shown that, apart from the appearance of a band of fully occupied interface states close to the valence band, no other important effects are expected to appear for a small relaxation. In particular, the energy discontinuities between both bulk bands are almost independent of this amount of relaxation.

The homopolar heterojunctions are studied assuming that the charge neutrality points of both semiconductors are lined up. This approach is supported by the results obtained for heteropolar heterojunctions and we have therefore concentrated on the important problem of interface relaxation. We have obtained the interface band of electronic states as a function of the relaxation and have found in this way that a relaxation of 4 % is needed to raise 1/8 of the band of interface states above the valence band. Recently, a tight-binding model has been used to study the (111) Ge–GaAs interface self-consistently (Djafari-Rouhani *et al* 1978a). The main conclusions of this approach are that the occupancy of the band of interface states is 7/8, and that there is an interface relaxation of roughly 10 %. This relaxation is an approximate value and can only be taken as an order of magnitude, similar to the one obtained above. The comparison between both methods (Djafari-Rouhani *et al* 1978b) shows that a relaxation of 4 % or a little more must occur at the interface between Ge and GaAs, which lends support to the simple approach developed in §3. Finally, it could be of interest to stress the similarity between our results for the (111) Ge–GaAs interface and the results of Baraff *et al* (1977) for the (100) Ge–GaAs interface. In both cases the interface relaxation gives an increase of 4 % for the Ge–Ga bond length, which is in good agreement with the length obtained for the Ge–Ga bond by means of simple chemical arguments (Baraff *et al* 1977).

Acknowledgments

We thank L Dobrzynski and F García-Moliner for many helpful discussions and the critical reading of the manuscript. The work was supported in part by a contract ATP International (CNRS) number 2589.

References

Anderson R L 1960 *PhD Thesis* Syracuse University New York
Baraff G A, Appelbaum J A and Hamann D R 1977 *Phys. Rev. Lett.* **38**, 237–40
Bertoni C M, Bortolani V, Calandra C and Nizzoli F 1973 *J. Phys. C: Solid St. Phys.* 6 3612–30
Cohen M L and Bergstresser T K 1966 *Phys. Rev.* 141 789–96
Djafari-Rouhani B, Dobrzynski L and Lannoo M 1978a *to be published*
Djafari-Rouhani B, Dobrzynski L, Flores F, Lannoo M and Tejedor C 1978b *Solid St. Commun.* 27 29–31
Dobrzynski L, Cunningham S L and Weinberg W H 1976 *Surface Sci.* 61 550–62
Elices M, Flores F, Louis E and Rubio J 1974 *J. Phys. C: Solid St. Phys.* 7 3020–32

Flores F and Tejedor C 1977 *J. Physique* **38** 949–60
Flores F, Tejedor C and Martin-Rodero A 1978 *Phys. Stat. Solidi* (b) **88** 591–7
Frensley W R and Kroemer H 1976 *J. Vac. Sci. Technol.* **13** 810–5
—— 1977 *Phys. Rev.* B **16** 2642–52
Garcia-Moliner F and Flores F 1978 *Introduction to the Theory of Solid Surfaces* (Cambridge: Cambridge University Press) in press
Garcia-Moliner F and Rubio J 1971 *Proc. R. Soc.* A**324** 257–73
Herman F and Kasowski R V 1978 *Phys. Rev.* B **17** 672–4
Louis E 1977 *Solid St. Commun.* **24** 849–52
Louis E, Yndurain F and Flores F 1976 *Phys. Rev.* B **13** 4408–18
Mach R, Ludwig W, Eichborn G and Arnold H 1970 *Phys. Stat. Solidi* a **2** 701
Milnes A G and Feutch D L 1972 *Heterojunctions and Metal–Semiconductor Junctions* (New York: Academic Press)
Pickett W E and Cohen M L 1978 *Solid St. Commun.* **25** 225–7
Pickett W E, Louie S G and Cohen M L 1977 *Phys. Rev. Lett.* **39** 109–12
Pickett W E, Louie S G and Cohen M L 1978 *Phys. Rev.* B **17** 815–28
Sakurai T, Cardillo M J and Hagstrum H D 1977 *J. Vac. Sci. Technol.* **14** 397–9
Smith J R 1969 *Phys. Rev.* **181** 522–9
Tejedor C and Flores F 1978 *J. Phys. C: Solid St. Phys.* **11** L19–23
Tejedor C, Flores F and Louis E 1977 *J. Phys. C: Solid St. Phys.* **10** 2163–77
Zeidenbergs G and Anderson R L 1967 *Solid St. Electron* **10** 113–23

PHYSICAL REVIEW B   VOLUME 35, NUMBER 12   15 APRIL 1987-II

Acoustic deformation potentials and heterostructure band offsets in semiconductors

Manuel Cardona

Max-Planck-Institut für Festkörperforschung, D-7000 Stuttgart 80, Federal Republic of Germany
and Xerox Corporation, Palo Alto Research Center, Palo Alto, California 94304

Niels E. Christensen

Max-Planck-Institut für Festkörperforschung, D-7000 Stuttgart 80, Federal Republic of Germany
(Received 14 October 1986)

It is argued that the absolute hydrostatic deformation potentials recently calculated for tetrahedral semiconductors with the linear muffin-tin-orbital method must be screened by the dielectric response of the material before using them to calculate electron-phonon interaction. This screening can be estimated by using the midpoint of an average dielectric gap evaluated at special (Baldereschi) points of the band structure. This dielectric midgap energy (DME) is related to the charge-neutrality point introduced by Tejedor and Flores, and also by Tersoff, to evaluate band offsets in heterojunctions and Schottky-barrier heights. We tabulate band offsets obtained with this method for several heterojunctions and compare them with other experimental and theoretical results. The DME's are tabulated and compared with those of Tersoff's charge-neutrality points.

I. INTRODUCTION

The matrix elements for the interaction between carriers and acoustic phonons at band extrema of semiconductors can be evaluated from the deformation potentials for uniform strain (dependence of band extrema on strain).[1] While this is straightforward for the shear (traceless) components of the strain, problems arise when handling the hydrostatic components which accompany *longitudinal* phonons.[2] The corresponding deformation potentials are defined, for an infinite solid, to an arbitrary constant which represents the variation of the arbitrarily chosen zero of energy with hydrostatic stress. This arbitrariness should, of course, disappear when dealing with the finite solids found in nature. It should, therefore, be possible to define absolute deformation potentials for a uniform hydrostatic strain with respect to a fixed energy, e.g., the energy at infinity or at a point sufficiently far from the sample. Such deformation potentials would correspond, for the bottom of the conduction band, to the variation of the electron affinity with strain and for the top of the valence band to that of the ionization potential (photoelectric threshold). They should, therefore, be affected by surface properties rather than being a bulk property. Their evaluation as surface dependent quantities represents a formidable theoretical problem. The deformation potentials required to evaluate the electron-phonon interaction for phonons of wavelength much smaller than the sample size should be, however, bulk quantities independent of surface details.

In a recent paper,[2] Vergés *et al.* suggested that the linear muffin-tin-orbital (LMTO) method[3] provides a natural way of overcoming this problem. In this method, the solid is broken up into atomiclike spheres and all potentials are referred to the reference level which is chosen so that the Hartree potential of a single atomic sphere is zero at infinity. The solid can be terminated at any sphere

while leaving the electronic charge distribution in this sphere equal to that it would have in the bulk. An attempt was made to evaluate in this manner the electron-phonon interaction constants relevant to longitudinal acoustic phonons.[2] In doing so, the problem of screening by the dielectric function of the solid was overlooked: unscreened hydrostatic deformation potentials were used.

While the perturbations produced by the shear components of phonons are only insignificantly screened, strong screening should take place for the hydrostatic strain of long-wavelength longitudinal phonons. The present paper addresses this problem. Using the one-dimensional Penn model for the dielectric function,[4] it is shown that the average of the hydrostatic deformation potentials of the valence and conduction states which form the Penn gap must be screened by the full dielectric function [we call the average of the conduction and valence energies at the Penn gap the dielectric midpoint energy (DME)]. Thus the deformation potential of the DME, a_D, must be partly compensated by the screening response $a_D[1 - \epsilon^{-1}(q)]$. This screening response must be subtracted from all deformation potentials calculated in Ref. 2 in order to obtain the appropriate electron—LA-phonon coupling constants.

In this paper, results obtained by this technique for the electron-phonon coupling constants of group-IV elements and III-V and II-VI compound semiconductors are tabulated and compared with the few experimental and some theoretical data available. The calculations are performed with the LMTO method at the first Baldereschi special point.[5] The relevance of the *screened* deformation potentials to the problem of the dependence of the lattice constant of semiconductors on doping with either donors or acceptors is also discussed.

The concept of a midgap energy has been recently introduced by Tejedor, Flores, and Louis[6,7] and by Tersoff[8−10] in connection with the lining up of the band structures across semiconductor-semiconductor (heterojunction) and semiconductor-metal interfaces (Schottky barriers). This midgap point has also been referred to as the charge-neutrality point.[7–9] We suggest that this midgap point is basically the same as the DME discussed here for the screening of the electron—LA-phonon interaction. We in fact use the DME's calculated with the LMTO method for the first Baldereschi special points to evaluate valence-band offsets in several lattice matched heterojunctions and compare them with other available experimental and theoretical results. In doing so, we discuss the value of the dielectric constant to be used for the screening, an average of that of both constituents somewhat reduced from that for $q = 0$ because of the abruptness of the junctions. We also present a tabulation of DME's with respect to the top of the valence band obtained with the LMTO special point method and compare it with calculations of the charge neutrality point performed by Tersoff.

II. THE DIELECTRIC MIDPOINT ENERGY (DME)

As discussed in Refs. 1, 2, and 11, the LA phonon produces a perturbation on electronic band edges equivalent to a sinusoidal potential. This perturbation is different for each band edge. For long-wavelength phonons, this perturbation can be easily obtained by multiplying the strain associated with the phonon by a deformation potential which gives the change of the band edge energy-per-unit strain. It is helpful to decompose the local strain into irreducible symmetry components. For a cubic crystal they are the hydrostatic strain (multiple of the unit matrix), and two traceless strains which correspond to shear deformation along the $\langle 100 \rangle$ and $\langle 111 \rangle$ axes. The former will be strongly screened by the dielectric response of the crystal while the screening of the latter should be insignificant. Here we discuss the screening of the hydrostatic component, which was neglected in Ref. 2. We shall argue that there is a band energy, obtained as an average of the upper valence band and the lowest conduction band, whose deformation potential must be divided by the zero-frequency intrinsic dielectric response function $\epsilon(q)$. This energy will be called the dielectric midpoint energy (DME or E_D), and its hydrostatic deformation potential $dE_D/d\ln V$ (V is the volume) will be called a_D. For the wave vectors q involved in standard transport processes, $\epsilon(q)$ will be practically equal to its value ϵ for $q = 0$. Large concentrations of free carriers will modify $\epsilon(q)$ by adding to it their Lindhard polarizability.[12,13] We shall not consider this case here since our discussion can be trivially extended to deal with it. The unscreened deformation potential of all band extrema must be corrected by addition of the screening potential which acts on the DME.

The screening of an external electrostatic potential, acting equally on all band states, is rather trivial: it is performed simply by dividing the potential by $\epsilon(q)$. In the case of the perturbations generated by the hydrostatic strain which accompanies an LA phonon, the situation is not so simple since this perturbation is different for each band edge. Thus the notion of singling out some energy

(the DME) which can be screened by division by $\epsilon(q)$ naturally arises. We give here a heuristic derivation of the DME used in this work. A more rigorous derivation is given in the Appendix.

Let us first discuss briefly the nature of the intrinsic dielectric response of zinc-blende-type semiconductors for $\omega \approx 0$ (the acoustic phonons of interest here have frequencies much smaller than any characteristic frequency of the dielectric response) and $q \approx 0$. This response is generated by direct virtual transitions from the filled electronic states to the empty conduction states plus a small correction for the lattice polarizability (phonon contribution) in the case of ionic materials. We shall neglect this ionic contribution for the time being.

The simplest model for the dielectric response of semiconductors is the isotropic Penn model.[4] In this model, the valence band is described in a Jones zone which is symmetrized in k space by making it spherical. An average isotropic gap is then introduced between this band and the conduction bands, produced by the crystal potential. Thus the model is basically one-dimensional and the transitions around the "Penn gap" dominate the dielectric response. These concepts, in spite of their highly simplified nature, have been successfully applied to interpret many features related to the dielectric response of semiconductors.[13,14] The main feature we want to use now is the existence of a group of filled states clustered around a given valence-band energy E_V and a corresponding group of conduction states (E_c) which mainly produce the dielectric response. Let us consider the unscreened perturbation induced by an LA phonon on these states, i.e., their hydrostatic deformation potentials. The valence electrons will polarize so as to partially screen this perturbation. One may, at first glance, think that this will take place by setting up an electrostatic potential whose effect will be to replace the hydrostatic deformation potential of the valence band a_V by $a_V/\epsilon(q)$, regardless of the value of the deformation potential of the conduction band. This is of course wrong, since filled valence states and empty conduction states must contribute symmetrically to the dielectric response. (An empty state is a hole. Electrons and holes must be treated on the same footing.) Hence we infer that the dielectric screening represented by division by $\epsilon(q)$ must be applied to the average of a_V and a_c, i.e., to the deformation potential of a fictitious midgap state situated halfway between the conduction and valence states. The question of how to determine this state will be considered next.

The dielectric function $\epsilon(\omega,q)$ can be calculated by performing a straightforward Brillouin-zone integration of energies and matrix elements for interband transitions.[15] An analysis of this integration (see, for instance, Fig. 6 of Ref. 16) suggests that the main contribution to ϵ is for transitions from the two top valence bands (spin degenerate in Ge and Si but not in the zinc-blende structure) and the bottom two conduction bands (with the same degeneracy properties). Thus we shall consider only these bands here. The Brillouin-zone integration can be replaced by a sampling over a small number of so-called Baldereschi special points.[5] For the sake of simplicity, we shall use here the first Baldereschi point (and implicitly the other 23 generated from it by the operations of the O_h point groups):

$$\mathbf{k}_B = (2\pi/a_0)(0.622, 0.295, 0) . \tag{1}$$

We have calculated the energies of the two top valence bands and the two bottom conduction bands at k_B for a number of group-IV elements and, III-V and II-VI com-

pound semiconductors with the LMTO method.[3,17] The results are given in Table I for the lattice constants at zero pressure and temperature. For completeness, we have added to this table the energies of these bands at the Γ, X, and L points of the Brillouin zone, points which are also of importance to the dielectric response.[14,15] The table also contains the deformation potentials a (i.e., the

TABLE I. Energies (in eV) of (a) the top of the valence band and (b) the bottom of the conduction band calculated with the fully relativistic LMTO method at the Γ, X, L, and B points. At the B (first Balltereschi) point, the average values of the inversion asymmetry split spin doublet are listed. The corresponding volume deformation potentials are also given.

(a)

	Valence bands				Deformation potentials a			
	Γ_8	X_7	$L_{4,5}$	B	Γ_8	X_7	$L_{4,5}$	B
C	3.73	−2.73	0.91	−0.70	−15.42	−8.77	−12.59	−11.02
Si	−0.85	−3.76	−2.02	−2.99	−7.95	−5.06	−6.97	−5.62
Ge	−0.79	−4.03	−2.18	−3.19	−8.09	−4.28	−6.45	−5.15
α-Sn	−1.39	−4.26	−2.59	−3.53	−7.34	−3.60	−5.68	−4.46
AlP	−1.78	−3.95	−2.53	−3.26	−7.67	−5.39	−7.02	−5.98
AlAs	−1.51	−3.84	−2.35	−3.18	−6.46	−4.13	−5.71	−4.73
AlSb	−1.65	−3.96	−2.57	−3.43	−7.35	−4.41	−6.21	−5.02
GaP	−1.59	−4.32	−2.68	−3.51	−8.07	−4.43	−6.57	−5.34
GaAs	−1.07	−3.85	−2.19	−3.08	−8.77	−4.92	−7.15	−5.89
GaSb	−1.46	−4.18	−2.62	−3.56	−7.95	−4.10	−6.21	−4.97
InP	−2.08	−4.39	−2.99	−3.68	−6.91	−3.77	−5.59	−4.63
InAs	−1.94	−4.56	−2.94	−3.72	−7.83	−4.18	−6.29	−5.20
InSb	−1.95	−4.41	−2.94	−3.79	−7.31	−3.67	−5.72	−4.60
ZnSe	−2.80	−5.08	−3.67	−4.33	−8.62	−4.79	−7.09	−6.11
ZnTe	−2.28	−4.64	−3.22	−4.05	−9.49	−4.79	−7.51	−6.13
CdTe	−2.94	−4.94	−3.71	−4.41	−8.16	−4.45	−6.61	−5.60
HgTe	−2.45	−5.04	−3.43	−4.25	−10.45	−4.69	−8.02	−6.79

(b)

	Conduction bands				Deformation potentials a			
	Γ_6	X_6	L_6	B	Γ_6	X_6	L_6	B
C	17.67	8.54	12.14	10.96	−39.71	−17.69	−39.92	−20.41
Si	2.15	−0.25	0.51	1.74	−20.97	−5.73	−11.49	−8.17
Ge	−1.11	−0.23	−0.85	1.66	−17.20	−6.55	−10.99	−8.71
α-Sn	−2.53[a]	−0.91	−1.81	0.50	−15.28[a]	−5.84	−9.18	−8.00
AlP	−1.38	−0.32	0.92	1.96	−16.81	−5.31	−11.51	−8.06
AlAs	+0.49	−0.30	−0.54	2.00	−13.83	−4.59	−9.96	−7.15
AlSb	−0.59	−0.73	−0.69	0.95	−15.78	−5.02	−9.91	−7.59
GaP	−0.03	−0.05	−0.11	1.78	−15.90	−5.82	−11.14	−9.11
GaAs	−0.94	+0.20	−0.35	2.04	−15.93	−6.59	−11.49	−10.06
GaSb	−2.01	−0.85	−1.47	0.75	−16.35	−6.51	−9.80	−9.08
InP	−1.58	−0.42	−0.76	1.26	−12.37	−5.23	−9.38	−9.92
InAs	−2.60	−0.85	−1.34	1.08	−14.49	−5.90	−10.18	−10.45
InSb	−2.72	−0.93	−1.82	0.30	−13.12	−6.51	−8.75	−9.39
ZnSe	−1.91	−0.02	−0.53	1.61	−13.26	−6.28	−10.59	−12.85
ZnTe	−1.68	−0.31	−0.44	0.93	−14.74	−6.54	−3.56	−11.21
CdTe	−2.73	−0.68	−1.56	0.18	−10.88	−5.88	−9.06	−11.10
HgTe	−3.66	−0.81	−2.18	−0.33	−12.88	−9.27	−10.24	−12.35

[a]This Γ_6 state is now below the top of the valence and in agreement with experiment, see Groves and Paul (Ref. 64). In other cases in which this happens in the tables (e.g., Ge), it is an artifact of the LDA.

volume derivatives) of all the energies mentioned above. The LMTO calculations were fully relativistic, thus including spin-orbit interaction. In the ionic materials (III-V and II-VI compounds), the gap states at k_B are split by spin-orbit (s.o.) interaction, the splittings being in all cases less than 0.3 eV. We have listed in Table I the average of the split bands since we feel that these are the values which should be used to determine the E_D. For the Γ, X, and L points, the top of the valence bands are s.o. split. We list the true top without spin-orbit averaging.

The LMTO calculations just mentioned were performed with the local-density approximation (LDA) to the exchange-correlation potential. This approximation is known to lead to large errors ($\sim 100\%$) in the gaps for direct excitations from the valence to the conduction bands.[18,19] These errors can be removed, in an "ad hoc" manner, by introducing additional potentials at the atomic cores.[17] We have not followed this procedure here since we do not know what its effect on the Baldereschi point states is. The energies listed in Table I are uncorrected LDA results.

The effect of LDA inaccuracies on the DME will be examined next. We list in Table II the values of the average dielectric gap or Penn gap calculated from the data of Table I at $k_B(E_B)$. We also list in this table the experimental values of the average dielectric gap E_g (page 42 of Ref. 20) and the strongest structure in the imaginary part of the dielectric function, usually labeled E_2 (page 169 of Ref. 20) for the materials considered here. We note that the calculated E_B's represent rather well the experimental E_g's and E_2's (deviations less than 10%). The absolute

errors due to the LDA are thus less than for the fundamental (lowest) gap at Γ. The relative errors are of course even smaller, actually insignificant within the semiquantitative nature of the present treatment.

We have also listed in Table I the hydrostatic deformation potentials of the various states under consideration, also calculated with the LMTO method. We should keep in mind that the residual LDA-induced errors seem to be rather small for these deformation potentials.[2,17]

We have listed in Table II the position of the DME with respect to the top of the valence band ($E_D - E_V$) and the corresponding value for the charge neutrality points ($E_T - E_V$) calculated by Tersoff.[10] We find an excellent agreement between these two quantities. This agreement is even more remarkable when one considers that E_T in Ref. 10 was obtained from first principles band structures after applying a rigid shift between valence and conduction bands so as to correct for the LDA error in the lowest gap (the so-called "scissors" operator). No such shift has been applied here. We have not investigated the source of this paradox.

III. SCREENED ELECTRON LA-PHONON INTERACTION

As already mentioned, the screening potential which accompanies the LA-phonon perturbation corresponding to an unscreened hydrostatic deformation potential a (listed in Table I for several extrema) is obtained from the deformation potential of the DMP a_D with the expression:

TABLE II. Representative values E_B, E_g, and E_2 (in eV), for the Penn gap of several group-IV elemental and III-V and II-VI compound semiconductors. E_B has been calculated from the top valence and the bottom conduction bands at the Balderschi point k_B. E_g, from the tabulation in Ref. 20, represents the average gap obtained from $\epsilon(0)$ with the Penn model. E_2 is the energy of the major structure in $\epsilon_i(\omega)$ (also from Ref. 20). We also have listed in this table the lattice constant of these materials, the dielectric midpoint energy E_D obtained from the Baldereschi point data, its difference to the top of the valence band ($E_D - E_V$), and the corresponding difference for Tersoff's charge neutrality level $E_T - E_V$ (from Ref. 10). a_0 is the lattice constant in Å.

	E_B	E_g	E_2	$E_D - E_V$	$E_T - E_V$	E_D	a_0
C	11.66	13.5	12.5	1.40		+ 5.13	3.57
Si	4.73	4.77	4.40	0.23	0.36	−0.625	5.43
Ge	4.85	4.31	4.3	0.03	0.18	−0.765	5.65
α-Sn	4.03	3.06	3.75	−0.12		−1.515	6.47
AlP	5.22	5.67		1.13	1.27	−0.65	5.47
AlAs	5.18	5.14	4.7	0.92	1.05	−0.59	5.66
AlSb	4.38	4.14	4.25	0.41	0.45	−1.24	6.13
GaP	5.29	5.75	5.27	0.73	0.81	−0.865	5.44
GaAs	5.12	5.20	4.85	0.55	0.50	−0.520	5.65
GaSb	4.31	4.12	4.1	0.06	0.07	−1.405	6.10
InP	4.94	5.16	4.8	0.87	0.76	−1.210	5.86
InAs	4.80	4.58	4.5	0.62	0.50	−1.32	6.05
InSb	4.09	3.73	4.08	0.20	0.01	−1.745	6.47
ZnSe	5.94	7.05	6.4	1.44	1.70	−1.36	5.65
ZnTe	4.98	5.74	5.3	0.73	0.84	−1.56	6.10
CdTe	4.59	5.79	5.0	0.83	0.85	−2.115	6.48
HgTe	3.92	5.0	5.0	0.16	0.34	−2.29	6.48

$$\Delta a_D = a_D[\epsilon^{-1}(\omega,q)-1] . \tag{2}$$

For acoustic phonons $\omega \approx 0$. For intraband phonon scattering, we can also take $q \approx 0$ and use the static, q-independent dielectric constant ϵ. In Ge and Si, ϵ is purely of electronic origin and thus the Penn gap or the first Baldereschi gap is its main source. The analysis of Sec. II applies to this electronic ϵ. In ionic (III-V, II-VI) materials, there is a small contribution below ω_{TO} (transverse optic frequency) which can be easily estimated from ω_{TO} and ω_{LO} (longitudinal optic frequency) with the Lyddane-Sachs-Teller relation.[21] It is not clear whether the DME analysis given in Sec. II also applies to the ionic contribution to ϵ. Nevertheless, its effect in Eq. (2) is rather small since it amounts typically to $\approx 10\%$ of ϵ and Eq. (2) is dominated by the -1 inside the brackets. We shall therefore neglect the ionic contribution of ϵ and use for ϵ only the ir, purely electronic contribution, sometimes called ϵ_{ir} or ϵ_∞.

This dielectric constant is listed in Table III (from Ref. 22, p. 114) for the materials of interest here. We also list in this table the screening deformation potential Δa_D obtained with Eq. (1) from the data of Tables I (a_D) and III (ϵ), the screened value of a_D ($\bar{a}_D = a_D + \Delta a_D$, screened values are represented by a bar over the corresponding unscreened ones), and the screened deformation potentials of the top valence extrema (Γ_8) and that of the lowest conduction valleys (Γ_6, Δ_6, or L_6 as indicated). We also list in this table values obtained recently by Tersoff[23] ($a_V{}^c$) by matching his charge neutrality points and Van de Walle et al.[24] ($a_V{}^e$) by calculating superlattices consisting of the same material stressed and unstressed. Since Tersoff's calculation implies infinite screening ($\epsilon = \infty$), we have listed under $a_V{}^d$ the values which result from adding Δa_D to $\bar{a}_V{}^b$ and thus should be closer to the correctly screened $\bar{a}_V$'s.

We note that all theoretically predicted values of $\bar{a}_V$ are small and rather similar in magnitude. The corrected values from the Tersoff data ($\bar{a}_V{}^d$) fall between our calculations and those of Van de Walle et al. ($\bar{a}_V{}^e$). The sign reversals which appear now and then between different calculations should not be taken too seriously: the absolute values are very small when compared with unscreened deformation potentials. Hence, even if the signs are different, the differences between the various estimates are small. If we add to these $\bar{a}_V$'s the deformation potential of a direct gap at Γ, we obtain in the cases in which the lowest conduction-band minimum is at Γ (all the materials under consideration with the exception of Ge, Si, AlP, AlAs, AlSb, GaP) the deformation potential of the lowest

TABLE III. Infrared dielectric constant ϵ and various hydrostatic deformation potentials for the materials under consideration. $\bar{a}_V$ and $\bar{a}_c$ represent the screened deformation potentials of the highest valence and the lowest conduction states, $\bar{a}_D$ that of the dielectric midgap point. Δa_D represents the effect of screening on the deformation potentials. All deformation potentials (in eV) were obtained as described in the text, unless otherwise indicated. In the cases of conduction-band minima along $\langle 100 \rangle$ (Si, AlP, AlAs, AlSb, GaP, C) we took the deformation potentials to be those at X_6 since these points are either the minima or very close to them.

	ϵ^a	Δa_D	$\bar{a}_D$	$\bar{a}_V{}^b$	$\bar{a}_V{}^c$	$\bar{a}_V{}^d$	$\bar{a}_V{}^e$	$\bar{a}_C$
C	5.7	13.0	−2.8	−2.4				−4.7
Si	12	6.3	−0.5	−1.6	−0.4	−1.0	+0.8	+0.6
Ge	16	6.5	−0.4	−1.6	+0.65	+0.2	+1.8	−4.5
α-Sn	20	5.9	−0.3	−1.5				
AlP	8	6.1	−0.9	−1.5				+0.8
AlAs	9.1	5.3	−0.6	−1.2	+0.4	−0.2		+0.7
AlSb	10.2	5.7	−0.6	−1.2				+0.7
GaP	9.1	6.6	−0.6	−1.5				+0.8
GaAs	10.9	7.1	−0.7	−1.6	+0.65	−0.1	+0.7	−8.8
GaSb	14.4	6.5	−0.5	−1.4				−9.8
InP	9.6	6.5	−0.7	−0.4				−5.9
InAs	12.3	7.2	−0.6	−0.6				−7.3
InSb	15.7	6.5	−0.5	−0.8				−6.6
ZnSe	5.9	7.9	−1.6	−0.7				−7.4
ZnTe	7.3	7.5	−1.2	−2.0				−7.3
CdTe	7.2	7.2	−1.1	−1.0				−3.7
HgTe	9.3	8.5	−1.1	−2.0				

[a]From Ref. 22.
[b]Present calculations. Note that Cardona and Christensen (Ref. 65) have found that the calculated unscreened a_V for diamond is larger than the experimental one. The screened one given here is smaller but leads to better agreement with experiment.
[c]Theoretical, from Ref. 23.
[d]Theoretical, from Ref. 23 after adding $\bar{a}_D$.
[e]Theoretical, from Ref. 24.

Γ conduction-band valley $\bar{a}_C$. Since the deformation potential of this gap is large (~ -9 eV), the differences just mentioned are not too important in giving the value of $\bar{a}_c$. We note that in Ref. 24, $\bar{a}_c = -7.6$ eV is given for GaAs, which compares well with our result ($\bar{a}_c = -8.8$ eV). The differences in $\bar{a}_V$ reflect themselves more strongly in the values of $\bar{a}_c$ for Ge (L_1 band): in Ref. 24, $\bar{a}_c = -1.0$ eV is found, as compared with our value of $\bar{a}_c = -4.5$ eV. For Si (Δ_1 band), we find $\bar{a}_c = +0.6$ while $+3.1$ is found in Ref. 24. We should point out that a calculation of $\bar{a}_V$ which was implied to include screening has been performed by Wiley[25] for a few group IV and III-V materials. It yields values of $\bar{a}_V$ around $+2.5$ eV. However, this calculation uses the vacuum level as reference and the empirical dependence of the ionization energy on lattice constant as a basis. Its connection with the $\bar{a}_V$'s required for the electron-phonon interaction problem is not obvious.

IV. COMPARISON WITH EXPERIMENT

We shall now compare the calculated values of $\bar{a}_c$ given in Table III with experimental data. The most precise data should be found for the cases in which the conduction band minimum is at Γ since then no shear deformation potentials contribute to the scattering by LA phonons. Still, polar optical phonon and impurity scattering must be removed from experimental electron transport data in order to obtain the $\bar{a}_c$'s, hence the accuracy in the experimental determination of $\bar{a}_c$ is not expected to be too large. This fact has been best illustrated by Zawadski[26] who has given a plot of the variation of the reported values of $\bar{a}_c$ versus calendar year for InSb. They fluctuate between 4 and 30 eV, averaging around 12 eV, a value which comes close to the calculated unscreened one (-13.1 eV) and thus must be too high. Two values are reported in the literature which are close to the calculated one for InSb (-6.6 eV). They were found by rather reliable methods: $|\bar{a}_c| = (4.5 \pm 0.5)$ eV was obtained by measuring the attenuation of an acoustic wave traveling through doped InSb in a magnetic field,[27] $|\bar{a}_c| = 8.2$ eV was obtained from thermoelectric power in the phonon drag region.[28] Hot electron transport data[29] have yielded $|\bar{a}_c| = 6.9 \pm 0.4$ eV, in excellent agreement with our calculations. We feel that other existing experimental determinations are more indirect and thus more subject to error than the ones just given, which bracket our calculated value of 6.6 eV. In the case of GaAs, there is by now also a considerable amount of data, especially since the discovery of the modulation doping technique[30] which enables one to dope GaAs by placing the impurities in an adjacent AlAs layer, thus partly avoiding impurity scattering. Analysis of low temperature mobility data for such AlAs-GaAs multiple heterojunctions yields $|\bar{a}_c| = 13.5$ eV,[31] a result which has been criticized in Ref. 32 as disagreeing with data for *single* heterojunctions which yield $|\bar{a}_c| = 7$ eV. Analysis of bulk mobility data in high-purity bulk GaAs also give the value $|\bar{a}_c| = 7.0$ eV, in acceptable agreement with our value of 8.8 eV. We point out that 8.6 eV has also been given by Rode.[33] His values, however, are simply meant to be the pressure coefficient of the gap and thus not very relevant to the problem at hand except for the nontrivial fact, proven here, that the Γ_{15} valence state, after screening, is affected very little by the hydrostatic strain of the LA phonon ($\bar{a}_V \simeq 0$). We should also point out that Vinter[34] has recently reinterpreted the data of Ref. 30 by using more accurate wave functions for the quantized electrons. He finds $|\bar{a}_c| = 12$ instead of 13.5 eV, as found in Ref. 31. We feel that our value of 8.8 eV is also sufficiently close to 12 eV, although the discrepancy between 12 eV and the value found for bulk GaAs (7 eV) cannot be accepted.

We note that analysis of infrared absorption data for GaAs, to which many scattering mechanisms contribute, yields $|\bar{a}_c| = 15.7$ eV.[35] We believe this value to be too high. High values ($|\bar{a}_c| = 17.5$ eV) were also found from transport measurements in Ref. 36.

Low-field transport data are also available for InP. Their analysis yields the values $|\bar{a}_c| = 14.5$ (Ref. 37) and 18 eV (Ref. 38) which would be compatible with our unscreened data ($|a_c| = 12.4$ eV) but cannot be reconciled with the screened value (5.9 eV). For InAs, the value $|\bar{a}_c| = 11.5$ has been reported in Ref. 37, also higher than the calculated (screened) one (7.3 eV). It is not very likely that quadrupole scattering, of the type discussed by Lawaetz,[39] will provide the additional scattering mechanism to harmonize the theoretical and experimental values of $|\bar{a}_c|$.

The uncertainties just described get even worse for electron valleys off $\mathbf{k} = 0$, such as found in Ge, Si, GaP, and the Al compounds, as one has to include in the analysis the shear components of both TA and LA phonons. The value $|\bar{a}_c| = 5.7$ eV found in Ref. 40 for Ge is in reasonable agreement with our calculations (4.5 eV). That given for Si in the same work $|\bar{a}_c| = 3$ eV seems a little high (ours is 0.6 eV) although it agrees with the predictions of Ref. 24 ($a_c = 2$ eV). Other experimental data are given in Ref. 2. We point out that a method to determine $\bar{a}_c$, including its sign, has been suggested in Ref. 41. It involves the measurement of LA-phonon self-energies versus $\bar{q}$ in heavily doped silicon with neutron scattering. The experimental data seemed to favor $\bar{a}_c \simeq -5$ eV. We have reevaluated these data for $\bar{a}_c = 0$. While the calculated curve seems to *deviate* from the experimental data twice as much as that obtained for $\bar{a}_c = -5$, we feel that the uncertainty of the data and the theoretical processing (which ignores electron mean-free path) is large enough to make $\bar{a}_c \simeq 0$ acceptable.

The value $|\bar{a}_c| = 9 \pm 1$ eV has been obtained by Kocsis for an analysis of transport data in GaP.[42] It is also much higher than that predicted here (0.8 eV).

V. DEPENDENCE OF LATTICE CONSTANT ON DOPING

Doping with electrically active atoms (donors or acceptors) is known to change the lattice constant of semiconductors.[2,43,44] We treat here the case of heavy doping, by "shallow" hydrogenic impurities, in which the excess electrons or holes have no ionization energy. As first suggested by Yokota,[43] the effect can be broken up into two components, one due to the cores of the dopant ions and the other to the hydrostatic deformation potential of the band edge occupied by the free carriers. We shall describe the effect by the parameter β:

$$(\Delta a_0)/a_0 = \beta N_i , \qquad (3)$$

where N_i is the dopant concentration and β will be given in units of 10^{-24} cm^3. Thus $\beta = \beta_{\text{size}} + \beta_{e,h}$, where β_{size} corresponds to the hard-core effect of the ions and β_e (β_h) is the deformation potential effect for electrons (holes) given by:

$$\beta_{e,h} = \pm(\bar{a}_{c,v})/3B , \qquad (4)$$

where B is the bulk modulus and the $-$ ($+$) sign corresponds to electrons (holes). It was shown in Ref. 2 that the unscreened deformation potentials a_c (a_v) give the correct sign of β_e (β_h) but too large a magnitude (a factor of 2). As we shall see below (Table IV), agreement is improved if the screened $\bar{a}_{e,h}$ are used. If no pinning of the Fermi energy at the surface would take place, the argument for using an unscreened deformation potential may be made since the corresponding strain would be uniform. In samples exposed to air, however, the Fermi energy is pinned at the surface, somewhere in the gap, and the strain produced by the free carriers will not be uniform, relaxing when the surface is approached to within a few tens of an angstrom (screening length). The material in this region will thus polarize and screen the deformation potential in the manner discussed in Sec. II. Thus we conjecture that the screened $\bar{a}_c$ ($\bar{a}_v$) should be used in Eq. (4).

We present in Table IV the total values of β determined experimentally (β_{expt}) and those of β_e^{expt}(β_h^{expt}) obtained from the experimental ones after subtracting the hard-core effect calculated from the ionic radii as discussed in Ref. 2 (see also Ref. 44) for Si, Ge, GaAs, and GaP with different dopants. With the exception of electrons in X valleys (Si and GaP), the agreement between $\beta_{e,h}^{\text{expt}}$ and the values calculated from the $\bar{a}_{c,v}$ with Eq. (4) is rather satisfactory, especially in view of the scatter in the experimental data. For the case of the X valleys, the opposite sign obtained for $\beta_{e,h}^{\text{expt}}$ and $\beta_{e,h}^{\text{calc}}$. We should keep in mind, however, that in this case $\bar{a}_c$ is very small. A slight decrease in the screening ϵ would suffice to reverse its sign and thus restore sign agreement between theory and experiment. In any case, the agreement in Table IV is considerably better than that shown in Table V of Ref. 2 for unscreened deformation potentials.

VI. VALENCE-BAND OFFSETS AT HETEROJUNCTIONS

A. Lattice-matched heterojunctions

As can be seen from the lattice constants a_0 in Table I, many lattice-matched heterojunctions can be constructed with the materials under consideration here. It will become obvious in Sec. VI B that mismatches in a_0 of less than 1% are negligible within the type of accuracy aimed at here (≈ 0.1 eV). We shall consider heterojunctions with a_0 mismatches of less than 1% to be lattice matched; for the materials under consideration, the list of such heterojunctions is given in Table V. For each pair of materials, we give first that with the deeper valence-band top (Γ_{15})

TABLE IV. Experimental values of β, the constant which expresses the doping-induced changes in a_0, for several semiconductors. The sources are given in Ref. 2 unless otherwise specified. $\beta_{e,h}^{\text{expt}}$ represents the electronic contribution to $\beta_{e,h}$ obtained as specified in the text while $\beta_{e,h}^{\text{calc}}$ was obtained with Eq. (4).

Material Dopant and type	Silicon					Germanium			GaAs			GaP	
	P:e	As:e	Sb:e	B:h	Ga:h	P:e	As:e	Sb:e	Si:e	Se:e	Te:e	Ge:h	Te:e
β^{expt}	-1.3[a]	-0.1[a]	$+2.8$[a]	-5.2	$+1.7$[b]	$+7.4$[d]	$+1.3$	$+11$	$+3$	$+9.4$	$+11.6$	$+1$	$+3.0$[c]
$\beta_{e,h}^{\text{expt}}$	$+3.7$	$+2$	$+1.6$	-4.6	-1.9		$+5.8$	$+12.4$	$+9.9$	$+16$	$+11$	-2.2	$+0$
$\beta_{e,h}^{\text{calc}}$	-0.7	-0.7	-0.7	-2.6	-2.6	$+9.0$	$+9.0$	$+9.0$	$+18$	$+18$	$+18$	-2.7	-2.9

[a]Reference 66.
[b]Reference 67.
[c]Reference 68.
[d]Reference 69.

TABLE V. Valence-band ($\Delta E_V^{A,B}$) offsets (in eV) for nearly lattice matched heterojunctions between several group-IV elements and III-V and II-VI compounds calculated by different methods compared with recent experimental data. The compound with the deeper valence band is listed first. A value of 0.7 eV has also been calculated *ab initio* by Ihm and Cohen for ZnSe-GaAs (Ref. 70).

	LMTO[a]	LMTO[b]	SCIC[c]	CNP[d]	Experiment
AlP/Si	0.92	0.91	1.03	0.91	
AlP/GaP	0.38	0.34	0.36	0.46	
AlAs/Ge	0.87	0.84	1.05	0.87	0.95[e]
AlAs/GaAs	0.43	0.43	0.37	0.55	0.55,[f] 0.42[g]
AlSb/GaSb	0.34	0.30	0.38	0.38	0.4[h]
GaP/Si	0.53	0.57	0.61	0.45	0.80[i]
GaAs/Ge	0.51	0.45	0.63	0.32	0.56[j]
InSb/α-Sn	0.34	0.39			
InAs/GaSb	0.55	0.54	0.38	0.43	0.51,[k] 0.57[l]
ZnSe/Ge	1.46	1.57	2.17	1.52	1.52,[m] 1.29[m]
ZnSe/GaAs	0.99	1.13	1.59	1.20	1.10[m]
CdTe/α-Sn	0.99	1.12			1.0[q]
CdTe/InSb	0.66	0.73		0.84	0.87[n]
CdTe/HgTe	0.64	0.61	0.23	0.51	0.35,[o] 0.12[p]

[a]Present calculations, Eq. (5) with $\bar\epsilon$ equal to ϵ of Table III.
[b]Present calculations, Eq. (5) with $\bar\epsilon = 3.5$.
[c]Self-consistent interface calculations (SCIC), from Refs. 45 and 46.
[d]Calculations based on charge-neutrality point, from Ref. 10.
[e]Reference 71.
[f]Reference 72.
[g]Reference 73.
[h]Reference 74.
[i]Reference 75.
[j]Reference 76.
[k]Reference 77.
[l]References 63 and 78.
[m]Reference 79.
[n]Reference 80.
[o]Reference 81.
[p]Reference 82.
[q]Reference 83.

after heterojunction formation.

The band offsets for the Γ_{15} states calculated by us, by Tersoff[10] and by Van de Walle and Martin[45,46] are given in Table V compared with the most recent (or reliable, as judged by the present authors) experimental data. Other theoretical and experimental data can be found in Refs. 10, 45, and 46.

The procedure we have used for our calculations is based on the calculation of the Γ_{15} valence bands using the LMTO method with respect to the reference level of the ASA which, except for surface dipoles, should represent the potential at infinity.[2,7] When bringing two materials together to form a heterojunction, a potential difference will appear which will be screened by the electronic polarizability in a way similar to that discussed in Secs. II and III for acoustic phonons. The phonons of relevance, however, are of wavelength much larger than the lattice constants while for the heterojunction the potential variation occurs in a region of a depth typically equal to about $a_0/2$.[4,46] It is therefore questionable whether it is legitimate to screen with the full dielectric constant. We thus use now an effective dielectric constant $\bar\epsilon$ and consider this question in more detail below. The expression for the band offset between two materials A and

B can thus be written:

$$\Delta E_V^{A,B} = E_V^B - E_V^A - (E_D^B - E_D^A)(\bar\epsilon - 1)/\bar\epsilon \qquad (5)$$

where E_V's represent the energies of the Γ_{15} top of the valence band (including s.o. splitting), which are listed in Table I. The sign of Eq. (5) has been chosen such that if A has a deeper valence band, $\Delta E_V^{A,B}$ is positive. The effective dielectric constant $\bar\epsilon$ can be taken to be an average of the $q = 0$, $\omega = 0$ data for both materials, listed in Table III. As already mentioned, however, this probably overestimates the screening. A possible approach to correct this deficiency would be to estimate or assume a one-dimensional variation of E_B with z at the interface, decompose it into one-dimensional Fourier components and screen each according to the calculated $\epsilon(q)$ averaged, of course, for both materials.[47] For an interface with a transition region of width $\approx a_0/2$, as expected for [100] heterojunction planes, and a linear variation of the potential within this region, we find a maximum in the Fourier component of the potential for $q_M \simeq 5.6 \times d^{-1}$. Figure 7 of Ref. 47 shows that for Ge, GaAs, and ZnSe, $\epsilon(q_M) \simeq 2.3$. One may, therefore, be tempted to use this value for $\bar\epsilon$ in Eq. (5), regardless of material. This is a point of view similar to that adopted in Ref. 7, where it was suggested, as a result of microscopic calculations for the interface, that $\bar{\epsilon}=3.5$ regardless of material. The work of Tersoff, however, requiring exact lineup of E_T^A and E_T^B, implies $\bar{\epsilon}=\infty$ in the spirit of Eq. (5). $(E_T\approx E_D$ according to Table II.) *(See Note added in proof.)*

Fortunately, the band offsets $\Delta E_V^{A,B}$ calculated with our formulation do not depend critically on the value of $\bar{\epsilon}$ because the values of E_D are very similar for materials which yield well-ordered heterojunctions (see Table II). In order to illustrate the differences in band offset estimates produced by the uncertainty of $\bar{\epsilon}$, we present in Table V calculations for lattice-matched pairs of materials performed by replacing into Eq. (1) the energies of Tables I and II, with two values of $\bar{\epsilon}$, the average of those listed in Table III for both components ($q=0$ assumption), and $\bar{\epsilon}=3.5$ as suggested in Ref. 7. The difference between both predictions is small and no trends are apparent that may help us to describe which ansatz is preferable for $\bar{\epsilon}$. Table V clearly exposes the fallacy of the so-called common anion rule[48,49] (small band offset for common anions), a fact which has been also recently recognized by Tersoff.[50] The band offsets for pairs of materials with common anions are not particularly smaller than for other cases in which the anions are not common (e.g., InAs-GaSb).

B. Lattice-mismatched heterojunctions

Several heterojunctions with constituents differing in their lattice constants up to $\sim 7\%$ can be prepared. We discuss here the pairs Ge-Si ($\delta=\Delta a_0/a_0=0.04$), GaAs-Si ($\delta=0.04$), and GaAs-InAs ($\delta=0.07$). If the thickness of one of the materials is small (<50 Å for the cases mentioned) and that of the other large (substrate), the lattice constant of the thin component along the interface will match that of the substrate: the thin material will thus be strained. As it becomes thicker, the lattice mismatch (strain) is relieved through misfit dislocations. In the latter case, the theory of the previous subsection is applicable. In the former, one must correct for the hydrostatic and shear components of the strain. We thus decompose the strain tensor of the strained component as follows:

$$\delta \begin{bmatrix} 1 & 0 & 0 \\ 0 & 1 & 0 \\ 0 & 0 & -2C_{12}/C_{11} \end{bmatrix} = \delta_H \begin{bmatrix} 1 & 0 & 0 \\ 0 & 1 & 0 \\ 0 & 0 & 1 \end{bmatrix} + \delta_S \begin{bmatrix} -1 & 0 & 0 \\ 0 & -1 & 0 \\ 0 & 0 & 2 \end{bmatrix},$$

$$\delta_H = \tfrac{2}{3}\delta(1 - C_{12}/C_{11}), \quad \delta_S = -\tfrac{1}{3}\delta(1 + 2C_{12}/C_{11}),$$

$$(6)$$

where C_{12} and C_{11} are elastic stiffness constants. Equation (6) is valid for a (001) interface; generalization to other interfaces is trivial.

The correction of the values of $\Delta E_V^{A,B}$ obtained with Eq. (5) from the energies of the unstrained components (Tables I and II) for the hydrostatic component of the strain is straightforward. If A is the strained component of the heterojunction, one must add to $\Delta E_V^{A,B}$:

$$\Delta_H^{A,B} = (-a_V^A + a_D^A(\bar{\epsilon}-1)/\bar{\epsilon})\delta_H .$$

The values of the volume coefficients of the Γ_{15} valence band $a_V^{A,B}$ and those for the DME $a_D^{A,B}$ are listed in Tables I and III. We note that the a_V's and a_D's have the same sign and about the same magnitude for all materials. Hence, effects of δ_H on Γ_{15} and the DME are nearly the same and compensate each other. Actually, the a_V's are somewhat larger in magnitude than the a_D's and this effect is accentuated through multiplication by $(\bar{\epsilon}-1)/\bar{\epsilon}$. Hence a residual effect remains which tends to lower the Γ_{15} valence edge for the A material in the case $\delta_H > 0$ (i.e., $a_0^A > a_0^B$). We consider next the cases of special interest.

1. Silicon-germanium

Let us take material A to be silicon (strained) and B germanium (unstrained). We find

$$\Delta E_H^{A,B} = 1.6\delta_H = +0.02 \text{ eV} .$$

Thus in this case δ_H slightly increases the band offset between Si and Ge which for the unstrained materials is calculated to be 0.19 eV. If the strain is in the germanium side (compression), we find similarly:

$$\Delta E_H^{A,B} = -1.6\delta_H = -0.02 \text{ eV} , \qquad (7)$$

thus the magnitude of the effect is the same in both cases.

The treatment of the pure shear component of the strain is more complicated, especially if the strained component is silicon. In this case, the Γ_8 valence band splits into two and the strain couples it to its spin-orbit-split component Γ_7.[51] Since the coupling energy is larger than the spin-orbit splitting of Si (0.04 eV), the resulting shifts are strongly nonlinear in strain. These complications can be eliminated by treating the band offset for the average of the six Γ_{15} valence bands, the four Γ_8's and two Γ_7's, as discussed in Refs. 46 and 52: the effect of the pure shear component of the strain δ_S on the valence bands then disappears. The offset for the average valence bands of Ge and Si thus becomes $\Delta E_{V,av}^{A,B}=0.12$ eV, much smaller than the values calculated by Van de Walle and Martin (0.54 eV) (Refs. 46 and 52) with an "ab initio" pseudopotential method which is expected to give a better representation of the interface than the calculations performed here.

The experimental situation is, as for most heterojunctions, somewhat confused. Maybe the most reliable relevant data are those recently obtained for $Si_{1-x}Ge_x$ multiple quantum wells, for x up to $\simeq 0.5$.[53,54] An analysis of transport and other data enables the authors to deduce the relative positions of the lowest conduction bands in the two components, now split by the uniaxial stress. It is concluded in Refs. 46 and 52 that these data require band offsets which extrapolate for $x \to 1$ to $\Delta E_{V,av}^{A,B} > 0.5$ eV. Most other experimental data, however, yield somewhat smaller values of this offset, although neither the nature of the interface nor the strain is usually specified.[55-57] The photoemission work of Margaritondo et al.[56] yields $\Delta E_V^{A,B} = 0.2$ eV which, assuming that no uniaxial strain is present, would correspond to $\Delta E_{V,av}^{A,B} = 0.1$ eV, a result which would agree with ours. However, $\Delta E_V^{A,B} = 0.4 \pm 0.1$ eV was obtained in Ref. 57, using the same technique.

We note that Tersoff's calculations[10] yield a value of $\Delta E_V^{A,B} = 0.25$ eV for Si-Ge. Since these calculations do not include any uniaxial stress, they correspond to $\Delta E_{V,av}^{A,B} = 0.15$ eV, a number which agrees with our estimates. In view of the reliability of the calculations of Van de Walle and Martin,[52] we should examine the possibility that the difference between data based on midgap points (Tersoff's, ours) and theirs may be due to the uniaxial component of the stress.

For a (001) interface between materials with different lattice constants, the first Baldereschi points [Eq. (1)] are not all equivalent after the strain of Eq. (6) appears: they split into three groups of eight each. The question then arises of which of the split points must be matched with that of the unstrained material. We have not calculated the splitting of these points because of the complications which arise in the LMTO method when shear strain is present. We give, however, a simple model which enables us to make a crude estimate of the splitting and its effects on the matching across the interface.

Let us consider the Penn model of the electronic polarizability and the changes induced by strain, of both types described in Eq. (6), on the Penn gap and on E_D. The changes induced by δ_H and δ_S are, in principle, independent of each other. If we assume, however, that the change along a direction $\mathbf{k}$ of $\mathbf{k}$ space is given solely by the compression along this direction, the hydrostatic and the shear change become related through the expression (for E_D):

$$\Delta E_D(\mathbf{k}) = (\Delta a_D)(\mathbf{k} \cdot \underline{e} \cdot \mathbf{k}) \quad (8)$$

where $\underline{e}$ represents the strain tensor. Thus, for a shear strain E_D will depend on the direction of $\mathbf{k}$ and we must consider how to average the various $E_D(\mathbf{k})$ so as to obtain the one to be matched across a (001) interface. We note that a similar model was successfully used in Ref. 14 to explain the sign and magnitude of the long-wavelength stress-induced birefringence.

We call θ the angle between $\mathbf{k}$ and the (001) direction. $\Delta E_D(k)$ can thus be written, for the pure shear component of the strain of Eq. (6):

$$\Delta E_D(\mathbf{k}) = (\Delta a_D)\delta_s(2\cos^2\theta - \sin^2\theta) . \quad (9)$$

In order to decide which value of $\Delta E_D(\mathbf{k})$ to use for the matching across the interface, we consider the fact that in the Penn model, the states along $\mathbf{k}$ only contribute to the polarizability along $\mathbf{k}$. Hence, if we want to consider the effect of the dielectric response on the $\Delta E_D(\mathbf{k})$ of Eq. (9),

we must multiply this equation by $\cos^2\theta$ and average it for all directions of k. We find

$$\langle \Delta E_D \rangle_{(001)} = (\Delta a_D)\delta_s(2\langle \cos^4\theta - \sin^2\theta\cos^2\theta \rangle)/\langle \cos^2\theta \rangle$$
$$= \tfrac{4}{5}(\Delta a_D)\delta_s . \quad (10)$$

The average of $\cos^2\theta$ in the denominator of Eq. (10) has been introduced to take into account that such average appears for a pure hydrostatic strain. For a Si-Ge interface with Si under strain and Ge unstrained, we find with Eq. (6) that $\delta_S = -0.024$. With $\Delta a_D = 6.3$ eV (Table III), we finally obtain $\langle \Delta E_D \rangle_{001} = -0.12$ eV. The sign of $\langle \Delta E_D \rangle_{001}$ just found is such that it increases the band offset between Si and Ge and thus brings our calculated value closer to those of Van de Walle and Martin. Nevertheless, our calculated value will now be 0.22, still too small when compared with that of those authors (0.54 eV). Van de Walle and Martin also calculated a Si-Ge heterojunction between cubic Si and Ge with the same lattice constant. They found a band offset of 0.40 eV. The difference to the case in which Si was allowed to expand along the z axis, 0.14 eV,[52] does agree with the estimate of the effect of the uniaxial stress performed above. Hence the discrepancy between our result of 0.27 eV and that of Ref. 52 (0.54 eV) is still unanswered.

A reason for this discrepancy may be found in the fact that the spectrum of $\epsilon_2(\omega)$ of Si differs considerably from that of Ge.[20] This is the spectrum of virtual transitions which contribute to the polarizability for $\omega = 0$. Both Ge and Si have a peak in $\epsilon_2(\omega)$ at ~ 4.3 eV (listed as E_2 in Table II). They also have a peak, usually labeled E_1, at 2.1 in the case of Ge and at 3.4 for Si. Hence the shape of $\epsilon(\omega)$ is considerably changed in Ge with respect to that of Si. The increase in E_1 in the latter is responsible for the fact that the dielectric gap E_g of Si is $\simeq 4.8$ eV, as compared to 4.3 for Ge. The E_1 peak, due to transitions along $\langle 111 \rangle$, is not well sampled by the first Baldereschi point (see Table II; actually $E_D = 4.73$ eV for Si, smaller than the value of 4.85 calculated for Ge!). Inclusion of two Baldereschi points, $(2\pi/a_0)(\tfrac{3}{4}, \tfrac{1}{4}, \tfrac{1}{4})$ and $(2\pi/a_0)(\tfrac{1}{4}, \tfrac{1}{4}, \tfrac{1}{4})$, would remedy this problem as they include the transitions along the $\langle 111 \rangle$ directions responsible for E_1. Since the valence bands of both materials are very similar, the discrepancy just discussed must reflect itself in a higher E_D for Si than for Ge, in excess of the difference between the E_D calculated at the $\mathbf{k}_B$ (Table II). We estimate this additional increase in the E_D of Si to be $\simeq 0.3$ eV, which would bring our estimate of the offset to 0.52 eV, now in agreement with the data of Ref. 52.

2. GaAs-Si

We now discuss the GaAs-Si heterojunction ($\delta = 0.04$). Using the method described above, we find for a lattice-mismatched cubic heterojunction an average band offset $(\Delta E_{av}^{A,B})$ of 0.48 eV. For lattice-matched heterojunctions, the hydrostatic strain lowers this value to 0.35 eV if only the GaAs is strained and to 0.34 eV if only Si is strained. The shear correction of Eq. (10) brings this value down to 0.21 eV in the former case and to 0.22 eV in the latter. This is in reasonable agreement with Van de Walle's calculations, which yield 0.12 eV for strained GaAs and 0.14 eV for strained Si.[46]

We note that an offset of 0.05 eV has been measured for this system by Margaritondo[58] for layers of GaAs deposited on Si. The state of the strain in this layer, however, was not known.

3. GaAs-InAs

We discuss next the GaAs-InAs heterojunction for which $\delta = 0.07$. Our calculation without taking into account the strain yields actually an average Γ_{15} valence band for InAs lower than that for GaAs ($\Delta E_{av}^{A,B} = -0.15$ eV). The hydrostatic correction brings this value up to -0.12 eV. The uniaxial correction of Eq. (10) raises this value by 0.30 eV, up to $\Delta E_{av}^{A,B} = +0.18$ eV, with the GaAs side now deeper than InAs. This is in rather good agreement with the value of 0.11 eV calculated by Martin and Van de Walle.[46,59] The same value ($+0.18$ eV) is found with our method for a strained GaAs layer. It agrees even better with the results of Refs. 46 and 58 ($+0.21$ eV).

We note that for this system, an offset $\Delta E_V = 0.17 \pm 0.07$ eV was measured by Kowalczyk et al.[60] The heterojunction was formed by a thin layer of InAs deposited on a GaAs substrate. Because of the large spin-orbit splittings of the Γ_{15} valence states ($\Delta_0 = 0.34$ for GaAs, 0.38 for InAs), the nonlinear contributions to the strain splittings of the top of the valence band are negligible. So is the difference in the spin-orbit splittings. The linear splitting of the Γ_8 valence band of InAs by the uniaxial strain, however, should lift the top of the valence band of InAs by about 0.34 eV with respect to that of unstrained GaAs (we have used for this estimate the strain deformation potential $b = -1.8$ eV given in Ref. 61), thus increasing the band offset estimated here to $\Delta E^{A,B} = 0.52$ eV, now much higher than the experimental one [0.17 ± 0.07 eV (Ref. 60)]. A plausible explanation for the discrepancy is partial relaxation of the large shear component of the strain in the deposited InAs layer.

VII. CONCLUSIONS

We have shown that the valence- and conduction-band edges calculated at the first Baldereschi special points ($\mathbf{k}_B$) for group-IV elemental and III-V and II-VI compound semiconductors can be successfully used to estimate the effect of screening on the hydrostatic deformation potentials used to calculate the electron-phonon interaction. This procedure introduces the concept of a dielectric midgap point, somewhat similar to Tersoff's charge-neutrality point, but arrived at in a rather different, simpler manner. The results so obtained for the electron-phonon coupling constant yield reasonable agreement with available experimental data when the most reliable ones of the many available and conflicting data are chosen. They probably can be used as a guide to choose among such conflicting data. The screened deformation potentials so obtained also improve agreement of experimental and calculated values of the change in lattice constant with doping in heavily doped semiconductors.

The ideas used for the evaluation of screening effects on electron-phonon interaction constants can also be used to calculate the effect of screening on the band offset at heterojunctions obtained from the absolute energies calculated with the LMTO method. This procedure is straightforward in the case of nearly lattice-constant-matched pairs of materials (less than 1% mismatch). For strongly mismatched pairs, the effects of the hydrostatic and the uniaxial components of the resulting strain must be included. The former is straightforward to evaluate. The latter is estimated on the basis of a generalized Penn model which considers the inequivalence of the various points of the star of $\mathbf{k}_B$ after a shear strain is applied. Reasonable agreement with ab initio pseudopotential calculations for the same interfaces is obtained. The screening of the shear strain should also contribute small additional terms to the electron-phonon interaction constants which are similar to the octopole terms discussed by Lawaetz.[39] They have not been considered here any further.

In a recent paper, Priester et al. have performed a self-consistent tight binding calculation of the Ge-GaAs interface.[62] They use as reference sp^3 hybrids instead of the DME utilized here. They find for that system a band offset of 0.65 eV, only slightly higher than ours (0.51 eV). It has also come to our attention that Claessen et al.[63] have measured the dependence of the valence-band offsets on pressure for InAs-GaSb superlattices. They find this offset to increase at the rate 4.2 meV/kbar. Using the method discussed here, we also find an increase of this offset with pressure, but at a smaller rate, namely, 1.5 meV/kbar.

Note added in proof. W. A. Harrison and also J. C. Durán, F. Flores, C. Tejedor, and A. Munoz (unpublished) have recently calculated that the effective dielectric constant $\bar{\epsilon}$ of Eq. (5) should be rather close to that for $q = 0$. If we accept this conclusion, the LMTO[a] data of Table V should be preferred to the corresponding LMTO[b] data. A value of $\bar{a}_c = 11 \pm 1$ eV has been recently obtained by K. Hirakawa and H. Sakaki [Appl. Phys. Lett. 49, 14 (1986)] by investigating electron relaxation processes in AlGaAs-GaAs heterojunctions.

ACKNOWLEDGMENTS

This work originated as a result of several conversations with R. M. Martin and J. Tersoff. Thanks are due to them and others, including O. K. Andersen, W. A. Harrison, C. Herring, J. Menéndez, D. Olego, and C. Van de Walle, for many discussions and for communicating to us unpublished data. The paper was partly written up during the stay of one of the authors (M.C.) at the Xerox Palo Alto Research Center (PARC). Thanks are due to the Xerox Corporation and to the staff of Xerox PARC for making this stay possible.

APPENDIX

The dielectric response of a semiconductor to a low-frequency potential $V(r) = V_q e^{iqr}$ (time dependence omitted), such as that created by longitudinal acoustic phonons, can be approximately obtained with the Penn model.[4] This model replaces the Jones zone by a sphere of radius k_p and introduces an isotropic gap, the Penn gap E_p, between the occupied and unoccupied states at the boundary of the spherical Jones zone. The model is basically equivalent to that of a one-dimensional semiconductor with a gap at the edge of the Brillouin (Jones) zone. In order to obtain the dielectric response, we must evaluate the matrix element of $V(r)$ for virtual transitions between the valence and the conduction band in the neighborhood of E_p (we implicitly neglect umklapp processes in the dielectric response). For $q \to 0$, this matrix element tends to zero, hence we evaluate it for q small but finite. The

gap then takes place between valence states at $k_p - q/2$ and conduction states at $k_p + q/2$. Using $\mathbf{k} \cdot \mathbf{p}$-perturbation theory, the corresponding eigenstates can be written as

$$|c, k_p + q/2\rangle = (|c, k_p\rangle + (Pq/2E_p)|v, k_p\rangle)e^{iqr/2},$$
$$|v, k_p - q/2\rangle = (|v, k_p\rangle + (Pq/2E_p)|c, k_p\rangle)e^{-iqr/2}, \tag{A1}$$

where c and v denote the conduction and valence bands, respectively, and P is the interband matrix element of $\mathbf{p}$. The interband matrix element of $V(r)$ is readily obtained from Eqs. (A1):

$$\langle c, k_p + q/2 | V(r) | v, k_p - q/2 \rangle = \tfrac{1}{2} P_q / E_p [\langle c, k_p | V_q | c, k_p \rangle + \langle v, k_p | V_q | v, k_p \rangle]. \tag{A2}$$

V_q must be understood as the self-consistent sum of the applied unscreened potential and the dielectric response. If the former is an external electrostatic potential, both matrix elements in the rhs of Eq. (A1) become equal and the dielectric response reduces the external potential $V_{ex,q}$ to a total potential $V_q = V_{ex,q}/\epsilon(q)$, where $\epsilon(q)$ is the static, q-dependent dielectric function. For q smaller than about 0.2 times $2\pi/a_0$ ($a_0 =$ crystallographic lattice constant), $\epsilon(q)$ is nearly equal to the value at $q = 0$. At $q = 2\pi/a_0$, it reduces to a smaller value, close to that obtained by Tejedor and Flores[7] for the screening of band offsets in heterojunctions.

In the case of the unscreened potential induced by the hydrostatic component of the strain of a longitudinal acoustic phonon, $\langle c, k_p | V_q | c, k_p \rangle$ and $\langle v, k_p | V_q | v, k_p \rangle$ are not equal since the corresponding deformation poten-

tials usually differ (the Penn gap depends on hydrostatic strain). Nevertheless, Eq. (A2) indicates that the calculation of the screened response is basically the same as that for an electrostatic potential provided one replaces this potential by the average of the potentials seen by the valence and the conduction bands. This *average* must be screened by dividing by $\epsilon(q)$. Hence the deformation potentials obtained in Ref. 2 must be corrected for the effect of screening on the deformation potential of the *dielectric midpoint energy* (average of the conduction and valence bands which form the Penn gap) before using them to evaluate the electron-phonon interaction. A similar correction should be applied to the difference in absolute band energies in order to obtain band offsets in heterojunctions.

[1]J. Bardeen and W. Shockley, Phys. Rev. **80**, 72 (1950); C. Herring and E. Vogt, *ibid.* **101**, 944 (1956).

[2]J. A. Vergés, D. Glötzel, M. Cardona, and O. K. Andersen, Phys. Status Solidi B **113**, 519 (1982).

[3]O. K. Andersen, Phys. Rev. B **12**, 3060 (1975).

[4]D. Penn, Phys. Rev. **128**, 2093 (1962).

[5]A. Baldereschi, Phys. Rev. B **7**, 5212 (1973).

[6]C. Tejedor, F. Flores, and E. Louis, J. Phys. C **10**, 2163 (1977).

[7]F. Flores and C. Tejedor, J. Phys. C **12**, 731 (1979).

[8]J. Tersoff, Phys. Rev. Lett. **52**, 465 (1984).

[9]J. Tersoff, Phys. Rev. B **30**, 4874 (1984); see also W. A. Harrison and J. Tersoff, J. Vac. Sci. Technol. B **4**, 1068 (1986).

[10]J. Tersoff, in *Heterojunctions: A Modern View of Band Discontinuities and Applications*, edited by G. Margaritondo and F. Capasso (North-Holland, Amsterdam, in press).

[11]F. S. Khan and P. B. Allen, Phys. Rev. B **29**, 3341 (1984); S. Rodriguez and E. Kartheuser, *ibid.* **33**, 772 (1986).

[12]J. Bok and M. Combescot, in *Proceedings of the International Conference on the Physics of Semiconductors, Stockholm, 1986* (World Science, Singapore, 1987).

[13]L. Vina, S. Logothetidis, and M. Cardona, Phys. Rev. B **30**, 1979 (1984).

[14]M. Cardona, in *Atomic Structure and Properties of Solids*, edited by E. Burstein (Academic, New York, 1972), p. 514.

[15]See, for instance, J. R. Chelikowski and M. L. Cohen, Phys.

Rev. B **14**, 556 (1976).

[16]F. H. Pollak, M. Cardona, C. W. Higginbotham, F. Herman, and J. P. Van Dyke, Phys. Rev. B **2**, 352 (1971).

[17]N. E. Christensen, Phys. Rev. B **30**, 5753 (1984).

[18]G. B. Bachelet and N. E. Christensen, Phys. Rev. B **31**, 879 (1985).

[19]M. S. Hybertsen and S. G. Louie, Phys. Rev. Lett. **55**, 1418 (1985).

[20]J. C. Phillips, *Bonds and Bands in Semiconductors* (Academic, New York, 1973).

[21]R. H. Lyddane, R. G. Sachs, and E. Teller, Phys. Rev. **59**, 673 (1941).

[22]W. A. Harrison, *Electronic Structure and the Properties of Solids* (Freeman, San Francisco, 1980).

[23]J. Tersoff (private communication).

[24]C. G. Van de Walle, C. Herring, and R. M. Martin, Phys. Rev. B (to be published).

[25]J. D. Wiley, Solid State Commun. **8**, 1865 (1970).

[26]W. Zawadzki, in *Physics of Narrow Gap Semiconductors*, edited by E. Gornik, H. Heinrich, and L. Palmetshofer (Springer, Heidelberg, 1982), p. 8.

[27]K. W. Nill and A. L. McWhorter, J. Phys. Soc. Jpn. Suppl. **21**, 755 (1966).

[28]S. M. Puri, Phys. Rev. **139**, A995 (1965).

[29]H. Kahlert and G. Bauer, Phys. Rev. B **7**, 2670 (1973).

[30]H. L. Störmer, A. C. Gossard, W. Wiegmann, R. Blondel, and K. Baldwin, Appl. Phys. Lett. **44**, 139 (1984).

[31]E. E. Mendez, P. J. Price, and M. Heilblum, Appl. Phys. Lett. **45**, 294 (1984).

[32]W. Walukiewicz, H. E. Ruda, J. Lagewski, and H. C. Gatos, Phys. Rev. B **32**, 2645 (1985); P. J. Price, *ibid.* **32**, 2643 (1985).

[33]D. L. Rode, in *Semiconductors and Semimetals,* edited by R. K. Willardson and A. C. Beer (Academic, New York, 1975), Vol. 10, p. 84. The argument that the deformation potential of the Γ_{15} valence state should be taken to be zero was early given by H. Ehrenreich, J. Phys. Chem. Solids **2**, 131 (1957).

[34]B. Vinter, Phys. Rev. B **33**, 5904 (1986); see also K. Hirakawa and H. Sakaki, Appl. Phys. Lett. **49**, 889 (1986). These authors report $|a_c|=11\pm1$ eV.

[35]P. Pfeffer, I. Gorczyca, and W. Zawadski, Solid State Commun. **51**, 179 (1984).

[36]H. J. Lee, J. Basinski, L. Y. Juravel, and J. C. Woolley, Can. J. Phys. **57**, 419 (1978).

[37]D. L. Rode, Phys. Rev. B **3**, 3287 (1971).

[38]B. R. Nag and G. M. Dutta, J. Phys. C **11**, 119 (1978).

[39]P. Lawaetz, Phys. Rev. **183**, 730 (1969).

[40]K. Murase, K. Enjouji, and E. Otsuka, Jpn. J. Appl. Phys. **29**, 1255 (1970).

[41]L. Pintschovius, J. A. Vergés, and M. Cardona, Phys. Rev. B **26**, 5658 (1982).

[42]S. Kocsis, Phys. Status Solidi A **28**, 133 (1975).

[43]J. Yokota, J. Phys. Soc. Jpn. **19**, 1487 (1964).

[44]V. Pietsch and K. Unger, Phys. Status Solidi A **80**, 165 (1983).

[45]C. G. Van de Walle and R. M. Martin, Mater. Res. Soc. Symp. Proc. **63**, 21 (1985); J. Vac. Sci. Technol. B **4**, 1055 (1986).

[46]C. G. Van de Walle, Ph.D. thesis, Stanford University, Palo Alto, California, 1986 (unpublished).

[47]G. Srinivasan, Phys. Rev. **178**, 1248 (1969).

[48]J. O. McCaldin, T. C. McGill, and C. A. Mead, Phys. Rev. Lett. **36**, 56 (1976).

[49]W. R. Frensley and H. Kroemer, Phys. Rev. B **16**, 2642 (1977).

[50]J. Tersoff, Phys. Rev. Lett. **56**, 2755 (1986).

[51]L. D. Laude, F. H. Pollak, and M. Cardona, Phys. Rev. B **3**, 2623 (1971); M. Chandrasekhar and F. H. Pollak, *ibid.* **15**, 2127 (1977).

[52]C. G. Van de Walle and R. M. Martin, Phys. Rev. B **34**, 5621 (1986).

[53]G. Abstreiter, H. Brugger, T. Wolf, H. Jorke, and H. J. Herzog, Phys. Rev. Lett. **54**, 2441 (1985).

[54]D. V. Lang, R. People, J. C. Bean, and A. M. Sargent, Appl. Phys. Lett. **47**, 1333 (1985).

[55]T. F. Kuech, M. Mäenpää, and S. S. Lau, Appl. Phys. Lett. **39**, 245 (1981).

[56]G. Margaritondo, A. D. Katnani, N. G. Stoffel, R. R. Daniels, and T. X. Zhao, Solid State Commun. **43**, 163 (1982).

[57]P. H. Mahowald, R. S. List, W. E. Spicer, J. Woicik, and P. Pianetta, J. Vac. Sci. Technol. B **3**, 1252 (1985).

[58]G. Margaritondo, Phys. Rev. B **31**, 2526 (1985).

[59]C. G. Van de Walle and R. M. Martin (unpublished).

[60]S. P. Kowalczyk, W. J. Schaffer, E. A. Kraut, and R. W. Grant, J. Vac. Sci. Technol. **20**, 705 (1982).

[61]A. Blacha, H. Presting, and M. Cardona, Phys. Status Solidi B **126**, 11 (1984).

[62]C. Priester, G. Allan, and M. Lanoo, Phys. Rev. B **33**, 7386 (1986).

[63]L. M. Claessen, J. C. Maan, M. Altarelli, P. Wyder, L. L. Chang, and L. Esaki, Phys. Rev. Lett. **57**, 2556 (1986).

[64]S. Groves and W. Paul, Phys. Rev. Lett. **11**, 194 (1963).

[65]M. Cardona and N. E. Christensen, Solid State Commun. **58**, 412 (1986).

[66]P. Becker (unpublished).

[67]W. H. Appel, Ph.D. thesis, University of Stuttgart, 1985 (unpublished).

[68]H. G. Bruhl, K. Jacobs, W. Seiffert, and J. Maege, Krist. Tech. **14**, K29 (1979); see also Ref. 44.

[69]G. Contreras, L. Tapfer, A. K. Sood, and M. Cardona, Phys. Status Solidi B **131**, 475 (1987).

[70]J. Ihm and M. L. Cohen, Phys. Rev. B **20**, 729 (1979).

[71]M. K. Kelly, D. W. Niles, E. Colavita, G. Margaritondo, and M. Henzler (unpublished), quoted in Ref. 59.

[72]J. Batey and S. L. Wright, J. Appl. Phys. **59**, 1200 (1986).

[73]J. Menéndez, A. Pinczuk, D. J. Werder, A. C. Gossard, and J. H. English, Phys. Rev. B **33**, 8863 (1966).

[74]J. Menéndez, A. Pinczuk, D. J. Werder, and J. P. Valladares, Solid State Commun. (to be published).

[75]P. Perfetti, F. Patella, F. Settle, C. Quaressima, F. Capasso, A. Savoia, and G. Margaritondo, Phys. Rev. B **30**, 4533 (1984).

[76]J. R. Waldrop, E. A. Kraut, S. P. Kowalczyk, and R. W. Grant, Surf. Sci. **132**, 513 (1983).

[77]J. Sakaki, L. L. Chang, R. Ludeke, C. A. Chang, G. A. Sai-Halasz, and L. Esaki, Appl. Phys. Lett. **31**, 211 (1977).

[78]G. A. Sai-Halasz, L. L. Chang, J.-M. Welter, C.-A. Chang, and L. Esaki, Solid State Commun. **27**, 935 (1978).

[79]S. P. Kowalczyk, J. T. Cheung, E. A. Kraut, and R. W. Grant, Phys. Rev. Lett. **56**, 1605 (1986).

[80]K. J. McKey, P. M. G. Allen, W. G. Herrender-Harker, R. H. Williams, C. R. Whitehouse, and G. M. Williams, Appl. Phys. Lett. **49**, 354 (1986).

[81]S. P. Kowalczyk, E. Kraut, J. R. Waldrop, and R. W. Grant, J. Vac. Sci. Technol. **21**, 482 (1982).

[82]D. J. Olego, J. P. Faurie, and P. M. Racah, Phys. Rev. Lett. **55**, 328 (1985); see also J. M. Berroir, Y. Guldner, and M. Voos, IEEE J. Quantum Electron. **QE-22**, 1793 (1986).

[83]S. Takatani and Y. W. Chung, Phys. Rev. B **31**, 2290 (1985).

PHYSICAL REVIEW B   VOLUME 18, NUMBER 2   15 JULY 1978

Self-consistent electronic structure of (110) Ge-ZnSe

Warren E. Pickett* and Marvin L. Cohen

*Department of Physics, University of California, Berkeley, California 94720
and Materials and Molecular Research Division, Lawrence Berkeley Laboratory, Berkeley, California 94720*
(Received 10 March 1978)

Self-consistent pseudopotential calculations of the electronic structure of the (110) Ge-ZnSe interface (IF) indicate a density of localized IF states that may be experimentally detectable. The full spectrum is presented and the character of the IF states is discussed. Evaluations of the bond charges indicate little charge transfer parallel to the IF. A proposed relaxation of atoms at the IF is also presented. The detailed electronic structure of Ge-ZnSe is found *not* to be a simple extension of that of Fe-GaAs.

I. INTRODUCTION

Since the appearance of reports[1,2] that abrupt GaAs-Al$_x$Ga$_{1-x}$As interfaces (IFs) could be fabricated by molecular-beam-epitaxial (MBE) techniques, there have been a number of theoretical studies by various methods of abrupt semi-conductor-semiconductor IFs. These include the (110), (100), and (111) IFs of GaAs-AlAs[3-6] and of Ge-GaAs.[3,7-10] Only very recently there have been reports[11,12] of MBE fabrication of *abrupt* Ge-GaAs IFs. The obvious implication is that other abrupt IFs can also be made.

In this paper we present results of a self-consistent pseudopotential calculation for an abrupt (110) Ge-ZnSe IF. This system is interesting for both experimental and theoretical reasons. Various crystallographic IFs have been fabricated from Ge and ZnSe by liquid- and vapor-phase epitaxy, and the resulting heterojunctions have found workable applications, including transitors and photovoltaic converters. Although all Ge-ZnSe heterojunctions to date have been graded to some extent, it seems likely that abrupt counterparts will soon be made. From a theoretical point of view, the study of this IF, following work[3,7] on the same crystallographic IF in AlAs-GaAs and Ge-GaAs using identical methods, should lead to a better understanding of the physics of IF behavior, as well as providing detailed results for a particular system. Some aspects of the work described here have been reported elsewhere.[13]

Localized IF states are found in various regions below the valence-band maximum, as was the case in Ge-GaAs. In Sec. II we discuss the dispersion and character of the localized states and present their local state density near the IF. The method of calculation allows a direct inspection of the self-consistent potential and its influence on the electrostatic IF dipole and band discontinuities. The charge density is analyzed with a view toward understanding the IF chemistry and possibly predicting relaxation which may occur at this IF. The conclusions are summarized in Sec. III.

II. RESULTS AND DISCUSSION

The method of calculation has been described in detail elsewhere.[3] The effects of the ion cores on the valence electrons are represented by local pseudopotentials. With the effects of exchange and correlation included in the local density approximation, the valence electrons are allowed to re-adjust until the charge density is consistent with the potential from which it is determined. The Ge potential was used previously[3,7] in the study of the Ge-GaAs IF; for Zn and Se the potentials were the same as were used in surface studies.[14] In using a local potential for Zn we are neglecting effects due to the $3d$ levels at ~ -10 eV; this is not expected to be a serious limitation.

The ideal (110) IF geometry is shown in Fig. 1. Within the IF unit cell there are two bonding chains directed across the IF, one containing the Ge-Zn bond ($ABAB$ in Fig. 1) and the other ($CDCD$) containing the Ge-Se bond. In addition, in each atomic layer, bonding chains (Ge-Ge or Zn-Se) run parallel to the IF. The lattice constant of the parent Ge and ZnSe bulk crystals was taken to be 5.658 Å and the lattice mismatch, $\sim 0.2\%$, was ignored.

The superlattice configuration was used here, as in previous studies, to introduce three-dimensional periodicity and allow the use of Fourier-space techniques. The superlattice consisted of five layers of Ge and five layers of ZnSe, repeated periodically in the direction z perpendicular to the IF. To check that the five-layer–five-layer (denoted 5-5) superlattice is sufficient to reproduce the properties of a single IF, a calculation for a Ge-GaAs 5-5 superlattice was compared with the previously published[3,7] 9-9 superlattice calculation.

Ge-ZnSe (110) Interface

○ Zn   ◉ Ge   ● Se

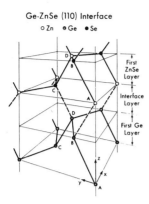

(110) Ge-ZnSe Self-Consistent Potential (eV)

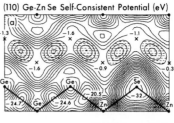

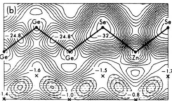

FIG. 1. Atomic positions near the Ge-ZnSe (110) interface. Heavy solid lines denote bonding directions, except for bonds across the interface which are denoted by dashed lines. The bonding chains $ABAB$ and $CDCD$ contain the Ge-Zn and Ge-Se bonds, respectively.

FIG. 3. Contour plot of the self-consistent potential in the (a) $ABAB$ and (b) $CDCD$ bonding planes perpendicular to the interface. The zero of potential is the valence-band maximum of Ge, and the hatched areas denote regions of *positive* potential. Interstitial positions are marked by x's.

The self-consistent charge density in the IF region was virtually identical in the two calculations, indicating a very small interaction between neighboring IFs even in the 5-5 superlattice. However, it can be expected that a smaller superlattice will make the identification of IF states less straightforward; this is discussed further below.

A. Self-consistent potential

In Fig. 2 we present a plot of the potential averaged parallel to the IF, $\overline{V}(z)$, across the interfacial region. The electrostatic dipole, equal to the difference in average potential across the IF, is 0.25 ± 0.1 eV, with ZnSe having the higher average potential. The uncertainties quoted have arisen from interaction of neighboring IFs in the superlattice geometry. Close inspection of $\overline{V}(z)$ in Fig. 2 reveals that the potential away from the

IF has not become exactly periodic. Averaging parallel to the IF, of course, eliminates most of the variations in the potential which are important in inducing localized states. In Fig. 3 we show a contour plot of the potential, measured relative to the valence-band maximum of bulk Ge, in the two planes containing the Ge-Zn and Ge-Se bonds across the IF. On the scale of the variation of the potential in the unit cell (~ 30 eV), the transition from the Ge to ZnSe potential occurs entirely in the interface layer; the potential in the Ge-Ge and and Ge-Se bonding regions, even in the first layer away from the IF, is nearly bulklike.

The most interesting feature of the potential is its behavior in the regions where it is positive (hatched regions in Fig. 3). These regions constitute the "channels" in the diamond (or zinc blende) lattice within which the charge density vanishes. From the figure it is evident that the potential in the "channels" is similar in Ge, in ZnSe, or at the IF. This is the basis of the model of Frensley and Kroemer[15] for band discontinuities at an IF, in which they assumed the average of the potentials at the two interstitial positions in the bulk unit cell to be continuous across the IF. These interstitial positions are marked by x's in Fig. 3, and the values of the self-consistent potential at the interstitial positions are labeled. It is not possible to make a clear assessment of the accuracy of the Frensley-Kroemer *ansatz* because of the small but nonnegligible interaction of the two IFs in the unit cell; however, it does seem to

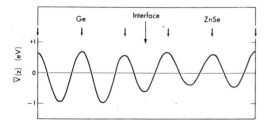

FIG. 2. Self-consistent potential $\overline{V}(z)$, averaged parallel to the interface, plotted perpendicular to the interface. The long arrow denotes the geometrical interface and the short arrows mark the positions of atomic planes.

be satisfied at least to within 0.5 eV.

The results of the self-consistent calculation give a valence-band discontinuity $\Delta E_v = 2.0 \pm 0.3$ eV, with this uncertainty reflecting possible inaccuracies arising from the use of local potentials as well as that due to interaction of the IFs. In the interpretation of experiments the "electron affinity rule" value[16] of 1.90 eV is usually assumed, since no attempt at measurement has been made. The model of Frensley and Kroemer[15,17] gives $\Delta E_v = 1.84$ eV, in reasonable agreement with our result. The model of Harrison[18] gives the significantly smaller value of $\Delta E_v = 1.46$ eV. An experimental determination of ΔE_v for this IF would provide a test for the various theories.

B. Interface states

One of the most fundamental aspects in understanding the electronic structure of an IF is the spectrum of IF states. In Fig. 4 we show the (110) projected band structures (PBS) of Ge and ZnSe, with relative alignment determined[3] from the self-consistent potential. True IF states, with a charge density which decays rapidly away from the IF, are allowed only in mutual gaps of the PBSs. Resonances which have an enhanced density near the IF may occur within either or both of the PBSs.

We find no IF states in the thermal gap, in agreement with the experimental consensus.[16] The localized states in the valence-band region are shown in Fig. 4 in relation to the PBSs. Clearly

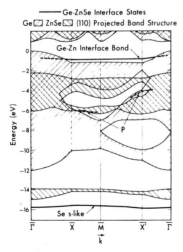

FIG. 4. Spectrum of interface states in (110) Ge–ZnSe, in relation to the projected band structures of Ge and ZnSe. Resonances have not been included in this graph, but are discussed in the text.

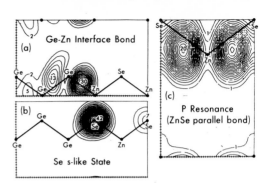

FIG. 5. Contour plots of the charge density of the interface states shown in Fig. 4. (a) and (b) are plotted in planes perpendicular to the interface while (c) is plotted in the ZnSe atomic layer adjacent to the interface. At least 90% of the charge of each of these states is confined to the plane which is shown, and the charge of each state is normalized to unity in the unit cell.

identifiable as true IF states are (a) a Se-derived s-like band at ~ −16 eV, well separated from the bulk Se s-like band throughout the Brillouin zone (BZ), and (b) a Ge-Zn p-like bonding state at the IF near −0.8 eV. Both of these bands are rather flat, showing ≳0.5 eV dispersion throughout the BZ. In addition there is a well-localized Zn-Se bonding state parallel to the IF, denoted P in Fig. 4. The charge densities of these states are shown in Fig. 5.

Before proceeding to the discussion of these states, it should be noted that the (5-5) geometry imposes some limitations on the identification of IF states and resonances. As mentioned above, the potential midway between IFs, in our case only $2\frac{1}{2}$ layers from the geometrical IF, has not quite become bulklike. One result is that the bulk band edges are not well defined, with an estimated uncertainty of ~0.3 eV in this calculation. Thus, although the position of the P "resonance" is shown in Fig. 4 to lie within one of the PBSs, its proximity to the band edges and degree of localization near the IF suggest that, for a single isolated IF, P is a true IF state.

A more quantitative estimate of the limitations imposed by the superlattice geometry is possible. Since the unit cell contains two IFs, well-defined IF states and resonances appear in (nearly degenerate) pairs. If the IFs are sufficiently close, these states overlap, introducing an energy splitting which gives a measure of the interaction of the IFs. For the very localized Se s-like states, the splitting varied from 0.02 to 0.04 eV over the BZ and is negligible. Likewise, the Zn-Se bonding state P is confined almost entirely to the first ZnSe atomic layer, and the splitting is ≳ 0.02 eV

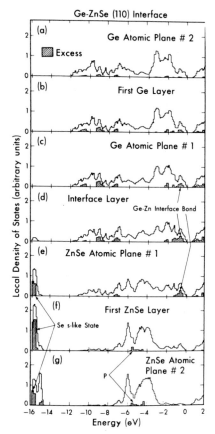

FIG. 6. Local density of states near and at the (110) Ge-ZnSe interface. The "excess" denotes localized states. The notation and interpretation is given in the text.

at $\overline{X}' = (0, \frac{1}{2})$ (in reduced units) and at $\vec{k} = (\frac{1}{2}, \frac{1}{4})$, the two points where it was clearly identifiable. The much larger splitting of 0.4–0.6 eV of the Ge-Zn bonding state is due to this state being directed perpendicular to the IF and decaying rather slowly into the Ge. This state is well-formed only because it is sufficiently split off from the bulk spectrum.

The local density of states (LDOS) in various layers near the IF is shown in Fig. 6. The "interface layer," "first Ge layer," etc., are as shown in Fig. 1. "Ge atomic plane No. 1," etc., denotes a slab also one layer thick but centered at the atomic plane rather than between atomic layers. The "excess" denotes the amount by which the LDOS exceeds that of Ge and of ZnSe, and which must represent states localized in that layer. The Se s-like state and the Zn-Se p state are localized on the ZnSe side of the IF, while the Ge-Zn bond-

ing state is confined near the geometrical IF. Also apparent are resonances on the Ge side of the IF at −11 to −7 eV (below the ZnSe p bands) and at −2.5 to −1 eV (above the ZnSe p bands). Examination of the charge density of these states indicate that they arise from bulk Ge states which peak near the IF before decaying quickly into the ZnSe gaps, similarly to the "metal-induced gap states"[19] found in Schottky-barrier studies. The density of localized states is large enough to be studied experimentally by photoemission spectroscopy, as has been done for the Ga-GaAs IF,[20] if abrupt IFs can be prepared.

It is important to recognize that the states "localized" at an IF are in general completely bandlike in the plane parallel to the IF and are localized only in the dimension perpendicular to the IF. The physics of this kind of localization is not yet well understood[13]; however, consideration of this one-dimensional localization can help understand the dispersion of the IF states. The Se s-like state (respectively, Ge-Zn bonding state) is strongly confined to only the CDCD (ABAB) bonding chain in the (two-dimensional) IF unit cell, and thus is very weakly coupled to the same state in the next unit cell. This accounts for the very small dispersion of these states (see Fig. 4), and to a good approximation they can be considered to be localized in three dimensions, similarly to the higher atomic core states in a crystal. The Zn-Se bonding state, on the other hand, is strongly coupled to the neighboring unit cells in the chain direction, resulting in ~2-eV dispersion. This state is truly localized in one direction only.

The Se s-like state can be considered as resulting from a bulk Se s-state responding to the more strongly attractive potential of Ge (relative to Zn) and lowering its energy. The Ge-Zn IF bond can be regarded variously as (a) bulk Ge p responding to the weaker attractive potential of Zn (relative to Ge), (b) a bulk Zn p state responding to the weaker attractive potential of Ge (relative to Se), or, more symmetrically, (c) the Ge and Zn dangling-bond surface states near the top of the gap overlapping and forming a bonding state. Each of these viewpoints is qualitatively in accord with a Ge-Zn bonding state near the bottom of the fundamental gap.

These two states would have been expected by analogy with previous calculations[3,7] on the Ge-GaAs (110) IF. In addition a Zn s-like state and a Ge-Se bonding state would be expected, as the analogs of strongly localized states found in Ge-GaAs. We have found no evidence for either of these states. It is unlikely that the interaction between IFs in this calculation (discussed above) could account for the disruption of the otherwise well-localized states that are anticipated. It is more likely that these anticipated states are reduced to (perhaps weak) resonances due to the rather poor mutual overlap of the "stomach" gaps (in the range −7 to −4 eV) in Ge-ZnSe (the overlap is essentially perfect in Ge-GaAs), but the details are not understood at present.

C. Charge density

In Fig. 7 we show contour plots of the valence pseudocharge density in the planes containing the two bonding chains perpendicular to the IF. The charge density in the Ge-Zn bond peaks at a significantly lower value than occurs in the Ge-Ge or Zn-Se bond; in the Ge-Se bond the contours are elongated in the Ge direction, indicating more charge than in a Zn-Se bond. In spite of the marked ionic character of the Zn-Se bonds it is not difficult to divide the unit cell into volumes containing the various bonding regions, integrate the charge density, and thus assign "bond charges." It should be emphasized that these bond charges in total contain all the charge in the unit cell and are not *directly* related to various "bond charge models" introduced previously in semiconductor physics; for example, with the definition used here, all bulk tetrahedral semiconductors would have a bond charge of exactly two electrons.

The result of this division of charge is that, with the exception of the Ge-Zn and Ge-Se bonds, each bond in the unit cell contains 2.000 ±0.005 electrons. The Ge-Zn (Ge-Se) bond contains 1.54 (2.46) electrons. In bulk, the Zn, Ge, and Se atoms contribute (0.50, 1.00, and 1.50) electrons to each of their four bonds in the tetrahedral struc-

ture, and evidently this occurs in each Ge-Ge and Zn-Se bond up to the IF. Without any charge transfer, the Ge-Zn and Ge-Se bonds would be then expected to contain 1.50 and 2.50 electrons. Thus 0.04 electrons are transferred from the Ge-Se bond to the Ge-Zn bond, resulting in saturated bonds (i.e., all bonding states below a well-defined gap are occupied).

These observations on bond characteristics have implications for the atomic geometry at the polar IFs as well as for the nonpolar (110) IF. Baraff et al.[8] have given general arguments showing that unreconstructed low-index polar IFs cannot be semiconducting. Their argument was "global" in nature, using the fact that a unit cell with an odd number of electrons must have partially occupied bands, and hence be metallic. Our results give a "local" picture, with the metallic behavior arising from unsaturated (or oversaturated) bonds. The (110) IF is semiconducting because charge transfer can result in saturated bonds. Although both the Ge-Zn and Ge-Se bonds at this IF are saturated, it is likely that the latter, with almost one more electron, is considerably "stronger" than the former. This suggests a contraction of the Ge-Se bond with a concomitant elongation of the Ge-Zn bond but such an effect would be difficult to detect experimentally. This effect was suggested originally[3] for the Ge-GaAs (110) IF.

III. CONCLUSIONS

The results and interpretation of a self-consistent calculation of the electronic structure of the (110) Ge-ZnSe IF have been presented in Sec. II. Localized states are found to occur at this IF, as was the case in similar calculations on (110) Ge-GaAs, but a comparison of these two IFs, as well as (110) AlAs-GaAs, will be presented elsewhere.[13]

There are two principal results of this work. First, the density of localized states at the IF is large enough to be experimentally measurable by photoemission spectroscopy, if abrupt IFs can indeed be fabricated. In principle, angle-resolved photoemission could be used to verify the specific character of the IF states, although this would be a more difficult experiment. Second, the character of the bonds at the IF indicate that there is little charge transfer parallel to the IF.[21] The bond charges themselves suggest relaxation, with the Ge-Se bond contracting while the Ge-Zn bond stretches. Simple considerations indicate that this relaxation would result in the IF states becoming more highly localized.

(110) Ge-ZnSe
Valence Charge Density

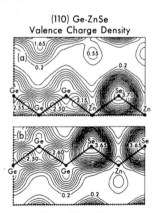

FIG. 7. Contour plots of the valence charge density in the two bonding planes perpendicular to the (110) Ge-ZnSe interface. The charge density is normalized to unity in the unit cell. The relative maxima in the charge density are quoted to the nearest 0.05 units.

ACKNOWLEDGMENT

One of the authors (W.E.P) would like to acknowledge financial support from IBM.

*Present address: Dept. of Physics, Northwestern University, Evanston, Ill. 60201.

[1] R. Dingle, Crit. Rev. Solid State Sci. 5, 585 (1975).

[2] L. Esaki and L. L. Chang, Crit. Rev. Solid State Sci. 6, 195 (1976).

[3] W. E. Pickett, S. G. Louie, and M. L. Cohen, Phys. Rev. B 17, 815 (1977).

[4] E. Caruthers and P. J. Lin-Chung, Phys. Rev. Lett. 38, 1543 (1977).

[5] E. Caruthers and P. J. Lin-Chung, J. Vac. Sci. Technol. (to be published).

[6] J. N. Schulman and T. C. McGill, in Ref. 5.

[7] W. E. Pickett, S. G. Louie, and M. L. Cohen, Phys. Rev. Lett. 39, 109 (1977).

[8] G. A. Baraff, J. A. Appelbaum, and D. R. Hamann, Phys. Rev. Lett. 38, 237 (1977); J. Vac. Sci. Technol. 14, 999 (1977).

[9] E. Louis, Solid State Commun. 24, 849 (1977).

[10] F. Herman and R. V. Kasowski, in Ref. 5.

[11] R. S. Bauer and J. C. McMenamin, in Ref. 5.

[12] R. W. Grant, J. R. Waldrop, and E. A. Kraut, in Ref. 5.

[13] W. E. Pickett and M. L. Cohen, in Ref. 5.

[14] J. R. Chelikowsky and M. L. Cohen, Phys. Rev. B 13, 826 (1976).

[15] W. R. Frensley and H. Kroemer, J. Vac. Sci. Technol. 13, 810 (1976).

[16] A. G. Milnes and D. L. Feucht, *Heterojunctions and Metal-Semiconductor Junctions* (Academic, New York, 1972).

[17] W. R. Frensley and H. Kroemer, Phys. Rev. B 16, 2642 (1977).

[18] W. A. Harrison, J. Vac. Sci. Technol. 14, 1016 (1977).

[19] S. G. Louie and M. L. Cohen, Phys. Rev. B 13, 2461 (1976); S. G. Louie, J. R. Chelikowsky, and M. L. Cohen, J. Vac. Sci. Technol. 13, 790 (1976).

[20] R. Z. Bachrach, in Ref. 5.

[21] We have recently found that the charge transfer at the Ge-GaAs interface, reported in Refs. 3 and 7, is in error. In fact, only ~0.02 electrons are transferred from the Ge-As bond to the Ge-Ga bond at that interface.

PHYSICAL REVIEW B   VOLUME 35, NUMBER 15   15 MAY 1987-II

Theoretical study of band offsets at semiconductor interfaces

Chris G. Van de Walle*

Stanford Electronics Laboratories, Stanford, California 94305
and Xerox Corporation, Palo Alto Research Center, 3333 Coyote Hill Road, Palo Alto, California 94304

Richard M. Martin

Xerox Corporation, Palo Alto Research Center, 3333 Coyote Hill Road, Palo Alto, California 94304
(Received 11 August 1986)

We present a first-principles approach to deriving the relative energies of valence and conduction bands at semiconductor interfaces, along with a model which permits a simple interpretation of these band offsets. Self-consistent density-functional calculations, using *ab initio* nonlocal pseudopotentials, allow us to derive the minimum-energy structure and band offsets for specific interfaces. Here we report results for a large number of lattice-matched interfaces, which are in reasonable agreement with reported experimental values. In addition, our systematic analysis leads to the important conclusions that, for the cases considered, the offsets are independent of interface orientation and obey the transitivity rule, to within the accuracy of our calculations. These are necessary conditions for the offsets to be expressible as differences between quantities which are intrinsic to each of the materials. Based on the information obtained from the full interface calculations, we have developed a new and simple approach to derive such intrinsic band offsets. We define a reference energy for each material as the average (pseudo)potential in a "model solid," in which the charge density is constructed as a superposition of neutral (pseudo)atomic densities. This reference depends on the density of each type of atom and the detailed form of the atomic charge density, which must be chosen consistently for the different materials. The bulk band structures of the two semiconductors are then aligned according to these average potential positions. For many cases, these model lineups yield results close to those obtained from full self-consistent interface calculations. We discuss the comparison with experiments and with other model theories.

I. INTRODUCTION

It has become technologically possible to grow high-quality epitaxial interfaces between two different semiconductors, using techniques such as molecular-beam epitaxy. The most important parameters characterizing such heterojunctions are the valence- and conduction-band discontinuities. These discontinuities can form a barrier for carrier transport across the interface; the knowledge of these quantities is therefore essential for calculating the transport properties of the interface, or the electrostatic potential in a heterojunction device. Examples of such novel semiconductor structures include quantum-well lasers, high-speed modulation-doped field-effect transistors, and superlattice photodetectors. Measured experimental values for band lineups are not well established yet, even though considerable progress has been made in growth and analysis techniques. In this paper, we will present a theoretical approach to deriving the band offsets.

Let us suppose we know the band structures of the semiconductor bulk materials A and B. We now want to figure out what the band structure looks like around an interface A/B. It is only in a very narrow region around the junction that the potential will be changed from its shape in the respective bulk materials, as we will show. Band bending caused by space-charge layers occurs on a length scale that is much larger than the atomic distances

over which the band offsets occur; therefore, the bands can be considered to be flat on this scale, except for the sharp discontinuity at the interface. We are then confronted with the problem of how to line up these bulk bands with respect to one another, which amounts to determining the lineup of electrostatic potentials. This type of information cannot be obtained from regular bulk calculations alone. For an infinite solid, no absolute energy reference is provided by the calculations (i.e., no "vacuum zero" is present to which other energies could be referred).[1] Therefore one cannot compare separate calculations on different solids. The fundamental reason for this is the long range of the Coulomb interaction: the charge distribution at a surface or an interface will determine the position of the energy levels deep in the bulk.

A number of model theories[2-6] have been developed which attempt to predict the lineups from information on the bulk alone; they necessarily rely on certain assumptions to establish an absolute energy scale, to which values for different materials can be referred. The electron affinity rule[2] assumed that the energy difference between the conduction band and the vacuum level, as measured at a surface, would be fixed, and derived conduction-band discontinuities in this fashion. Frensley and Kroemer[3] attempted to identify a reference level in each semiconductor that would correspond to the vacuum level. Harrison's theory of natural band lineups[4] established an absolute energy scale by referring everything to energy eigenvalues of the free atom. A very different approach has been developed by Tejedor and Flores,[5] and more recently, by Tersoff.[6] Their model is based upon simple screening arguments to define a "neutrality level" for each semiconductor, which will be aligned when an interface is formed.

All of these model theories rely on information about the bulk alone, and do not provide a complete description of the electron distribution at the interface. The only way to obtain a full picture of this effect is to perform a calculation in which the electrons are allowed to adjust to the specific environment created by the interface. This can be accomplished by performing self-consistent calculations, which will correctly describe the electrostatic potential shift that determines the lineups. Density-functional theory provides a fundamental theoretical framework to address this problem, and has the advantage that one can use the same methods which have been applied to a wide variety of solid-state problems.[7] Pickett et al.[8] and Kunc and Martin[9] have performed calculations which followed this approach; however, they used empirical pseudopotentials instead of the more recent ab initio pseudopotentials, and they only studied a small number of interfaces.

In this paper, we will carry out a systematic study of the band offset problem for a large number of heterojunctions; preliminary results for some of these systems have been reported elsewhere.[10] Our calculations are performed on a superlattice geometry, and based on local-density-functional theory,[11] applied in the momentum space formalism,[12,13] and using nonlocal norm-conserving pseudopotentials.[14] From the self-consistent potentials we obtain information about potential shifts at the interface. Combining this with bulk band-structure calculations will allow us to derive values for valence- and conduction-band discontinuities. Spin-orbit splitting effects in the valence bands are added in a posteriori. We also need to address the "band-gap problem,"[15-17] and examine to what extent the local-density approximation (LDA) is able to produce a reliable description of these heterojunction systems. Our discussion will indicate that for the semiconductors studied here the lineup of the bands should not be greatly modified by the known corrections to the local-density approximation.

Self-consistent calculations such as those performed in the present study provide a way to take all the effects of the electronic structure of the interface into account. This also implies that the results do not immediately tell us what physical mechanisms are dominant in determining the lineups. It is therefore important to systematically analyze a large number of interfaces, which will allow us to extract some general features of the lineup mechanism. In particular, we study the dependence of the lineups on interface orientation, and also examine to what extent the lineup mechanism can be considered to be linear. Linearity can be tested by checking whether transitivity is obeyed; it implies that the lineups can be obtained as a difference between quantities which are intrinsic to each semiconductor.

Based upon the information obtained from the full self-consistent calculations, we have developed a simple model to derive the lineups. We divide the problem into one part which can be expressed as the difference between quantities which are intrinsic to each of the materials, and another which involves corrections due to the detailed electronic charge density at the interface. To define appropriate intrinsic quantities, we choose to describe the bulk solids by a superposition of neutral atoms. The average potential in such a "model solid" can be found on an absolute scale from atomic calculations, and is not influenced by boundary effects. At the junction between two model solids, a shift in the average potentials occurs, which we take as the reference with respect to which any additional dipole corrections will be measured. The bulk band structures of the two materials are then aligned according to these average potential positions. A short description of this model was given elsewhere.[18] For non-polar interfaces, the model lineups yield results close to those obtained from full self-consistent interface calculations, and to reported experimental values. This indicates that for these interfaces the additional dipole contributions are small. Furthermore, these lineups are independent of interface orientation and obey the transitivity rule, corresponding to what was found from the ab initio calculations.

We have applied our methods to both lattice-matched and strained-layer interfaces between pairs of group-IV elements and III-IV and II-VI compound semiconductors. Interfaces between materials which are lattice mismatched are receiving considerable attention nowadays; the strains which are present in such strained-layer structures have important effects on the electronic structure.[19] We have performed extensive calculations for such systems, in particular for the Si/Ge interface, the results of which have been reported elsewhere.[18,20,21] In this paper we will concentrate upon lattice-matched systems. In the next section, we will describe the self-consistent calculations, and illustrate them with the example of a GaAs/AlAs interface. In Sec. III, we will give an overview of the broad range of lattice-matched systems that we studied, and derive some important and general conclusions. Section IV contains a description of the model solid approach that allows us to determine the lineups in a simpler way. We present a comparison with other theories and with experiment in Sec. V. Section VI concludes the paper.

II. SELF-CONSISTENT CALCULATIONS

A. Derivation of band lineups

In this paper, we will be reporting results for lattice-matched interfaces. We consider two semiconductors to be matched if the difference in lattice constant is less than 0.5%. We then fix the materials to have the same lattice constant in the interface calculation; the values we have used are listed in Table I. The geometry we use for the interfaces in this study is an ideal structure, in which the zinc-blende (or diamond) structure is continued throughout the system, with an abrupt change in the type of material right at the junction, and no displacements of the atoms from their ideal positions. We have performed density-functional total energy calculations for representative cases (GaAs/AlAs, and closely related checks[21] on

TABLE I. Lattice constant a, spin-orbit splitting (Ref. 25) and configuration (Ref. 35) used in atomic calculations for selected semiconductors.

Semiconductor	a (Å)	Spin-orbit splitting (eV)	Configuration			
Si	5.43	0.04		$s^{1.46}p^{2.54}$		
Ge	5.65	0.30		$s^{1.44}p^{2.56}$		
AlAs	5.65	0.28	Al,	$s^{1.11}p^{1.89}$;	As,	$s^{1.75}p^{3.25}$
AlP	5.43	0.04	Al,	$s^{1.11}p^{1.89}$;	P,	$s^{1.75}p^{3.25}$
AlSb	6.08	0.65	Al,	$s^{1.11}p^{1.89}$;	Sb,	$s^{1.75}p^{3.25}$
GaAs	5.65	0.34	Ga,	$s^{1.23}p^{1.77}$;	As,	$s^{1.75}p^{3.25}$
GaP	5.43	0.08	Ga,	$s^{1.23}p^{1.77}$;	P,	$s^{1.75}p^{3.25}$
GaSb	6.08	0.82	Ga,	$s^{1.23}p^{1.77}$;	Sb,	$s^{1.75}p^{3.25}$
InAs	6.08	0.38	In,	$s^{1.38}p^{1.62}$;	As,	$s^{1.75}p^{3.25}$
ZnSe	5.65	0.43	Zn,	$s^{1.02}p^{0.98}$;	Se,	$s^{1.86}p^{4.14}$

Si/Ge) to examine the validity of this assumption. We found that the ideal structure is very close to the minimum-energy configuration, with very small forces acting on the atoms. More importantly, we studied what effect the small displacements that might occur (on the order of 0.05 Å) would have on the band offsets. For nonpolar interfaces as studied here we found the effects to be negligible. However, we should note that displacements of charged atoms at polar interfaces would be expected to produce dipole shifts, as was indeed found in the studies in Ref. 9.

A major problem that has to be faced in calculating the electronic structure of an interface is the loss of translational symmetry, which is essential for using a reciprocal

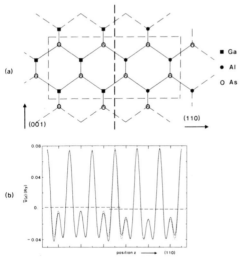

FIG. 1. (a) Schematic representation of a GaAs/AlAs (110) interface. The supercell used in the interface calculations is indicated in dotted lines; it contains 12 atoms and 2 identical interfaces. (b) Variation of the $l=1$ component of the total potential $\bar{V}(z)$ [as defined in Eq. (1)] across the (110) interface. The dashed lines represent the corresponding potentials for the bulk materials. These coincide with $\bar{V}(z)$ in the regions far from the interfaces. However, the average levels of the two bulk potentials (dashed horizontal lines) are shifted with respect to one another.

space formulation of the problem. The actual calculations are therefore performed on a superlattice, consisting of slabs of the respective semiconductors in a particular orientation. A typical (110) interface between two semiconductors, GaAs and AlAs, is sketched in Fig. 1(a). We also indicate a supercell appropriate for calculating the properties of this interface; it contains 12 atoms and 2 identical interfaces. Of course, what we emphasize here are the results for an isolated interface. These can be derived from our calculations to the extent that the interfaces in the periodic structure are well separated. We will establish a posteriori that this is the case, by examining charge densities and potentials in the intermediate regions, and showing them to be bulklike.

The self-consistent calculations are performed within the framework of local-density-functional theory,[11] applied in the momentum space formalism.[12,13] We use nonlocal, norm-conserving, ab initio pseudopotentials;[14] this term indicates that these potentials are generated using only theoretical calculations on atoms, without introducing any type of fitting to experimental band structures or other properties. All elements are therefore treated in the same way, which is particularly important when we want to include different materials in the same calculation, as for an interface. This is not true for the empirical pseudopotentials which have been used in previous interface calculations.[8,9] For Zn, the pseudopotential includes the $3d$ electrons as part of the core. We obtain self-consistent solutions for the charge density and the total potential, which is the sum of ionic, Hartree, and exchange-correlation potentials. The latter is calculated using the Ceperley-Alder form.[22] The first cycle requires a trial potential, a possible choice for which is the ionic potential screened by the dielectric function of a free-electron gas. An even better choice in many instances is the potential corresponding to a superposition of free-atom charge densities. Convergence of the self-consistent iterations is obtained with the help of the Broyden scheme.[23]

We include plane waves with kinetic energy up to 6 Ry in the expansion of the wave functions (corresponding to more than 650 plane waves in some cases). A set of 4 special points was used for sampling k space.[24] We will show later that these choices are sufficient for deriving the quantities we are interested in here. In the final self-consistent solution, a redistribution of electrons occurs in the interface region. The resulting self-consistent potential across the supercell is plotted in Fig. 1(b), for the example of GaAs/AlAs. Because the *ab initio* pseudopotentials are nonlocal, the total potential consists of different parts corresponding to different angular momenta l. We only show the $l = 1$ part of the potential here; this is the most important one in determining the lineup of the *p*-like valence bands. In the plot, the variation of the space coordinate **r** is limited to the component perpendicular to the interface, and values of the potentials are averaged over the remaining two coordinates, i.e., in the plane parallel to the interface:

$$\bar{V}(z) = [1/(Na^2)] \int \int V(\mathbf{r}) dx \, dy \ . \tag{1}$$

In the regions far from the interface, the crystal should recover properties of the bulk. Therefore we also plot (broken lines) the potentials determined separately from calculations on bulk GaAs and AlAs. One sees that already one layer away from the interface the potential assumes the form of the bulk potential. Similar results hold for the charge density. This confirms, *a posteriori*, that the two interfaces in our supercell are sufficiently far apart to be decoupled, at least as far as charge densities and potentials are concerned. The average levels of the potentials which correspond to the bulk regions are also indicated in Fig. 1(b). We denote these average levels by $\bar{V}_{\text{GaAs}}$ and $\bar{V}_{\text{AlAs}}$, and define the shift $\Delta \bar{V} = \bar{V}_{\text{GaAs}} - \bar{V}_{\text{AlAs}}$.

To get information about band discontinuities, we still have to perform the band calculations for the bulk materials. These were carried out with a 12-Ry cutoff; tests have shown that the choice of this cutoff is not critical for deriving the valence-band lineups. We find that the valence-band maximum in GaAs is 9.60 eV above the average potential $\bar{V}_{\text{GaAs}}$. In AlAs, the valence band occurs at 9.29 eV above $\bar{V}_{\text{AlAs}}$. From Fig. 1(b), we find $\Delta \bar{V} = 0.035$ eV. Figure 2 illustrates the resulting band lineups; we find a discontinuity in the valence band of $\Delta E_v = 0.34$ eV (upward step in going from AlAs to GaAs). We did not include spin-orbit splitting in our density-functional calculations. These effects can be added in *a posteriori*, by using experimental values for spin-orbit splittings.[25] For GaAs/AlAs, this brings the discontinuity to a value of $\Delta E_v = 0.37$ eV.

B. Accuracy

We estimate the inaccuracy of our calculations to be on the order of 0.05–0.10 eV. We have checked the convergence with respect to energy cutoff by increasing the cutoff to 9 Ry; this caused a change in $\Delta \bar{V}$ of less than 0.03 eV, in the direction of shifting ΔE_v towards higher values. We have also performed test calculations, using a local potential, to determine whether the interfaces in our supercell are sufficiently well separated. Increasing the number of atoms in the supercell to 16 resulted in a negligible change (less than 0.02 eV) in $\Delta \bar{V}$, thus confirming that a cell with 12 atoms suffices for our purposes. We also checked how good an assumption it is to put the atoms in the ideal structure. We calculated the forces on the atoms, for a GaAs/AlAs(110) interface, with 12

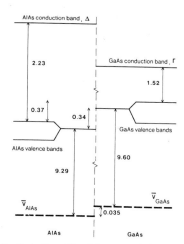

FIG. 2. Derivation of band lineups: relative position of the average potentials $\bar{V}_{\text{AlAs}}$ and $\bar{V}_{\text{GaAs}}$, and of the AlAs and GaAs valence and conduction bands. All values shown are derived with the $l = 1$ angular momentum component chosen as the reference potential; the band lineups, however, are unique and independent of this choice. Valence-band splittings due to spin-orbit splitting are indicated separately. Experimental band gaps were used to derive conduction-band positions.

atoms in the unit cell; they turned out to be smaller than 0.03 mdyne. This would lead to changes in the atomic positions smaller than 0.03 Å. We have checked that displacements of this size have a negligible effect on the band lineups. All this confirms that our choice of parameters allows us to obtain a numerical accuracy in deriving the lineups of 0.05–0.10 eV. Similar results were found for the Si/Ge interface.[21]

It is also appropriate to consider what effects the use of the local-density approximation (LDA) has upon the accuracy of our results. It is well known that the LDA severely underestimates the magnitudes of band gaps in semiconductors. More generally, the positions of the bulk bands with respect to the reference potential $\bar{V}$ can be subject to significant corrections, which can only be obtained by going beyond the LDA. This has been the subject of extensive recent theoretical investigations.[15–17] Precise information about the required corrections to the LDA for all semiconductors is not yet available at this time. Such corrections would need to be taken into account in the derivation of ΔE_v and ΔE_c. We expect, though, that for many of the systems that we studied the value of ΔE_v will not be significantly affected. From our comparison of LDA eigenvalues with experimental band structures, and from theoretical analysis,[17] there is evidence that the corrections needed to bring the conduction bands into agreement with experiment are fairly uniform for all conduction-band points (except for the Γ point, which, however, bears little relation to the conduction band as a whole, and has little weight in the Brillouin zone). As long as these corrections to the LDA are similar for the two materials on either side of the interface, the *relative* positions of valence and *representative* conduction bands are still given reliably by our lineup scheme. This seems to be true for most of the materials in Table I. The discrepancies tend to be larger between narrow-gap and wide-gap semiconductors (such as Ge and ZnSe), in which case somewhat larger errors (up to 0.3 eV) may occur.

In addition, the change in reference potential $\Delta \bar{V}$ contains long-range electrostatic dipole terms. Since these are given strictly in terms of the ground-state charge density, they would be correctly given by the exact density functional. However, the LDA may introduce errors, which one would expect to affect the interface dipole if the errors are different on the two sides of the interface. We have argued[21] that our results for $\Delta \bar{V}$ for Si/Ge are not greatly affected because the LDA errors are similar in the materials. Thus, just as in the previous paragraph, we conclude that corrections to the dipole terms should be small for interfaces between similar materials, such as most of the cases studied here, but may be larger effects for interfaces between more dissimilar materials.

In terms of deriving values for ΔE_c, we are confronted with the problem that many of the materials we are studying are direct gap semiconductors. The conduction band at Γ is not representative for the conduction-band structure as a whole, and may show large discrepancies; it is also quite sensitive to the energy cutoff, and to the inclusion of relativistic effects.[26] Because of these uncertainties in the gap at Γ, we will use experimental information about band gaps[25] to include conduction bands into the picture. We thus report our *ab initio* results for valence-band offsets, and find the conduction-band lineup by subtracting the valence-band discontinuity from the ex-

perimental band-gap difference. For the GaAs/AlAs interface this leads to $\Delta E_c = 0.34$ eV (higher in AlAs, with the lowest conduction band in AlAs situated at Δ).

III. RESULTS FROM SELF-CONSISTENT CALCULATIONS

A. Overview of results for lattice-matched interfaces

We have studied a variety of lattice-matched (110) interfaces, the results for which are given in Table II. In all cases, the convention is used that a positive value for the discontinuity at a junction A/B corresponds to an upward step in going from A to B. For interfaces between a group-IV element and a III-V compound, the (110) orientation is the only one which avoids charge accumulation without the need for mixing at the interface.[27] Our values have been adjusted to include spin-orbit splitting, the experimental values for which are listed in Table I. The correction to ΔE_v due to spin-orbit splitting is typically smaller than 0.05 eV. The only case in which it is really sizable is InAs/GaSb, where it increases ΔE_v by 0.15 eV. For GaSb and AlSb, there is some uncertainty in the value of the spin-orbit splitting. The result $\Delta E_v = 0.38$ eV in Table II was derived using the spin-orbit values from Table I. If the spin-orbit splittings in these two materials were equal, the value of ΔE_v would be 0.32 eV.

The column "empirical pseudopotentials" in Table II contains values derived by performing self-consistent density-functional calculations very similar to ours, but with empirical pseudopotentials.[8] We notice a significant

TABLE II. Heterojunction band lineups for lattice-matched (110) interfaces, obtained by self-consistent interface calculations (SCIC), and by the model solid approach. Other theoretical and experimental results are listed for comparison.

Heterojunction	SCIC	Model solid	Empirical pseudopotential[a]	ΔE_v (eV) Harrison[b] "Natural"	"Pinned"	LMTO[c]	Tersoff theory[d]	Experiment
AlAs/Ge	1.05	1.19		0.70	0.78	0.73	0.87	0.95[e]
GaAs/Ge	0.63	0.59	0.35	0.67	0.66	0.24	0.32	0.56[f]
AlAs/GaAs	0.37	0.60	0.25	0.03	0.12	0.49	0.55	0.55[g]
AlP/Si	1.03	1.16		0.87	0.79	0.93	0.91	
GaP/Si	0.61	0.45		0.86	0.69	0.75	0.45	0.80[h]
AlP/GaP	0.36	0.70		0.01	0.10	0.18	0.46	
ZnSe/GaAs	1.59	1.48	2.0±0.3	1.42	1.35	1.75	1.20	1.10[i]
ZnSe/Ge	2.17	2.07	2.0±0.3	2.09	2.01	1.99	1.52	1.52[i]
InAs/GaSb	0.38	0.58		0.72	0.42	0.36	0.43	0.51[j]
AlSb/GaSb	0.38	0.49		0.09	0.18	0.17	0.38	0.45[k]

[a]Reference 8.
[b]Reference 48.
[c]Reference 49.
[d]J. Tersoff, J. Vac. Sci. Technol. B **4**, 1066 (1986).
[e]M. K. Kelly, D. W. Niles, E. Colavita, G. Margaritondo, and M. Henzler (unpublished); quoted in G. Margaritondo, Phys. Rev. B **31**, 2526 (1985).
[f]J. R. Waldrop, E. A. Kraut, S. P. Kowalczyk, and R. W. Grant, Surf. Sci. **132**, 513 (1983).
[g]Reference 39.
[h]P. Perfetti, F. Patella, F. Sette, C. Quaresima, A. Capasso, A. Savoia, and G. Margaritondo, Phys. Rev. B **30**, 4533 (1984).
[i]S. P. Kowalczyk, E. A. Kraut, J. R. Waldrop, and R. W. Grant, J. Vac. Sci. Technol. **21**, 482 (1982).
[j]Reference 43.
[k]J. Menéndez and A. Pinczuk (private communication).

difference with our values, due to our use of *ab initio* pseudopotentials, compared to their empirical pseudopotentials (fitted to reproduce experimental band structures). If we use those same pseudopotentials in our calculations, we reproduce their result (within the numerical accuracy of 0.05 eV). This indicates that the essential difference is in the choice of the pseudopotential—the *ab initio* pseudopotential providing a better justified starting point. We will discuss the other entries in the table after we have presented the model solid approach.

B. Dependence on interface orientation

For the GaAs/AlAs system, we have also studied other interface orientations. In particular, for the (100) interface we find a valence-band discontinuity of 0.37 eV, the same as the value for the (110) interface. For (111), we find $\Delta E_v = 0.39$ eV. This indicates that ΔE_v does not depend on interface orientation, a result that was also found experimentally.[28] Let us note that this is not necessarily valid for pseudomorphic strained-layer systems, in which different strains associated with different interfaces can have sizable effects on the lineups, as discussed in Refs. 20 and 21. It also has been shown that rearrangements of atoms at polar interfaces can change the offsets.[9] Within such limitations, we believe that the result that the offset is orientation independent can be considered an important general result for suitably chosen lattice-matched interfaces.

C. Pressure dependence of the lineups

We have also performed self-consistent interface calculations for GaAs/AlAs interfaces under hydrostatic pressure. Two groups[29] have performed photoluminescence experiments on GaAs/Ga$_{1-x}$Al$_x$As heterojunctions, in order to vary the band offsets and to use this information to determine their magnitudes at zero pressure. In the interpretation of the experimental results, it was assumed that ΔE_v remains constant under pressure. It is appropriate to examine the validity of that assumption. Since the bulk moduli of the two materials are very similar (784 kbar for GaAs, and 733 kbar for AlAs), it is safe to assume that the only effect of hydrostatic pressure will be to decrease the lattice constant of the overall system, according to the relation:

$$P = -B \, \Delta V/V = -3B \, \Delta a/a \, ,$$

where P is the pressure, B is the bulk modulus, V is the volume, and a is the lattice constant. We have therefore performed interface calculations at four different lattice constants, ranging from 5.65 to 5.50 Å, as well as the cor-

responding bulk calculations for the compressed solids. We found that

$$\Delta E_v = \Delta E_v^0 - 0.64 \, \Delta V/V$$
$$= \Delta E_v^0 + 0.82 \times 10^{-3} P \, ,$$

where ΔE_v^0 is the valence-band discontinuity at zero pressure, and P is expressed in kbar. This is to be compared, for instance, with the change in the GaAs direct band gap under pressure, which we calculate to be

$$\Delta E_g = \Delta E_g^0 - 8.33 \, \Delta V/V$$
$$= \Delta E_g^0 + 10.6 \times 10^{-3} P \, .$$

We see that the change in ΔE_v is more than an order of magnitude smaller than the change in the gap; a pressure change of 10 kbar will increase the gap by 0.1 eV, but only change ΔE_v by less than 0.01 eV.

D. Transitivity

It is interesting to examine our results to establish the extent to which theory supports the proposition that the band offsets for any pair of semiconductors can be expressed as a difference of numbers *intrinsic* to each material. This has been observed from experiment,[30,31] and is an implicit assumption in theories such as Refs. 2–6. It is clear that our full interface calculations do not assume linearity, i.e., we do *not* postulate that our heterojunction lineups be given by the difference of two numbers which would each be characteristic of a particular semiconductor, independent of which heterojunction it is used in. *A posteriori*, however, we can check how close our results are to linearity, by examining transitivity, i.e., whether the following equation is satisfied:

$$\Delta E_v(A,B) + \Delta E_v(B,C) = \Delta E_v(A,C) \qquad (2)$$

where

$$\Delta E_v(A,B) = E_v(B) - E_v(A) \, .$$

In Table III, we list values for these quantities, which allow us to conclude that the transitivity rule [Eq. (2)] is satisfied to better than 0.06 eV, which is on the order of the numerical accuracy of the calculations. It is interesting to note that transitivity also holds for strained-layer interfaces, taking the appropriate strains into account to construct pseudomorphic interfaces. We have checked this for Si/Ge/GaAs (results for Si/Ge and Si/GaAs were reported in Ref. 18) where Eq. (2) turned out to be satisfied to within 0.01 eV. The fact that transitivity is satisfied shows that the deviations from linearity are small. Together with the orientation independence, we believe that

TABLE III. Examination of transitivity [Eq. (2)] for various sets of systems. ΔE_v values are from Table II. The values in the last two columns are equal to within the numerical accuracy of the calculations, showing that transitivity is satisfied.

A	B	C	$\Delta E_v(A,B)$	$\Delta E_v(B,C)$	$\Delta E_v(A,B) + \Delta E_v(B,C)$	$\Delta E_v(A,C)$
AlAs	GaAs	Ge	0.37	0.63	1.00	1.05
AlP	GaP	Si	0.36	0.61	0.97	1.03
ZnSe	GaAs	Ge	1.59	0.63	2.22	2.17

this is indicative of the *intrinsic* nature of the band offsets for large classes of lattice-matched systems.

Our general conclusions regarding orientation independence and linearity indicate that in principle it is possible to derive the lineups by determining a reference level for each semiconductor, and lining up the band structures according to these reference levels. In the next section, we will describe how we define an appropriate level for each material.

IV. DEFINITION OF A REFERENCE MODEL SOLID

As we already pointed out, a pure bulk calculation cannot provide information about absolute energy positions. An absolute energy scale only enters into the problem if one does not deal with an infinite solid, but instead the crystal is terminated—i.e., by a surface. A particular choice of reference surface must be made, which will then allow us to express all energies with respect to the vacuum level. Our choice for terminating the solid should correspond as closely as possible to the situation at an interface; this immediately excludes using the structure of a real surface, which might involve complicated relaxation and reconstruction. Also, we do not want to perform a complete self-consistent calculation for a surface—since that would be computationally even harder than an interface calculation. We have therefore developed a model theory, which allows us to calculate the reference energy for a particular choice of reference surface. The model corresponds to a superposition of atomic charged densities, which is known to give reasonable results for a number of bulk properties. Mattheiss, for instance, used it to study energy bands of transition metals.[32] Here it turns out to be particularly suited to the derivation of semiconductor interface properties. The model will be used *only* to find a value for the average electrostatic potential (on an absolute scale) for each semiconductor. The positions of the bands with respect to this average potential are still obtained from self-consistent calculations for the bulk crystals, as was described in the last paragraph of Sec. II A. Within this model we can thus line up the band structures for different crystals without the need for a self-consistent interface calculation of the type described in Sec. II A.

We construct the model solid by taking a superposition of neutral atomic spheres. The potential outside each such sphere goes exponentially to (an absolute) zero; this will be the zero of energy for the model solid. When we use such neutral, spherical objects to construct a semi-infinite solid, the presence of a surface will not induce any shift in the average potential, since no dipole layers can be set up. This feature of the model was also stressed in earlier work that used the overlapping spherical atomic charge-density approximation, for instance to calculate work functions.[33] This also means that the potential shift between two solids will only depend on "bulk" properties, and not on the specific arrangement of atoms at the interface.

One has to check, of course, that such a model solid can adequately represent the real crystal. This is not difficult to imagine in the case of elemental semiconductors, but

somewhat harder to understand for materials in which the bonds have more of an ionic character, such as the III-V or even the II-VI compounds. Apart from the *a posteriori* justification that the obtained results are quite good, we can also rely on information obtained from pseudopotential[34] or tight-binding[35] calculations on bulk materials. Examination of the distribution of electrons in the bonds shows that the number of electrons around each atom is approximately equal to its nuclear charge, i.e., one can still talk about "neutral spheres."

Full information about the atomic potential can be obtained by performing an atomic calculation (of the Herman-Skillman type). Since all our calculations for the solid are based on pseudopotentials, we actually perform the atomic calculations on the "pseudoatom." The choice of pseudopotential for this purpose is arbitrary, so long as the same ionic potential is used throughout the calculations. We now must find the average potential in the model solid, which is a superposition of atomic charge densities. The total potential is the sum of ionic, Hartree, and exchange and correlation potentials:

$$V^l = V^{\text{ion},l} + V^H + V^{\text{xc}} . \tag{3}$$

The superscript l on $V^{\text{ion},l}$ reflects the fact that we are working with nonlocal pseudopotentials.[14] The choice of angular momentum component does not influence the final results, so long as we consistently use the same angular momentum component of the pseudopotential as our reference. The first two terms in (3) are linear in the charge density, and can therefore also be expressed as a superposition of atomic potentials. Their average value in the solid is

$$\bar{V}^{\text{ion},l} + \bar{V}^H = \sum_i (1/\Omega) \int (V_i^{\text{ion},l} + V_i^H) d\tau , \tag{4}$$

where Ω denotes the volume of the unit cell, and the index i runs over all atoms in the unit cell. Convergence is no problem in the numerical integration, since for each neutral atom the long-range part of the ionic potential (which is the same for each l) is canceled by the Hartree potential. The exchange and correlation potential V^{xc} is not linear in the charge density, and can therefore not be expressed as a superposition of atomic potentials. This contribution, however, is local in nature and does not depend upon the specific way in which we terminate the solid. It can easily be calculated for a bulk solid, and added in afterwards.

We illustrate the procedure with the example of an AlAs/GaAs interface. To perform the atomic calculations, we have to choose a configuration, i.e., the occupation x and y of the s and p orbitals: $s^x p^y$ (the d character of the bonds is small in the semiconductors studied here). Naturally, we want this choice to be as close as possible to the configuration that an atom would have in the solid. It is not easy to extract this type of information from pseudopotential calculations on the bulk crystal. Since angular momentum is not a good quantum number in the solid, there is no straightforward way to distinguish between s or p character of wave functions. We therefore extract these values from tight-binding calculations,[35] in which the choice of basis set provides a natural separation between s and p states. We used $s^{1.23}p^{1.77}$ for Ga, $s^{1.11}p^{1.89}$ for Al, and $s^{1.75}p^{3.25}$ for As. The atomic charge density does not vary much when the configuration is changed; still, the average potentials tend to be rather sensitive to the choice of configuration. Although a change in configuration causes only small shifts in the long-range tail of the wave function, these changes at large r values may have a significant effect on the average. For Ga, going from an sp^2 to an s^2p configuration shifts the average potential up by 0.72 eV; the variation is close to linear. An analogous change for Al introduces a shift of 0.70 eV. This indicates that the uncertainties become less severe when we look at potential *differences*. For example, the lineup in the GaAs/AlAs system will be determined by the difference in average potential between Ga and Al; it is to be expected that the configurations will be similar for these atoms in GaAs and AlAs. Making the *same* change in the configuration on both sides will have no effect on the potential difference. We establish as our convention that atomic configurations will be used which are obtained from tight-bonding theory[35] for all systems that we study.

We then carry out the atomic calculations on the pseudoatom in the configuration $s^x p^y$, and obtain the charge density and potentials. Next, we proceed with the superposition scheme. Figure 3 shows the shape of the charge density for an AlAs/GaAs(110) interface between two model solids. For plotting purposes, we have averaged the charge density in planes parallel to the interface in a fashion similar to the potential in Eq. (1). Note that, within the model, there is a certain amount of "spillover" between the charge densities in the region near the interface, with tails of the wave function from AlAs extending into the GaAs and vice versa. This reflects the fact that we do not model the interface by a discontinuous charge density, but rather a smooth variation over a region of atomic dimensions, which is expected to closely mimic the situation at a real interface. The main difference between

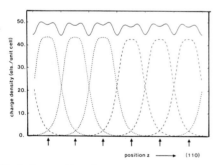

FIG. 3. Superposition of atomic charge densities to form a model solid interface. We show the plane-averaged charge density for pairs of GaAs (on the left, dotted lines) and AlAs atoms (on the right, dashed lines) in (110) planes. The units are (electrons/unit cell), for a supercell with 48 electrons. The arrows indicate the positions of the atomic planes. The solid line represents the superposition, which corresponds to the charge density in the model solid. Notice that the model solid is *not* cut off abruptly at the plane of the interface.

the model solid and the self-consistent charge density is that in the real solid some charge is drawn away from the regions near the atoms and piled up in the bonds. The qualitative aspect of the charge distribution near the interface is fairly well represented by the model solid, however.

Next, we evaluate the integral in Eq. (4). From that equation, it also follows that the average ionic and Hartree potentials are proportional to Ω^{-1}. Using the values of the volume of the unit cell in AlAs and GaAs, and summing over the two atoms in the bulk unit cell, we can derive the average potentials. Choosing the $l=1$ angular momentum component, as before, this leads to the following values of $(\overline{V}^{\text{ion},l=1}+\overline{V}^H)$: -7.82 eV for GaAs, and -8.08 for AlAs. The exchange and correlation contributions to the average potentials are $\overline{V}^{xc}_{\text{GaAs}}=\overline{V}^{xc}_{\text{AlAs}}=-8.71$ eV. Finally, we add up the contributions for the individual materials, and find the shift in the total potential on either side of the interface: $\overline{V}_{\text{GaAs}}-\overline{V}_{\text{AlAs}}=(-16.53$ eV$)-(-16.79$ eV$)=0.26$ eV. This is to be compared with the value obtained from the full self-consistent calculations on the interface, using the supercell technique: $\Delta\overline{V}=0.03$ eV. The deviation here is actually larger than it will be in most other cases. Once we know $\Delta\overline{V}$, we can line up the band structures of materials, which are obtained from self-consistent bulk calculations. These band structures are significantly more accurate than those which would correspond to a model solid of superimposed atomic charge densities. We will assume that they are referred to the average electrostatic potential that we calculated for the model solid. Since the charge density of the model solid is not quite the same as that in the real bulk, the corresponding average potential can only be an approximation to the actual quantity. The model solid, however, enables us to obtain this average electrostatic potential on an absolute scale, and we will see by examining the results that the approximation is a good one. We have specified above the conventions which are used in deriving such a value; it is uniquely defined by the choice of pseudopotential, local-density approximation, and atomic configurations.

We have studied a variety of other lattice-matched (110) interfaces. The configurations[35] that were used in the free-atom calculations are listed in Table I, and the results for ΔE_v (including spin-orbit splitting) are given in Table II. For lattice-matched systems, the model solid approach will yield the same value for the band alignment, irrespective of the interface orientation. This corresponds to what we found above from the self-consistent interface calculations on GaAs/AlAs. Table II only contains results for the (110) orientation. For interfaces between a group-IV element and a III-V or II-VI compound, or between compounds which do not have any elements in common, the (110) orientation is the only one which avoids charge accumulation without the need for mixing at the interface.[27] It has been shown[9] that for polar interfaces different types of mixing can lead to different dipoles at the interface, which significantly alter the band lineups. This effect cannot be described by the present model solid approach, in which the neutral spheres cannot generate any net dipole across the interface, and it is clearly beyond the scope of *any* theory[2–6] which assumes the dipole to be fixed by consideration of the bulk alone. Other limitations of the model solid approach will be discussed in the next section.

V. DISCUSSION

A. Comparison with experiment

In Table II, we also list experimental data from various sources. At the present time, not all of these values are equally reliable. A striking example is that of the GaAs/AlAs interface, for which "Dingle's 85/15 rule"[36] had become widely accepted: $\Delta E_v = 0.15\Delta E_g$, where ΔE_g is the difference in direct band gaps. Since last year, however, this value has been challenged and new results now indicate that more than 30% of the discontinuity is in the valence band.[37-42] This example shows that even for this most widely used heterojunction the correct value could only be established by performing many experiments on high-quality interfaces, using a variety of different techniques. Since most of the heterojunctions listed in Table II have not received such careful attention, one should be very cautious when referring to these reported valence-band discontinuities.

We will attempt to give a brief overview of the experimental techniques which, at present, we regard to be the most reliable ones for deriving the band offsets, and illustrate them by references to work on GaAs/AlAs. Photoluminescence experiments on quantum wells can give very accurate results, but should be limited to cases in which the band offsets can be derived without having to rely on the precise knowledge of additional quantities, such as effective masses or exciton binding energies. Structures in which a crossover of bands can be observed are most appropriate, e.g., in the AlGaAs/AlAs heterojunctions as a function of composition[38] or pressure.[29] I-V and C-V measurements may or may not be reliable, depending on the system and the procedure used. The reason is that heterojunctions often contain charges at or near the interface, which may cause significant band bending. One should therefore either eliminate these charges,[39] or use a measurement procedure that is insensitive to these effects, such as C-V profiling through the junction.[40] A promising new approach is that of charge-transfer measurements at single heterojunctions[41] or in modulation-doped superlattices.[42] Finally, we have noticed that photoemission spectroscopy, while in principle providing a direct measurement of the valence-band discontinuity, has produced widely varying results by different groups for the same system. A possible reason is the technological difficulty involved in producing high-quality epitaxial interfaces. Measurements on lower-quality heterojunctions can lead to a ΔE_v value which is not representative of an ideal system. For a more detailed evaluation of current experimental techniques, we refer to the critical review by Duggan.[37]

For GaAs/AlAs, our model solid result is very close to the present experimental value; closer, indeed, than the self-consistent calculation. Another very interesting case is that of InAs/GaSb, in which experimentally a "broken-gap lineup" was detected,[43] meaning that the conduction band in InAs is lower in energy than the

valence band in GaSb. From the self-consistent interface calculations, we find that $\Delta E_v = 0.38$ eV. The band gap of InAs is 0.41 eV at 0 K, and 0.35 eV at room temperature,[25] which means that the conduction band of InAs almost lines up with the valence band of GaSb. This result is close to that obtained from an earlier self-consistent calculation on a InAs/GaSb(100) interface.[44] The model solid result for ΔE_v is 0.51 eV, which even more clearly leads to a "broken gap" lineup.

B. Comparison with other model theories

In Table II, we have also given results from a number of other models. We should point out that these numbers [for Harrison's theory, the linear muffin-tin orbitals (LMTO) calculations, and Tersoff's approach] do not include spin-orbit splitting. However, as we have remarked before, these corrections are usually smaller than 0.05 eV. We will briefly describe these models here, and point out similarities and differences with our model solid approach. We will discuss the electron affinity rule,[2] the Frensley-Kroemer theory,[3] Harrison's theory of natural band lineups,[4] and the model developed by Tejedor and Flores,[5] and independently by Tersoff.[6] We will also devote some attention to a line-up scheme that occurs naturally in the context of LMTO calculations.[45]

Our model solid approach is in spirit related to the electron affinity rule,[2] in that it derives the band discontinuities as a difference between quantities which are defined for each semiconductor individually. In the case of electron affinities, the problem is that these quantities are measured experimentally for a specific surface, and therefore depend on orientation, relaxation, reconstruction, etc., which can all introduce extra dipoles that shift the energy bands in the bulk. One could try to define an "electron affinity" which would only take the "intrinsic" contribution due to the bulk into account, and ignore the surface effects. Van Vechten[46] has argued that such quantities would predict the lineups reliably. The main problem with such an approach is that the separation between the bulk and the "surface" part is not unique, so that it is not clear how to derive an "intrinsic electron affinity" from experimental information alone.

Our model solid approach defines a reference level corresponding to a well specified "model surface," which, by construction, cannot introduce any "extra dipoles." The reference potential can therefore in principle be considered to be a quantity intrinsic to the bulk material (its actual value being determined by the conventions regarding pseudopotentials and configurations that we specified above). Furthermore, the fact that the results are so close to those from self-consistent calculations shows that this model provides a good description of the charge distribution at the heterojunction. Since we do not allow any *additional* charge rearrangement, we should always expect some deviation between the model solid results and the full self-consistent calculations. The comparison of our results for valence-band offsets shows that these deviations are fairly small, however.

Additional dipoles may be due to several different sources. Displacements of atoms around the interface may in certain cases set up dipoles which shift the energy levels. We have argued that at (110) interfaces the deviations from the ideal structure will be small, and will have small effects on the lineups. In some cases, like ZnSe/Ge, however, the sizable difference in ionicity may introduce more significant displacements, and consequently larger dipoles. Also, we do not consider the model to be applicable to polar interfaces. As explained above, additional dipoles which depend on the type of mixing at the interface can occur in such cases.

Small deviations from our model may also be caused by the fact that we are using *neutral* atoms as our building blocks. Use of charged objects, however, would destroy the simple, "dipole-free" picture of the reference surface that our superposition of neutral atoms provides. Frensley and Kroemer have actually constructed a model in which they superimposed spherical ions to construct the solid.[3] They chose the mean interstitial potential in the diamond or zinc-blende structure as the electrostatic reference potential for each semiconductor. If the crystal were viewed as a superposition of spherical charges, this reference potential would correspond to the vacuum potential, provided the charges were so localized that the charge density in the interstitial region was negligible. These potentials were then lined up, taking a dipole shift into account, which was expressed in terms of charges on the atoms, and subsequently in terms of electronegativity differences. It turned out that these dipole shifts were quite small in most cases, indicating that the intrinsic lineups should be close to the true result. We came to the same conclusion in the present work, using a better justified value for the intrinsic potential.

Frensley and Kroemer used empirical pseudopotentials to generate values for the reference potentials. To really test how good this procedure is, one should use the better quality pseudopotentials which are available nowadays, as we have done in our studies. Since we had the results from bulk calculations at our disposal, we could examine the potential values in the interstitial regions. It turned out that the values we obtained (without the dipole correction) were quite different from Frensley and Kroemer's original results, and also different from the results from self-consistent interface calculations (by more than 0.25 eV, on the average). We assumed that the qualitative result that the additional dipole shifts are small remains valid, such that these corrections would not significantly affect the lineups; in any case, adding the dipole shifts suggested by Frensley and Kroemer made the agreement with our values even worse. Inspection of the potential in the interstitial region showed us why the results would not be reliable. We found that the potential does not really flatten out near the interstitial site, as Frensley and Kroemer assumed, and still shows significant structure. This is true both for elemental and compound semiconductors. Under these circumstances, it is hard to determine what the appropriate value for the reference potential is; is it the value at the interstitial point itself, or an average over some region? This can make a difference of up to 1 eV. Frensley and Kroemer themselves acknowledged that their electrostatic potential inside the interstices of the diamond structure was only flat within about 1 eV.[47] To

make things worse, it turned out that the value of the potential at the interstitial point was only converged at a much higher cutoff than we needed for the other aspects of our calculations (e.g., larger than 18 Ry for Si). This would require one to do the bulk calculation with a much higher accuracy than is typically required for deriving energy eigenvalues. This is to be expected, if one insists on deriving an accurate value at one point, instead of dealing with averaged quantities, or properties which depend only on the total charge density. We therefore conclude that the Frensley-Kroemer scheme *in principle* offers a very attractive approach, but turns out to be unsuitable for generating accurate values for the lineups.

Our approach has in common with Harrison's theory of natural band lineups[4] that a reference energy level for each material is derived from atomic information. However, a key difference should be emphasized: Harrison's model was based on the atomic term values, which he assumed to carry over from atom to solid. This is clearly different from our model solid approach, in which all electronic energy levels are shifted by the superposition of atomic potentials. This choice to define the average potential of the model solid is better justified by self-consistent calculations and seems to be in better agreement with experiment. Harrison has recently developed a new point of view, which is closely related to Tersoff's approach, but applied in a tight-binding context.[48] Here, the averaged hybrid energy is identified as the "neutrality level," that will be pinned at a heterojunction. Values from both the "natural band lineup" scheme and the new "pinned" theory are given in Table II. It is not clear to what extent this new method actually improves the natural band lineups. It might be that the tight-binding formalism, while very appealing because of its simplicity and generality, is not accurate enough to predict values on a scale that is necessary for these applications.

Another scheme which establishes an absolute energy reference level for tetrahedral semiconductors has been developed in the context of the LMTO all-electron method, by Vergés *et al.*,[45] and Christensen and Andersen.[49] They interested in deriving absolute deformation potentials for individual semiconductors, but their approach can also be used to line up band structures of pairs of different semiconductors. Their reference level is the zero of electrostatic potential in the infinite crystal, as evaluated with the atomic-spheres approximation, i.e., with point charges placed at the atomic and tetrahedral interstitial sites. This turns out to be a reasonable *ansatz* for deriving the band lineups. Values for ΔE_v obtained by this approach are listed in Table II.

The last source of additional dipoles to be discussed here is due to screening effects, of the type that play the dominant role at a metal-metal interface. Such dipoles are clearly not present in our model solid approach, which is therefore not applicable to metallic interfaces or Schottky barriers. Tersoff,[6,50,51] and before him Tejedor and Flores,[5] have argued that such screening will also be the dominant effect that determines the lineups at a semiconductor-semiconductor interface: dipoles will be set up which will drive the system towards alignment of the "neutrality levels" of the materials (as would be the case at the junction between two metals, where the Fermi levels line up). Although this picture seems to be contrary to the assumptions that underlie the model solid approach, the two points of view may actually be not that far removed from one another. We have remarked before that the superposition of atomic charge densities effectively deals with a model surface for which there is significant overlap of the charge densities, with the tails of the bulk charge density of one material sticking out into the other side. This charge distribution may actually incorporate much of the dipole that Tejedor and Flores and Tersoff consider to be the dominant effect.

It is essential to point out, in this context, that the concept of "dipole" at an interface is not uniquely defined—its magnitude depends on the choice of "reference surfaces" that are brought together to create an interface within a specific model. It is therefore possible for different models to obtain good results, while claiming to deal with dipoles of very different magnitude. The reference surfaces that we have chosen here are clearly a good "ansatz:" the "additional dipoles," due to charge redistribution at the interface are small. Tejedor and Flores and Tersoff do not need to make an ansatz for the charge density since they consider only the final lineup that the system would attain *if* the screening effects were as strong as they are at a metal/metal interface. We can recognize two problems with this approach. First, assuming that a unique neutrality level exists, no convincing evidence has been given so far that the induced dipoles are actually strong enough to drive the system towards the "neutral" lineup. Second, the assignment of a neutrality level to each material is not straightforward. Tersoff has suggested two possibilities: a simple average of the indirect gap,[50] or a branch point derived from a Green's-function approach;[6,51] neither of these is rigorous. The Green's-function approach itself involves a number of approximations and assumptions.[51] For instance, a specific choice of orientation has to be made in the Green's function $G(\mathbf{R},E)$, and only the $\langle 110 \rangle$ direction produced reasonable results. Also, the branch point energy E_b depends on the value of $\mathbf{R}$. Still, the success of the theory clearly depends on how accurately these numbers can be generated, and it may be somewhat fortuitous that the particular choice that was made produces values that are reasonably close to the self-consistent results.

From this overview, it should be clear that none of the model theories is able to adequately deal with *all* effects of electronic rearrangement at the interface. Our superposition of atomic charge densities model is the only one which is based upon, and has been directly compared with, results from self-consistent interface calculations. This places our approach on a strong footing, particularly since the values we obtain are so close to the self-consistent results.

VI. CONCLUSIONS

We have described our first-principles approach to deriving band offsets at semiconductor interfaces. Density-functional theory and *ab initio* pseudopotentials were used to perform self-consistent calculations, and derive valence-band discontinuities for a large number of lattice-matched interfaces. The calculations were illustrated with the example of a GaAs/AlAs interface, and we presented an analysis of the sensitivity of our results to the procedures used; the numerical precision is on the order of 0.05–0.10 eV. Values for valence-band discontinuities were summarized in Table II. We also listed reported experimental values in that table; we have pointed out which measurements we consider reliable, and have given a critical overview of the available experimental techniques.

Our systematic analysis has allowed us to draw some general conclusions: for suitably chosen heterojunction systems, the lineups are independent of orientation, and they obey the transitivity rule. This indicates that the offsets can be described by a linear theory, in which the lineups are obtained as the *difference* between quantities which are *intrinsic* to each material. To establish the reference levels for each solid, we constructed a model based on superposition of (pseudo)atomic charge densities. The average (pseudo)potential in such a model solid can be derived from atomic calculations; the atomic configurations were taken from tight-binding theory to simulate as well as possible the solid. This uniquely defines the reference potentials. The band structures obtained from self-consistent bulk calculations are then aligned according to these reference levels. The resulting lineups are close to those obtained from full self-consistent interface calculations, and to reliable experimental values. We consider this to be evidence that our *ansatz* is close to the real situation at the interface. Extra dipoles may be present in certain cases; however, in the cases studied these amount to only small corrections, and at this point we believe there is no simple universal theory that describes the exact screening mechanism.

We have also discussed the relationship with other model theories. In particular, we compared our approach to Tersoff's theory, which seems to produce good results, even though it relies on certain untested assumptions, and the prescription for finding the reference level is not rigorous. The advantages of our superposition of neutral atomic charge densities are that it provides a well defined, physical model, that the numerical work is straightforward, and that, even though it should only be considered as an *ansatz*, the results are close to those obtained from self-consistent interface calculations.

ACKNOWLEDGMENTS

We are indebted to K. Kunc, O. H. Nielsen, and R. J. Needs, whose developments we have used. Conversations with O. K. Andersen, M. Cardona, D. J. Chadi, W. A. Harrison, W. C. Herring, W. E. Spicer, J. Tersoff, and D. J. Wolford are gratefully acknowledged. We thank M. S. Hybertsen and S. G. Louie for providing us with a copy of their work prior to publication, and G. Schwartz, J. Menéndez, and A. Pinczuk for communicating their results on AlSb/GaSb interfaces. This work was partially supported by U.S. Office of Naval Research (ONR) Contract No. N00014-82-C0244.

*Present address: IBM Thomas J. Watson Research Center, Yorktown Heights, New York 10598.

[1] L. Kleinman, Phys. Rev. B **24**, 7412 (1981).

[2] R. L. Anderson, Solid-State Electron. **5**, 341 (1962).

[3] W. R. Frensley and H. Kroemer, Phys. Rev. B **16**, 2642 (1977).

[4] W. A. Harrison, *Electronic Structure and the Properties of Solids* (Freeman, San Francisco, 1980), p. 253.

[5] C. Tejedor and F. Flores, J. Phys. C **11**, L19 (1978); F. Flores and C. Tejedor, *ibid.* **12**, 731 (1979).

[6] J. Tersoff, Phys. Rev. B **30**, 4874 (1984).

[7] For a review, see R. M. Martin, in *Festkörperprobleme (Advances in Solid State Physics)*, edited by P. Grosse (Vieweg, Braunschweig, 1985), Vol. XXV, pp. 3–17.

[8] W. E. Pickett, S. G. Louie, and M. L. Cohen, Phys. Rev. B **17**, 815 (1978); J. Ihm and M. L. Cohen, *ibid.* **20**, 729 (1979); W. E. Pickett and M. L. Cohen, *ibid.* **18**, 939 (1978).

[9] K. Kunc and R. M. Martin, Phys. Rev. B **24**, 3445 (1981).

[10] C. G. Van de Walle and R. M. Martin, in *Computer-Based Microscopic Description of the Structure and Properties of Materials*, Materials Research Society Symposia Proceedings, edited by J. Broughton, W. Krakow, and S. T. Pantelides (Materials Research Society, Pittsburg, 1986), Vol. 63, p. 21.

[11] P. Hohenberg and W. Kohn, Phys. Rev. **136**, B864 (1964); W. Kohn and L. J. Sham, *ibid.* **140**, A1133 (1965).

[12] J. Ihm, A. Zunger, and M. L. Cohen, J. Phys. C **12**, 4409 (1979).

[13] O. H. Nielsen and R. M. Martin, Phys. Rev. B **32**, 3792 (1985).

[14] D. R. Hamann, M. Schlüter, and C. Chiang, Phys. Rev. Lett. **43**, 1494 (1979). Here we use the tabulated potentials given by G. B. Bachelet, D. R. Hamann, and M. Schlüter, Phys. Rev. B **26**, 4199 (1982).

[15] C. S. Wang and W. E. Pickett, Phys. Rev. Lett. **51**, 597 (1983).

[16] L. J. Sham and M. Schlüter, Phys. Rev. Lett. **51**, 1888 (1983); Phys. Rev. B **32**, 3883 (1985).

[17] M. S. Hybertsen and S. G. Louie, Phys. Rev. Lett. **55**, 1418 (1985); Phys. Rev. B **32**, 7005 (1985); **34**, 5390 (1986).

[18] C. G. Van de Walle and R. M. Martin, J. Vac. Sci. Technol. B **4**, 1055 (1986).

[19] G. C. Osbourn, J. Appl. Phys. **53**, 1586 (1982).

[20] C. G. Van de Walle and R. M. Martin, J. Vac. Sci. Technol. B **3**, 1256 (1985).

[21] C. G. Van de Walle and R. M. Martin, Phys. Rev. B **34**, 5621 (1986).

[22] D. M. Ceperley and B. J. Alder, Phys. Rev. Lett. **45**, 566 (1980); J. Perdew and A. Zunger, Phys. Rev. B **23**, 5048 (1981).

[23] P. Bendt and A. Zunger, Phys. Rev. B **26**, 3114 (1982).

[24] A. Baldereschi, Phys. Rev. B **7**, 5212 (1973); D. J. Chadi and M. L. Cohen, *ibid.* **8**, 5747 (1973); H. J. Monkhorst and J. D. Pack, *ibid.* **13**, 5188 (1976); A. H. MacDonald, *ibid.* **18**, 5897 (1978).

[25] *Landolt-Börnstein, Numerical Data and Functional Relationships in Science and Technology* (Springer, New York, 1982), Group III, Vol. 17 a-b.

[26] G. B. Bachelet and N. E. Christensen, Phys. Rev. B **31**, 879 (1985).

[27] W. A. Harrison, E. A. Kraut, J. R. Waldrop, and R. W. Grant, Phys. Rev. B **18**, 4402 (1978); R. M. Martin, J. Vac. Sci. Technol. **17**, 978 (1980).

[28] W. I. Wang, T. S. Kuan, E. E. Mendez, and L. Esaki, Phys. Rev. B **31**, 6890 (1985).

[29] D. J. Wolford, T. F. Kuech, J. A. Bradley, M. Jaros, M. A. Gell, and D. Ninno, Bull. Am. Phys. Soc. **31** (3), 557 (1986); J. Vac. Sci. Technol. B **4**, 1043 (1986); U. Venkateswaran, M. Chandrasekhar, H. R. Chandrasekhar, B. A. Vojak, F. A. Chambers, and J. M. Meese, Phys. Rev. B **33**, 8416 (1986).

[30] A. D. Katnani and G. Margaritondo, J. Appl. Phys. **54**, 2522 (1983); Phys. Rev. B **28**, 1944 (1983).

[31] A. D. Katnani and R. S. Bauer, Phys. Rev. B **33**, 1106 (1986).

[32] L. F. Mattheiss, Phys. Rev. **134**, 970 (1964).

[33] M. Weinert and R. E. Watson, Phys. Rev. B **29**, 3001 (1984), and references therein.

[34] O. H. Nielsen and R. M. Martin, in *Proceedings of the 17th International Conference on the Physics of Semiconductors*, edited by D. J. Chadi and W. A. Harrison (Springer-Verlag, New York, 1985), p. 1162.

[35] D. J. Chadi (private communication); the tight-binding method is described in Phys. Rev. B **19**, 2074 (1979).

[36] R. Dingle, in *Festkörperprobleme/Advances in Solid State Physics*, edited by H. J. Queisser (Vieweg, Braunschweig, 1975), Vol. 15, p. 21.

[37] G. Duggan, J. Vac. Sci. Technol. B **3**, 1224 (1985).

[38] P. Dawson, B. A. Wilson, C. W. Tu, and R. C. Miller, Appl. Phys. Lett. **48**, 541 (1986).

[39] J. Batey and S. L. Wright, J. Appl. Phys. **59**, 200 (1986).

[40] H. Kroemer, W.-Y. Chien, J. S. Harris, Jr., and D. D. Edwall, Appl. Phys. Lett. **36**, 295 (1980).

[41] W. I. Wang and F. Stern, J. Vac. Sci. Technol. B **3**, 1280 (1985).

[42] T. J. Drummond and I. J. Fritz, Appl. Phys. Lett. **47**, 284 (1985).

[43] J. Sakaki, L. L. Chang, R. Ludeke, C.-A. Chang, G. A. Sai-Halasz, and L. Esaki, Appl. Phys. Lett. **31**, 211 (1977); L. L. Chang and L. Esaki, Surf. Sci. **98**, 70 (1980).

[44] J. Ihm, P. K. Lam, and M. L. Cohen, Phys. Rev. B **20**, 4120 (1979).

[45] J. A. Vergés, D. Glötzel, M. Cardona, and O. K. Andersen, Phys. Status Solidi B **113**, 519 (1982).

[46] J. A. Van Vechten, J. Vac. Sci. Technol. B **3**, 1240 (1985).

[47] W. R. Frensley and H. Kroemer, J. Vac. Sci. Technol. **13**, 810 (1976).

[48] W. A. Harrison and J. Tersoff, J. Vac. Sci. Technol. B **4**, 1068 (1986).

[49] N. E. Christensen and O. K. Andersen (private communication).

[50] J. Tersoff, Phys. Rev. B **32**, 6968 (1985).

[51] J. Tersoff, Surf. Sci. **168**, 275 (1986).

PHYSICAL REVIEW B   VOLUME 35, NUMBER 18   15 JUNE 1987-II

Interface phenomena at semiconductor heterojunctions: Local-density valence-band offset in GaAs/AlAs

S. Massidda, B. I. Min,* and A. J. Freeman

Materials Research Center and Department of Physics and Astronomy, Northwestern University, Evanston, Illinois 60201

(Received 10 March 1987)

The valence-band offset ΔE_v at the lattice-matched GaAs/AlAs(001) interface is derived from highly precise self-consistent all-electron local-density band-structure calculations of the $(GaAs)_n(AlAs)_n(001)$ superlattices (with $n \leq 3$). We calculate ΔE_v by using the core levels —available uniquely from an all-electron approach—as reference energies. Since these are experimentally accessible quantities, a direct comparison with experiment is, in principle, possible. We find that $\Delta E_v = 0.5 \pm 0.05$ eV, in very good agreement with recent experimental results ($\Delta E_v = 0.45$–0.55 eV). Calculated core-level shifts are also compared to experiment. These results, which are closely related to changes in the charge-density distribution at the interface, contribute to understanding the underlying mechanism of the band discontinuity.

In a recent review, Bauer and Margaritondo[1(a)] have emphasized that understanding interface phenomena at semiconductor heterojunctions is essential for the design of novel devices. To this end, a precise knowledge of the band structure and especially of the band alignment at a semiconductor heterojunction—as probably the single most important property of the interface—is necessary. Particular attention has been devoted to the almost perfectly lattice-matched GaAs/AlAs heterojunction, both from the experimental and the theoretical points of view.[2] However, there is still no general agreement regarding the value of, and microscopic mechanism causing, the band discontinuities at this interface. The experimental valence-band offsets (ΔE_v), as measured by several techniques, range[2-5] from 0.19 to 0.65 eV. Until recently, a partitioning of the valence- and conduction-band gap contributions into a ratio,[5] $\Delta E_c : \Delta E_v = 85.15$, was universally accepted; the results of several recent experiments,[2] however, have indicated a larger $\Delta E_v \sim 0.45$–0.55 eV, and a ratio $\Delta E_c : \Delta E_v \sim 60 : 40$. Most of these results derive from extrapolation of the measurements at the $Al_x Ga_{1-x} As$/GaAs interface, with $0.2 \leq x \leq 0.6$.

Following Anderson's[6] early effort with an electron affinity rule, a few theoretical models[1,7-11] were proposed to calculate the valence-band offset at the interface of semiconductors. These models also fail to agree for the GaAs/AlAs interface: Harrison's[7] tight-binding approach gave too small a valence-band offset (~ 0.04 eV) [the difficulties of this approach with Al-containing compounds have been related[8] to the observed anomaly of the Al-X ($X = P$, As, Sb) bond lengths entering this model]; Frensley and Kroemer[9] first found $\Delta E_v = 0.26$ eV, and later, in a revised version, $\Delta E_v = 0.69$ eV; the model recently proposed by Tersoff[10] gives $\Delta E_v = 0.35$ eV; Van de Walle and Martin[11] constructed the model solid by superposing neutral atomic spheres to estimate $\Delta E_v = 0.60$ eV. A common feature of all these theoretical approaches is that they evaluate the offset by the alignment of certain "reference levels" which are characteristic of the bulk semiconductors. In the same spirit, the absolute energy

positions of the deep-level impurities have been proposed[12] as reference energies; this yields $\Delta E_v = 0.42$ eV. The empirical rule proposed by Bauer and Margaritondo[1(a)] and Katnani and Margaritondo[1(b)] gives 0.43 eV. The question has been raised if, indeed, the band offset can be determined by knowing only the properties of the separate bulk materials, i.e., without performing a calculation on the interface. The first published report of the AlAs/GaAs(110) valence-band offset based on a self-consistent study of the interface is the pioneering work of Pickett, Louie, and Cohen.[13] Using an empirical local-pseudopotential approach, they found $\Delta E_v = 0.25$ eV, when the band offset is calculated by the relative alignment of the average potential on the two sides of the interface. Recently, the self-consistent *ab initio* nonlocal-pseudopotential method was applied by Van de Walle and Martin[11] to various heterojunctions; they obtained $\Delta E_v = 0.37$ eV for AlAs/GaAs(110).[11]

In this Rapid Communication, we report a theoretical determination of the valence-band offset for the GaAs/AlAs interface based on the self-consistent all-electron energy-band-structure calculations for the $(GaAs)_n(AlAs)_n(001)$ ($n = 1, 2, 3$) superlattice. We use the core levels as reference energies to determine the relative alignment of the valence-band edges. We find a valence-band offset $\Delta E_v = 0.5 \pm 0.05$ eV, in very good agreement with the recent experimental results. Since our determination of the band offset relies on quantities that are—in principle—experimentally accessible [e.g., by x-ray photoemission spectroscopy (XPS)[3]] a direct comparison with experiment is possible, which is different from the reference level used in the pseudopotential calculation.

Since GaAs and AlAs have almost the same lattice constant (the experimental mismatch is about 0.1%),[14] we used the same experimental value in all our calculations. The effect of strain,[15] which may be expected to be small for this closely matched heterojunction, is neglected in first approximation in this work. We have performed self-consistent full-potential linearized augmented-plane-wave (FLAPW)[16] calculations on the two bulk semiconductors and on the $(GaAs)_n(AlAs)_n(001)$ superlattices, with $n \leq 3$. The superlattices have a tetragonal structure (space group D_{2d}^5), with lattice parameters $a = a_0/\sqrt{2}$ and $c = na_0$ and $4n$ atoms per unit cell (and a_0 is the zinc-blende lattice parameter).

Band energies were calculated semirelativistically, while the core states were treated fully relativistically and updated at each iteration. The Hedin-Lundqvist[17] form of the exchange-correlation potential was employed. The calculations on bulk GaAs and AlAs were performed by using ten special k points[18] in the irreducible wedge of the Brillouin zone, while for the $n = 2, 3$ superlattices we used three special k points[19] in the two-dimensional- (2D-) like wedge of the Brillouin zone. (Calculations with more than three special k points showed that sufficient precision was obtained.) Inside the muffin-tin spheres, angular momenta up to $l = 8$ are used in the expansion of the wave functions, and up to $l = 4$ for the charge density and potential. In the interstitial regions, the wave functions are expanded in terms of all the plane waves with wave vector $k \leq k_{max} = 2.48$ a.u. The resulting convergence, determined by the parameter $k_{max}\langle R_{MT}\rangle \sim 5.7$, where $\langle R_{MT}\rangle$ is the average muffin-tin radius, is sufficient to lead to stable band eigenvalues and charge densities.

We have performed six independent self-consistent calculations in order to test the stability of the results with respect to the parameters entering the calculations. Particular attention was devoted to the treatment of the "semicore" Ga $3d$ states. About 0.15 electrons spill out of each ($R_{MT} = 2.3$ a.u.) muffin-tin sphere. Since the potential profile across the interface is very sensitive to the correct distribution of the electronic charge, we describe the spillout-core charge density by using the overlapping-charge method. A less precise treatment, such as the uniform spreading of this charge in the interstitial regions, results in an artificial charge transfer between the two sides of the interface and a remarkable alteration of the band lineup. This result indicates that the band lineup depends critically on the charge transfer at the interface. We have used different sets of sphere radii and also treated the Ga $3d$ electrons as band states. The corresponding deviations in the ΔE_v values are ~ 0.01 and ~ 0.03 eV, respectively. The remarkable consistency of these results provides evidence for the precision of our calculations.

Figure 1 shows a diagram of the energy levels near the interface. The binding energies of the selected core levels relative to the top of the valence bands (E_{c1}^b and E_{c2}^b) are obtained from the self-consistent band structure of the bulk semiconductors. The superlattice calculation gives the binding-energy differences (ΔE_B) of the same core levels on the two sides of the interface, and finally, ΔE_v is calculated from

$$\Delta E_v = E_{c1}^b - E_{c2}^b - \Delta E_B . \tag{1}$$

This approach assumes that in a heterostructure the binding energies of the core levels, E_{c1}^b and E_{c2}^b with respect to the corresponding valence-band maxima, are equal to their values in the bulk compounds when atoms are far enough away from the interface. However, since one deals with a finite-thickness superlattice in any model calculation, the concept of "local band structure," which was

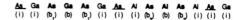

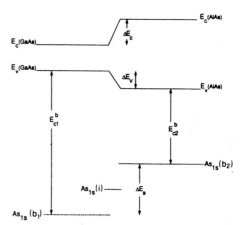

FIG. 1. Schematic diagram of the energy levels in the $(GaAs)_3(AlAs)_3$ superlattice showing various quantities described in the text.

implicit in the previous assumption, is now lost. We therefore need to make the further assumption that in the superlattice the value of ΔE_B is the same as in a real heterojunction. While the first assumption can be easily accepted, the second will be verified on the basis of our calculations.

To evaluate the band offset we have chosen the following representative core levels as the reference energies: As $1s$, As $3d_{5/2}$ and Ga $1s$, Al $1s$. As seen from Fig. 1, in the $(GaAs)_3(AlAs)_3$ superlattice two independent Ga (Al) sites exist, one (corresponding to two atoms) being "interface" [referred to as Ga(i) and Al(i)] and the other being "bulk" [Ga(b) and Al(b)]. On the other hand, three different As sites exist: (i) Two As atoms are on the GaAs side [As(b_1)], (ii) two As atoms are on the AlAs side [As (b_2)], and (iii) two As atoms are at the interface [As(i)] and share two bonds with Ga(i) and two bonds with Al(i).

The core-binding-energy differences ($E_{c1}^b - E_{c2}^b$) in the bulk compounds are given in Table I. A first remark on this data is related to the As core-level shifts in going from GaAs to AlAs: the change in binding energy of the As $3d$ states (~ 0.8 eV larger in GaAs) is consistent with its slightly lower ionicity compared to AlAs.[20,21] A direct

TABLE I. Core-energy differences and corresponding valence-band offset values ΔE_v in (i) the bulk compounds and (ii) the $(GaAs)_n(AlAs)_n$ superlattices with $n = 2, 3$. Energies are in eV.

		$n=2$		$n=3$	
	$E_{c1}^b - E_{c2}^b$	ΔE_B	ΔE_v	ΔE_B	ΔE_v
As $1s$	0.87	0.44	0.43	0.41	0.46
As $3d_{5/2}$	0.81	0.37	0.44	0.34	0.47

comparison with x-ray photoemission spectroscopy experiments is also possible; Ludeke, Ley, and Ploog[21] found an upward shift in the As $3d$ level of 0.6 eV going from GaAs to AlAs. The agreement of our result with their experiment (within the resolution of the measurement) supports the use of the local-density-approximation (LDA) core levels to calculate the valence-band offset.

Table I also lists the energy differences, ΔE_B, of the chosen core levels on the two sides of the interface for the $n=2,3$ superlattices, and the resulting ΔE_v values. If we use the As $1s$ and $3d_{5/2}$ levels we get a valence-band offset ~ 0.47 eV for the $n=3$ superlattice and $\Delta E_v \sim 0.44$ eV for the $n=2$ case. This difference shows that an $n=3$ superlattice is already thick enough to determine the band offset with good precision, and we can estimate the uncertainty due to the finite superlattice thickness to be of this magnitude. In this respect, a further test is provided by comparing the bulk and interface Ga (Al) core levels in $(GaAs)_3(AlAs)_3$. The Ga $1s$ and Al $1s$ core-energy difference is ~ 0.07 eV smaller for the bulk than for the interface Ga (Al) atoms. If we use these levels and the bulk Ga and Al atoms, however, we get a larger (0.07 eV) value for the band offset than that calculated by using the As levels. This difference can be attributed to a nonperfect cancellation of errors when different core levels are used. [For instance, a smaller $k_{max}=2.3$ cutoff gives different (~ 0.04 eV larger) Ga $1s$–Al $1s$ energy separations, but very stable values for the As $1s$ and As $3d_{5/2}$ energy differences.]

In order to correctly compare these calculated results with experiment, we need to first consider the effect of spin-orbit coupling. Its effect on ΔE_v can be expected to be small, since the top of the valence band is mainly As p-like in both GaAs and AlAs. We can now estimate the resulting corrections *a posteriori* using the known values of the spin-orbit splittings. The spin-orbit splitting Δ_0 shifts the top of valence bands by $\frac{1}{3}\Delta_0$, and we can therefore estimate the consequent change in ΔE_v to be one-third of the difference between the spin-orbit splittings in GaAs and AlAs. Using published values[22] gives a positive increase $=\frac{1}{3}(0.34-0.29)$ eV ~ 0.02 eV toward a *larger* offset. In the *worst* (highly conservative) case of adding this uncertainty to our earlier uncertainty from the difference between the $n=2$ and 3 results, our valence-band offset is 0.5 eV with an uncertainty ≤ 0.05 eV. Thus our result appears to be in very good agreement with the latest experimental results,[2] which give $\Delta E_v \sim 0.45$–0.55 eV.

The fact that the offest is already established after only a very few layers is consistent with the results of Pickett *et al.*[13] for the (110) GaAs/AlAs interface and of Van de Walle and Martin[15] for the Si/Ge interface. To further verify this conclusion, we have calculated the angular-momentum-decomposed charges Q_l inside the muffin-tin spheres; their values in the superlattice are compared with those in the bulk in Table II. The results show that (i) the Q_l values in the $n=3$ superlattice are very similar to those of the bulk compounds, and (ii) the two adjacent Ga (Al) atoms have (within our numerical precision) exactly the same Q_l values. Further, in order to prevent any numerical difference (such as different structure of k-point

TABLE II. l decomposition of the charge density inside muffin-tin spheres (radii are 2.3 a.u.).

	Q_s	Q_p	Q_d	Q_{tot}
	GaAs (bulk)			
Ga	0.84	0.74	0.09	1.69
As	1.35	1.88	0.04	3.28
	AlAs (bulk)			
Al	0.64	0.68	0.10	1.44
As	1.35	2.01	0.03	3.40
	$(GaAs)_3(AlAs)_3$			
Ga(b)	0.84	0.73	0.09	1.68
Ga(i)	0.84	0.73	0.09	1.68
Al(i)	0.65	0.66	0.10	1.43
Al(b)	0.65	0.66	0.10	1.43
As(b_1)	1.36	1.87	0.04	3.27
As(i)	1.35	1.94	0.04	3.33
As(b_2)	1.35	2.00	0.03	3.39

meshes) from affecting our conclusions, we performed self-consistent calculations on the $n=1$ superlattice, and on the two bulk compounds in the $n=1$ superlattice structure, D_{2d}^1. The l-decomposed integrated charge differences inside the Ga and Al muffin-tin spheres are smaller than 0.8%, while the As Q_l value is halfway be-

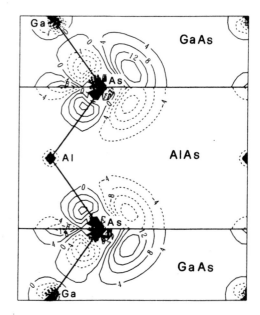

FIG. 2. Difference between the charge density of $(GaAs)_1(AlAs)_1$ and those of the bulk semiconductors. Contours are given in units of $10^{-4}e/a_0^3$, i.e., ~ 0.06 electrons per unit cell.

tween the GaAs and AlAs bulk values. In order to demonstrate the interface effects, we show in Fig. 2 contour plots of the difference between the charge density of the $(GaAs)_1(AlAs)_1$ superlattice and those of the pure compounds. These charge-density deformations, which give rise to the induced interface dipole moment at an abrupt interface, are seen to be quite small (note the scale in Fig. 2), and to fall off very rapidly away from the As interface atom. Surprisingly, we recover almost bulklike properties already in the first Ga and Al atoms away from the interface.

Finally, a question concerning the validity of our results could arise from the use of the local-density approximation. However, since we only use the LDA to derive the valence-band discontinuity, the well-known band-gap problem should not affect our results. Furthermore, although the energies of localized states such as the core states are usually poorly described by the LDA, we believe that relative energy differences are meaningful. In this context, let us look at the XPS measurements. Waldrop et al.[23] reported $\Delta E_v \sim 0.4$ eV for GaAs grown on AlAs(110) and $\Delta E_v \sim 0.15$ eV for the reverse sequence,

which raised the question of the commutativity of the offset. More recent XPS measurements,[3] however, gave a commutative $\Delta E_v = 0.38-0.39$ eV. Unfortunately, the absolute value of the offset in these experiments relies on an accurate knowledge of the binding energies of the core levels in the bulk semiconductors; a precise value of the Al $2p$ binding energy is, however, lacking.[3] Thus, comparison between our results and experiment (while agreeing within the uncertainty of the experiment) requires a more precise measurement of the core binding energy.[1,3]

In conclusion, using the first-principles FLAPW band-structure method, we have obtained the valence-band offset for the GaAs/AlAs interface. Using the core levels as reference energies produces a very good value compared to experiment and may turn out to be an important tool for predicting the band offset of semiconductor heterojunctions.

This work was supported by the National Science Foundation through the Northwestern University Materials Research Center Grant No. DMR 852028.

*Present address: Argonne National Laboratory, Argonne, IL 60439.

[1](a) R. S. Bauer and G. Margaritondo, Phys. Today 40 (No. 1), 27 (1987), and references therein; (b) A. D. Katnani and G. Margaritondo, Phys. Rev. B 28, 1944 (1983).

[2]For recent critical reviews, see H. Kroemer, J. Vac. Sci. Technol. B 2, 433 (1984); G. Duggan, ibid. 3, 1224 (1985). The most recent results, to our knowledge, are reported in A. D. Katnani and R. S. Bauer, Phys. Rev. B 33, 1106 (1986); B. A. Wilson, P. Dawson, C. W. Tu, and R. C. Miller, in Layered Structures and Epitaxy, Materials Research Symposium Proceedings, Vol. 56, edited by J. M. Gibson, G. S. Osbourn, and R. M. Tromp (Materials Research Society, Pittsburgh, PA, 1986), p. 307; Appl. Phys. Lett. 48, 541 (1986); J. Batey and S. L. Wright, J. Appl. Phys. 59, 200 (1985).

[3]Katnani and Bauer, Ref. 2.

[4]Wilson et al. and Batey and Wright, Ref. 2.

[5]R. Dingle, W. Wiegmann, and C. H. Henry, Phys. Rev. Lett. 33, 827 (1974).

[6]R. L. Anderson, Solid State Electron. 5, 341 (1962); see also, C. Mailhiot and C. B. Duke, Phys. Rev. B 33, 1118 (1986); J. L. Freeouf and J. M. Woodall, Surf. Sci. 168, 518 (1986).

[7]W. A. Harrison, J. Vac. Sci. Technol. 14, 1016 (1977). Recently, W. A. Harrison and J. Tersoff [J. Vac. Sci. Technol. B 4, 1068 (1986)] reported $\Delta E_v = 0.12$ eV, including the effect of screening. See also P. Vogl, H. P. Hjalmarson, and J. D. Dow, J. Phys. Chem. Solids 44, 365 (1983).

[8]S. Gonda, Solid State Commun. 60, 249 (1986).

[9]W. R. Frensley and H. Kroemer, J. Vac. Sci. Technol. 13, 810 (1976); Phys. Rev. B 16, 2642 (1977).

[10]J. Tersoff, Phys. Rev. B 30, 4874 (1984); Phys. Rev. Lett. 56, 2755 (1986).

[11]C. G. Van de Walle and R. M. Martin, J. Vac. Sci. Technol. B 4, 1055 (1986).

[12]J. M. Langer and H. Heinrich, Phys. Rev. Lett. 55, 1414 (1985); M. J. Caldas, A. Fazzio, and A. Zunger, Appl. Phys. Lett. 45, 671 (1984).

[13]W. E. Pickett, S. G. Louie, and M. L. Cohen, Phys. Rev. B 17, 815 (1978); see also a recent calculation by A. C. Ferraz and G. P. Srivastava, Semicond. Sci. Technol. 1, 169 (1986).

[14]Handbook of Chemistry and Physics, edited by R. W. Weast (Chemical Rubber Company, Boca Raton, FL, 1985).

[15]C. G. Van de Walle and R. M. Martin, Phys. Rev. B 34, 5621 (1986).

[16]H. J. F. Jansen and A. J. Freeman, Phys. Rev. B 30, 561 (1984).

[17]L. Hedin and B. I. Lundqvist, J. Phys. C 4, 2064 (1971).

[18]D. J. Chadi and M. L. Cohen, Phys. Rev. B 8, 5747 (1973).

[19]S. L. Cunningham, Phys. Rev. B 10, 4988 (1974).

[20]J. Ihm and J. D. Joannopoulos, Phys. Rev. 24, 4191 (1981).

[21]R. Ludeke, L. Ley, and K. Ploog, Solid State Commun. 28, 57 (1978).

[22]J. C. Phillips, Bonds and Bands in Semiconductors (Academic, New York, 1973), cf. Table 7.3.

[23]J. R. Waldrop et al., J. Vac. Sci. Technol. 19, 573 (1981).

A universal trend in the binding energies of deep impurities in semiconductors

M. J. Caldas, A. Fazzio, and Alex Zunger

Solar Energy Research Institute, Golden, Colorado 80401

(Received 7 June 1984; accepted for publication 28 June 1984)

Whereas the conventional practice of referring binding energies of deep donors and acceptors to the band edges of the host semiconductor does not produce transparent chemical trends when the same impurity is compared in different crystals, referring them to the vacuum level through the use of the photothreshold reveals a remarkable material invariance of the levels in III-V and II-VI semiconductors. It is shown that this is a consequence of the *antibonding* nature of the deep gap level with respect to the impurity atom-host orbital combinations.

Whereas, since the early days of atomic and molecular physics, electronic energy levels have been naturally referred to the vacuum state (vacuum referred binding energies, or VRBE), in impurity physics, it has long become customary to refer acceptor or donor levels to either the valence (v) or the conduction (c) host band edges (hereafter denoted as host referred binding energies, or HRBE). This latter choice has been motivated not only by the paradigms of effective mass theory (which associated the generic evolution of shallow levels from these band edge states), but also by the obvious relation of electron/hole emission and capture processes to such band edges. The organization of a large body of observed electric levels with reference to E_v or E_c (see for example, recent compilations in Refs. 1–3) had unravelled many well-known chemical trends in the binding energies of *shallow* impurities (central cell effects). At the same time, this traditional choice of HRBE has revealed obscure trends in the material dependence of the HRBE for *deep* impurities,[1–3] which have since been accepted as part of the complex reality of the physics of deep centers. We show here that the VRBE is a more natural reference system and that it organizes many of the puzzling material-dependent trends in terms of the different positions of the *host states* relative to vacuum (photothreshold).

Recently, first-principles self-consistent electronic structure calculations have become possible[4] for series (e.g., V through Cu) of deep transition atom centers in semiconductors such as Si[5] and GaP.[6] Among others, they have shown that the impurity levels are not[7] "pinned" to the host vacancy level (as previously suggested[8]), nor are the host band edges the physically relevant states that determine generically the *position* of such levels. Instead, it was found that many of the results of the detailed calculations could be qualitatively understood in terms of a model of three "effective levels." This can be used here to illustrate that the physical invariant is the VRBE and not the HRBE. For impurity levels of a given symmetry Γ (say, t_2) associated with a certain site S in the host crystal (say, cation substitutional), these zero-order states are the $3d\Gamma$ orbitals of the effective impurity (I) ion[5] with energy ϵ_I, and the host (H) states ϵ_H^v and ϵ_H^c showing up as the largest peaks in the S-centered, Γ-projected local density of states in the valence (v) and conduction (c) bands, respectively. When ϵ_I is not too close to ϵ_H^v, a two-level model (ϵ_I and ϵ_H^v) suffices.[5] Having the same symmetries (Γ around S), they will interact (Fig. 1) through the

coupling element $V = \langle I |\Delta V| H \rangle$. This will result in an impurity-like bonding state in the valence band (t_2^{CFR} of Refs. 5 and 6), and in an antibonding gap level (the t_2^{DBH} dangling bond hybrid of Ref. 5) at the energy $E_{imp} = \bar{\epsilon} + [\Delta^2 + V^2]^{1/2}$. Here $\bar{\epsilon} = (\epsilon_I + \epsilon_H^v)/2$ and $\Delta = (\epsilon_I - \epsilon_H^v)/2$ are the zero-order centroid and level separation, respectively. The energy E_{VBM} of the valence-band maximum (VBM) is at a fixed distance from ϵ_H^v. All energies are referred to the vacuum. The antibonding character of the gap level implies that its energy E_{imp} is decided by two opposing and partially cancelling effects. This is illustrated in Fig. 1. If ϵ_H^v is well above ϵ_I [host anion with shallow p orbital, such as in ZnTe, Fig. 1(a)] then $\bar{\epsilon}$ is shallow, but Δ is large, leading to a weak level repulsion $R = [\Delta^2 + V^2]^{1/2}$. If, on the other hand, ϵ_H^v is closer to ϵ_I [host anion with deeper p orbitals, such as ZnS, Fig. 1(b)], then $\bar{\epsilon}$ is deeper, but Δ is smaller, increasing thereby the level repulsion R. Hence, the positions of E_{imp} relative to vacuum in cases (a) and (b) (VRBE, depicted as shaded areas in Fig. 1) are expected to be considerably closer to one another than the HRBE's (cross-hatched area in Fig. 1). The cancellation is particularly effective when ϵ_I of the *effective* impurity ion (where all s electrons occupy the d shell[5]) is above ϵ_H^v. *No cancellation is expected to occur if the gap level is not antibonding*. Such is the case in hydrogenic impurities that merely split an *already existing* host state into the gap. Hence, their energy will follow the HRBE. Note that since the coupling V and the Coulomb repulsion energies U vary considerably in going from one class of crystals to the other[3] (e.g., the more ionic II-VI relative to the III-V), this invariance may be restricted to one class of materials at a time.

To check the idea of material invariance of the VRBE's, we have used our calculated levels of GaP:Fe and InP:Fe,

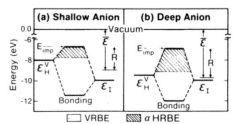

FIG. 1. Schematic two-level model for a deep impurity in a semiconductor with (a) shallow anion (e.g., ZnTe) and (b) deep anion (e.g., ZnS).

284

TABLE I. Observed[13] photothreshold values Φ, band gaps E_g, and single acceptor (A), single donor (D), and double acceptor (AA) energies (in eV, relative to the valence-band maxima) of deep transition atom impurities.[1-3,15]

Host	E_g (eV)	Φ (eV)	Impurity							Type
			V	Cr	Mn	Fe	Co	Ni		
ZnS	3.85	7.5	2.11	1.0	−0.6ᵃ	1.75	...	0.75ᵇ		D
			...	2.78ᶜ				2.48		A
ZnSe	2.80	6.82	1.6ᵃ	0.44	−0.86ᵃ	1.25	0.3	0.16		D
			...	2.24ᶜ	...	...	...	1.85		A
CdSe	1.98	6.62ᵈ	1.4	0.64	...	0.64	0.15	0.3		D
			...	...	...	...	...	1.81		A
CdTe	1.48	5.78	0.74	...	...	...	...	...		D
			...	1.3	...	...	...	...		A
InP	1.41	5.69	...	0.94	0.2	0.7	0.24	...		A
GaP	2.35	5.9ᵉ	...	1.12	0.4	0.85	0.41	0.5		A
			...	1.85		2.25		1.55		AA
GaAs	1.50	5.49	1.29	0.81	0.1	0.46	0.16	0.22		A

ᵃ Predicted in Ref. 3.
ᵇ Tentative, see Ref. 3.
ᶜ Optical value.
ᵈ Wurtzite structure, Ref. 15.
ᵉ Extrapolated from Fig. 10–13 in Ref. 14.

referring all one-electron energy levels to the *electrostatic potential of each host crystal at its interstitial site*. The potential at the empty interstitial site has been shown[9] to be a reasonable approximation to an internal (surface independent) vacuum level, and had produced reasonable predictions for band alignments at interfaces.[9] We found that whereas their HRBE differ substantially ($E_c − 1.22$ and $E_c − 0.28$ eV, for GaP:Fe and InP:Fe, respectively), their VRBE are much closer ($− 2.98$ and $− 2.86$ eV, respectively). A similar insight has been derived by Jaffe and Zunger[10] by analyzing the band-gap anomaly in ternary chalcopyrites.

Motivated by the above considerations, we follow recent suggestions[11,12] and refer the experimentally determined HRBE of transition atom acceptors [first (o/ −) and second (− / =)] levels in III-V and II-VI semiconductors and donors [(o/ +)] levels in III-V and II-VI semiconductors to an approximate vacuum level, taken as the experimentally determined[13] photothreshold Φ for the (110) surface (Table I). We neglect material variations in surface corrections because (i) the experimental precision for surface corrections is poorer[13] than for Φ, (ii) Φ has proven to correlate well[14] with E_{VBM} for a given sequence of common-cation compounds, (iii) only relative *shifts* in the vacuum level from one material to the other are needed here. The reliability of this approximation has already been verified in the study of vacancies in III-V materials[15] where empirical band structures of III-V materials were fit to Φ. We show in Fig. 2 the results for eight semiconductors for which reliable data exist,[1-3,16] indicating the oxidation states that exist at each region of the gap. Note that the $1 +$, $2 +$, and $3 +$ oxidation states shown in Fig. 2 correspond to the $A^=$, A^-, and A^0 charge states in III-V semiconductors, and to A^-, A^0, and A^+ charge states in II-VI's. The remarkable result is that *within a class of compounds, the VRBE of each impurity are nearly constant, despite significant variations in HRBE.*

Few chemical trends become apparent. (i) Shallow acceptors in CdTe and ZnTe (e.g., Cu, with[1,2] $E_A = E_v + 0.15$ eV) become deep acceptors in ZnS and ZnSe (around[17] E_v

$+ 1.3$ and $E_v + 0.7$ eV, respectively for Cu in ZnS and ZnSe) merely because the VBM in the latter systems recedes, decreases Δ, and repells E_{imp} upwards, deep inside the gap. This is why CdTe can be made low resistivity p type by cation substitution, whereas sulphides cannot.[10] On the other hand, isovalent substitutional elements lacking a deep ϵ_l (e.g., Li, Na) can form shallow acceptors in II-VI's, having hence similar HRBE and different VRBE. The same is true for Mn acceptors in III-V's: they are deep in GaP but shallow in GaAs because $\Phi_{GaP} > \Phi_{GaAs}$. (ii) Cr, Co, and Ni impurities that exist as deep donors in ZnS, ZnSe, and CdSe, but were not observed in CdTe, are indeed predicted here to be inside the CdTe valence band. (iii) Iron impurity forms a midgap (semi-insulating) level in InP, but Cr is needed to form a midgap level in GaAs (despite the similarities in band gaps), since the VBM of InP is lower than that of GaAs. (iv) Impurities in CdTe and ZnTe have similar HRBE (hence only one is

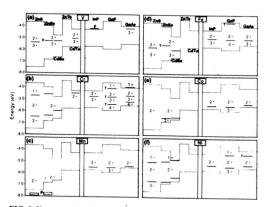

FIG. 2. Vacuum related binding energies of six $3d$ impurities in eight host semiconductors (cf. Table I), showing that the regions of stability of the $1 +$, $2 +$, and $3 +$ oxidation states are very similar in materials of the same class. (T) tentative experimental value; (P) predicted.

672   Appl. Phys. Lett., Vol. 45, No. 6, 15 September 1984

Caldas, Fazzio, and Zunger   672

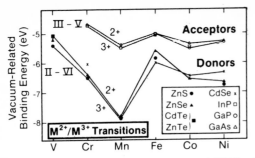

FIG. 3. Universal binding energy curve for deep acceptors in III-VI's and deep donors in II-VI's.

free-ion ionization energies[18]; the jump is larger in the more ionic II-VI systems since the impurity Mott–Hubbard Coulomb repulsion energies U are larger ($U \sim 10$–20 eV in free ions, 2–3 eV in II-VI's, and 1–2 eV in III-V's[3]). The overall width of the distribution of VRBE is dictated by the host covalency. One hopes that the universality of VRBE could be used to predict the approximate location of unknown deep centers in crystals and alloys from the knowledge of Φ and the level position in related semiconductors.

shown in Fig. 2), since their Φ's are nearly identical. Other common-anion semiconductors (e.g., InP and GaP) show variations in HRBE of deep level, since their Φ's are different. (v) The failure to detect a V acceptor in InP (despite its existence in GaAs) is consistent with the prediction that it lies just above the CBM; however, an experimental search for the V acceptor level in GaP would be important to shed light on its position GaAs. (vi) Cr in GaP can appear in the $1+$ oxidation state, whereas it does not exist in GaAs and InP (but could be forced into the gap by applying pressure) since the conduction-band minima of the latter materials are lower than in GaP. (vii) we predict that the VRBE of transition atom impurities in mixed alloys (e.g., $ZnS_x Se_{1-x}$ or $GaAs_x P_{1-x}$) will follow the variations with x in Φ and not the HRBE (e.g., the CBM or any CB in particular).

Figure 3 shows the universal trends in the VRBE of donors (M^{2+}/M^{3+}) in II-VI's, and of acceptors (M^{3+}/M^{2+}) in III-V's. (Similar trends are obtained for acceptors in II-VI's except that the jump is between Cr and Mn.) The overall trend, including the local minima in Mn parallels that in

[1]"Landolt–Börnstein Numerical Data and Functional Relationships in Science and Technology", edited by O. Madelung, M. Schultz, and H. Weiss (Springer, Berlin, 1982), Vols. 17a and b.
[2]U. Kaufmann and J. Schneider, Adv. Electron. Electron Phys. **58**, 81 (1983).
[3]A. Fazzio, M. Caldas, and A. Zunger, Phys. Rev. B **29**, 5999 (1984); **30**, 3449 (1984).
[4]U. Lindefelt and A. Zunger, Phys. Rev. **26**, 846 (1982).
[5]A. Zunger and U. Lindefelt, Phys. Rev. **27**, 1191 (1983); **26**, 5989 (1982); Solid State Commun. **45**, 343 (1983).
[6]V. Singh and A. Zunger (unpublished).
[7]V. Singh, U. Lindefelt, and A. Zunger, Phys. Rev. **25**, 2781 (1982).
[8]H. P. Hajalmarson, P. Vogl, D. J. Wolford, and J. D. Dow, Phys. Rev. Lett. **44**, 810 (1980).
[9]W. R. Frensley and H. Kroemer, Phys. Rev. **16**, 2642 (1977).
[10]J. E. Jaffe and A. Zunger, Phys. Rev. **29**, 1882 (1984), see Sec. III and Fig. 9.
[11]L. A. Ledebo and B. K. Ridley, J. Phys. **15**, L961 (1982).
[12]P. Rojo, P. Leyral, A. Nouailhat, G. Guillot, B. Lambert, B. Deveaud, and R. Coquille, J. Appl. Phys. **55**, 395 (1984).
[13]R. K. Swank, Phys. Rev. **153**, 844 (1967); T. E. Fischer, Phys. Rev. **142**, 519 (1966); G. W. Gobeli and F. G. Allen, Phys. Rev. **137**, 245 (1965).
[14]W. A. Harrison, Electronic Structure and the Properties of Solids (W. H. Freeman and Co., San Francisco, 1980), p. 254.
[15]S. Das Sarma and A. Madhukar, Solid State Commun. **38**, 183 (1981); Phys. Rev. B **24**, 2051 (1981).
[16]J. M. Baranowski and Phuong An, Phys. Status Solidi B **122**, 331 (1984).
[17]I. Broser, K. H. Franke, and H. J. Schulz, in II-VI Semiconducting Compounds, edited by D. Thomas (Benjamin, New York, 1967), p. 81 and ibid., G. B. Stringfellow and R. H. Bube, p. 1315.
[18]J. S. Griffith, The Theory of Transition Metal Ions (Cambridge, Cambridge, 1971), p. 101.

Semiconductor Heterojunction Interfaces: Nontransitivity of Energy-band Discontinuities

J. R. Waldrop and R. W. Grant

Electronics Research Center, Rockwell International, Thousand Oaks, California 91360

(Received 18 September 1979)

A direct experimental test has revealed that heterojunction energy-band discontinuities are nontransitive. This result was obtained by an x-ray photoemission-spectroscopy investigation of abrupt (110) interfaces in the heterojunction series Ge/CuBr, CuBr/GaAs, and GaAs/Ge. The sum of the valence-band discontinuities for these inteaces is 0.64 ± 0.05 eV, a large deviation from the zero sum expected by transitivity.

A fundamental feature of an abrupt semiconductor heterojunction is the discontinuity in the valence band and conduction band, ΔE_v and ΔE_c, that arises from the bandgap change ΔE_g across the interface. Theoretical models[1-3] have been proposed to predict ΔE_v (or ΔE_c); these models have as a common feature a transitive relationship for the band discontinuities. In general, such models express a band discontinuity as the difference in an energy associated with each individual semiconductor. The widely used electron-affinity rule,[1] whereby $\Delta E_c(A/B) = |\chi^A - \chi^B|$, is an example of a transitive model; χ is the respective electron affinity of semiconductors A and B which form the junction A/B. Transitivity, if true, is appealing for the relative simplicity

brought to the resulting models; implied is that interface properties *per se* need not be investigated to predict ΔE_γ and ΔE_c.

A transitive model has the property that if $\Delta E_v(A/B)$, $\Delta E_v(B/C)$, and $\Delta E_v(C/A)$ are the valence-band discontinuities associated with heterojunction interfaces from semiconductors A, B, and C, the relationship

$$\Delta E_v(A/B) + \Delta E_v(B/C) + \Delta E_v(C/A) = 0 \qquad (1)$$

must be valid. Since $\Delta E_v + \Delta E_c = \Delta E_g$, any conclusions drawn for ΔE_v can always be expressed in terms of ΔE_c. An experimental test of Eq. (1) is thus a test of transitivity.

The electronic properties of relatively few abrupt heterojunctions have been studied experi-

1686   © 1979 The American Physical Society

mentally. As a result, data are not available to
determine whether heterojunction band-edge discontinuities are transitive. Semiconductors in
row four of the periodic table, Ge, GaAs, ZnSe,
and CuBr, are all lattice matched, have tetrahedral crystal structures, and range from covalent
to highly ionic. Thus, if these semiconductors
can be grown epitaxially to form abrupt heterojunctions, characterization of at least three appropriate interfaces would test transitivity.

We report the first experimental results which
demonstrate that no general transitive relationship exists for heterojunction band discontinuities.
Specifically, by using x-ray photoemission spectroscopy (XPS) the (110) nonpolar abrupt interfaces in the series Ge/CuBr, CuBr/GaAs, and
GaAs/Ge have been found to exhibit a large deviation from transitivity. To study this series of
junctions, CuBr epitaxial layers were grown on
Ge and GaAs; this to our knowledge is the first
reported characterization of a heterojunction involving a I-VII compound.

A generalized band diagram is given in Fig. 1
for a heterojunction interface between semiconductors A and B. Shown are the valence- and
conduction-band edges E_v and E_c, $\Delta E_v(A/B) \equiv E_v^{\,B}$
$- E_v^{\,A}$, $\Delta E_c(A/B) \equiv E_c^{\,B} - E_c^{\,A}$, and the binding-energy separation, $\Delta E_B(A/B) \equiv \Delta E_b^{\,B} - E_b^{\,A}$, between
arbitrary core levels b which have binding ener-
gy $E_b^{\,A}$ and $E_b^{\,B}$ in semiconductors A and B, respectively. By inspection of the figure, $\Delta E_v(A/B)$
can be expressed as

$$\Delta E_v(A/B) = \Delta E_B(A/B) + (E_b^{\,A} - E_v^{\,A})$$

$$- (E_b^{\,B} - E_v^{\,B}). \quad (2)$$

Since A/B is any heterojunction, similar expressions can be written for $\Delta E_v(B/C)$ and $\Delta E_v(C/A)$.
Upon substitution of these expressions into Eq.
(1), terms of the form $(E_b^{\,A} - E_v^{\,A})$ cancel to yield

$$\Delta E_B(A/B) + \Delta E_B(B/C) + \Delta E_B(C/A) = 0. \quad (3)$$

These ΔE_B quantities can be measured with high
accuracy at appropriate heterojunctions by XPS;
thus, Eq. (3) provides a sensitive and direct experimental test of Eq. (1). In our experiment, A
= Ge, B = CuBr, and C = GaAs. Although the binding energies in Fig. 1 and in XPS measurements
are referenced to the Fermi energy E_F such that
$E_B \equiv 0 = E_F$, Eqs. (1)–(3) involve only energy differences. Thus, knowledge of the actual position
of the Fermi level is not required and bulk doping
differences or interface states resulting in band

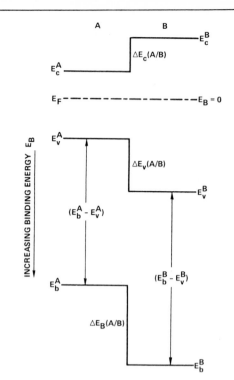

FIG. 1. Generalized energy-band diagram for a thin
abrupt A/B heterojunction interface.

bending do not affect the analysis or the XPS
measurement.

The experimental apparatus consists of a Hewlett-Packard 5950A XPS spectrometer combined
with an ultrahigh-vacuum sample preparation
chamber. This system also includes LEED (low-
energy electron diffraction), a rastered sputter
ion gun, a sample heater, and a CuBr sublima-
tion source. System base pressure is $\sim 2 \times 10^{-10}$
Torr. The XPS x-ray source is Al $K\alpha$ ($h\nu = 1486.6$
eV) radiation.

Epitaxial CuBr films were grown on (110)GaAs
and (110)Ge substrates by vacuum sublimation of
CuBr. Sublimation has frequently been used to
prepare polycrystalline films of CuBr with zinc-
blende structure.[4]

The GaAs substrate was etched in 4:1:1 H_2SO_4:
H_2O_2:H_2O solution and was cleaned under vacuum
by heating ($\sim 620\,^\circ$C) until no O or C was detectable
by XPS. At room temperature this surface ex-
hibited the (1×1) LEED pattern which is observed
on the cleaved, stochiometric (110) surface. The
Ge substrate was etched in a dilute HF solution

1687

and then cleaned under vacuum by ~1-keV Ar^+-ion sputtering and 550 °C annealing cycles until no O or C was detectable and a LEED pattern was obtained. The room-temperature (110)Ge LEED pattern was complex and strongly resembled the reported $c(8 \times 10)$ pattern.[5]

The CuBr film growth proceeded at a deposition rate of ~3 Å/sec on room-temperature substrates; growth was stopped at a layer thickness of 25-30 Å. Chamber vacuum during deposition was 2×10^{-9} Torr. No O or C was detectable in the XPS spectra of the CuBr films.

Examination by LEED was used to confirm the epitaxy of the CuBr films. The CuBr overlayer on GaAs exhibited a sharp LEED pattern that appeared to contain only integral-order spots and have lattice vectors parallel to the corresponding substrate vectors. Only the electron energy maximizing the pattern spot intensities distinguished the overlayer and clean substrate patterns; this suggests a (1×1) surface structure for the CuBr on (110)GaAs. Deposition of CuBr on room-temperature (110)Ge did not result in a LEED pattern; however, slow incremental heating of the sample to ~150 °C caused a pattern to appear. No evidence of higher-order spots was observed in the CuBr LEED pattern and, as with GaAs, the lattice vectors were parallel to the corresponding substrate vectors. This suggests that CuBr epitaxially grown on (110)Ge also forms a (1×1) surface structure.

For heterojunction samples consisting of a thin (on the order of the 25-Å XPS sampling depth)

overlayer of one semiconductor on a thick substrate of another, photoelectrons originating from each side of the interface can be observed in the same XPS spectrum. The upper half of Fig. 2 shows the core-level XPS binding-energy spectrum in the vicinity of the As 3d and Br 3d core-level peaks for the (110)CuBr/GaAs junction; similarly, the lower half of Fig. 2 shows a spectrum that includes the Ge 3d and Br 3d peaks for the (110)Ge/CuBr junction. For both junctions, a core-level peak originating from each side of the interface is evident. The ΔE_B indicated in the figure is that needed to test Eq. (3). To accurately determine ΔE_B, a background function which is proportional to the integrated peak area was subtracted from the data to correct for the effect of inelastic scattering. Core-level energies were consistently measured at the center of the peak width at half-height; this eliminated the necessity of resolving spin-orbit splitting to obtain high-precision peak positions.

Interface abruptness was assessed by comparing core-level peaks from pure samples of Ge, GaAs, and CuBr with the corresponding core-level peaks from the thin heterojunction samples. No evidence of interfacial chemical reactions was found (interfacial chemical effects would produce XPS peak broadening or splitting). In addition, the reduction of substrate core-level peak intensities with coverage and the accompanying appearance of the overlayer LEED pattern was consistent with uniform film growth and abrupt junction formation.

Table I gives the ΔE_B values measured for Ge/CuBr, CuBr/GaAs, and GaAs/Ge. The core level used in Ge, GaAs, and CuBr was Ge 3d, As 3d, and Br 3d, respectively. The ΔE_B value for the GaAs/Ge heterojunction was obtained from previously reported (110)Ge/GaAs data[6] which used the Ga 3d core level in GaAs. In an independent measurement on clean (110)GaAs [surfaces which exhibited (1×1) LEED patterns] the energy separation, determined as described above, of

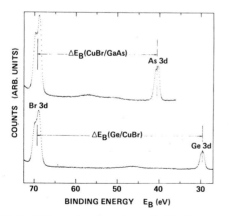

FIG. 2. XPS core-level spectra in the binding-energy region of the Br 3d, As 3d, and Ge 3d levels obtained from thin, abrupt (110)CuBr/GaAs and (110)Ge/CuBr heterojunctions.

TABLE I. XPS core-level binding-energy difference ΔE_B for abrupt interfaces which involve Ge, GaAs, and CuBr.

Interface	ΔE_B (eV)
(110)Ge/CuBr	39.85 ± 0.03
(110)CuBr/GaAs	-28.77 ± 0.03
(110)GaAs/Ge	-11.72 ± 0.02

the Ga $3d$ and As $3d$ core levels was found to be 21.92 ± 0.01 eV.[7] This value was used to compute the ΔE_B(GaAs/Ge) quoted in Table I.

Substitution of the ΔE_B values in Table I into Eq. (3) shows that the sum is nonzero and that Eq. (3), and therefore Eq. (1), is clearly not satisfied: ΔE_v(Ge/CuBr) + ΔE_v(CuBr/GaAs) + ΔE_v(GaAs/Ge) = -0.64 ± 0.05 eV. This result provides the first direct experimental proof that semiconductor-heterojunction band discontinuities are nontransitive quantities.

For perspective, this transitivity deviation can be compared to the magnitude of the ΔE_v's involved. By use of valence-band XPS data from pure samples of Ge, GaAs, and CuBr, and approximate value of the parameter $E_b - E_v$ for each material was estimated by inspection. From Eq. (2) the ΔE_v's of Ge/CuBr, CuBr/GaAs, and GaAs/Ge are found to be in the range: $0.4 \lesssim \Delta E_v \lesssim 0.9$ eV. Thus, the 0.64-eV transitivity deviation is comparable in magnitude to the individual ΔE_v values.

An interesting consequence of nontransitivity would appear in a repeating slab structure of, for example, Ge/CuBr/GaAs/Ge, etc. If the bulk semiconductor doping is chosen so that a flat-band condition is expected, the electrostatic potential would have to change by 0.64 eV for each repeat of three interfaces. Therefore, the potential across a repeating structure would become arbitrarily large. As this is unreasonable, nontransitivity of energy-band discontinuities implies that charge accumulation and/or space-charge formation must occur at one or more of the interfaces in each three-junction sequence to result in band bending that cancels the potential change.

A primary objective for a theoretical model of semiconductor heterojunctions should be a quantitative prediction of the interface band discontinuities. Models[1-3] which have been developed for this purpose have a transitive relationship for the band discontinuities. The widely used electron-affinity rule[1] depends on the difference in a surface property of semiconductor materials (this approach has been reviewed in detail[8]). Models developed in Refs. 2, 3 express band discontinuities in terms of bulk-material properties. The explicit calculation of interface electronic structure has been used to obtain energy-band discontinuities for a few selected heterojunctions.[9,10] The self-consistent pseudopotential calculations[10] for (110) interfaces of Ge/GaAs, GaAs/ZnSe, and ZnSe/Ge suggest that these band discontinuities may be nontransitive; however, the reported error limits do not allow an unambiguous conclusion. The large deviation from transitive behavior for semiconductor-heterojunction energy-band discontinuities that we report suggests that heterojunction models need to explicitly treat true interface properties associated with reconstruction and charge redistribution and should not be inherently transitive if ΔE_v and ΔE_c are to be accurately predicted.

The authors acknowledge helpful discussions with Dr. E. A. Kraut and Dr. S. P. Kowalczyk. This work was supported by the U. S. Office of Naval Research, Contract No. N00014-76-C-1109.

[1]R. L. Anderson, Solid-State Electron. **5**, 341 (1962).

[2]W. A. Harrison, J. Vac. Sci. Technol. **14**, 1016 (1977).

[3]W. R. Frensley and H. Kroemer, Phys. Rev. B **16**, 2642 (1977).

[4]See, for example, S. F. Lin, W. E. Spicer, and R. S. Bauer, Phys. Rev. B **14**, 4551 (1976).

[5]B. Z. Olshanetsky, S. M. Repinsky, and H. A. Shklyaev, Surf. Sci. **64**, 224 (1977).

[6]R. W. Grant, J. R. Waldrop, and E. A. Kraut, Phys. Rev. Lett. **40**, 656 (1978).

[7]R. W. Grant, J. R. Waldrop, S. P. Kowalczyk, and E. A. Kraut, unpublished data.

[8]H. Kroemer, Crit. Rev. Solid State Sci. **5**, 555 (1975).

[9]G. A. Baraff, J. A. Appelbaum, and D. R. Hamann, Phys. Rev. Lett. **38**, 237 (1977).

[10]W. E. Pickett, S. G. Louie, and M. L. Cohen, Phys. Rev. Lett. **39**, 109 (1977); W. E. Pickett and M. L. Cohen, Phys. Rev. B **18**, 939 (1978); J. Ihm and M. L. Cohen, Phys. Rev. B **20**, 729 (1979).

1689

PHYSICAL REVIEW B   VOLUME 28, NUMBER 4   15 AUGUST 1983

Microscopic study of semiconductor heterojunctions: Photoemission measurement of the valance-band discontinuity and of the potential barriers

A. D. Katnani and G. Margaritondo

Department of Physics, University of Wisconsin, Madison, Wisconsin 53706

(Received 4 January 1983)

We report on synchrotron-radiation photoemission measurements of the valence-band discontinuity and of the Fermi-level position for 25 different interfaces involving group-IV, III-V, and II-VI semiconductor substrates and Ge or Si overlayers. A comparison is made between our measured discontinuities and the predictions of current theoretical models. We find the best agreement with empirically corrected versions of the models of Harrison and of Frensley and Kroemer. However, we present a new empirical rule based on our present results and on those of other authors which yields even more accurate predictions of band discontinuities. The measured Fermi-level-pinning position of each substrate is the same for both Ge and Si overlayers. This result is discussed in terms of the "defect model" of Fermi-level pinning, originally developed for Schottky barriers.

I. INTRODUCTION

Photoemission experiments have been performed in recent years[1-5] on a limited number of prototypical semiconductor-semiconductor interfaces to test the current theoretical models for heterojunction-band discontinuities. These experiments have emphasized the need for a systematic study on a large number of heterojunctions. In fact, photoemission discontinuity measurements on a few prototypical interfaces did not provide conclusive evidence in favor or against any one theory. We report here the results of a systematic photoemission study of 25 interfaces involving Ge or Si overlayers on group-IV, III-V, and II-VI semiconductor compounds. These results were compared to the predictions of the electron-affinity rule,[6] the pseudopotential approach of Frensley and Kroemer,[7] Harrison's linear combination of atomic orbitals (LCAO) model,[8] Adam and Nussbaum's model,[9] and the Von Ross model.[10] The measured valence-band discontinuities, ΔE_v, show the best overall correlation with the predictions of empirically corrected versions of the models of Harrison and Frensley and Kroemer. However, even for those models the accuracy in predicting ΔE_v, 0.15–0.2 eV, is not sufficient for most practical applications. A somewhat more accurate, empirical approach to predict valence-band discontinuities was developed based on our present data and on those of other authors. This approach calculates ΔE_v as the difference between the valence-band-edge positions of the two semiconductors, empirically deduced from the measured discontinuities between each semiconductor and Si or Ge.

We also measured the interface Fermi-level position, E_F, relative to the substrate valence-band maximum, E_v. The most relevant result of these measurements is that we obtained the same value of E_F for a given substrate, both for Ge and Si overlayers. This result suggests that the defect model originally proposed by Spicer and co-workers[11] for metal-semiconductor interfaces could be applicable in certain cases to heterojunction interfaces.

Figure 1 shows a schematic energy-band diagram of two semiconductors forming a heterojunction. The transport properties of all heterojunction devices strongly depend on three interface characteristics[12-14]: band discontinuities, interface states, and potential-barrier height. The change in the forbidden gap across the interface is distributed between a valence-band discontinuity, ΔE_v, and a conduction-band discontinuity, ΔE_c. These discontinuities may form barriers for the charge carriers crossing the interface and dramatically influence the operation of heterojunction devices.[12] Interface states, including defect states, also influence the heterojunction-device behavior by acting as charge traps or recombination centers.[13,14] Finally, the position of E_F at the interface determines the barrier height on the two sides of the interface, V_{D1} and V_{D2}.

Interface characteristics, such as ΔE_v, ΔE_c, and the interface states, were the subject of many theories and experiments over the past 20 years. In 1962, Anderson[6] formulated a semiempirical rule to calculate ΔE_c, based on free-surface properties of the two semiconductors. In his model ΔE_c is given by the difference between the electron affinities of the semiconductors. The lack of reliable measurements of the band discontinuities and the spread in

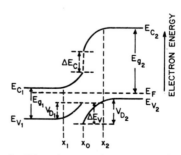

FIG. 1. Schematic energy-band diagram for a semiconductor-semiconductor interface. The two semiconductors have band gaps E_{g1} and E_{g2}. The difference between the two gaps gives rise to a conduction-band discontinuity, ΔE_c, and to a valence-band discontinuity, ΔE_v.

the experimental values of the electron affinity made it hard to test the accuracy of Anderson's model. On the other hand, this "electron-affinity rule" has been—and still is—very widely used in heterojunction-device research. The Anderson model was criticized on theoretical grounds by Kroemer[15] because it uses a free-surface parameter, the electron affinity, to describe interface properties.

Two different kinds of approaches were used in later theoretical works. The first kind of theory tried to calculate the band lineup from bulk crystal properties. The second kind of approach calculated the local electronic structure of the interface in detail, leading in particular to a direct estimate of the discontinuities. Examples of the first approach are the potential-matching model of Frensley and Kroemer,[7] the tight-binding approach of Harrison,[8] the continuous—intrinsic-Fermi-level model by Adams and Nussbaum,[9] and the continuous—conduction-band model by Von Ross.[10] Examples of the second approach are the self-consistent calculation of Baraff et al.[16] and Pickett et al.,[17] the cluster approach of Swart et al.,[18] and the tight-binding approach of Pollman and Pantelides.[19] In principle, some of the models of the second kind can be developed to any degree of accuracy and therefore they are ideal methods for the estimation of discontinuities. However, the practical present accuracy in estimating ΔE_v and ΔE_c is still limited. Furthermore, these approaches imply lengthy and expensive calculations. As a consequence, "general" approaches such as the electron-affinity rule and the other models mentioned above[7-10] are still very widely used in heterojunction research. In turn, it is necessary to perform more extensive tests of the accuracy of these approaches.

The validity and the limits of accuracy of the theoretical models for band discontinuities cannot be tested without direct and reliable measurements of these parameters. Transport techniques provide only indirect, "macroscopic" estimates of ΔE_v and ΔE_c. Futhermore, these measurements rely heavily on specific assumptions about the distribution of dopants at the interface and about the spacial distribution of the interface states. In 1978 the first results were reported on measurements of the valence-band discontinuity using photoemission spectroscopy.[4,20,21] At the same time, photoemission provided an insight into the microscopic electronic structure of the interface and a *local* measurement of ΔE_v. Other surface-sensitive techniques such as Auger electron spectroscopy and electron-energy-loss spectroscopy have also been used to study heterojunction interfaces.[22] However, the most extensive results were obtained using photoemission spectroscopy and in particular photoemission with synchrotron radiation. For example, studies of the interface states and in general of the evolution of the local electronic structure during the interface formation were made possible by the use of angle-resolved photoemission.[23,24]

The detailed theoretical calculations available for the GaAs-Ge interface[16-19] stimulated many experimentalists to investigate the microscopic characteristics of that interface by photoemission spectroscopy. A few other interfaces of fundamental and technological importance have also been studied with surface-sensitive techniques.[25-28] Those pioneering experiments were not sufficient to test the discontinuity models and to assess their limits. For example, the experimental values for the GaAs-Ge valence-band discontinuity ranged between 0.25 and 0.65 eV. This range of values reflects in part the experimental uncertainty—but is also a result of the dependence of ΔE_v on experimental variables such as annealing or substrate orientation. The wide range of reported values made it impossible to test the discontinuity models based on this interface only. The situation did not improve much when photoemission measurements of ΔE_v became available for a few other prototypical interfaces.[25-28] This suggested to us that the best way to test the discontinuity models was to investigate their ability to reproduce the general dependence of ΔE_v on the properties of the two semiconductors forming the heterojunction. This test required systematic, time-consuming measurements for a large number of interfaces under similar experimental conditions. We present here the results of the first systematic investigation of this kind.

The theoretical problems concerning the Fermi-level pinning and the potential-barrier formation are similar to those found for metal-semiconductor interfaces. The Schottky barrier, which is the equivalent for metal-semiconductor interfaces of the heterojunction potential barriers, has been widely studied.[11,29-31] Experimental results demonstrated that for many metal-semiconductor interfaces the Fermi-level-pinning position is obtained at small metal coverages and is independent of the metal overlayer. This observation led Spicer and co-workers[11] to propose in 1979 the "defect model" for Fermi-level pinning at III-V—semiconductor-metal interfaces, which relates this effect to native defects of the semiconductor surface created during the interface formation. Several theoretical studies recently tried to understand the nature of the Fermi-level-pinning defects. Daw et al.[32] identified the defects as vacancies created at the surface during the metal deposition. However, Dow et al.[33] argue that antisite defects are the most energetically favorable. In general, the nature of the local defects at metal-semiconductor interfaces and their role in pinning E_F remains a rather controversial issue. Recent experiments[34] suggested that the defect model could be extended to certain kinds of semiconductor-semiconductor interfaces. Our present systematic results on the Fermi-level-pinning position strengthen that conclusion.

The remainder of this article will be organized as follows. Sections II and III will discuss the experimental procedure and present the experimental results. These results will be analyzed and discussed in Sec. IV, and our conclusions will be summarized in Sec. V.

II. EXPERIMENTAL PROCEDURE

The experimental procedure consisted of taking photoemission spectra on clean, cleaved semiconductor substrates and then on the same substrates covered by Ge or Si overlayers of increasing thickness. The spectra were then analyzed to deduce the value of the valence-band discontinuity and the pinning position of the Fermi level at the interface. The experiments were performed under ultrahigh vacuum [operating pressure $(4-60)\times10^{-11}$ Torr, including evaporation]. The substrates were cleaved *in situ*. Table I lists the samples we studied, their source, doping, and dopant. The initial position of the Fermi lev-

TABLE I. Characteristics of the semiconducting substrates used in our experiment.

Semiconductor	Dopant	N_D, cm^{-3}	d, Å	E_g, eV	Source[a]
Si	B,n	10^{15}	2.35	1.11	ESPI
Ge	Sb,n	10^{15}	2.44	0.67	GTI
GaAs	Te,n	$10^{17}-10^{18}$	2.45	1.40	CLC
GaP	S,n	$10^{17}-10^{18}$	2.36	2.25	CI
GaSb	Te,n	$10^{17}-10^{18}$	2.65	0.67	MS
InAs	S,n	$10^{17}-10^{18}$	2.61	0.36	MS
InP	Sn,n	$10^{17}-10^{18}$	2.54	1.34	CC
InSb	S,n	$10^{17}-10^{18}$	2.81	0.17	MS
CdS	n[b]	10^{16}	2.53	2.42	CLC
CdSe	n[b]	10^{16}	2.63	1.70	CLC
CdTe	n[b]	10^{16}	2.81	1.44	JW
ZnSe	n[b]	10^{16}	2.45	2.67	CLC
ZnTe	n[b]	10^{16}	2.64	2.26	CLC

[a]ESPI indicates Electronic Space Product Inc.; GTI is Glass Technology Inc.; CI is Cambridge Instruments Company; MS is Metal Specialties Company; CC is Crystal Company; CLC is Cleveland Crystal Company; JW indicates courtesy of Professor J. D. Wiley.
[b]Nominally undoped.

el E_F indicated a flat band condition, i.e., no band bending and therefore a low density of cleavage steps, except for GaP, Si, and Ge which exhibit band bending due to intrinsic states in the forbidden gap.

Ge was evaporated from a tungsten basket and Si from a homemade, miniature electron-bombardment source where electrons with 3-KeV energy were directed against a Si single crystal. The overlayers were deposited on room-temperature substrates and their thicknesses were monitored with a quartz-crystal oscillator. The evaporation rates ranged from 0.3 to 1.5 Å per minute. Photons of energy 40–200 eV were used to probe the freshly-cleaved surfaces and the overlayer-covered surfaces. The photons were emitted by the University of Wisconsin Synchrotron Radiation Center storage ring Tantalus and monochromatized by a grazing-incidence "Grasshopper" monochromator. The photoelectrons were analyzed by a double-pass cylindrical-mirror analyzer. The overall experimental resolution (analyzer and monochromator) was 0.2–0.5 eV. Data acquisition was controlled by a Tektronics 4051 minicomputer.

For each interface we measured the photoelectron energy-distribution curves (EDC's) of the valence band and of several core levels on the clean substrate and then on the overlayer-covered surface for different overlayer thicknesses. We selected the photon energy so as to minimize the photoelectron escape depth for maximum surface sensitivity. The top of the valence band was measured by a linear extrapolation of the leading edge of the valence-band EDC's. The energy position of all spectral features was referred to the Fermi level of the system, deduced from the leading EDC edge of a thick metal film, gold or aluminum.

The accuracy of our experimental measurements of E_v and E_F, which are discussed in the next section, is primarily limited by the determination of the valence-band edge from the experimental spectra. This factor is more important than other factors such as the accuracy in determining the core-level peak position and the Fermi

edge. Kraut et al.[28] recently proposed a method to estimate E_v with very high accuracy based on the calculated density of states in the valence-band region. However, this method could not be systematically applied to all our interfaces since the required theoretical density of states is only available for a few substrates. Therefore, in our systematic study we preferred to use the most common approach to determine the valence-band edge, i.e., linear extrapolation of the leading edge of the EDC's. This method, although somewhat less accurate than that proposed by Kraut et al.,[28] could be consistently used for all the substrates we investigated, and therefore it was more appropriate for our systematic measurements. Repeated tests on a large number of systems indicate that ±0.05 eV is the typical estimated accuracy in determining E_v's with this approach. The corresponding accuracy of our measurements is ±0.1 eV for ΔE_v and for the Fermi-level-pinning position.

III. EXPERIMENTAL RESULTS

Measurements of the valence-band discontinuity from the EDC's were performed with two different methods. One of these methods was direct but it could only be used for a small number of interfaces. The other method was less direct, but it could be applied to all interfaces. Whenever possible, both methods were applied and their results were compared to test the reliability of the less direct method.

The large ΔE_v's at certain interfaces, e.g., ZnSe-Ge, CdS-Ge, and CdS-Si, enabled us to spectrally resolve the valence-band edges of both semiconductors, substrate and overlayer. Figure 2 shows the valence-band EDC's for a clean, cleaved CdS(10$\bar{1}$0) surface (bottom curve) and for increasing Si coverages of the same surface. Notice that at intermediate coverages both valence-band leading edges are visible. A direct measurement of ΔE_v is possible in this case by linear extrapolation of the two edges, and it gives a discontinuity of 1.55 ± 0.1 eV. The edge positions

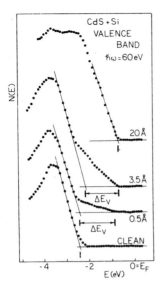

FIG. 2. EDC's of the valence band of freshly cleaved and Si-covered CdS substrate taken with 60-eV photon energy. Notice that at intermediate coverages one is able to resolve the valence-band edges of both semiconductors. This makes it possible to perform a direct measurement of ΔE_v. The two valence-band-edge positions are estimated by linear extrapolation as shown in the figure and discussed in the text. The estimated accuracy of ΔE_v is ± 0.1 eV, and its average value deduced from Si overlayers of different thicknesses on CdS is 1.55 eV.

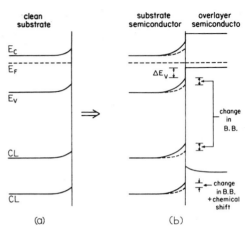

FIG. 3. Energy-band diagram for the interface between clean substrate and vacuum (a) and between clean substrate and overlayer (b). The solid lines in (b) represent the final position of the valence-band edge, of the conduction-band edge E_c, and of two different core levels, CL, while the dashed lines reproduce the same positions as in (a). The difference between the solid lines and dashed lines for E_v and E_c is due to the change in band bending (B.B.). One of the core levels is primarily affected by the band-bending changes, while the other core level is also strongly affected by changes in the chemical shift, as discussed in the text. ΔE_v is equal to the distance in energy between the *final* positions of the E_v's of the substrate and of the overlayer.

deduced by linear extrapolation were cross-checked for different systems with those deduced from the energy positions of the corresponding core levels. For Ge and Si core levels this was done by using as a reference the distance in energy between the bulk Ge $3d$ or Si $2p$ levels and features in the valence band, including the edge. For the substrate core levels the test is part of the indirect method of estimating ΔE_v discussed in the next paragraph.

The above direct method of estimating ΔE_v could not be applied to most interfaces since the two edges were too close in energy to be resolved. For those interfaces we used the more indirect approach to estimate ΔE_v. This method consisted of measuring the substrate valence-band-edge position, correcting it for band-bending changes during the interface formation, and estimating ΔE_v from the distance in energy between this corrected substrate valence-band edge and the measured valence-band edge of the overlayer. Figure 3(a) shows the energy-band diagram of the interface between a clean substrate and vacuum. Photoemission probes the surface region and enables one to measure the top of the valence band at the vacuum-substrate interface. Figure 3(b) shows the energy-band diagram for the semiconductor-semiconductor interface. As the overlayer is deposited, the top of the valence band of the substrate moves from the dashed-line position to the solid-line position. This change in position is due to the change in band bending caused by the changes in the local charge distribution. Also, notice that the substrate core-level positions change with increasing coverage. Typical-

ly, at thicknesses ≥ 5 Å one already observes the top of the valence band of the overlayer (solid line on the overlayer side). ΔE_v in Fig. 3(b) is the distance between the two solid lines representing the substrate and overlayer valence-band edges at the interface. This is equal to the distance in energy between the top of the valence band of the clean surface, corrected for the change in band bending, and the top of the valence band for the overlayer. In practice, the shift of the substrate valence-band edge due to change in band bending is obscured by the overlayer valence-band signal. However, the band-bending changes can be deduced from the shift in energy of the substrate core-level peaks with increasing coverage. In general, this shift is due to changes both in the band-bending and in the core-level chemical shift. The chemical-shift changes are due to the formation of interface chemisorption bonds. We found that the substrate *cation* core-level peaks are primarily affected by the band-bending changes during the early stages of interface formation. In fact, for all interfaces we observed at small coverages a correlation between the cation core-level shift and the substrate valence-band-edge shift. We found a similar correlation with the shift of other substrate features in the valence-band EDC's. Also, for small and intermediate (< 30 Å) overlayer thicknesses the cation core-level line shape showed no broadening with increasing coverage and this again indicated a negligible change in chemical shift. For some interfaces, for example, ZnSe-Si, ZnSe-Ge, and CdS-Ge, a broadening of the cation core-level peak was observed at high coverages (≥ 30 Å). However, this broadening did not affect our estimate of the band-bending changes which occurred at much lower coverages as shown by the initial

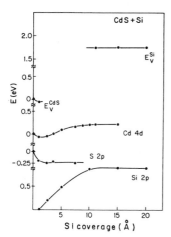

FIG. 4. Energy shift of the valence-band edge and of the Cd $4d$, S $2p$, and Si $2p$ core levels with increasing Si coverage on a CdS substrate. The zero for each plot is the clean surface energy positions. Notice the low-coverage correlation between E_v and Cd $4d$ shifts. This plot was used to estimate ΔE_v as explained in the text. The estimated ΔE_v is 1.60 eV, in good agreement with the direct method.

shift of the peak. In summary, the above observations enabled us to estimate the band-bending changes from the substrate cation core-level shift. In turn, we estimated ΔE_v by subtracting the shift of the cation core-level peak from the distance in energy between the leading edges of the EDC's taken before and after interface formation.

An example of the above method is the estimate of ΔE_v for the CdS-Si interface. Figure 4 shows the shift with coverage of the top of the valence band and of the core-level peaks. Notice the correlation between the Cd $4d$ shift

and the shift of the top of the valence band at small coverages. Instead, the S $2p$ shift is not correlated with those of E_v and Cd $4d$, and therefore is due to a combination of change in chemical shift and change in band bending. As discussed above, ΔE_v is estimated from the distance in energy between the initial and final valence-band edges after subtracting from it the total shift of the cation core-level peak. The ΔE_v deduced in this way from the data of Fig. 4 is 1.6 eV. This value is in agreement with the result of the direct method, 1.55 eV. The agreement demonstrates the reliability of the "indirect" method of measuring ΔE_v. Similar positive tests of the indirect method were performed for all the interfaces for which the direct method could be used. Table II summarizes our results on ΔE_v for the different interfaces.

An investigation of the interdiffusion across the interface is important for a meaningful comparison with theory because all theoretical models calculate the band discontinuity for an abrupt interface. The intensity attenuation of the substrate and of the overlayer core-level peaks during the interface formation was used to monitor possible interdiffusion process. In Fig. 5 we plot the normalized Cd $4d$ and S $2p$ intensities versus coverage for Si overlayers on CdS. The plot is consistent with an exponential attenuation of both core-level peaks with increasing coverage. The exponential attenuation length deduced from this plot, ~ 9 Å, is close to the escape depth for photoelectrons of this energy across the Si overlayer.[35] This indicates the formation of an abrupt interface.

The interface-pinning position of the Fermi level within the substrate gap was derived from the position of the top of the valence band in the clean-substrate EDC's after correction for the changes in band bending. The absolute position of E_F for our spectrometer was deduced from a linear interpolation of the leading spectral edge of a thick film of freshly evaporated Al or Au. The change in band bending was again estimated from the cation core-level shift. Figure 6 shows the Fermi-level shift at the interface as a function of coverage for CdS with Si and Ge over-

TABLE II. Experimental valence-band discontinuities measured from our spectra and corresponding theoretical predictions.[a]

Substrate	Experimental		EA		AN		Von Ross		Harrison		FK	
	Si	Ge	Si	Ge	Si	Ge	Si	Ge	Si	Ge	Si	Ge
Ge	−0.17		−0.31		−0.21		−0.42		−0.38		0.09	
Si		0.17		0.31		0.21		0.42		0.38		−0.09
GaAs	0.05	0.35	0.27	0.70	0.15	0.37	0.30	0.73	0.03	0.35	0.80	0.71
GaP	0.95	0.80	0.33	0.64	0.58	0.79	1.15	1.57	0.50	0.88	0.96	0.87
GaSb	0.05	0.20	−0.37	−0.07	−0.21	0.00	−0.42	0.00	−0.81	−0.31	0.73	0.64
InAs	0.15	0.33	0.15	0.46	−0.37	−0.16	−0.74	−0.31	−0.29	0.09	1.22	1.13
InP	0.57	0.64	0.55	0.85	0.12	0.34	0.24	0.67	0.14	0.64	1.42	1.33
InSb	0.00	0.00	−0.34	−0.03	−0.47	−0.25	−0.93	−0.50	−1.09	−0.71		
CdS	1.55	1.75	1.30	1.61	0.66	0.88	1.32	1.75	1.62	2.00	2.26	2.17
CdSe	1.20	1.30	0.49	0.80	0.30	0.52	0.60	1.03	0.85	1.23	2.13	2.04
CdTe	0.75	0.85	0.64	0.94	0.17	0.39	0.34	0.77	−0.18	0.20	1.74	1.69
ZnSe	1.25	1.40	1.68	1.99	0.79	1.00	1.57	2.00	1.08	1.46	1.91	1.82
ZnTe	0.85	0.95	0.64	0.96	0.58	0.80	1.16	1.59	0.00	0.38	1.58	1.49

[a]Values in eV, uncertainty ±0.1 eV. The theories are the following: EA indicates Anderson's electron-affinity rule (Ref. 6), AN is the Adam-Nussbaum model (Ref. 9), Von Ross's model (Ref. 10), Harrison's tight-binding model (Ref. 8), and FK is the Frensley-Kroemer model (Ref. 7). ΔE_v is taken to be positive when the overlayer valence-band edge is *above* the substrate valence-band edge.

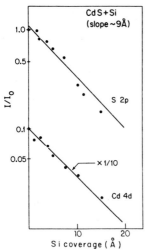

FIG. 5. Intensity attenuation of the Cd $4d$ and S $2p$ with increasing Si coverage of a CdS substrate. The exponential attenuation length estimated from this plot is 9 Å. This is close to the escape depth from Si for photoelectrons at these energies and it indicates that the interface is reasonably abrupt.

layers. Notice that the final pinning position of E_F is the same for CdS($10\bar{1}0$)-Si, CdS($10\bar{1}0$)-Ge, and CdS($11\bar{2}0$)-Ge. Table III summarizes our results on the Fermi-level-pinning position for different interfaces. These results generalize the above observation—the pinning positions for a given substrate are independent, within the experimental uncertainty, of the nature of the overlayer, Ge or Si.

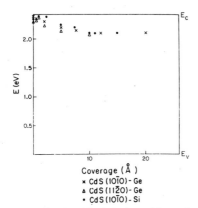

FIG. 6. Interface Fermi-level-pinning position vs the nominal thickness of the overlayer during the formation of interfaces between CdS($10\bar{1}0$) or CdS($11\bar{2}0$) and Si or Ge. The Fermi-level position in the CdS gap was estimated from the distance in energy between the substrate E_v corrected for band-bending changes and the Fermi level of the photoelectron spectrometer. In turn, the Fermi level of the spectrometer was deduced by linear interpolation of the leading spectral edge of a freshly evaporated thick film of metal. The band-bending changes were estimated from the Cd $4d$ shift.

IV. DISCUSSION

The discussion of the experimental results will be organized in Secs. IV A—IV D. First, we shall briefly present the current theoretical approaches to predict band discontinuities for all heterojunction interfaces[6–10] and discuss their characteristics. Second, we shall estimate the general theoretical accuracy limits of these approaches. The estimated accuracy limits underlying all models will then be compared with the specific accuracy limits of each model. Third, we shall describe our new empirical method to predict band discontinuities. Finally, we shall discuss our experimental results on the Fermi-level interface pinning position.

A. Theoretical discontinuity models

We have shown that the general theoretical approaches developed to calculate ΔE_v for *any* heterojunction interface include Anderson's electron-affinity rule,[6] Harrison's LCAO model,[7] the Frensley-Kroemer pseudopotential model,[7] the Adam-Nussbaum continuous—intrinsic-E_F rule,[9] and the continuous—conduction-band-edge rule by Von Ross.[10] As already mentioned, the Anderson model expresses ΔE_c as the difference between the electron affinities of the two semiconductors. Harrison used a tight-binding approach to calculate the absolute position of the valence-band maximum, while Frensley and Kroemer used a pseudopotential approach to calculate the valence-band maximum relative to an average interstitial potential. ΔE_v in both cases is simply given by the difference between the calculated valence-band maxima of the two semiconductors. In both approaches terms calculated from the bulk crystal parameters replace the electron affinities used in Anderson's model. Two other general discontinuity models were proposed in recent years. Adam and Nussbaum[9] calculated the valence-band discontinuity by aligning the intrinsic Fermi levels of the two semiconductors, while Von Ross[10] simply estimated ΔE_c to be zero, and therefore ΔE_v to be equal to the difference between the forbidden gaps.

A common characteristic of all the above models is that they express the band discontinuities as the difference between two terms characteristic of the two semiconductors. Therefore, ΔE_v and ΔE_c are linearly related to these terms (notice that the sum of the two discontinuities is equal to the difference between the forbidden gaps). This "linearity" is a powerful simplification and at the same time a limiting factor. For example, all linear models ignore the peculiar *microscopic* properties of each interface. In fact, most of them give a band discontinuity which is independent of the crystallographic faces involved in the interface and of the general interface morphology. This implies, for example, that the predicted band discontinuities must be the same for different surface orientations of a given substrate combined with a given overlayer. It also implies that the discontinuities are not different for ordered and disordered overlayers. Two other general consequences of the linearity of the above models are the commutativity and the transitivity of the predicted discontinuities. The commutativity rule implies that the valence- (or conduction-) band discontinuity for the interface between a substrate of material A and an overlayer of material B

(*A-B* interface) is equal in magnitude and opposite in sign with respect to that for the *B-A* interface. The transitivity rule implies, for example, that the sum of the valence- (or conduction-) band discontinuities for the three interfaces formed by different combinations of three given semiconductors is zero, i.e., the valence-band discontinuities for the *A-B*, *B-C*, and *C-A* interfaces add up to zero.

B. General accuracy limits and specific accuracy limits of the linear discontinuity models

The peculiar microscopic properties of each interface, such as the charge distribution on each side of the interface due to the formation of chemical bonds, in principle, affect the band discontinuity. A realistic band-discontinuity model should take these effects into consideration. Therefore, all the above linear models, which essentially ignore the peculiar microscopic properties of each interface, have intrinsic accuracy limits. An estimate of these general accuracy limits and of the specific accuracy of each model can be obtained from our results. The general accuracy limits for all linear models can be estimated by testing the predictions discussed in Sec. IV A. The specific accuracy of each model can be estimated by a direct comparison between our results and its predicted band discontinuities.

1. General accuracy limits

The underlying accuracy limits arising from the linearity of the models were empirically estimated by analyzing the extent to which our data and those of other authors agree with the predicted independence of substrate orientation and of overlayer ordering, with the commutativity rule and with the transitivity rule. In particular, we did find the measured ΔE_v to be independent of the surface orientation for substrates with Ge overlayers. In fact, we measured the same ΔE_v's for CdS(10$\bar{1}$0)-Ge and CdS(11$\bar{2}$0)-Ge interfaces. However, earlier experiments by Fang et al.[36] and by Grant et al.[4] revealed non-negligible substrate surface-orientation effects. For example, Grant et al.[4] measured discrepancies of the order of 0.2 eV between the ΔE_v's of Ge-covered GaAs substrates with different orientations.

The independence of overlayer ordering was tested for Ge overlayers on[25] Si without detecting significant changes in ΔE_v when the overlayer was ordered by annealing. A difference in ΔE_v of the order of 0.2–0.3 eV was reported for ordered and disordered Ge overlayers on[34] GaAs and on epitaxial ZnSe substrates,[27] while no difference was observed for ordered and disordered ZnSe overlayers on Ge.[27] Our préliminary tests did not show any ΔE_v difference between ordered and disordered Ge overlayers on cleaved ZnSe. Systematic data on the effects of overlayer ordering are not yet available, but from the above preliminary results they do not appear to affect the ΔE_v's by more than a few tenths of an electronvolt per interface—0.1–0.15 eV on the average. This point is relevant since in our systematic study we tried to use similar experimental conditions for all interfaces, and in particular room-temperature substrates during deposition which give disordered overlayers—while some of the discontinuity models apply to ordered systems. However,

TABLE III. Interface Fermi-level-pinning position.[a]

Substrate	Si	Ge
Si		0.40
Ge	0.30	
GaAs	0.80	0.75[b]
GaP	1.05	1.05
InP	0.75	0.80
CdS	2.10	2.10
CdSe	1.30	1.40
CdTe	1.00	0.95
ZnSe	1.80	1.80
ZnTe	1.25	1.20

[a]Measured with respect to the substrate valence-band edge. Uncertainty ±0.1 eV.
[b]See Ref. 34.

the limited-overlayer ordering effects mentioned above do not jeopardize the overall comparison between our data and those models, and do not significantly affect our tests of the models and the corresponding conclusions.

The commutativity implied by all linear models was tested in three experimental measurements. One is our own present experiment and the other two are x-ray photoelectron spectroscopy (XPS) measurements by Waldrop et al.[5] and by Kowalczyk et al.[27] We have not observed a dependence of ΔE_v on the growth sequence for the Ge-Si combination. In fact, we have measured the same ΔE_v, 0.17 eV, for Ge-Si and Si-Ge interfaces. Instead, the other two experiments have measured deviations from the commutativity rule. Waldrop et al.[5] found a deviation of 0.25 eV for GaAs-AlAs and AlAs-GaAs and Kowalczyk et al.[27] found deviations of 0.32–0.54 eV for ZnSe-Ge and Ge-ZnSe.

We analyzed eleven different groups of three semiconductors each to test the transitivity rule. Each group includes Si, Ge, and a third material, X. Table IV lists the difference between ΔE_v's of X-Ge and X-Si interfaces. This difference should be equal to the valence-band

TABLE IV. Test of the transitivity rule.[a]

X (substrate)	$\Delta E_v^{X\text{-}Ge}$-$\Delta E_v^{X\text{-}Si}$
GaAs	0.20
GaP	0.32
GaSb	0.15
InAs	0.20
InP	0.07
InSb	0.00
CdS	0.20
CdSe	0.10
CdTe	0.10
ZnSe	0.15
ZnTe	0.10

[a]Reported in the table is the sum of the ΔE_v's for the two heterojunctions involving a given substrate X and Si or Ge. The transitivity rule implied by all linear models requires this sum to be equal to the discontinuity of the Si-Ge interface, 0.17 eV.

discontinuity of the Si-Ge interface, 0.17 eV, if the transitivity rule holds. In fact, the difference is equal to the measured ΔE_v of the Si-Ge interface within the combined experimental uncertainty—except for GaP, where a discrepancy of the order of 0.15 eV occurs. Previous experiments by Waldrop and Grant[3] on GaAs-Ge, CuBr-GaAs, and CuBr-Ge interfaces revealed a 0.64-eV deviation from zero of the sum of the corresponding ΔE_v's. Also, a recent XPS study[27] of ΔE_v for Ge-ZnSe and ZnSe-GaAs when combined with previous ones[3-5] for GaAs-Ge reveals a deviation of 0.20 eV from the zero sum predicted by the transitivity rule.

In conclusion, several of the above tests revealed deviations from the general predictions of all linear models beyond the combined experimental uncertainty. From the magnitude of these deviations, we conclude that the effects ignored by the linear models are not negligible, but they do not affect each band discontinuity by more than 0.25 eV. In fact, their average magnitude, corresponding to the average accuracy limit underlying all linear models, appears close to 0.15 eV.

2. Specific accuracy limits of each theoretical model

The above tests confirmed that the accuracy of any linear model is necessarily limited. However, the predictions of each model do not necessarily achieve even the above accuracy limits. Therefore, specific tests of the predictions of each model are necessary to select the most accurate among them. Table II lists, together with our results, the ΔE_v's predicted by the different linear models for the interfaces we studied.

The most widely used band-discontinuity model is Anderson's electron-affinity rule. Recent photoemission experiments[37] in which the two electron affinities and ΔE_v were measured in the same system demonstrated the

failure of this model in predicting the band discontinuity. In fact, the discrepancy between ΔE_v and the measured difference in the electron affinities was of the order of 0.5 eV. One routine difficulty in using Anderson's model is selecting the appropriate electron affinities from the wide range of values found in the literature for each semiconductor. Table II shows that even after making a "biased" selection of the published electron affinities that best fit our data the average accuracy is not better than 0.25 eV.

To analyze the correlation between theory and experiment in the case of the Frensley and Kroemer model, a plot of our results versus the theoretical predictions is shown in Fig. 7. The solid line is the line of perfect agreement. The correlation between our results and the model is not excellent, although the model does give reasonable predictions for some interfaces, e.g., GaP-Si. The average accuracy is about 0.4 eV, i.e., worse than the experimental uncertainty and above the underlying accuracy limit of all linear models. Corrections for the interface dipoles do not improve the agreement with our results. However, the model does reproduce recent XPS measurement of ΔE_v's for[26] GaAs-InAs and ZnSe-GaAs.[27] This suggests that the discrepancy between this theory and our data primarily arises from errors in the predicted E_v's for Ge and Si. In fact, if we move the predicted valence-band-edge energy position by 0.70 eV for Si and by 0.40 eV for Ge we improve the overall accuracy of the model bringing it to ~0.20 eV. Figure 8 shows a plot similar to that of Fig. 7 after including this empirical correction, and emphasizes the better correlation between theory and experiment.

Figure 9 shows a comparison between our results and Harrison's model. The correlation between our results and the predictions of the model is reasonable. However, the overall accuracy of the model is about 0.4 eV, i.e., again worse than the experimental uncertainty. Notice that the

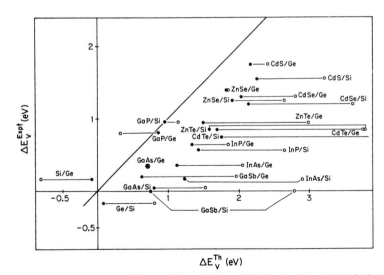

FIG. 7. Comparison between our experimental data and the predictions of the Frensley-Kroemer model (closed circles) (Ref. 7). The solid line is the line of perfect agreement. We also include in the plot a version of the model corrected for the local dipoles (open circles) (Ref. 7). In both cases there is limited correlation with our results.

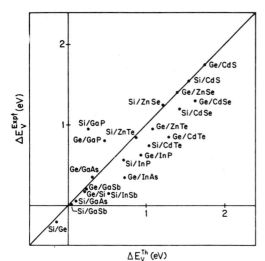

FIG. 8. Comparison between our data and an empirically corrected version of the Frensley-Kroemer model. The corrections are discussed in the text. Notice that the overall accuracy of the model is improved after the correction.

model successfully predicts the band discontinuities for lattice-matched interfaces, while it becomes much less accurate for lattice-mismatched interfaces. For example, the predicted ΔE_v's for GaAs-Ge and ZnSe-Ge, which exhibit good lattice matching, are in excellent correlation with the

experimental findings. On the contrary, the predictions for InSb-Ge, GaSb-Ge, and CdTe-Ge which have very severe lattice mismatch are very far from the experimental results. This observation leads us to introduce a simple correction for interface relaxation[2] to compensate for the lattice mismatch. We assumed that the overlayer interatomic distance approaches the substrate interatomic distance, d, near the interface. As a result the calculated E_v for the overlayer changes at the interface because of the dependence of the interatomic matrix elements on the interatomic distances. This empirical correction substantially improved the accuracy of Harrison's model. In fact, similar improvements can be obtained by replacing the overlayer interatomic distance with the average of substrate and overlayer interatomic distances. Figure 10 shows a comparison between our experimental findings and the predictions of the model after substituting the overlayer interatomic distance with the substrate interatomic distance (open circles), or with the average of the overlayer and substrate interatomic distances (closed circles). The improvement with respect to Fig. 9 is evident. Notice in particular that the correction is successful in improving the model for lattice-mismatched interfaces, e.g., for InSb-Ge, GaSb-Ge, CdTe-Ge, CdTe-Si, and ZnTe-Si. The average accuracy of the model, after including either one of the above corrections, is improved to 0.15–0.2 eV.

Finally, we compared our results to the predictions of the Adam-Nussbaum model[9] and of the Von Ross model.[10] The average accuracy in reproducing our data is of the order of 0.4 eV for both models. This accuracy is comparable to the average accuracies of the Frensley-

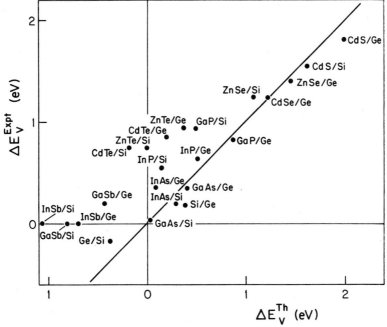

FIG. 9. Comparison between our results and the prediction of Harrison's model (Ref. 8). The average accuracy of the model is ~0.4 eV. Notice that the accuracy is better for lattice-matched interfaces, e.g., GaAs-Ge, than for lattice-mismatched interfaces, GaSb-Ge.

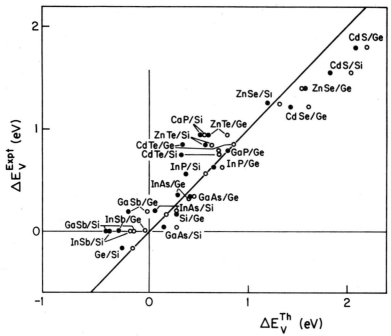

FIG. 10. Comparison between our results and empirically corrected versions of Harrison's model. The corrections are discussed in the text. Notice the improvement in accuracy for lattice-mismatched interfaces. The average accuracy of the model is close to 0.15 eV after the corrections.

Kroemer and Harrison models, but it is worse than the average accuracies of their empirically modified versions.

In summary, most current discontinuity models do not reproduce our results with an accuracy close to the empirically estimated general limits of all linear models. The best overall agreement is given by the empirically corrected versions of the models of Harrison and Frensley and Kroemer, which both reach an average accuracy of the order of 0.15–0.2 eV.

C. Empirical table to predict valence-band discontinuities

The accuracy limits estimated in the preceding section for the current discontinuity models are not sufficient for most applications in heterojunction-device research. In particular, the widely used electron-affinity rule is among the least accurate models. Even the most sophisticated theoretical calculations[16-19] do not provide the required accuracy. This led us[38,39] to develop an empirical method to estimate ΔE_v's, based on a table of experimentally deduced valence-band-edge positions of the semiconductors we studied. Table V lists empirical E_v's referred to the valence-band edge of Ge. The distance in energy between the valence-band edge of Ge and that of a given material X was estimated by taking the average of the ΔE_v of X-Ge and of the sum of the ΔE_v's of X-Si and Si-Ge. Whenever available, discontinuities measured with photoemission methods by other authors were considered, and an average

TABLE V. Empirical position in energy of the valence-band edge.[a]

Semiconductor	E_v
Ge	0.00
Si	−0.17
GaAs	−0.33[b]
GaP	−0.96
GaSb	−0.21
InAs	−0.33
InP	−0.69
InSb	−0.11
CdS	−1.73
CdSe	−1.33
CdTe	−0.88
ZnSe	−1.41[c]
ZnTe	−0.98

[a]Position in eV, referred to the valence-band edge of Ge. These positions were empirically estimated from the experimental ΔE_v's as discussed in the text (Ref. 38). The valence-band discontinuity at the interface between any two semiconductors listed in the table can be simply estimated by taking the difference of their empirical E_v's.
[b]Average value deduced from the data of Refs. 2, 20, 28, 34, and 37.
[c]Average value deduced from the data of Refs. 27 and 2.

of all the experimental ΔE_v's for each interface was used in estimating ΔE_v.[38] The value of ΔE_v for the interface between two materials listed in Table V can be simply estimated by taking the difference of their E_v's. We tested the accuracy of our empirical approach by reversing the procedure to develop the table, i.e., by using it to predict our results (notice that this is not a trivial test since the table is based on results by other authors as well as on our data). The accuracy was always better than 0.16 eV, and the average accuracy was of the order of 0.05 eV per interface. Therefore, the use of our empirical table gives a better average accuracy than all the models we tested. Attempts to use the table to predict band discontinuities between compound semiconductors[38] indicate that our empirical approach is close to the underlying average accuracy limits of all linear models as estimated above, 0.15 eV.

It should be emphasized that our approach is *not* a new theory. It is simply an optimized empirical table based on the "linearity" assumption. While it can be useful for practical uses, it does not provide an insight into the nature of the band discontinuities. This insight must be provided by theories based on physical assumptions, and the discussion in Secs. IV A–IV C clarifies to some extent what the important factors are influencing the band discontinuities. The success of the modified versions of the tight-binding and pseudopotential models, although limited, indicates that the absolute position of the bulk valence-band edges *is* an important factor in ΔE_v. The empirically determined accuracy limits which underlie all linear models indicate that "local" effects contribute to ΔE_v by no more than a few tenths of an eV. Of course, effects of this magnitude are important in a number of practical problems, and a satisfactory theory of the band discontinuities should be able to describe and predict them. Futher refinements of the "realistic" calculations of the interface electronic structure[16–19] are the only hope to solve this problem. The linear, i.e., "nonlocal" approaches—including our own empirical rule—cannot be improved beyond the estimated ∼0.15-eV accuracy limit, which is not satisfactory for many applications. Further experiments are also necessary to detect the nature and magnitude of the local contributions to ΔE_v, thereby guiding the theoretical efforts to include these contributions in a satisfactory description of the band discontinuities.

D. Potential-barrier heights

The interface position of the Fermi level in the gaps of the two semiconductors determines the band bending on each side of the junction and therefore the potential-barrier height seen by carriers crossing the junction region. As already mentioned, there is a correspondence between the establishments of these barriers and the creation of the Schottky barrier at a metal-semiconductor interface.[39] Extensive experimental and theoretical work has clarified several important features of the Schottky-barrier formation process—but also created some controversy. There is general agreement that for interfaces between silicon and simple metals the Fermi level is pinned in its interface position by localized interface states, as indicated by photoemission and energy-loss experiments. For III-V metal interfaces the experiments are yielding apparently contradic-

tory results.[39] On one hand, several interface properties exhibit a general dependence on the chemical parameters of the metal and of the semiconductor. For example, the Schottky-barrier heights on InP can assume either one of two possible values, depending on the interface reactivity. On the other hand, the Schottky barrier is generally established at a very early stage of formation of the metal overlayer, and it appears related to a limited number of pinning positions for E_F. These features were explained by Spicer and co-workers[11] in terms of their "defect model," which attributes the pinning of E_F to surface defects created by the metal chemisorption process on the semiconductor surface. The experimental basis and the theoretical implications of the above two results have been discussed in detail in a number of recent reviews,[39] which also propose possible ways to reconcile them. Therefore, we shall not give here a full discussion of those issues. The relevant points to our present results are that some of the above results find their counterparts in our present data on heterojunction interfaces.

Similar to silicon simple-metal interfaces, localized electronic states have been detected at the Si-Ge interface and theoretically explained in terms of Si-Ge chemisorption bonds.[25] However, this is the only heterojunction interface for which chemisorption-induced interface states are easily detected with angle-integrated photoemission. For the other interfaces we studied, the information on the local electronic states responsible for the Fermi-level pinning is indirect, and primarily given by the study of E_F as a function of the overlayer thickness.

One important point which was raised in support of the defect model[11] is that the pinning position of E_F for several III-V compounds appears independent of the nature of the overlayer. For example, for n-type GaAs several different kinds of overlayers were reported to pin the Fermi level ∼0.75 eV above the top of the valence band. Mönch and Gant[40] found that the adsorption of Ge gives the same pinning positions of the Fermi level on GaAs as that of metal atoms, suggesting an extension of the defect model to heterojunction interfaces. This hypothesis is strengthened and generalized by our present results. The results of Table III show that the pinning positions for Ge or Si on n-type GaAs—as measured by Mönch et al.[34] for ordered or disordered Ge overlayers and by ourselves—are coincident with the above value of 0.75 eV within the experimental uncertainty. This observation is generalized by the results of Table III, which shows that the pinning position is the same for a given substrate, independent of the overlayer. This is consistent with the predictions of the defect model (which was originally developed for interfaces involving III-V compounds). However, the chemical properties of Ge and Si are too close to each other to consider this a very strong argument in favor of the defect model. The above similarity of the pinning positions on GaAs for different classes of overlayers remains the strongest indication from our work that the defect model could be extended to semiconductor-semiconductor interfaces.

Other features of the data in Table III are related to the defect model. It was suggested that the study of the chemical trends upon varying the substrate is an effective approach to study the nature of the Fermi-level-pinning defects.[41] Recently, Allen and Dow calculated the different energy levels for surface antisite defects on GaAs and GaP.[41] The calculated acceptor levels are not far from the pinning positions of E_F shown in Table III for n-type substrates. Much more interesting, however, is the fact that the distance between E_v and E_F changes on going from GaAs to GaP as qualitatively predicted by the theory.[41] These results, therefore, appear consistent with a role of surface antisite defects in the Fermi-level pinning. One interesting fact is that the theoretical results given by *interface* antisite defects show a much poorer correlation with the experimental data.

The Fermi-level pinning at heterojunction interfaces raises interesting questions about its correlation with the establishment of the band discontinuities. The pinning positions of E_F in the two gaps are trivially related to ΔE_v and ΔE_c. One could, therefore, propose a *gedanken* experiment in which the clean surfaces of the two semiconductors chemisorb just enough foreign atoms to reach the final pinning position of E_F (the real experiments show that less than 0.1 monolayer of foreign atoms are sufficient for most III-V substrates), and then they are brought together by aligning the E_F's to form the interface. In this approach, ΔE_v would be given, at least in first approximation, by the difference of the distances between the Fermi level and the valence-band edge for the two surfaces. There are of course some very fundamental problems with this *gedanken* experiment. For example, the defect levels—and therefore the pinning positions of E_F in the two gaps—could be substantially changed by the formation of the interface. However, many experiments have revealed that the pinning position of E_F, established at submonolayer coverage, does not change much in many cases when the overlayer grows thicker.[39] Interestingly enough, the difference between the pinning positions for GaAs and ZnSe reported in Table III gives ~1 eV, which coincides with the measured ΔE_v for the ZnSe-GaAs interface. Without further experimental tests, it is impossible to decide if this is just a coincidence. Otherwise, it could indicate that the above *gedanken* experiment is valid within reasonable limits, and it would explain why the empirical accuracy limit underlying all the linear discontinuity models, ~0.15 eV, is after all so surprisingly good. In that case, however, one would have to find an explanation for the agreement between our data and the ΔE_v's given by the modified tight-binding and pseudopotential models, as discussed in Sec. IV B.

V. SUMMARY

The main results of our systematic study are the following. The valence-band discontinuity and the interface Fermi-level-pinning positions were measured for Ge and Si overlayers on Si and Ge substrates and on 11 compound-semiconductor substrates. The measured ΔE_v's were used to empirically estimate an underlying, average accuracy limit of ~0.15 eV for all the linear discontinuity models. There are the models which express ΔE_v and ΔE_c as the difference of two terms related to the two semiconductors—and in particular the models most widely used in heterojunction research such as the electron-affinity rule.

We then used the experimental data to estimate the accuracy of each linear model without questioning *a priori* its theoretical soundness. The best average accuracy, 0.15–0.2 eV, was found for empirically modified versions of the Harrison model and of the Frensley-Kroemer model. We also proposed an empirical optimization of all linear models, based on the use of the terms in Table V which, in turn, were deduced from experimental ΔE_v's. This approach appears able to reach the general accuracy limits for linear models.

Some features in the measured pinning positions of E_F indicate that the basic assumption of a local-defect-related pinning mechanism could be extended from III-V—semiconductor-metal interfaces to heterojunction interfaces. We also raised the problem of the correlation between the band discontinuities and the pinning of E_F in the two gaps. However, a satisfactory treatment of this problem requires more experimental tests and a considerable amount of fundamental theoretical work.

ACKNOWLEDGMENTS

This research was supported by the National Science Foundation under Grant No. DMR-82-00518. The experiments were made possible by the friendly and expert assistance of the staff of the University of Wisconsin Synchrotron Radiation Center (supported by the National Science Foundation, Grant No. DMR-80-20164). We are grateful to N. G. Stoffel for participating in these experiments and for critically reading the manuscript. R. R. Daniels and Te-Xiu Zhao actively helped us in the experimental work, and we had several illuminating scientific interactions with H. Kroemer, W. Harrison, W. R. Frensley, A. Nussbaum, R. S. Bauer, W. Mönch, R. W. Grant, E. A. Kraut, S. P. Kowalczyk, and P. Perfetti.

[1]A. D. Katnani, R. R. Daniels, Te-Xiu Zhao, and G. Margaritondo, J. Sci. Technol. 20, 662 (1982); A. D. Katnani, N. G. Stoffel, R. R. Daniels, Te-Xiu Zhao, and G. Margaritondo, J. Vac. Sci. Technol. A 1, 692 (1983).

[2]G. Margaritondo, A. D. Katnani, N. G. Stoffel, R. R. Daniels, and Te-Xiu Zhao, Solid State Commun. 43, 163 (1982).

[3]J. R. Waldrop and R. W. Grant, Phys. Rev. Lett. 26, 1686 (1978).

[4]R. W. Grant, J. R. Waldrop, and E. A. Kraut, Phys. Rev. Lett. 40, 656 (1979).

[5]J. R. Waldrop, S. P. Kowalczyk, R. W. Grat, E. A. Kraut, and D. L. Miller, J. Vac. Sci. Technol. 19, 573 (1981).

[6]R. L. Anderson, Solid State Electron. 5, 341 (1962).

[7]W. R. Frensley and H. Kroemer, Phys. Rev. B 15, 2642 (1977).

[8]W. Harrison, J. Vac. Sci. Technol. 14, 1016 (1977).

[9]M. J. Adam and Allen Nussbaum, Solid State Electron. 22, 783 (1979).

[10]Oldwig Von Ross, Solid State Electron. 23, 1069 (1980).

[11]W. E. Spicer, P. W. Chye, P. Skeath, C. Y. Su, and I. Lindau, J. Vac. Sci. Technol. 16, 1422 (1979); W. E. Spicer, I. Lindau,

P. Skeath, and C. Y. Su, *ibid.* 17, 1019 (1980).

[12]A. G. Milnes and D. L. Feucht, *Heterojunctions and Metal-Semiconductor Junctions* (Academic, New York, 1972).

[13]W. G. Oldham and A. G. Milnes, Solid State Electron. 7, 153 (1964); J. Jerhot and V. Snejdar, Phys. Status Solidi A 34, 505 (1976).

[14]H. C. Card, J. Appl. Phys. 50, 2822 (1979).

[15]H. Kroemer, CRC Crit. Rev. Solid State Sci. 5, 555 (1975).

[16]G. A. Baraff, Joel A. Appelbaum, and D. R. Hamann, Phys. Rev. Lett. 38, 237 (1977).

[17]Warren Pickett, Steven G. Louis, and Marvin Cohen, Phys. Rev. Lett. 39, 109 (1977); Phys. Rev. B 17, 815 (1978).

[18]C. A. Swarts, W. A. Goddard, and T. G. McGill, J. Vac. Sci. Technol. 19, 551 (1981).

[19]J. Pollman and S. Pantelides, J. Vac. Sci. Technol. 16, 1498 (1979).

[20]P. Perfetti, D. Denley, K. A. Mills, and D. A. Shirley, Appl. Phys. Lett. 33, 66 (1978).

[21]R. S. Bauer and J. C. McMenamin, J. Vac. Sci. Technol. 15, 1444 (1978).

[22]W. Mönch and H. Gant, J. Vac. Sci. Technol. 17, 1094 (1980).

[23]D. Denley, K. A. Mills, P. Perfetti, and D. A. Shirley, J. Vac. Sci. Technol. 16, 1501 (1979).

[24]P. Zurcher, G. J. Lapeyre, J. Anderson, and D. Frankel, J. Vac. Sci. Technol. 21, 476 (1982).

[25]P. Perfetti, N. G. Stoffel, A. D. Katnani, G. Margaritondo, C. Quaresima, F. Patella, A. Savoia, C. M. Bertoni, C. Calandra, and F. Manghi, Phys. Rev. B 24, 6174 (1981); G. Margaritondo, N. G. Stoffel, A. D. Katnani, H. S. Edelman, and C. M. Bertoni, J. Vac. Sci. Technol. 18, 290 (1981); G. Margaritondo, N. C. Stoffel, A. D. Katnani, and F. Patella, Solid State Commun. 36, 215 (1980).

[26]S. P. Kowalczyk, W. J. Schlaffer, E. A. Kraut, and R. W. Grant, J. Vac. Sci. Technol. 20, 705 (1981).

[27]S. P. Kowalczyk, E. A. Kraut, J. R. Waldrop, and R. W. Grant, J. Vac. Sci. Technol. 21, 482 (1982).

[28]E. A. Kraut, R. W. Grant, J. R. Waldrop, and S. P. Kowalczyk, Phys. Rev. Lett. 44, 1620 (1980). A comparison of the results of these authors with those obtained by linear extrapolation confirms the accuracy of the latter method quoted in the text.

[29]R. H. Williams, R. R. Varma, and V. Montegomery, J. Vac. Sci. Technol. 16, 1143 (1979).

[30]P. Skeath, I. Lindau, P. W. Chye, C. Y. Su, and W. E. Spicer, J. Vac. Sci. Technol. 16, 1143 (1979).

[31]L. J. Brillson, C. F. Brucker, N. G. Stoffel, A. D. Katnani, and G. Margaritondo, Phys. Rev. Lett. 46, 838 (1981).

[32]M. S. Daw and D. L. Smith, Phys. Rev. B 20, 5150 (1979); Appl. Phys. Lett. 36, 690 (1980).

[33]J. D. Dow and R. E. Allen, J. Vac. Sci. Technol. 20, 659 (1982).

[34]W. Mönch, R. S. Bauer, H. Gant, and R. Murschall, J. Vac. Sci. Technol. 21, 498 (1982).

[35]I. Lindau and W. E. Spicer, J. Electron Spectrosc. Relat. Phenom. 3, 409 (1974).

[36]F. F. Fang and W. H. Howard, J. Appl. Phys. 35, 3 (1964).

[37]P. Zurcher and R. S. Bauer, J. Vac. Sci. Technol. A 1, 695 (1983).

[38]A. D. Katnani and G. Margaritondo, J. Appl. Phys. 5, 2522 (1983), and references therein.

[39]For a complete discussion of these problems and for further references see G. Margaritondo, Solid State Electron. 26, 499 (1983), or L. J. Brillson, Surf. Sci. Rep. 2, 123 (1982).

[40]W. Mönch and H. Gant, Phys. Rev. Lett. 48, 512 (1982).

[41]A. D. Katnani, G. Margaritondo, R. E. Allen, and J. D. Dow, Solid State Commun. 44, 1231 (1982).

Understanding and controlling heterojunction band discontinuities

D. W. Niles, E. Colavita, and G. Margaritondo[a]

Synchrotron Radiation Center and Department of Physics, University of Wisconsin–Madison, Stoughton, Wisconsin 53589

P. Perfetti, C. Quaresima, and M. Capozi

Istituto di Struttura della Materia, Via E. Fermi, 00044 Frascati, Italy

(Received 23 August 1985; accepted 30 September 1985)

We discuss two recent results on the microscopic nature and control of the band lineup at semiconductor–semiconductor interfaces. First, we identified a correlation between measured heterojunction band discontinuities and Schottky barrier heights of the corresponding semiconductors, as predicted by several theoretical models. Second, we found that ultrathin metal intralayers modify the band lineup of polar interfaces by several tenths of an electron volt. At least in principle, this degree of freedom can be exploited to tailor heterojunction devices.

I. INTRODUCTION

The band discontinuities which accommodate the difference in forbidden gap are the most important characteristic of a semiconductor–semiconductor interface.[1] They determine the transport properties of the interface. They are a fundamental factor in the ultimate performance of heterojunction devices ranging from simple photon detectors to superlattices.

Substantial progress was recently made in understanding the microscopic nature of the band discontinuities. The main reason for this progress was the extensive application of photoemission techniques as direct probes of the interface formation process.[2] These studies also stimulated renewed theoretical efforts to clarify this crucial issue.[3–13]

We present here two recent contributions to a better understanding of heterojunction band discontinuities and to their ultimate control. The first result is a positive test of the correlation between band discontinuities and Schottky barriers, predicted by several theories.[4,5] The second result is the successful use of ultrathin metal intralayers to modify band lineups.

II. CORRELATION BETWEEN BAND DISCONTINUITIES AND SCHOTTKY BARRIERS

The recent months have produced a series of developments in heterojunction theory.[4–6,13] Most interesting are the results produced by two fundamental classes of theories, those based on the concept of midgap energy,[4] and those related to the electron affinity rule.[5,13] Both approaches predicted the correlation between band discontinuities and Schottky barriers which is discussed in this section.

A general midgap-energy heterojunction theory was developed by Tersoff.[4] He proposed the minimization to zero of the interface dipoles as the band lineup criterion for semiconductor–semiconductor interfaces. This implied the alignment of the midgap energy points of the two semiconductors, and therefore the same calculations could be used to derive Schottky barriers and band discontinuities.[14] One interesting aspect of Tersoff's results is their remarkable accuracy. When tested with extensive photoemission data,[15] Tersoff's predicted discontinuities appear close to the underlying general accuracy limit of all linear discontinuity models ± 0.15 eV.[16]

The correlation between Schottky barriers and valence band discontinuities is an elementary by-product of Tersoff's midgap-energy approach. Given the Schottky barrier heights between two given semiconductors and the same metal E_B^1 and E_B^2, and the difference between their forbidden gaps ΔE_g, Tersoff's valence band discontinuity ΔE_v for the interface between those two semiconductors must satisfy the equation

$$\Delta E_v = \Delta E_g - (E_B^1 - E_B^2) = \Delta E_g - \Delta E_B .\qquad(1)$$

The above equation, however, is not unique to the midgap-energy approach. It is also predicted by the Schottky model for metal–semiconductor interfaces, combined with the electron affinity rule. The electron affinity rule originally developed by Anderson,[17] is the oldest theoretical model of heterojunction band lineup. The interest on this kind of theories was recently renewed by several works.[3,5,6,13] For example, Freeouf and Woodall[13] emphasized the importance of metallurgical effects by applying the "effective work function" model to the problem of III–V heterojunctions. Deep impurity levels were proposed[6] as substitute reference energies instead of the vacuum level, thereby by-passing the accuracy problems caused by the experimental determination of the electron affinities. Tejedor and Flores[3] attacked the crucial problem of the microscopic interface dipoles in the general framework of the electron affinity rule. Along the same lines, Duke and Mailhiot[5] recently estimated that the interface-dipole corrections to the electron affinity rule are smaller than 100 meV unless they are enhanced by interface atomic relaxation.

A test of the validity of Eq. (1) cannot discriminate between Tersoff's model[4] and the electron-affinity approaches. It is, nevertheless, a necessary test for both classes of theories, and in general, for the hypothesis that microscopic interface dipole contributions to the band discontinuities have limited magnitude on the average.[16]

The best opportunity for the test is offered by the extensive data on Schottky barriers involving Au[1] and by the equally extensive photoemission discontinuity measurements re-

304

TABLE I. Measured Schottky barrier heights and forbidden gaps.

Semiconductor	Au/semiconductor Schottky barrier height (eV)[a]	E_g (eV)
Si	0.81	1.11
Ge	0.45	0.67
GaAs	0.90	1.35
GaP	1.30	2.24
GaSb	0.60	0.67
InP	1.27	0.49
CdSe	0.49	1.74
CdTe	0.60	1.44
ZnSe	1.36	2.58

[a] Reference 1.

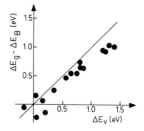

FIG. 1. The correlation between heterojunction valence band discontinuities and Schottky barrier heights, predicted by Eq. (1) (which corresponds to the straight line with slope 1). Notice that the data points tend to be on one side of the line, suggesting that microscopic interface corrections systematically tend to increase ΔE_v with respect to the predictions of Eq. (1).

ported, e.g., in Ref. 16. The input data and the results of the test are shown in Tables I and II and in Fig. 1. The correlation is quite evident. The average magnitude of the discrepancy between the two columns in Table II is 0.2 eV, which is close to the empirically estimated general accuracy limit of all linear discontinuity models.[16] This confirms that the interface dipole contributions are small on the average as predicted both by Tersoff[4] and by Duke–Mailhiot[5]; and also from an empirical point of view by Katnani and Margaritondo.[16] The peculiar correlation emphasized by Fig. 1 suggests that the dipole contributions tend to increase the valence band discontinuity with respect to the predictions of Eq. (1).

III. EFFECTS OF ULTRATHIN METAL INTRALAYERS

The above upper limit for the effects of the microscopic interface dipoles, although small from the point of view of theory, is still large for device technology. Discontinuity changes of the order of 0.1 eV correspond to huge modifications of the transport properties.[1] Furthermore, while the

TABLE II. Correlation between heterojunction valence band discontinuities and Schottky barrier heights.

Interface	$\Delta E_g - \Delta E_B$ (eV)	Experimental ΔE_v (eV)[a]
Ge–Si	− 0.08	− 0.17
GaAs–Si	0.15	0.05
GaP–Si	0.64	0.80
GaSb–Si	− 0.23	0.05
InP–Si	0.48	0.57
CdSe–Si	0.95	1.20
CdTe–Si	0.54	0.75
ZnSe–Si	0.92	1.25
Si–Ge	0.08	0.17
GaAs–Ge	0.23	0.35
GaP–Ge	0.72	0.80
GaSb–Ge	− 0.15	0.20
InP–Ge	0.56	0.64
CdSe–Ge	1.03	1.30
CdTe–Ge	0.62	0.85
ZnSe–Ge	1.00	1.40

[a] Reference 16.

dipole effects are limited on the average,[16] unusually large effects can be found for particular systems, e.g., for interfaces involving polar semiconductors. Thus, the interface dipoles still offer a possible degree of freedom in tailoring the properties of heterojunction devices.

A recent and very interesting example of this approach was the use of a doping interface dipole to tune the AlGaAs–GaAs discontinuity by Capasso et al.[18] We present here, empirical evidence that ultrathin metal intralayers are an alternative approach to modify interface dipoles and therefore, the band discontinuities.

The search for intralayer-induced effects was conducted on interfaces involving polar semiconductors, for which large dipole effects are probable. In a first series of experiments,[19] we explored the effects of ultrathin Al intralayers at the CdS–Ge and CdS–Si interfaces. The experiments were performed with intralayer thicknesses of 0.5–1 Å.

The valence band discontinuity was measured using synchrotron-radiation photoemission spectroscopy, following the procedure discussed in Refs. 2 and 16. The above interfaces offer an important advantage; the valence band discontinuity is directly visible in the photoemission spectra as a double edge. Therefore, the intralayer effects could be directly monitored without need for an indirect analysis based on core-level shifts.[16] The intralayer effects were quantitatively studied by estimating the intralayer and overlayer induced energy shifts of the substrate photoemission features. This estimate was based on a least-square realignment in energy of those features.[19] The analysis revealed a systematic intralayer-induced increase of the valence band discontinuity, by 0.15 eV on the average.

We recently extended the above tests to the prototypical lattice-matched interface between cleaved ZnSe and Ge. For this interface we used intralayer thicknesses up to 2 Å of Al. Once again, the intralayer causes an increase of the valence band discontinuity. The magnitude of the effect in this case is large enough to be directly visible without a computer analysis of the substrate spectral features.

This fact is evident, for example, for the spectra of Fig. 2, which were taken on cleaved ZnSe covered by 4 Å of Ge, with or without a 2 Å thick Al intralayer. The intralayer-

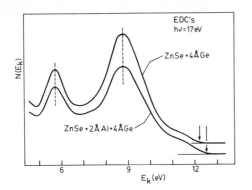

FIG. 2. Synchrotron-radiation photoemission spectra in the region near the valence band edge of cleaved ZnSe covered by 4 Å of Al, with and without a 2 Å thick Al intralayer. Notice the shift of the Ge-related leading edge, which corresponds to an intralayer-induced increase of ΔE_v by 0.2–0.3 eV.

induced changes in substrate band bending were compensated in this case by a simple visual realignment of the CdS-related features. This procedure is sufficient to detect the large intralayer-induced shift of the leading spectral edge. This edge corresponds to the top of the Ge valence band, and its shift corresponds to an increase of the valence band discontinuity. In the specific case of Fig. 2, the estimated magnitude of the increase (see vertical arrows) is approximately 0.2–0.3 eV. We emphasize that the weak Al-induced signal in this spectral region cannot account for the above effect. A detailed analysis which rules out other spurious factors can be found in Ref. 19.

At present, we do not have a microscopic explanation of the intralayer induced increase in the discontinuity. We can rule out simple models based on the formation of chemical bonds between substrate and intralayer. In fact, the expected direction of the corresponding dipoles would correspond to a decrease rather than to an increase of the discontinuity. On the other hand, Al was found to strongly influence the microdiffusion processes at II–VI interfaces.[20] Thus, the microscopic dipole responsible for the change in ΔE_v could be due to microdiffusion.[8] This possibility must be tested with

systematic studies of the microdiffusion processes at these interfaces, which are currently being performed.

ACKNOWLEDGMENTS

These experiments were supported by the National Science Foundation Grant No. DMR-84-21292. They were performed at the University of Wisconsin Synchrotron Radiation Center (supported by the NSF, Grant No. DMR-83-13523) and at the Frascati Synchrotron Radiation Project (PULS). We are glad to acknowledge stimulating discussions with C. Duke, A. Zunger, and J. Tersoff.

[a] Author to whom all correspondence should be addressed.
[1] A. G. Milnes and D. L. Feucht, *Heterojunctions and Metal–Semiconductor Junctions* (Academic, New York, 1972).
[2] For a recent review of the research in this area, see G. Margaritondo, Surf. Sci. (in press).
[3] C. Tejedor and F. Flores, J. Phys. C **11**, L19 (1978); F. Flores and C. Tejedor, *ibid.* **12**, 731 (1979).
[4] J. Tersoff, Phys. Rev. B **30**, 4874 (1984).
[5] C. B. Duke and C. Mailhiot, J. Vac. Sci. Technol. **3**, 915 (1985), and (private communication).
[6] M. J. Caldas, A. Fazzio, and A. Zunger, Appl. Phys. Lett. **45**, 671 (1984); A. Zunger, Annu. Rev. Mater. Sci. **15**, 411 (1985), and (private communication); J. M. Langer and H. Heinrich, Phys. Rev. Lett. **55**, 1414 (1985).
[7] W. A. Harrison, J. Vac. Sci. Technol. **14**, 1016 (1977).
[8] H. Kroemer, in *Proceedings of the NATO Advanced Study Institute on Molecular Beam Epitaxy and Heterostructures*, Erice, 1983, edited by L. L. Chang and K. Ploog (Martinus Nijhoff, The Netherlands) (in press).
[9] H. Unlu and A. Nussbaum, Solid State Electron. (in press).
[10] F. Bechstedt, R. Enderlein, and O. Heinrich, Phys. Status Solidi B **126**, 575 (1984).
[11] M. L. Cohen, Adv. Electron. Electron Phys. **51**, 1 (1980), and references therein.
[12] J. Pollman, Festkörperprobleme **20**, 117 (1980).
[13] J. L. Freeouf and J. M. Woodall, Surf. Sci. (in press); Appl. Phys. Lett. **39**, 727 (1981).
[14] J. Tersoff, Phys. Rev. Lett. **52**, 465 (1984).
[15] G. Margaritondo, Phys. Rev. B **31**, 2526 (1985).
[16] A. D. Katnani and G. Margaritondo, Phys. Rev. B **28**, 1944 (1983).
[17] R. L. Anderson, Solid State Electron. **5**, 341 (1934).
[18] F. Capasso, A. Y. Cho, K. Mohammed, and P. W. Foy, Appl. Phys. Lett. **46**, 664 (1985); F. Capasso, K. Mohammed, and A. Y. Cho, J. Vac. Sci. Technol. B **3**, 1245 (1985).
[19] D. W. Niles, G. Margaritondo, C. Quaresima, M. Capozi, and P. Perfetti (unpublished).
[20] L. J. Brillson, Surf. Sci. Rep. **2**, 123 (1982).

J. Vac. Sci. Technol. A, Vol. 4, No. 3, May/Jun 1986

306

VOLUME 58, NUMBER 12   PHYSICAL REVIEW LETTERS   23 MARCH 1987

Role of Virtual Gap States and Defects in Metal-Semiconductor Contacts

W. Mönch

Laboratorium für Festkörperphysik, Universität Duisburg, D-4100 Duisburg, Federal Republic of Germany
(Received 20 November 1986)

Chemical trends of barrier heights reported for metal- and silicide-silicon contacts are analyzed. The data are easily explained when both virtual gap states of the complex band structure of the semiconductor and electronic levels of defects created in the semiconductor close to the interface during its formation are considered. The virtual gap states determine the barrier heights when either the defect density is low or the defects are completely charged or all neutral.

PACS numbers: 73.30.+y, 73.20.−r, 73.40.Ns

The rectifying properties of metal-semiconductor contacts, which were discovered by Braun,[1] are caused by depletion layers on the semiconductor side of the interface, as was first shown by Schottky.[2] The fundamental parameter which characterizes such a junction is its barrier height, i.e., the energy distance from the Fermi level to the bottom of the conduction band at the interface when the semiconductor is doped n type. A basic understanding of Schottky contacts thus needs a model which explains the chemical trends of the barrier heights observed with different metal-semiconductor pairs.

Schottky[3] and Mott[4] proposed the barrier height to equal the difference of the work function of the metal and the electron affinity of the semiconductor. Although for a given semiconductor the barrier heights are generally found to increase when the work function of the metal in contact becomes larger, the simple Schottky-Mott rule is not obeyed by the experimental data. Bardeen[5] attributed this discrepancy to the presence of interface states. They could accommodate charge which is transferred between the metal and the semiconductor because of their generally different electronegativities. This means that a dipole layer exists at the interface. Since the work function of metals and their electronegativities were found to be linearly related, such interface states intuitively explain that the barrier heights are increased by metals with larger work functions but do not follow the Schottky-Mott rule. Two basically different models on the physical nature of such interface states have been suggested. In the following, they will be briefly reviewed.

The first model, which was introduced by Heine,[6] assumes that within the band gap of the semiconductor the wave functions of the metal electrons are tailing into the virtual gap states (VGS) of the complex band structure of the semiconductor. Since the virtual gap states are split off from the valence and the conduction band, their character varies across the gap from mostly donor type close to the top of the valence band to mostly acceptor type close to the bottom of the conduction band. The charge transferred between the metal and the semiconductor then pins the Fermi level above, at, or below the charge-neutrality level E_0 of the virtual gap states when

the electronegativity of the metal is smaller, equal to, and larger than, respectively, the one of the semiconductor. In the following, three different and independent results will be presented which support the VGS model of Schottky contacts.

For the column-IV elemental and the III-V compound semiconductors Tersoff[7] has calculated the charge-neutrality levels of the VGS. He has obtained good agreement between $(E_{cs} - E_0)$ and the barrier heights Φ_{Bn} experimentally determined with gold Schottky contacts on samples doped n type. This finding is supporting the VGS model since the electronegativities of gold and of the semiconductors only differ slightly. Second, the adsorption of cesium[8] and of chlorine[9] was found to pin the Fermi level above and below, respectively, the charge-neutrality level of the VGS at cleaved GaAs-(110) surfaces. Since the electronegativities of cesium and of chlorine are smaller and larger, respectively, by almost the same amount than the value of gallium arsenide the results mentioned are again in support of the VGS model.

The third indication is represented by the data plotted in Fig. 1. When interface states are assumed to be present in a metal-semiconductor junction the barrier height Φ_{Bn} increases proportionally to the work function Φ_M of the metal.[13,14] The slope parameter $S = d\Phi_{Bn}/d\Phi_M$ only depends on the product of the density of states $D_{vs}(E_0)$ around the charge-neutrality level of the interface states and the width δ of the related dipole layer as

$$S = [1 + e_0^2 D_{vs}(E_0)\delta/\varepsilon_0]^{-1}. \qquad (1)$$

In the VGS model, this product $D_{vs}\delta$ is determined by the average band-gap energy of the semiconductor which, on the other hand, is related to the electronic polarizability $(\varepsilon_\infty - 1)$ of the semiconductor.[12] Although in some cases the experimental slope parameters are not well defined,[10,11] the S values of nineteen different semiconductors follow a pronounced chemical trend[12] when $(1/S - 1)$ is plotted over $(\varepsilon_\infty - 1)$ as shown in Fig. 1. A least-squares fit to the data yields

$$(1/S - 1) = 0.1(\varepsilon_\infty - 1)^2 \qquad (2)$$

and a regression coefficient $r = 0.91$. This result again

© 1987 The American Physical Society

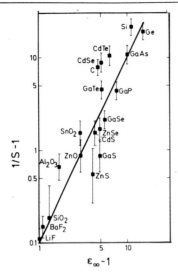

FIG. 1. Slopes $S = d\Phi_{Bn}/d\Phi_M$ plotted vs the electronic contribution ε_∞ of the dielectric constant of the semiconductor. The data were taken from Refs. 10 and 11 in the manner of Ref. 12.

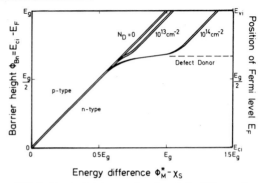

FIG. 2. Barrier height as a function of the difference between the internal part of the metal work function Φ_M^* and the electron affinity X_s of the semiconductor for three different densities of defects (shown schematically in the manner of Refs. 21 and 22).

strongly supports the VGS model of metal-semiconductor contacts.

The second model which was proposed by Wieder[15] and Spicer et al.[16] identifies the interface states in Schottky contacts as electronic states of native defects which are created during the formation of the junctions (see Mönch[17] for a review). The defect model was motivated by the observations that Schottky barriers on III-V compound semiconductors were found to be insensitive to within 0.2 eV to the metals used and to follow no apparent chemical trend. Up until now, no spectroscopic evidence has emerged for any defect such as vacancies or antisite defects which have been discussed theoretically.[18,19] This might indeed prove to be difficult since in many cases chemical reactions and intermixing were observed.[20] Therefore, interfaces between most metals and III-V compound semiconductors are difficult to characterize with respect to local variations in chemical composition.

The influence of defect levels on the barrier height of metal-semiconductor junctions was theoretically studied by Zur, McGill, and Smith[21] and by Duke and Mailhiot.[22] They placed defects 5–10 Å apart from the interface into the semiconductor. The metals were described by a jellium model. The results of these calculations are schematically explained in Fig. 2. When the area density N_D of defects is kept below 10^{13} cm^{-2} the position of the Fermi level moves across the energy gap of the semiconductor to the same extent as the internal work function Φ_M^* of the jellium metal is increased. For larger defect densities, the Fermi level gets transitionally pinned

at the defect level, until the defects are all charged, and then further moves towards the top of the valence band at the interface as a function of metal work function.

These models have been developed in parallel to many experimental studies of metal-semiconductor junctions which have provided a huge body of data on electronic, structural, and chemical properties of such interfaces. In the following, the analysis of experimental results and the search for chemical trends shall be restricted to metal-silicon contacts for the following reason. The interfaces of Schottky contacts on III-V compound semiconductors were found to be intermixed in many cases[20] and they are thus difficult to characterize chemically. Metal-silicon junction, on the other hand, can be prepared with quite abrupt interfaces since the controlled formation of silicides, which are mostly metallic, is a well-established technique.[23] Cross-sectional pictures obtained with high-resolution transmission-electron microscopy have proven that, for example, epitaxial films of NiSi$_2$, NiSi, and Pd$_2$Si may be grown on silicon substrates (see, e.g., the work of Liehr et al.[24] and Ho[25]).

In searching for chemical trends of the barrier heights measured now with metal-silicon contacts, the main difficulty arises with the ordering of the metals. The first choice, which was motivated by the early Schottky-Mott rule, has still remained the metal work function which, however, contains an internal part plus a surface dipole contribution. The internal part of the work function, which is of interest in interfaces, may be approximated by the electronegativity of the metal. Here, the most popular scale has been the one developed by Pauling.[26] He has designed his set of values to describe the partly ionic character of covalent bonds, and it is this field where Pauling's electronegativities have their merits in semiconductor bulk and surface physics, too (see, e.g., the work of Mönch[27]). In metal-semiconductor con-

1261

tacts, on the other hand, the ionicity of metallic bonds comes into play. Chemical trends in the properties of metal alloys and intermetallic compounds have been successfully described by another set of electronegativities which were derived by Miedema, el Châtel, and de Boer.[28] In the present paper, the further analysis will use the electronegativities proposed by Miedema.

In Fig. 3 barrier heights measured with metal- and with silicide-silicon junctions are plotted over electronegativities based on the Miedema scale. For silicides M_mSi_n, the geometric mean $(X_M^m X_{Si}^n)^{1/(m+n)}$ of the metal and the silicon electronegativities were taken.[26,31] In this respect the plot differs from a similar one by Schmid[29] who has introduced Miedema's electronegativities in the discussion of metal-silicon junctions but has plotted Φ_{Bn} vs X_M only. In Fig. 3, the data points are obviously arranged in two groups. The straight line drawn in full is a least-squares fit to fifteen data points and is given by

$$\Phi_{Bn} = 0.17 \langle X_M \rangle - 0.04 \text{ eV},$$

with a regression coefficient $r = 0.98$. The marked data point labeled CNL represents the charge-neutrality level of the virtual gap states of the complex band structure in silicon ($X_{Si} = 4.7$ eV) as calculated by Tersoff.[7] Obviously, the charge-neutrality level of the VGS fits exactly into that straight line. This finding implies that those barrier heights, which define that straight line in Fig. 3, are determined by the VGS of silicon. The broken line

connecting another eighteen data points resembles the shape of the curve shown in Fig. 2 which was obtained for heavily defected metal-semiconductor contacts.

Considering the VGS and the defect model of metal-semiconductor junctions as outlined above, the data plotted in Fig. 3 suggest the following explanation. Those metal-silicon junctions, the barrier heights of which are found close to the straight-line fit, are exhibiting a density of defects below approximately 10^{13} cm^{-2} and their barrier heights are thus determined by the tailing of the metal electron wave functions into the virtual gap states of the silicon bond structure. The other silicon Schottky contacts contain a large defect density of approximately 10^{14} cm^{-2}. As the inflection of the dashed line indicates, one defect level is located at approximately 0.62 eV below the bottom of the conduction band. That defect level was already concluded by Schmid[29] from his Φ_{Bn}-vs-X_M plot for the silicide-silicon junctions. For barrier heights less than about 0.6 eV the dashed curve is running in parallel to the straight line, which is determined by the VGS of silicon, but is shifted by 0.1 eV to lower values. This indicates the presence of another defect level at or above $1.12 - 0.37$ eV $= 0.75$ eV above the top of the valence band. The explanations just given for the data plotted in Fig. 3 are strongly supported by results of a study on nickel-silicide-silicon interfaces recently published by Liehr et al.,[24] which will be discussed in the following.

Both groups of data points in Fig. 3 contain results

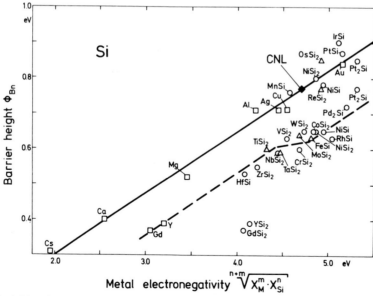

FIG. 3. Barrier heights of metal- and silicide-silicon contacts vs the effective Miedema electronegativities. The barrier heights were adopted from Refs. 24, 29, and 30.

VOLUME 58, NUMBER 12   PHYSICAL REVIEW LETTERS   23 MARCH 1987

from the study of Liehr *et al.* for NiSi$_2$ and NiSi which differ by $\Delta\Phi_{Bn} \approx 0.15$ eV. With both nickel silicides the larger values were found when epitaxial silicides were grown, which in the case of the disilicide consisted of either pure type-*A* or type-*B* interfaces. With a mixture of both types, which only differ in the stacking sequence when passing the interface, the lower value of the barrier height was observed regularly. Cross-sectional TEM pictures always revealed such interfaces exhibiting the lower Φ_{Bn} values to be less perfect, containing phase domain boundaries with faceted or stepped structures.[24] With the nickel monosilicide, large barrier heights were also observed only when the interfaces were of the same high degree of perfection as found with single-type disilicide. "The degree of perfection of the interfacial structure is more important than specific epitaxy in determining the barrier height," as was pointed out by Schmid *et al.*[32] From preliminary evaluations of further capacitance spectroscopy studies they also computed approximately 10^{12} to 10^{13} interface states per square centimeter for the single-type NiSi$_2$-Si interfaces but about 1 order of magnitude more for the mixed-type interfaces. These experimental findings by Schmid *et al.* are in excellent agreement with the explanations of the data plotted in Fig. 3 which were given above.

The results of the present paper may be summed up as follows. The analysis of the chemical trend of the barrier heights reported for 31 different metal- and silicide-silicon interfaces has revealed that both VGS and defects are needed to explain the complete set of experimental data. When, however, the experimental conditions during the preparation of Schottky contacts can be controlled such as to reduce the defect density to below approximately 10^{13} per square centimeter then the barrier height is determined by the virtual gap states of the semiconductor band structure only. In highly defected Schottky barriers the virtual gap states also determine the barrier heights when all the defects are either charged or neutral. A preliminary analysis of barrier heights observed experimentally with metal-GaAs(110) contacts show that the same concepts apply to these interfaces, too.

[1]F. Braun, Pogg. Ann. **153**, 556 (1874).

[2]W. Schottky, Naturwissenschaften **26**, 843 (1938).

[3]W. Schottky, Phys. Z **41**, 570 (1940).

[4]N. F. Mott, Proc. Cambridge Philos. Soc. **34**, 568 (1938).

[5]J. Bardeen, Phys. Rev. **71**, 717 (1947).

[6]V. Heine, Phys. Rev. **138**, A1689 (1965).

[7]J. Tersoff, Phys. Rev. Lett. **52**, 465 (1984), and Phys. Rev. B **30**, 4874 (1984).

[8]W. E. Spicer, P. E. Gregory, P. W. Chye, I. A. Babalola, and T. Sukegawa, Appl. Phys. Lett. **27**, 617 (1975).

[9]D. Troost, L. Koenders, Y.-L. Fan, and W. Mönch, to be published.

[10]S. Kurtin, T. C. McGill, and C. A. Mead, Phys. Rev. Lett. **22**, 1433 (1970).

[11]M. Schlüter, Phys. Rev. B **17**, 5044 (1978).

[12]W. Mönch, in *Festkörperprobleme: Advances in Solid State Physics*, edited by P. Grosse (Vieweg, Braunschweig, 1986), Vol. 26, p. 67.

[13]A. M. Cowley and S. M. Sze, J. Appl. Phys. **36**, 3212 (1965).

[14]C. Tejedor, F. Flores, and E. Louis, J. Phys. C **10**, 2163 (1977).

[15]H. Wieder, J. Vac. Sci. Technol. **15**, 1498 (1978).

[16]W. E. Spicer, P. W. Chye, P. R Skeath, and I. Lindau, J. Vac. Sci. Technol. **16**, 1422 (1979).

[17]W. Mönch, Surf. Sci. **132**, 92 (1983).

[18]M. S. Daw and D. L. Smith, Phys. Rev. B **20**, 5150 (1979).

[19]R. E. Allen and J. D. Dow, J. Vac. Sci. Technol. **19**, 383 (1981).

[20]L. N. Brillson, Surf. Sci. Rep. **2**, 123 (1982).

[21]A. Zur, T. C. McGill, and D. L. Smith, Phys. Rev. B **28**, 2060 (1983).

[22]C. B. Duke and C. Mailhiot, J. Vac. Sci. Technol. B **3**, 1170 (1985).

[23]S. P. Murarka, J. Vac. Sci. Technol. **17**, 775 (1980).

[24]M. Liehr, P. E. Schmid, F. K. Le Goues, and P. S. Ho, Phys. Rev. Lett. **54**, 2139 (1985).

[25]P. S. Ho, J. Vac. Sci. Technol. A **1**, 745 (1983).

[26]L. Pauling, *The Nature of the Chemical Bond* (Cornell Univ. Press, Ithaca, New York, 1960).

[27]W. Mönch, Solid State Commun. **58**, 215 (1986).

[28]A. R. Miedema, P. F. el Châtel, and F. R. de Boer, Physica (Amsterdam) **100B**, 1 (1980).

[29]P. E. Schmid, Helv. Phys. Acta **58**, 371 (1985).

[30]W. Mönch, Surf. Sci. **21**, 443 (1970).

[31]J. L. Freeouf, Solid State Commun. **33**, 1059 (1980).

[32]P. E. Schmid, M. Liehr, F. K. Legoues, and P. S. Ho, in *Thin Films—Interfaces and Phenomena*, edited by R. J. Nemanich, P. S. Ho, and S. S. Lau (Materials Research Society, Pittsburgh, 1986), p. 469.

Pressure Dependence of Band Offsets in an InAs-GaSb Superlattice

L. M. Claessen,[a] J. C. Maan, M. Altarelli, and P. Wyder

Max Planck Institut für Festköperforschung, Hochfeld Magnetlabor, F38042 Grenoble Cedex, France

and

L. L. Chang and L. Esaki

IBM Thomas J. Watson Research Center, Yorktown Heights, New York 10598

(Received 2 June 1986)

Using magneto-optical methods, we have measured the pressure dependence of the energy difference between subbands in an InAs-GaSb superlattice associated with the GaSb valence and the InAs conduction bands, respectively. The experimental results allow a determination of the pressure dependence of the energy separation between the InAs conduction band and the GaSb valence band which is found to decrease at a rate of 5.8 meV/kbar. This result shows that both the conduction- and the valence-band offsets are pressure dependent. Therefore these experiments constitute a critical test for different theories of band lineup.

PACS numbers: 73.40.Lq, 62.50.+p, 78.20.Ls

Usually, the relative positions of the energy bands within a single bulk semiconductor are well known. However, the positions of the band edges in one semiconductor relative to those in another when they are in contact with each other (band lineup) provide a problem in solid-state physics which is neither experimentally nor theoretically well understood. Yet this problem has become particularly relevant, and at the same time experimentally accessible, through the possibility of the growth of high-quality interfaces and heterojunctions by modern growth techniques like molecular-beam epitaxy (MBE).[1,2]

Conceptually the band-lineup problem can be divided into two parts: (i) Which energy level must be lined up at the interface in order to determine the band offsets, and (ii) where does this level lie with respect to the band edges? There exist several band-lineup theories, but the accuracy of both experimental and theoretical values is not sufficient to distinguish clearly between them. The essential difference between these theories is their choice of this energy level. As hydrostatic pressure has a strong effect on the relative positions of the energy bands in a solid, and therefore in general on the positions of the bands with respect to this common energy, it is of considerable importance to investigate the band lineup in a semiconductor interface as a function of pressure and to compare the results with existing band-lineup models. For this purpose we present experimental results of the pressure dependence of the lineup of the bands at the InAs-GaSb interface, by use of magneto-optical methods.

We have chosen the InAs-GaSb interface because it has been studied experimentally very carefully before[3-6] and because this system provides one of the most severe tests for any band-lineup theory. The peculiarity of this system is that the conduction-band (CB) edge of InAs is at a lower energy than the valence-band (VB) edge of GaSb. This fact leads to a strong dependence of the electronic properties of InAs-GaSb heterostructures, e.g., superlattices (periodic alternate thin GaSb and InAs layers), on the exact value of this energy overlap. Several results of optical experiments on this system can be explained with a value of 150 meV for this difference, with an experimental error of 50 meV.[4-6] These experimental values are probably the most accurately known in the literature; note that, for instance, the lineup of the most extensively studied GaAs-Ga$_{1-x}$Al$_x$As system is still controversial.[7] Hydrostatic pressure has a strong effect on the energies of the bands in these semiconductors. The energy gap, E_G, increases by 10 meV/kbar and 14 meV/kbar for InAs and GaSb, respectively.[8] In particular, at easily attainable pressures (10 kbar), the band-gap variation is comparable to the energy overlap between the valence and the conduction bands.

The main features of the electronic band structure of the investigated superlattice (consisting of many layers of alternate 12-nm InAs and 8-nm GaSb, grown on a GaSb ⟨100⟩ substrate) is illustrated in Fig. 1. This superlattice shows an electronlike level (E_1, at higher energy than the InAs bulk CB edge because of the confinement in the InAs layer) and a holelike level (H_1, at a lower energy than the GaSb VB edge because of the confinement in the GaSb layer). An extensive review of the electronic properties of this kind of superlattices can be found in Ref. 6. In a simplified manner, appropriate for the understanding of the present experiments, the energy difference between the E_1 and the H_1 subband edge at zero wave vector is given by the InAs-CB GaSb-VB discontinuity Δ minus the confinement energy for the electrons (the shift of the subband with respect to the InAs band edge) minus the hole confinement energy. Therefore a measurement of $E_1 - H_1$ as a function of pressure provides direct information about the pressure dependence of the band lineup. Previous measure-

2556   © 1986 The American Physical Society

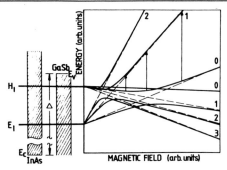

FIG. 1. One period of an InAs-GaSb superlattice showing the band lineup of the InAs conduction- and the GaSb valence-band edges, and the positions of the electronlike subband (E_1) and the holelike subband (H_1). In the right-hand part of the figure the holelike (moving downward) and electronlike (moving upward) Landau levels of these subbands in the absence of (dashed lines) and in the presence of (solid lines) coupling between them for a magnetic field perpendicular to the layers of the superlattice are shown schematically. The arrows indicate transitions which have been observed experimentally. Because of the coupling between the electronlike and holelike Landau levels these transitions have an interband character in the vicinity of, and an intraband (cyclotron resonance) character further away from, the crossing.

ments[4,5] on the same sample by use of far-infrared magneto-optical transmission at zero pressure have determined E_1 to be 40 meV lower than H_1, and subsequent theoretical calculations[9] have shown that these experiments can be explained with band-structure calculations using a value of 150 meV for Δ. Here we report the results of the same experiment, i.e., measurement of $E_1 - H_1$ by use of far-infrared absorption in a magnetic field, for different hydrostatic pressures. This is illustrated in Fig. 1, which shows schematically the holelike and the electronlike Landau levels of the sample as a function of a magnetic field perpendicular to the layer. As usual, the continuum of states for motion in the plane of the layer is split into a set of equidistant linearly field-dependent levels, with hole levels moving downward and electron levels moving upward in energy. However, a small interaction between the holelike and the electronlike Landau levels leads to an anticrossing between the two as indicated in the figure. Transitions which can be observed in the present experiment are also shown. The experiments were done at $T = 4.2$ K in a commercial Cu-Be liquid pressure cell, with mineral oil as the pressure-transferring medium (see Ref. 8, p. 184). An optically pumped cw molecular gas laser was used as radiation source.

Representative transmission curves at different radiation energies as a function of the magnetic field at fixed pressure and a plot of the observed transmission minima as a function of radiation energy are shown in Fig. 2. If,

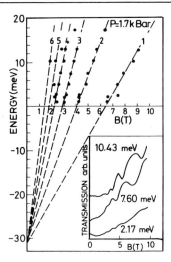

FIG. 2. Observed transition energy between subbands of an InAs-GaSb superlattice as a function of magnetic field at a pressure of 1.7 kbar. The inset shows the experimental spectra. The dashed lines show the linear extrapolation (i.e., with the assumption of pure interband transitions) indicating a material with a negative energy gap $E_1 - H_1$. The transitions are labeled according to the noninteracting model in which the quantum number is that of the two participating Landau states.

in a qualitative way, one assumes unperturbed, equidistant, linearly field-dependent Landau levels (no anticrossing), a linear extrapolation to zero field (the dashed lines) leads to a negative energy gap, i.e., $E_1 - H_1$. In this way the applied magnetic field is used to obtain the zero-field properties of the sample. Figure 3 shows the pressure dependence of the last high-field transition of

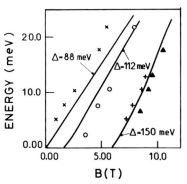

FIG. 3. Pressure dependence of the high-field transition at (triangles) 0, (plusses) 1.7, (circles) 6.6, and (crosses) 10.7 kbar in an InAs-GaSb superlattice. The lines are theoretical calculations of these transitions, with assumption of a linear pressure dependence of the band-lineup parameter Δ.

Fig. 2 with increasing pressure. This transition moves to lower magnetic fields, while at the same time the slope of the energy-versus-field dependence decreases by 30%. These results are a direct consequence of the pressure dependence of the lineup at the interface: If the energy difference between the GaSb VB and the InAs CB decreases, the energy separation between the E_1 and the H_1 subbands will also decrease. If we assume no interaction between the hole and electronlike Landau levels (i.e., simple interband Landau-level transitions obeying the selection rule $\Delta n = 0$, no anticrossing), the data can be analyzed by the drawing of straight lines through the transitions as shown in Fig. 2 and evaluation of the intercept with the energy scale at zero field as $E_1 - H_1$ at a given pressure. In this way, one finds that this quantity decreases linearly with pressure at roughly 4 meV/kbar. Obviously such a simplified analysis disregards the coupling between the energy levels and, in particular, does not explain the change of slope which is observed. As can be seen from Fig. 1, inclusion of the coupling between the levels has two effects. First, as $E_1 - H_1$ decreases with increasing pressure, the transitions at a fixed energy (i.e., $0 \rightarrow 1$, as indicated in the figure) move to lower magnetic field, and second, the field at which anticrossing occurs decreases. This latter effect results in a gradual change in character of this particular transition, i.e., a changeover from a more interbandlike transition with a steeper slope to a more intrabandlike (cyclotron resonance) transition with a steeper slope to a more intrabandlike (cyclotron resonance) transition with a lesser slope. To analyze this subtle band-structure behavior in more detail, we calculate the pressure dependence of the full band structure within the framework of a six-band $\mathbf{k \cdot p}$ model as described elsewhere.[9] In the calculation, the full VB-CB coupling is taken into account, both within each material and between the adjacent unstrained materials, by use of proper boundary conditions for the wave functions at the interface. Standard values for the band parameters are used and the only adjustable variable is Δ which is assumed to be linearly pressure dependent. The results are shown in Fig. 3 as the lines. The best agreement with the experiments is obtained by use of a decrease of Δ of 5.8 meV/kbar. This slightly stronger pressure dependence is obviously a consequence of the inclusion of all other effects of the band structure neglected in the more simple analysis (nonparabolicity, subband coupling, effects of this coupling on the confinement energies, etc.). The calculation also shows the tendency of the slope to decrease with pressure as experimentally observed. The essential experimental result therefore is that the offset between the InAs CB and the GaSb VB reduces at a rate of 5.8 meV/kbar. Note that because of a slight lattice mismatch between GaSb and InAs, the band offset one measures in superlattice experiments will be affected by strain.[10-12] However, since the compressibilities of InAs

and GaSb are nearly equal,[8] no additional strain is induced by the pressure, and hence the pressure dependence of the offset is not affected.

As the InAs and GaSb energy gaps increase by 10 meV/kbar and 14 meV/kbar, respectively, it is therefore clear that our experimental results imply that if the pressure is increased, neither the valence bands (which would lead to a decrease of Δ of 10 meV/kbar) nor the conduction bands (which similarly give 14 meV/kbar) in both materials remain constant. It is evident that if the criterion which determines the band offset is the lineup of a reference level, this will be the same for all pressures. The position of the energy bands with respect to that level, however, will in general be pressure dependent. Band-lineup models should be able to explain consistently this pressure dependence of the bands and the band offsets. In this connection we will briefly discuss different theoretical approaches. The most recent suggestion, proposed by Langer and Heinrich,[13] derives the valence-band offsets by use of transition-metal impurity levels as the common energy. To be consistent with our experimental results the position of these deep-level impurities must show a pressure dependence with respect to both the VB and the CB. For GaAs this is indeed so (the Cr level increases by 4.8 meV/kbar with respect to the VB.[14] Unfortunately, in InAs and GaSb deep-level impurity levels are not studied in sufficient detail to make a more quantitative statement. If, as recently postulated,[15,16] the charge-neutrality level is used as the reference energy, the pressure dependence can be estimated crudely from the work of Tersoff[16] by the assumption that the relative position of this level in the energy gap is not pressure dependent. This estimate gives 2 meV/kbar for Δ, which is close to the experiment result. It should be noted that the pioneering Harrison[17] model, which measures the position of the valence bands relative to the average atomic potential in the semiconductors, would predict that Δ varies as the InAs gap (the relative positions of the valence bands being almost pressure independent), which is not in agreement with the experiment. From other methods, such as the electron affinity rule,[18] the theory by Frensley and Kroemer,[19] and ab initio calculations,[20] it is rather difficult to extract predictions about the pressure dependence of the band lineup.

In summary, we have measured the pressure dependence of the InAs-GaSb band lineup. Our data show that offset between the GaSb VB and the InAs CB decreases by 5.8 meV/kbar. This value cannot be explained by the pressure dependence of the energy gaps in the bulk materials alone. In addition, the experiments show a pressure-dependent gradual change from interband to intraband transitions, effects which can be explained by our taking into account the full band structure of the system. We believe that the study of the pressure dependence of the band offset may be a useful

2558

tool for the test of heterojunction lineup theories, the main point being that since the band structure of each material at the interface is strongly pressure dependent, the comparison of the band offsets with and without pressure is in some sense equivalent to a comparison of different samples.

It is a great pleasure to acknowledge many useful discussions about this work with G. Martinez.

[a]Also at Research Institute for Materials, University of Nijmegen, Toernooiveld, NL-6525 ED Nijmegen, The Netherlands.

[1]H. Kroemer, Surf. Sci. 132, 543 (1983).

[2]H. Kroemer, J. Vac. Sci. Technol. B 2, 433 (1984).

[3]G. A. Sai-Halasz, L. L. Chang, J.-M. Welter, C.-A. Chang, and L. Esaki, Solid State Commun. 27, 935 (1978).

[4]Y. Guldner, J. P. Vieren, P. Voisin, M. Voos, L. L. Chang, and L. Esaki, Phys. Rev. Lett. 45, 1719 (1980).

[5]J. C. Maan, Y. Guldner, J. P. Vieren, P. Voisin, M. Voos, L. L. Chang, and L. Esaki, Solid State Commun. 39, 683 (1981).

[6]L. L. Chang, J. Phys. Soc. Jpn. 49, Suppl. A, 997 (1980).

[7]G. Duggan, J. Vac. Sci. Technol. B 3, 1224 (1985).

[8]G. Martinez, in Optical Properties of Solids, edited by M. Balkanski, Handbook on Semiconductors Vol. 2 (North-Holland, Amsterdam, 1980) p. 194.

[9]A. Fasolino and M. Altarelli, Surf. Sci. 142, 322 (1984).

[10]Y. Guldner, G. Bastard, J. P. Vieren, M. Voos, J. P. Faurie, and A. Million, Phys. Rev. Lett. 51, 907 (1983).

[11]S. P. Kowalczyk, J. T. Cheung, E. A. Kraut, and R. W. Grant, Phys. Rev. Lett. 56, 1605 (1986).

[12]G. Y. Wu and T. C. McGill, Appl. Phys. Lett. 47, 634 (1985).

[13]J. M. Langer and H. Heinrich, Phys. Rev. Lett. 55, 1414 (1985).

[14]A. M. Hennel and G. Martinez, Phys. Rev. B 25, 1039 (1982).

[15]F. Flores and C. Tejedor, J. Phys. C 12, 731 (1979).

[16]J. Tersoff, Phys. Rev. B 30, 4874 (1984).

[17]W. A. Harrison, J. Vac. Sci. Technol. 14, 1016 (1977).

[18]R. L. Anderson, Solid State Electron. 5, 341 (1962).

[19]W. R. Frensley and H. Kroemer, Phys. Rev. B 16, 2642 (1977).

[20]M. L. Cohen, Adv. Electron. Electron Phys. 51, 1 (1980).

Tunable barrier heights and band discontinuities via doping interface dipoles: An interface engineering technique and its device applications

F. Capasso, K. Mohammed, and A. Y. Cho

AT&T Bell Laboratories, Murray Hill, New Jersey 07974

(Received 14 February 1985; accepted 4 March 1985)

We present, for the first time, a technique to effectively tune barrier heights and band discontinuities at semiconductor heterojunctions using doping interface dipoles (DID). The DID consists of two ultrathin ionized donors and acceptor sheets *in situ* grown within 100 Å of the heterointerface by MBE. Using a DID the photocollection efficiency of an AlGaAs/GaAs abrupt heterojunction has been increased by one order of magnitude. Detailed investigations show the importance of nonequilibrium and hot electron transport effects and tunneling in heterostructures with DIDs. Several new applications of this concept are discussed. A DID can be used to convert a type I heterojunction in a type II or staggered heterojunction. A DID at the heterointerfaces of superlattice and staircase avalanche photodiodes can further enhance the ionization of electrons at the conduction band steps. The speed of heterojunction photodetectors such as the ones employing $InP/Ga_{0.47}In_{0.33}As$ can be enhanced by a DID without requiring grading of the interface. The effects of doping fluctuations in the charge sheets of the DIDs on transport properties across the heterointerface are also briefly discussed. The DID represents a new interface engineering technique in the sense that it allows one to selectively engineer the energy band diagram of a heterostructure within less than 100 Å from the heterointerface and as such has important implications for the physics of interfaces and the design of novel devices.

I. DOPING INTERFACE DIPOLES

Barrier heights and band edge discontinuities at heterojunctions are playing an increasingly important role in the physics of semiconductor–semiconductor interfaces[1] and in the design of novel heterojunction devices.[2] Band edge discontinuities are usually treated as basic properties of a given heterojunction. However, considerable evidence indicates that the effective discontinuities in "real world" heterojunctions can be a function of the substrate crystal orientation,[3] of the starting surface stoichiometries and reconstructions[4] and, possibly, of the order of growth of the two layers.[5]

Thus, the question has been raised if band discontinuities can in some way be tuned in a given heterojunction by properly controlling the interface chemistry and nanostructure over an atomic scale during the growth.[4]

Compositional grading at the interface is an effective way to control barrier heights, and is extensively used in semiconductor devices. This method, however, eliminates the abruptness of the heterojunction. In many cases one would like to preserve such a feature while simultaneously being able to tune the barrier height. In this paper we demonstrate for the first time that the barrier heights and effective band discontinuities at an abrupt, intrinsic heterojunction can be artificially tuned via the use of a doping interface dipole (DID) grown by molecular beam epitaxy (MBE). This concept is illustrated in Fig. 1. Figure 1(a) represents the band diagram of an abrupt heterojunction. The material is assumed to be undoped (ideally intrinsic) so that we can neglect band-bending effects over the short distance (a few hundred Å) shown here.

We next assume to introduce *in situ*, during the growth of a second identical heterojunction, one sheet of acceptors and one sheet of donors of identical doping concentrations, at the same distance $d/2$ ($\lesssim 100$ Å) from the interface [Fig. 1(b)].

The doping density N is in the 1×10^{17}–$1 \times 10^{19}/cm^3$ range, while the sheets' thickness t is kept small enough so that both are depleted of carriers ($t \lesssim 100$ Å). The DID is therefore a microscopic capacitor. The electric field between the plates is σ/ϵ, where $\sigma = eNt$. There is a potential difference $\Delta\Phi = (\sigma/\epsilon)d$ between the two plates of the capacitor. Thus the DID produces abrupt potential variations across a heterojunction interface by shifting the relative positions of

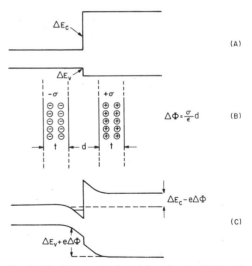

FIG. 1. (a) Band diagram of an intrinsic heterojunction. (b) Schematics of doping interface dipole. σ is the sheet charge density and $\Delta\Phi$ the dipole potential difference. (c) Band diagram of an intrinsic heterojunction with doping interface dipole. For simplicity of illustration, the potential drop across each charge sheet $[1/2(\sigma/\epsilon)t]$ is assumed small compared to $\Delta\Phi$.

315

the valence and conduction bands in the two semiconductors outside the dipole region [Fig. 1(c)]. This is done without changing the electric field outside the DID.

The valence band barrier height at the heterojunction is increased by the DID to a value $\Delta E_v + e\Delta\Phi + (\sigma/\epsilon)t$. If $\Delta\Phi$ is dropped over a distance of a few tens of Å and the total potential drop across the charge sheets $[= (\sigma/\epsilon)t]$ is small compared to $\Delta\Phi$, the valence band discontinuity has effectively been increased by $e\Delta\Phi$.

The DID reduces the energy difference between the condition band edges on both sides of the heterointerface to $\Delta E_c - e\Delta\Phi$. On the low gap side of the heterojunction a triangular quantum well is formed. Since the electric field in this region is typically $\gtrsim 10^5$ V/cm and $e\Delta\Phi \approx 0.1$–0.2 eV, the bottom of the first quantum subband E_1 lies near the top of the well. Therefore, the thermal activation barrier seen by an electron on the low gap side of the heterojunction is reduced from ΔE_c to $\Delta E_c - e\Delta\Phi/2$.

Electrons can also tunnel through the thin ($\lesssim 100$ Å) triangular barrier; this further reduces the effective barrier height. In the limit of a DID a few atomic layers thick, having a potential $\Delta\Phi$, the triangular barrier is totally transparent and the conduction band discontinuity is lowered to $\Delta E_c - e\Delta\Phi$. By inverting the position of the donor and acceptor sheets one can instead increase the conduction band discontinuity and decrease the valence band one.

Note that experimental evidence suggests that "natural" dipoles may occur at polar heterojunction interfaces causing the orientation dependence of band discontinuities.[3] Interface defects may also produce dipoles capable of altering band discontinuities.[6]

II. PHOTOCOLLECTION MEASUREMENTS

To verify the barrier lowering due to the DID, we have grown by MBE[7] heterojunction AlGaAs/GaAs pin diodes

on p-type (100)GaAs substrates. Two types of structures were grown: one with and the other without dipole. The one with dipole consists of four GaAs layers first, $p^+ > 10^{18}$/cm^3 (5000 Å), undoped (5000 Å), $p^+ = 5 \times 10^{17}$/cm^3 (100 Å), forming the negatively charged sheet of the dipole, and undoped (100 Å), followed by four Al$_{0.26}$Ga$_{0.74}$As layers, undoped (100 Å), $n^+ = 5 \times 10^{17}$/cm^3 (100 Å), forming the positively charged sheet of the dipole, undoped (5000 Å), and $n^+ > 10^{18}$/cm^3 (5000 Å). The second type of structure is identical, with the exception that it doesn't have DID. They were grown consecutively in the MBE chamber without breaking the vacuum to ensure virtually identical growth conditions. It is important to note that the charge sheets were introduced by controlling the shutters of the doping ovens, without interrupting the growth of the GaAs and AlGaAs layers. This minimizes the formation of defects in the interface region.

Beryllium was used for the p-type dopant and silicon for the n type. The substrate temperature was held at 590 °C during growth. The background doping of the undoped layers is $\lesssim 10^{14}$ cm^{-3}.

The solid and dashed lines in Fig. 2(a) are, respectively, the band diagram of the diodes at zero applied bias, with and without dipole (not to scale). The depletion region width is $W = 1 \mu$m for both structures. In the structure with the DID the electric field inside the dipole layer is strongly increased while it slightly ($\simeq 10\%$) decreased outside the dipole (compared to the structure without dipole) since the potential drop across the depleted i layer is identical to that of the diodes without dipole. This can be understood rigorously as follows. The electric field in the i layer of the p^+in^+ diode without DID is created by two very thin depleted regions, adjacent to the i layer, of charge density per unit area $-\sigma'$ and $+\sigma'$ located in the p^+ and n^+ layers, respectively. In the diodes with DID the charge density in these sheets σ'' is

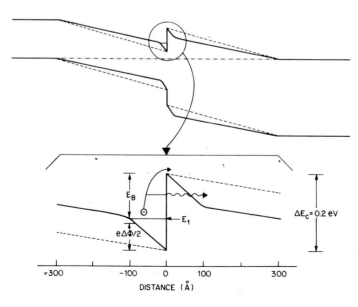

FIG. 2. (a) Solid and dashed lines represent, respectively, the band diagram of the pin diodes with and without interface dipole (not in scale). (b) Band diagram of the conduction band near the heterointerface of the diodes with and without dipole (in scale).

E_B

E_1

$e\Delta\Phi/2$

$\Delta E_c = 0.2$ eV

+300   −100   0   100   300

DISTANCE (Å)

smaller ($\sigma'' < \sigma'$), i.e., their thickness is reduced in order to achieve the same potential difference across the i layer [see Fig. 2(a)]. It follows that at zero bias one has:

$$\frac{e\sigma'}{\epsilon} W - \Delta E_c = E_{g,\text{GaAs}} \quad \text{(without DID)}, \tag{1}$$

$$e\frac{\sigma''}{\epsilon} W - \Delta E_c + e\Delta\Phi + e\frac{\sigma}{\epsilon}t = E_{g,\text{GaAs}} \quad \text{(with DID)}, \tag{2}$$

where W is the i layer thickness.

It is easily seen from Eqs. (1) and (2) that the electric field outside the dipole is reduced by

$$\Delta F = \frac{\sigma' - \sigma''}{\epsilon} = \frac{\Delta\Phi + (\sigma/\epsilon)t}{W}. \tag{3}$$

In the DIDs used in our experiments $\Delta\Phi$ ($= 0.14$ V) is twice the total potential drop across the two charge sheets of the dipole [$(\sigma/\epsilon)t$]. Thus, $\Delta F = 2.1 \times 10^3$ V/cm which is a small correction compared to the value of the electric field in the p^+in^+ without DID ($= 1.62 \times 10^4$ V/cm).

Figure 2(b) gives the conduction band diagram near the heterointerface at zero bias (to scale), drawn with the aid of Eqs. (1) and (2), for the cases with and without dipole. For ΔE_c we have used the value 0.2 eV, following the new band lineups for AlGaAs/GaAs.[8] The barrier height E_B is $\simeq 113$ meV which corresponds to about a factor of 2 smaller than in the case without dipole ($= \Delta E_c$).

We have measured the photocollection efficiency of the two structures; light chopped at 1 kHz and incident on the AlGaAs side of the diode was used and the short-circuit photocurrent was measured with a current sensitive 181 PAR preamplifier followed by a 5604 PAR lock in. The input impedance of the preamplifier was either 10 or 100 Ω depending on the scales.

The power of the incident radiation was kept low ($\lesssim 1$ nW) to minimize changes of the heterojunction potential profile due to possible charge accumulation in the conduction band notch. Absolute efficiency data were obtained by comparing the photoresponse to that of a calibrated Si photodiode. In Fig. 3 we have plotted the zero bias external quantum efficiency η as a function of wavelength for devices with and without dipole. In the ones without dipole, η is very small ($\lesssim 2\%$) for $\lambda \gtrsim 7100$ Å; this wavelength corresponds to the band gap of the $\text{Al}_{0.26}\text{Ga}_{0.74}$ As layer as determined by photoluminescence measurements. At wavelengths longer than this and shorter than ≈ 8500 Å photons are absorbed partly in the GaAs electric field region and partly in the p^+ GaAs layer within a diffusion length from the depletion layer. Thus most of the photoinjected electrons reach the heterojunction interface and have to surmount the heterobarrier of height $\Delta E_c = 0.2$ eV to give rise to a photocurrent. Thermionic emission and recombination, due to the unavoidable presence of interface states, limit therefore the collection efficiency. This explains the low quantum efficiency for $\lambda > 7100$ Å, since ΔE_c is significantly greater than kT.[9]

For $\lambda < 7100$ Å the light is increasingly absorbed in the AlGaAs as the photon energy increases and the quantum efficiency becomes much larger than for $\lambda > 7100$ Å, since most of the photocarriers don't have to surmount the heterojunction barrier to be collected. For $\lambda < 6250$ Å the quantum efficiency decreases since losses due to recombination of photogenerated holes in the n^+ AlGaAs layer and to surface recombination start to dominate. This wavelength dependence of the efficiency is typical of abrupt AlGaAs/GaAs heterojunctions without interface charges.[9,10]

The solid curve in Fig. 3 is the photoresponse in the presence of the DID. A striking difference is noted as compared to the case with no dipole. While the quantum efficiencies for $\lambda \lesssim 7100$ Å are comparable, at longer wavelengths it is enhanced by a factor as high as one order of magnitude in the structures with dipoles. This effect was reproduced in four sets of samples.

The physical interpretation is simple. The barrier height E_B has been lowered by $\simeq 87$ meV [Fig. 2(b)], which enhances thermionic emission across the barrier. Tunneling through the thin triangular barrier [illustrated in Fig. 2(b)] and the smaller reflection coefficient, which reduces carrier thermalization in the conduction band, will also contribute

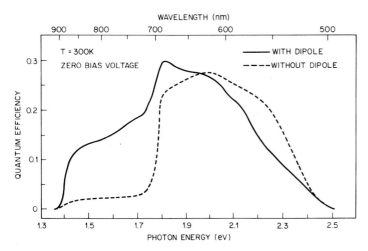

WAVELENGTH (nm)

FIG. 3. External quantum efficiency of the heterojunctions with and without dipole at zero bias vs photon energy. Illumination is from the wide gap side of the heterojunction.

J. Vac. Sci. Technol. B, Vol. 3, No. 4, Jul/Aug 1985

317

to the enhanced collection efficiency. In fact, a non-negligible source of leakage of carriers across the reduced barrier, is due to nonequilibrium energetic photocarriers created within an energy relaxation mean free path (≈ 3300 Å in a field of 10^4 V/cm for a 0.2 eV electron in GaAs) from the interface, an effect first proposed by Tansley.[11] There is clear evidence for this effect in Fig. 3. In the diodes with dipoles the efficiency increases significantly as the photon energy varies from 1.5 to 1.70 eV. In this range the initial energy ΔE of the excited photoelectron in the GaAs with respect to the bottom of the conduction band varies from 0.07 to 0.34 eV, since $\Delta E = [(h\nu - E_g)(m_h^*/m_e^* + m_{hh}^*)]$ and is comparable or greater than the barrier height $\{E_B = 0.11$ eV [Fig. 2(b)]\}. Thus the percentage of nonequilibrium carriers that traverse the interface without significant energy relaxation increases as photons are absorbed closer to the interface. In the diodes without dipoles, on the other hand, the efficiency is much less dependent on photon energy in the same range, since photoelectrons tend to thermalize before going over the barrier irrespective of where they are generated, because of the significantly greater barrier height ($\Delta E_c = 0.2$ eV).

This important effect illustrates clearly that at least in the heterojunctions with DID one cannot assume for the electrons photoexcited in the low gap side of the junction a simple Maxwell–Boltzmann distribution in equilibrium with the lattice. In fact, additional experimental results discussed further on in this paper show that this approximation, widely used in calculating photocollection efficiencies, is generally incorrect also for our pin heterojunctions without DID.

Figures 4 and 5 show the photocurrent spectral response at different reverse bias voltages for the structures without and with DID, respectively. The efficiency rapidly increased with reverse voltage in both structures and then saturated. Above 10 V the quantum efficiency in the energy range 1.5–1.7 eV are very similar in both structures and $\simeq$40%–50%. This is expected, since at fields $> 10^5$ V/cm the electrons acquire so much energy that the barrier height is no more a significant limiting factor to the efficiency.

Figure 6 illustrates the forward bias dependence of the

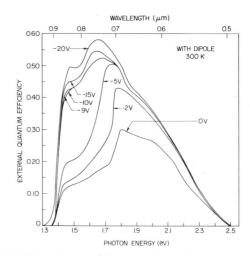

FIG. 5. External quantum efficiency vs wavelength at different reverse bias voltages for diodes with dipole.

external quantum efficiency of the structures with and without dipole, at 300 K at a wavelength of 8000 Å. The curves exhibit a relatively flat portion and a rapid (roughly exponential) decay at voltages greater than 0.6 and 0.8 V in the diodes without and with DID, respectively. This type of behavior has been observed before by several authors and is due to increased band flattening at high forward biases. As the forward bias is increased the conduction band spike projects above a greater portion of the conduction band in the depleted GaAs layer where electrons are photoexcited. Thus an increasingly large fraction of electrons does not gain suffi-

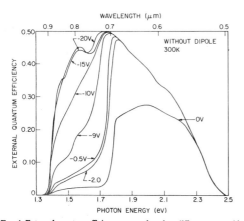

FIG. 4. External quantum efficiency vs wavelength at different reverse bias voltages for diodes without dipoles.

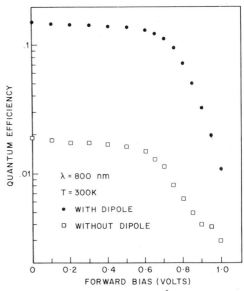

FIG. 6. External quantum efficiency at $\lambda = 8000$ Å vs forward bias voltage of the diodes with and without dipoles.

J. Vac. Sci. Technol. B, Vol. 3, No. 4, Jul/Aug 1985

318

cient energy to surmount the barrier and cannot be efficiently collected. Electrons that don't make it over the barrier then recombine with holes through the always present interface states or via bulk recombination. Recombination increases rapidly with forward bias because of the large increase in dark currents which injects holes (and electrons) in the depletion layer. Also, the electric field is strongly reduced at large forward bias thus increasing the electron transit time and enhancing the probability of recombination. In conclusion the rapid fall off of the efficiency at high forward bias results from the combined effects of the conduction band barrier and increased recombination.[9]

The cutoff voltage V_c, i.e., the voltage at which the efficiency starts to roll off exponentially with bias, is higher in the diodes with DID (0.8 V) than in those without ($V_c \approx 0.6$ V). This is intuitive since in the diodes with DID the conduction band spike projects above a smaller portion of the conduction band edge in the intrinsic GaAs layer. Thus a larger voltage is required to start cutting off the photocollection of carriers. The difference in the cutoff voltages for the two structures ($\Delta V_c \approx 0.2$ V) is comparable to twice the difference between the conduction band barriers of the two structures ($\Delta E_c - E_B \simeq 87$ mV, see Fig. 2), as expected from elementary energy band diagram considerations.

The rounding off of the efficiency curve before V_c is much more pronounced in the structure with DID. This is due to tunneling through the triangular barrier which makes the onset of the cutoff more gradual with voltage. Note that at high forward bias ($\gtrsim 0.8$ V) the difference between the efficiencies of the diodes with and without dipoles tends to decrease. This is because some of the channels of photocollection by which the structure with dipole is more efficient (tunneling through the barrier, nonequilibrium and hot electrons surmounting the barrier) are quenched as the conduction band in the GaAs flattens (see Fig. 2).

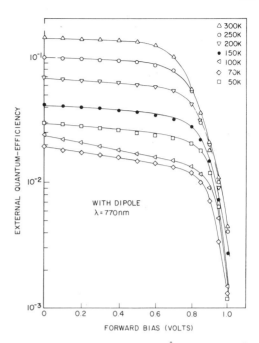

FIG. 8. External quantum efficiency at $\lambda = 7700$ Å vs forward bias at different temperatures for a diode with dipole.

More insight into the collection dynamics is gained by studying the temperature dependence of quantum efficiency vs forward bias (Figs. 7 and 8). The decrease of efficiency with decreasing lattice temperature (at temperatures > 100 K) is a manifestation of reduced thermionic emission. The contribution of reduced light absorption (due to the increasing band gap) is only a small fraction ($\lesssim 10\%$) of this decrease. We have not been able, however, to fit the curves of Figs. 7 and 8 by assuming, as routinely done in theories of photocollection in heterojunctions, that photocarriers are in a Maxwell–Boltzmann distribution at the lattice temperatures. The decrease of the efficiency with temperatures is significantly less pronounced than if photocarriers were all at the lattice temperature. The above suggests that there are basically two distributions of photocarriers in the low gap GaAs side of the heterojunction: one at the lattice temperature T, thermalized by reflections at the barrier and by phonon collisions, and another one with an average energy per electron greater than $\frac{3}{2}kT$. This interpretation is further supported by the observation that at low temperatures (< 100 K) the quantum efficiency increases with decreasing temperature. This occurs because at sufficiently low temperatures phonon scattering is strongly reduced, thus decreasing carrier thermalization and increasing the fraction of photocarriers in the hot-electron distribution and their energy. Strong experimental evidence for the existence of these two distributions has been found recently in the study of photocollection in InP/Ga$_{0.47}$In$_{0.33}$As heterojunctions.[12,13]

The cutoff voltage defined above tends to increase at lower temperatures as a result of the lower dark current and of the

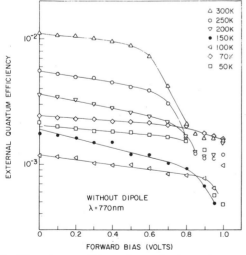

FIG. 7. External quantum efficiency at $\lambda = 7700$ Å vs forward bias at different temperatures for a diode without dipole.

J. Vac. Sci. Technol. B, Vol. 3, No. 4, Jul/Aug 1985

319

attendant decrease in carrier recombination.

One last point worth discussing is what is the smallest distance at which the donor and acceptor layers can be placed. This of course depends on the diffusion coefficient of the dopants, which in turn depends on the doping density, the substrate temperature, and on the growth time. For Si and Be in the AlGaAs heterostructure system one should be able to place the doping layers as close as 10 Å for substrate temperatures $\lesssim 600\,°C$ and growth times of < 1 h without incurring into significant interdiffusion.

III. INTERFACE ENGINEERING: DEVICE APPLICATIONS OF DIDs

Doping interface dipoles allow one to selectively engineer the band diagram near the interface. Thus barrier heights and band offsets can be artificially controlled and tailored to a specific device application. This method can be described as interface engineering and represents an important extension of the band gap engineering method and philosophy previously discussed by one of us.[2,14]

Many transport properties are exponentially dependent on barrier heights and band offsets. Thus small artificial variations of these quantities, as induced by the DIDs, can produce significant changes in those properties and allow a new degree of freedom in device design.

In this section we shall discuss several applications of the DID concept. One of the most intriguing possibilities is the conversion of a type I heterojunction into a type II or staggered heterojunction. Figure 9(a) represents the band dia-

gram of a type I heterojunction. The insertion of a DID of potential $\Delta\Phi$ greater than ΔE_c brings the conduction band edge of the wider gap semiconductor below that of the smaller gap material leading to an effectively staggered heterojunction [Fig. 9(b)]. Again, the dipole should be thin enough that electrons can tunnel through the conduction band spike.

DIDs may also prove very useful in increasing the speed of heterojunction photodetectors such as a long wavelength avalanche photodiodes with separate absorption ($Ga_{0.47}In_{0.53}As$) and multiplication regions (InP). This detector is one of the most promising for long wavelength (1.3 and 1.5 μm) fiber optic communication systems. One problem, however, is the large valence band discontinuity ($\Delta E_v = 0.43$ eV) the holes have to surmount to be injected from the $Ga_{0.47}In_{0.53}As$ layer into the InP gain region. Extensive studies have shown that holes pile up at the interface and are emitted thermionically with a long-time constant which leads to a long tail in the detector response time.[15] This tail can be reduced or eliminated by growing between the InP and the $Ga_{0.47}In_{0.53}As$ a quaternary GaInAsP layer of intermediate band gap[16] or an $InP/Ga_{0.47}In_{0.53}As$ variable gap superlattice which simulates a graded gap GaInAsP layer (pseudoquaternary).[17] An interesting alternative, probably the simplest, is the use of a DID (as shown in Fig. 10).

Another important application is the enhancement of impact ionization of one type of carrier in superlattice and staircase avalanche photodiodes. Recently, Capasso et al.[18] demonstrated that the difference between conduction and valence discontinuities in a heterojunction superlattice can lead to the enhancement of the ionization rates ratio. These experiments led to the conception of the staircase solid state photomultiplier.[19] Introducing DIDs at the band steps of these detectors can be used to further enhance the ionization probability of electrons at the steps, since carriers ballistically gain the dipole potential energy in addition to the energy ΔE_c obtained from the band step. In addition, the DID helps by promoting over the valence band barrier holes created near the step by electron impact ionization without sacrificing speed. Thus the dipole energy $e\Delta\Phi$ should equal or exceed the valence band barrier. This application is illustrated in Fig. 11. The top part of Fig. 11 gives the band diagram of the staircase solid state photomultiplier. Since the conduction band discontinuity exceeds the band gap after the step,

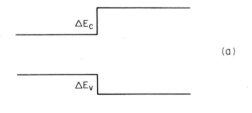

(a)

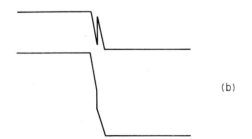

(b)

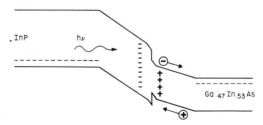

FIG. 9. (a) Band diagram of type I heterojunction. (b) Band diagram of the same heterojunction with an added DID. Since the dipole potential $\Delta\Phi$ is chosen larger than ΔE_c, the heterojunction is converted effectively in a staggered (or type II) one.

FIG. 10. Band diagram of a long wavelength heterojunction avalanche photodiode with separated multiplication (InP) and absorption ($Ga_{0.47}In_{0.73}As$) regions. The DID effectively reduces the valence band barrier, allowing high speed collection of photoinjected holes.

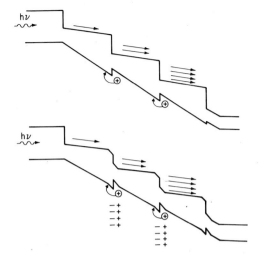

FIG. 11. (a) Band diagram of staircase APD without DID. (b) Band diagram of staircase APD with DID.

while the valence band step is of the opposite sign to assist hole ionization and the electric field is too small to cause hole initiated ionization, only electrons ionize and the steps are the analog of the dynodes in a photomultiplier. Adding dipoles at the step as shown in Fig. 11(b) will minimize hole trapping effects. In addition, in the event that in the structure of interest the conduction band step is not quite equal to the required ionization energy in the material after the step, the dipole can help to compensate the small energy deficit (0.1–0.2 eV).

DIDs can also be used at the interfaces of heterojunction and quantum well lasers to increase the confinement energies of carriers.

One last important point about DIDs is worth mentioning. In analyzing the electrostatics of the DID we neglected the statistical fluctuations of the doping concentration in the two charge sheets. For a doping density of $5 \times 10^{17}/\mathrm{cm}^3$ the average distance between impurities is $\simeq 125$ Å which is of the order of the sheet thickness (100 Å). In this limit the always present spatial fluctuations in the doping density become important. This means that the barrier or discontinuity lowering $e\Delta\Phi$ will fluctuate with position in the plane of the heterojunction interface. An electron approaching the interface will see a spatially fluctuating barrier height with relative minima and maxima. The minima act as "electrostatic potential pinholes"; in the sense that the thermionic emission probability over the barrier and tunneling probabil-

ity through the barrier at these pinholes can be greatly increased. The effects of doping fluctuations cannot be easily observed in the present experiments but may be observable in noise measurements particularly at high frequency.

The study of tunneling through such fluctuating barriers presents a considerable theoretical challenge because the standard one-dimensional treatment is obviously not applicable. Such investigations appear of considerable importance from both a basic and a device point of view.

In conclusion, DIDs represent a technique to effectively tune barrier heights and band offsets at heterojunctions and as such have great potential for new device applications.

[1]G. Margaritondo, Surf. Sci. **132**, 469 (1983); Solid State Electron. **26**, 499 (1983).
[2]F. Capasso, Surf. Sci. **132**, 527 (1983); **142**, 513 (1984).
[3]R. W. Grant, J. R. Waldrop, and E. A. Krawt, Phys. Rev. Lett. **40**, 656 (1983).
[4]R. S. Bauer and H. W. Sang, Jr., Surf. Sci. **132**, 479 (1983).
[5]J. R. Waldrop, S. P. Kowalczyk, and R. W. Grant, J. Vac. Sci. Technol. **19**, 573 (1981).
[6]A. Zur, T. C. McGill, and D. C. Smith, Surf. Sci. **132**, 456 (1983).
[7]A. Y. Cho, Thin Solid Films **100**, 291 (1983).
[8]R. C. Miller, D. A. Kleinman, and A. C. Gossard, Phys. Rev. B **29**, 7085 (1984).
[9]See, for example, T. S. Te. Velde, Solid State Electron. **16**, 1305 (1973); A. Ya. Shik, and Y. V. Shmartsev, Sov. Phys. Semicond. **15**, 799 (1984). These considerations are valid for nonideal heterojunctions with interface states and nonzero interface recombination velocity s. All grown heterojunctions belong to this class. In an ideal heterojunction diode without interface states ($s = 0$), under steady state conditions the collection efficiency would be unity, irrespective of the barrier height.
[10]S. F. Womac and R. H. Rediker, J. Appl. Phys. **43**, 4129 (1972). In previous work on p^+ (GaAs)-n(Al$_{0.33}$Ga$_{0.67}$As) MBE abrupt heterojunctions grown by one of us [H. Kroemer, Wu Yi Chien, H. C. Casey, and A. Y. Cho, Appl. Phys. Lett. **33**, 749 (1978)] a high quantum efficiency was found also for photon energies smaller than the AlGaAs band gap and attributed to a positive interface charge. The lack of such effects in our structures may be due to the greatly improved material quality. Note also that recent experiments by Nottenburg and Ilegems on p^+ (GaAs)-n(AlGaAs) heterojunctions found a quantum efficiency vs wavelength curve similar to ours with no evidence of interface charges (private communication).
[11]T. L. Tansley, Phys. Status Solidi **24**, 615 (1967).
[12]S. R. Forrest, D. K. Kim, and R. G. Smith, Solid State Electron. **26**, 951 (1983).
[13]Y. Takanashi and Y. Horikoshi, Jpn. J. Appl. Phys. **20**, 1271 (1981).
[14]F. Capasso, J. Vac. Sci. Technol. B **1**, 457 (1983).
[15]S. R. Forrest, O. K. Kim, and R. G. Smith, Appl. Phys. Lett. **41**, 95 (1982).
[16]J. Campbell, A. G. Dentai, W. S. Holden, and B. L. Kasper, Electron. Lett. **19**, 818 (1983).
[17]F. Capasso, H. M. Cox, A. L. Hutchinson, N. A. Olsson, and S. G. Hummel, Appl. Phys. Lett. **45**, 1193 (1984).
[18]F. Capasso, W. T. Tsang, A. L. Hutchinson, and G. F. Williams, Appl. Phys. Lett. **40**, 38 (1982).
[19]F. Capasso, W. T. Tsang, and G. F. Williams, IEEE Trans. Electron Devices, **ED-30**, 381 (1983).

J. Vac. Sci. Technol. B, Vol. 3, No. 4, Jul/Aug 1985

321

Heterojunction band discontinuity control by ultrathin intralayers

D. W. Niles and G. Margaritondo[a]

Department of Physics and Synchrotron Radiation Center, University of Wisconsin, Madison, Wisconsin 53706

P. Perfetti, C. Quaresima, and M. Capozi

Instituto di Struttura della Materia, Via E. Fermi, 00044 Frascati, Italy

(Received 5 July 1985; accepted for publication 27 August 1985)

We present evidence that the band lineup at a semiconductor-semiconductor heterojunction interface can be changed and potentially controlled by an ultrathin metal intralayer. Synchrotron-radiation photoemission experiments demonstrate that 0.5–2-Å-thick Al intralayers increase the valence-band discontinuity of CdS-Ge and CdS-Si heterojunctions by 0.15 eV on the average.

The nature of the band discontinuities caused by the difference of the forbidden band gaps is the most important open problem of the physics of semiconductor-semiconductor heterojunctions.[1–4] The most attractive goal is the possibility of controlling the band lineup and thus tailoring the properties of a wide variety of heterojunction devices.[5] Previous photoemission results have produced substantial progress in understanding the nature of the band discontinuities.[1–3,6] They did not provide, however, evidence that these parameters can be controlled by modifying the structure of the interface. The most important previous results in this area were negative—drastic changes in the interface preparation and structure failed to produce detectable changes in the valence-band discontinuity, ΔE_v, of GaAs-Ge.[7]

We present the first evidence obtained with photoemission spectroscopy that the band lineup can be changed by modifying the microscopic interface dipoles with an ultrathin (0.5–2 Å) intralayer. The study was performed with synchrotron-radiation photoemission on the interfaces between CdS and Si or Ge. We consistently found that an ultrathin Al intralayer at the heterojunction interface increases ΔE_v by 0.1–0.3 eV. The magnitude of these changes suggests that they are an effect of modifications in the microscopic interface dipoles.[6]

CdS-Ge and CdS-Si are ideal systems for this study. The ionicity of CdS makes it more likely to observe strong interface dipole effects than, for example, in interfaces involving III-V materials. The ΔE_v's for CdS-Ge and CdS-Si are large, and they correspond to a clearly visible double-edge structure in the valence-band spectra. This makes it possible to study ΔE_v without relying only on complicated and sometimes unreliable analyses of the core level spectra, as discussed in detail in Ref. 6. This is a crucial problem, since the main obstacle in the present experiments is the accuracy and reliability in measuring ΔE_v changes. Al intralayers were selected for these experiments since they are known to strongly influence the properties of other semiconductor interfaces, e.g., CdS-Au.[8]

The procedure followed in these experiments was similar to that described in Ref. 6, except for the presence of Al intralayers in some of the systems and for the more sophisticated data analysis. Preliminary experiments were performed at the University of Wisconsin Synchrotron Radiation Center, and then completed at the Synchrotron Radiation Facility of the Frascati National Laboratory. Ge and Al were deposited *in situ* by evaporation under ultrahigh vacuum conditions on roon-temperature substrates. This procedure gives amorphous Ge overlayers.[6] The photoelectron collection geometry corresponded to that of a double-pass cylindrical mirror analyzer, with the sample tilted at an angle of 30°–45° with respect to the analyzer axis.

The double-edge structure due to the lineup of the two valence-band edges is clearly visible in the photoemission spectra of Fig. 1, corresponding to a nominal thickness of 4 Å of Ge deposited on cleaved CdS (10$\overline{1}$0), with and without an Al intralayer. A similar structure is also visible in the upper two extended spectra of Fig. 2. From double-edge structures like those in Figs. 1 and 2 it is possible to deduce the magnitude of ΔE_v with an accuracy ranging between 0.1 and 0.2 eV, depending on the system. The above accuracy would have not been sufficient to detect the effects here de-

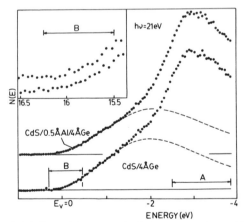

FIG. 1. Photoemission spectra taken in the region near the top of the valence band for Ge deposited on cleaved CdS, with or without a 0.5-Å Al intralayer. The two curves were shifted with respect to each other to obtain their best alignment in spectral region A, dominated by a CdS-derived peak. The dashed lines show the estimated Ge valence-band contributions to the spectra. The inset shows a portion of the same two curves, shifted with respect to each other to obtain the best alignment in the Ge-like spectral region B. The difference between the relative shifts in (A) and (B) reveals an intralayer-induced increase by 0.11 eV in the valence-band discontinuity ΔE_v. The energy scale is referred to E_v, the top of the Ge valence band for CdS,/Ge. The Fermi level also coincides with E_v.

[a] Author to whom all correspondence should be addressed.

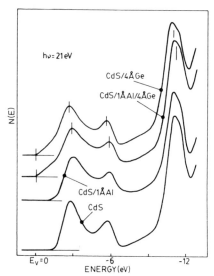

FIG. 2. Photoemission spectra taken on clean CdS (bottom), and on the same surface covered by 1 Å of Al and then by 4 Å of Ge. The top curve corresponds to CdS covered by 4 Å of Ge without Al intralayer. The comparison of the two top curves again reveals the Al-induced increase of ΔE_v.

scribed. One should not confuse, however, the accuracy in obtaining an *absolute* determination of ΔE_v with the accuracy in detecting *relative changes* in ΔE_v. As we shall now discuss, the latter can be much better than the former for data taken under the same conditions.

To reveal intralayer-induced changes in ΔE_v we used the following procedure. First, we shifted the two spectra of Fig. 1 with respect to each other to obtain the best alignment in the spectral region A. The peak in this region is due to the CdS substrate, and therefore this procedure will bring the substrate valence-band edges into alignment. This alignment was obtained by first normalizing each curve to its area, then calculating the area of their difference as a function of the relative shift and searching for a minimum of its magnitude. The best alignment obtained in this way corresponds to the curve positions shown in Fig. 1. The same alignment procedure was then repeated for the spectral region B, where the photoemission signal is almost entirely due to the Ge overlayer. At these coverages, the spectral edge already corresponds to the valence-band edge of Ge.[6] The results of the best alignment in region B are shown in Fig. 1 (inset).

In the absence of intralayer-induced changes of ΔE_v, the same relative shift should give the best alignment for both regions A and B. For the curves of Fig. 1, the relative shifts giving the best alignment for regions A and B differ from each other by -0.11 eV. This reveals an Al-induced increase of ΔE_v by the same magnitude.

The accuracy of our measurements of ΔE_v changes is *not* limited by the experimental resolution, as long as it stays constant for each series of experiments (this often neglected fact is well known, for example, to modulation spectroscopists). The accuracy limiting factor is, instead, the signal-to-noise ratio. This makes it possible to determine changes of ΔE_v with better accuracy than that achievable for the abso-

lute value of this parameter. From the experimental signal-to-noise ratio and from the dependence of the area of the difference curve on the relative shift, we estimated an accuracy of 0.005–0.01 eV in determining the best alignment shift for region A. The corresponding accuracy is worse for region B due to the smaller signal, and is 0.01–0.03 eV. The combined alignment accuracy, therefore, is better than 0.04 eV. Other possible factors affecting the accuracy in detecting the Al-induced ΔE_v changes were carefully analyzed and are discussed below (see Ref. 9). The analysis demonstrated that they did not jeopardize our capability to detect ΔE_v changes of this magnitude. We also analyzed other CdS-related features, e.g., the valence-band peak at ~ -5.8 eV, to test the shift deduced from region A. As one can see from the two top curves of Fig. 2, the shifts of the other CdS features are independently and quantitatively consistent with our estimated Al-induced changes in ΔE_v.[10]

Table I summarizes the results obtained for different thicknesses of the overlayer and of the intralayer. Al-induced increases of ΔE_v were consistently found, both for CdS-Ge and for CdS-Si. The scattering of the data in this table is too close to the experimental accuracy to clearly identify trends. There seems to be an indication, however, that thicker Al intralayers correspond to larger changes of ΔE_v. The average magnitude of the intralayer-induced changes listed in Table I is 0.15 eV, consistent with the upper limit deduced in Ref. 6 for the average dipole contribution to ΔE_v.

There are two kinds of interface dipoles that can influence ΔE_v and both of them can be affected by the Al intralayer.[11] The first kind corresponds to the charge distribution of interface chemical bonds. The second kind corresponds to the charge distribution arising from the migration of atomic species across the junction. The Al intralayer can obviously change the interface chemical bonds and the corresponding dipoles. Furthermore, it has been shown[8] to drastically change the microscopic diffusion processes, e.g., for CdS-Au. The experimental results obtained, e.g., by energy loss spectroscopy,[12] indicate that Al overlayers on CdS are reactive and that Al–S bonds are formed. The expected charge transfer for these bonds, however, would not be consistent with the sign of the Al-induced ΔE_v changes. Thus, these changes must be due to other modifications in the microscopic charge distribution, e.g., due to bonds at the Ge-Al interface, or to dipole modifications due to Al-induced diffu-

TABLE I. Estimated increase of the valence-band discontinuity ΔE_v (eV).

Overlayer	Thickness (Å)		Shift to align spectra		Al-induced increase in ΔE_v
	Al	overlayer	region A	region B	
Ge	0.5	4	0.13	0.02	0.11
Ge	1.0	4	0.27	0.06	0.21
Ge	0.5	10	0.12	-0.02	0.14
Ge	1.0	10	0.32	0.02	0.30
Ge	0.5	15	0.12	0.01	0.11
Ge	1.0	15	0.23	0.01	0.22
Si	1.0	4	0.14	0.02	0.12
Si	1.0	10	0.16	0.01	0.15

sion processes, or to a combination of the two kinds of mechanisms.

The analysis of core-level photoemission intensity as a function of coverage demonstrates that amorphous Si and Ge deposited on room-temperature CdS form sharp, nondiffusive interfaces.[6] Likewise, Al overlayers on CdS are very localized in space due to the high reactivity of Al.[8] For the same reason, the intralayer atoms are not likely to diffuse. The intralayer can, on the other hand, promote the diffusion of Cd, S, and overlayer atoms,[8] thus modifying the interface dipoles. This hypothesis must be tested with a complete study of the interface morphology which is currently being performed.

The detection of ΔE_v changes of the order of 0.1–0.2 eV has extremely important consequences in heterojunction research. Although apparently small and difficult to detect with photoemission methods, those changes correspond to dramatic changes in the transport properties across the heterojunction, and ultimately in the behavior and performances of the corresponding devices.[4,5] Our results open, at least in principle, the possibility of controlling the most important parameter in a heterojunction interface. Further research and, eventually, development will be required to test this possibility.

Note added in proof. Very recently, another successful method for controlling the discontinuities was reported by F. Capasso, A. Y. Cho, K. Mohammed, and P. W. Foy, Appl. Phys. Lett. **46**, 664 (1985).

This work was supported by the NSF grant DMR-84-21292. The international collaboration Frascati–Wisconsin is supported by the Italian National Research Council and by the NSF grant INT-81-22013. One of the authors (GM) is a Romnes Faculty Fellow.

[1] G. Margaritondo, Solid State Electron. **26**, 499 (1983); Surf. Sci. **132**, 469 (1983).

[2] R. S. Bauer, J. Vac. Sci. Technol. B **3**, 608 (1985).

[3] R. W. Grant, J. R. Waldrop, S. P. Kowalczyk, and E. A. Kraut, J. Vac. Sci. Technol. B **3**, 1295 (1985), and references therein.

[4] See, for example, A. G. Milnes and D. L. Feucht, *Heterojunctions and Metal-Semiconductor Junctions* (Academic, New York, 1972).

[5] F. Capasso, Surf. Sci. **142**, 513 (1984), and references therein.

[6] A. D. Katnani and G. Margaritondo, Phys. Rev. B **28**, 1944 (1983); for specific results on CdS/Ge also see A. D. Katnani, R. R. Daniels, Te-Xiu Zhao, and G. Margaritondo, J. Vac. Sci. Technol. **20**, 662 (1982).

[7] A. D. Katnani and R. S. Bauer, J. Vac. Sci. Technol. B **3**, 1239 (1985), and references therein.

[8] L. J. Brillson, G. Margaritondo, and N. G. Stoffel, Phys. Rev. Lett. **44**, 67 (1980). For a study of the CdS-Al interface see, for example, L. J. Brillson, R. S. Bauer, R. Z.Bachrach, and J. C. McMenamin, J. Vac. Sci. Technol. **17**, 476 (1980).

[9] The accuracy in determining the shift of the CdS valence-band features from region A could be affected (1) by the choice of its boundaries which affects, e.g., the normalization of two curves, (2) by the presence of a Ge signal together with the CdS peak, and (3) by other hypothetic Ge-induced or Al-induced changes in the density of states. The first factor was tested using different boundaries. All these tests gave the same optimum shift within 0.01–0.02 eV. The magnitude of the Ge signal was estimated from the spectral line shape in the region above the CdS edge and from the spectrum of amorphous Ge. The results are shown in Fig. 1 by dashed lines. The Ge signal is monotonically increasing with energy in region A. Therefore, it can only cause an *underestimate* of the corresponding shift, which corresponds to an underestimate of the Al-induced change in ΔE_v, and it cannot affect our qualitative conclusions. Furthermore, we found that the underestimate is not larger than a few 10^{-2} eV. The line shape analysis shows that the spectra in region A are essentially due to the superposition of the clean CdS and amorphous Ge spectra, and that the effects of factor (3) are very small. A quantitative estimate, ~ 0.01 eV, is given by the same test used for the first factor. The reliability of the shifts deduced for region A is of course increased by the analysis of other CdS-related features. The accuracy in determining the shift of the Ge valence-band edge from region B could be affected by Al-induced signal. Previous studies (e.g., Ref. 8) indicate that the Al-induced signal is very small there at these coverage levels. Figure 2, in fact, shows no detectable Al signal, and this corresponds to an absolutely negligible effect on the accuracy (the presence of Al is revealed by the Al 2p signal). We also emphasize that the spectral edge corresponds to that of the Ge valence band for these coverages (see Ref. 6).

[10] We estimate, however, that the accuracy obtained by analyzing other CdS features is worse than that obtained from region A, due to possible Al-induced spectral changes (Ref. 8). For example, the distance in energy between the center of gravity of the Cd 4d peak and the valence-band features could be affected by spectral contributions from dissociated Cd (see the two bottom spectra of Fig. 2).

[11] H. Kroemer, in *Proceedings NATO Advanced Study Institute on Molecular Beam Epitaxy and Heterostructures*, Erice 1983, edited by L. L. Chang and K. Ploog (Martinus Nijhoff, The Netherlands, 1983).

[12] L. J. Brillson, Surf. Sci. Rep. **2**, 123 (1982), and references therein.

324

Dipole-Induced Changes of the Band Discontinuities at the SiO$_2$-Si Interface

P. Perfetti and C. Quaresima

Istituto di Struttura della Materia, 00044 Frascati, Italy

C. Coluzza

Istituto di Fisica, Università di Roma "La Sapienza," 00100 Roma, Italy

C. Fortunato

Istituto di Elettronica dello Stato Solido del CNR, 00156 Roma, Italy

and

G. Margaritondo[a]

Synchrotron Radiation Center and Department of Physics, University of Wisconsin-Madison, Stoughton, Wisconsin 53589

(Received 8 May 1986)

We prove experimentally that the band lineup at the SiO$_2$-Si interface can be modified by means of an intralayer. Hydrogen and cesium intralayers produce modifications of 0.5 and 0.25 eV in opposite directions. Possible explanations of these dramatic changes are discussed.

PACS numbers: 73.40.Qv, 79.60.Eq

The application of photoemission techniques has produced substantial progress in understanding the nature of important semiconductor interface parameters.[1] The band discontinuities which accommodate the difference between the two forbidden gaps are the most important parameters of semiconductor (insulator)-semiconductor heterojunctions. Since the beginning of the research on heterojunctions, tailoring of these quantities with a controlled process has been the most ambitious goal. The achievement of such a goal would open the way for substantial improvement in the ultimate performance of all kinds of heterojunction devices, and for the development of entirely new devices.[2]

Our present results show that ultrathin Cs or H intralayers at the interface between SiO$_2$ and Si produce giant changes in the band lineup. Changes of the same magnitude have been independently observed in Si-SiO$_2$ interfaces by Grunthaner et al.[3] These results demonstrate the feasibility of changing band lineups by intervention in the local chemistry of the interface. Furthermore, their analysis helps clarify the nature of the observed intralayer effects for these and other interfaces.

Several previous studies[4-8] explored the effects on the band lineup of interface properties such as the overlayer morphology, the crystallographic orientation, and the growth sequence. Most of these investigations, however, revealed only small or negligible effects. In particular, no band-offset changes were detected from changes in the overlayer morphology in the ZnSe-Ge[4] and GaP-Si[5] heterojunctions, nor from changes in the stoichiometry and structure of the GaAs(100)-Ge interface,[6] in particular by addition of an Al intralayer.

Recently, experiments on Cd-Si, ZnSe-Ge, and SiO$_x$-Si interfaces suggested that the valence-band discontinuity can be changed by ultrathin intralayers of Al or H.[9,10] These results generated some controversy, since the observed changes were quite close to the experimental uncertainty in measurement of the band lineup. The experiments described in this article revealed changes beyond any conceivable experimental uncertainty, definitely demonstrating that the band lineup between two given materials can be significantly modified by ultrathin intralayers. Furthermore, they suggest that the interface dipole contribution to the band lineup can be as large as 0.5 eV or more. This conclusion is of fundamental relevance to the current controversy among different theoretical models for heterojunction band discontinuities.

At present, most theoretical approaches belong to two general categories: models based on the concept of midgap energy[11,12] and models related to the electron-affinity rule.[13-18] In the first case, the band discontinuities are determined by the alignment of the midgap energies of the two semiconductors. The midgap energy of each semiconductor is the level separating the valencelike and conductionlike interface gap states. This approach implies an important role of the interface dipoles in determining the band lineup.

The controversy among different kinds of band-lineup models is mostly related to the magnitude of the interface dipole contributions and to their nature. These questions also have a practical aspect, since interface dipoles provide, at least in principle, flexibility in modifying the band lineups. Our present work shows that the influence of the interface dipoles on the

2065

band lineup is relevant for the SiO_2-Si interface, and also that such dipoles can be changed in a controlled way.

Specifically, we found that the band discontinuities are dramatically changed by ultrathin interface intralayers. We used intralayers of two different materials with very different electronegativity, cesium and hydrogen. These materials produced giant changes of the valence-band discontinuity, ΔE_v, in opposite directions. ΔE_v increased by 0.25 eV with Cs, and decreased by 0.5 eV with hydrogen.

The experiments were performed with synchrotron radiation photoemission at the storage ring ADONE of the Frascati National Laboratory. For SiO_2-Si, photoemission is a straightforward probe of the valence-band discontinuity. In fact, for reasonable overlayer thicknesses, the substrate emission is still visible, and both valence-band edges appear in the spectra. The edge positions can be estimated by linear extrapolation. The accuracy of this method, although limited, is sufficient to detect the giant intralayer effects described here. Figure 1 shows the photoemission spectra in the region near the top of the valence bands, for intralayer-free SiO_2-Si and for SiO_2-Si with cesium (nominal thickness -0.5 Å) and hydrogen intralayers.

SiO_2 substrates were obtained by electron-bombardment deposition *in situ* of 50 Å of silicon on cleaved Si(111) in an oxygen atmosphere ($\sim 5 \times 10^{-5}$ mbar).

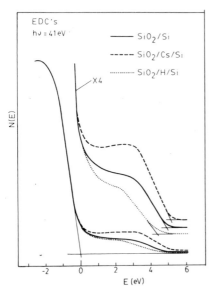

FIG. 1. Photoemission spectra taken in the region of the valence-band edges reveal the valence-band discontinuity at the SiO_2-Si interface. The spectral changes reveal the modifications of the valence-band discontinuities by cesium or hydrogen intralayers.

The cleaved Si substrate was kept at 250 °C and a 200-eV ion gun was used to bombard the system with oxygen ions during the SiO_2 growth. The good quality of the SiO_2 substrates obtained in this way was demonstrated by the valence-band photoemission spectra taken after growth. Subsequently, SiO_2-Si interfaces were obtained by electron-bombardment deposition of Si. Typical thicknesses of 10 Å were found ideal to observe both band edges with a well-developed overlayer valence band.

Cesium intralayers were obtained by *in situ* thermal evaporation. After depositing 0.5 Å of Cs, we observed no change in the line shape of the SiO_2 substrate photoemission spectrum. Hydrogen intralayers were obtained by exposure of the SiO_2 substrates for 2 min to a hydrogen atmosphere ($\sim 5 \times 10^{-4}$ mbar) in the presence of an incandescent filament. As in the case of cesium deposition, the line shape of the SiO_2 spectrum was not modified by the exposure to hydrogen.

We also obtained the same hydrogen-induced change revealed by Fig. 1 with two alternative procedures. Both procedures consisted of exposure of the SiO_2-Si interface to hydrogen, after deposition of the Si overlayer. In the first procedure the Si overlayer thickness was kept below 5 Å. In the second procedure the overlayer thickness was of the order of 10 Å, and the exposure to hydrogen was accompanied by hydrogen-ion bombardment with a 100-eV gun. The equivalency of ΔE_v changes obtained by the three above approaches indicates that hydrogen atoms located at the interface are always responsible for them. This is reasonable, since the SiO_2-Si interface has a large density of unsaturated Si bonds which can capture the H atoms.

The intralayer-induced modifications in the band lineup are evident in Fig. 1. Specifically, ΔE_v changed from 4.9 eV for the intralayer-free interface to 5.15 eV for the Cs intralayer and to 4.4 eV for the hydrogen intralayer. In the case of hydrogen, the question may arise of a possible simulation of the above effect by a regression of the silicon overlayer valence band due to hydrogenation. To rule out this possibility, we performed a similar photoemission study of the interface between amorphous silicon and hydrogenated amorphous silicon. We did not observe any measurable discontinuity at this interface, confirming earlier indications that the valence-band edges of a-Si and a-Si:H are aligned with respect to each other.[16]

The observed intralayer-induced changes in ΔE_v are due to modifications of the interface dipoles, and this demonstrates that such dipoles do play a major role in the band lineup. Several kinds of interface dipole terms can be affected by the intralayers: dipoles due to the formations of different kinds of interface bonds, dipoles due to the presence of dangling bonds, and dipoles due to diffusion of charged impurities. The interplay between different terms changes from interface to interface—for example, hydrogen intralayers produce opposite changes of ΔE_v in our interfaces and in those of Ref. 3, which were prepared with a completely different procedure. Thus, a complete theoretical treatment of these phenomena requires a complete description of all dipole terms.

Such description is a formidable theoretical task. We did, however, attempt to model the intralayer effects on one class of dipoles, i.e., those due to charge transfer upon formation of interface bonds. The amorphous character of our SiO_2-Si interface makes it impossible to use electronegativity approaches applicable to crystalline interfaces.[17] The approach we used bypasses this difficulty.

The model simulates the SiO_2-Si interface with two planes of spheres with densities equal to the surface density of crystalline Si, 1.36×10^{19} atoms/m², and to the average surface density of SiO_2, $\sim 8.4 \times 10^{18}$ molecules/m². The distance between the two planes is taken equal to the sum of the covalent radius of Si, $R_{Si} \sim 1.1$ Å, and an "equivalent radius" of the SiO_2 molecule, derived from the average SiO_2 bulk density, $R_{SiO_2} \sim (2.4 \times 10^{-2})^{-1/3}/2 \sim 1.7$ Å.

The electronegativity difference between the two species causes a charge transfer across the interface, and therefore an interface dipole. Also, the different surface densities of the Si and SiO_2 planes leaves unsaturated Si bonds at the interface. These unsaturated bonds provide natural bonding sites for the intralayer atoms. From the surface densities of Si and SiO_2, we estimate the density of Si dangling bonds to be of the order of 5×10^{18} m⁻².

The charge transfer can be estimated by the Sanderson criterion,[19-21] i.e., equal electronegativity value is reached when two or more species of different initial electronegativity react to form a stable chemical compound. For a molecule, this electronegativity, S_m, is given by

$$S_m = \left(\prod_{i=1}^{N} S_i \right)^{1/n},$$ (1)

where S_i is the electronegativity of the ith among the N atoms forming the molecule. The charge transfer (in electronic charges) affecting the ith atom is given by

$$\rho_i = (S_m - S_i)/\Delta S_i,$$ (2)

where $\Delta S_i = 2.08 S_i^{1/2}$ is a normalization factor. The values of S_i and of ΔS_i are tabulated in Ref. 18, and in Table I we show those relevant to this work.

In the first approximation, Eq. (2) can be used to calculate the charge transfer between adjacent Si and SiO_2 spheres in the two planes of atoms at the inter-

TABLE I. Parameters used for the Sanderson electronegativity model (from Ref. 18).

Species	S (eV)	ΔS (eV)
H	3.55	3.92
O	5.21	4.75
Si	2.84	3.51
Cs	0.28	1.1
SiO_2	4.25	4.29

face, neglecting the two bulks beyond it. This gives a charge transfer of $\rho_0 = 0.18$ electrons. A more realistic picture is obtained by inclusion of the charge transfers affecting the spheres in the planes beyond those at the interface. We accomplish this by modeling the bulk beyond each of the two interface planes with a series of equally spaced planes of spheres. The magnitude of the charge transfer for spheres in the nth plane from the interface is assumed to decrease exponentially with n:

$$\rho_n = A \exp(\alpha n).$$ (3)

The constants A and α (which is negative) are determined by the following conditions: (1) The charge transfer for $n = 1$ is obtained from Sanderson's rule. For the silicon side, this corresponds to Sanderson's estimate for the central atom in a hypothetical Si-Si-SiO_2 molecule. Equations (1) and (2) give $\rho_1 = 0.12$ electrons. (2) The total charge transfer across the interface is equal in magnitude to that calculated for two isolated planes, ρ_0. These conditions give $\alpha = \ln(\rho_1 / A)$ and $A = r_1 \rho_0 / (\rho_0 - \rho_1)$.

The estimated ρ_n's, combined with the average surface density of SiO_2, give the surface charge density for each plane of spheres. The corresponding dipole voltage drop is calculated with use of the appropriate interplanar distances, derived from the radii of the spheres, and the average of the Si and SiO_2 dielectric constants. The resulting total dipole voltage drop is 2.8 eV.

Let us now consider the effect of a hydrogen intralayer. We assume that the hydrogen atoms saturate the dangling silicon bonds at the interface, i.e., that their density is of the order of 5×10^{18} m⁻². This results in an additional interface dipole, due to transfer of charge from silicon to hydrogen. Such change, δV_H, can be estimated with a procedure similar to that used for the Si-SiO_2 interface dipole. The distance between the hydrogen and silicon planes is taken equal to the sum of the covalent radii of silicon and hydrogen, $1.1 + 0.32 = 1.42$ Å, and the dielectric constant is taken equal to that of amorphous Si, ~ 10. The result is $\delta V_H \approx 0.3$ eV. This corresponds to a *decrease* of ΔE_v of the same magnitude.

A similar approach can be used to estimate the effects of the Cs intralayer. The relevant differences

2067

between this case and hydrogen are that the density of intralayer atoms is given by the equivalent thickness of the intralayer, and that the charge transfer is from Cs to Si. We estimate an additional dipole which reduces the total interface dipole by $\delta V_{Cs} \simeq 0.2$ V, thereby increasing ΔE_v.

The above model, although very simple, gives surprisingly good results. We emphasize, however, that the model has severe limitations and is not intended to provide a detailed description of the interface structure. In particular, it takes into account only one class of interface dipole terms, and treats the charge transfers in a very approximate way. The differences between our results and those of Ref. 3 suggest that a more advanced description should take into account the presence of different interface oxides and perhaps microdiffusion processes. Furthermore, charges and defects at the Si-SiO$_2$ interface can be introduced by threefold coordinated Si atoms, strained Si—Si bonds, and valence-alternation pairs.[22,23] Impurities such as hydrogen and alkali metals interact with these factors. For example, Ngai and White[22] discussed the possible breakage of "weak" Si—O bonds by Na atoms, directly relevant to our Cs-intralayer results.

Clearly, our model does not account for all these potential contributions to the interface charges and dipoles. A realistic treatment requires additional information on the nature of the interface defects, which can be obtained with a coordinated use of photoemission and other transport and optical probes.[24] Our model, however, does show that at least one kind of possible interface-dipole changes produces effects of the same magnitude as those we observe. Thus, our conclusion, that the modulation of ΔE_v is due to changes in the interface dipoles, is reasonable.

Our results are very encouraging about the possibility of modulating the heterojunction band discontinuities by means of controlled interface doping. Together with the recent achievements of Capasso *et al.*,[25] these findings could deeply affect the technology of heterojunction devices. In particular, the current methods of producing interfaces between SiO$_2$ and hydrogenated amorphous silicon are likely to give discontinuities which are strongly affected by the mechanism we discovered.

This work was supported in part by the National Science Foundation, Grant No. DMR-84-21292, and by the Italian National Research Council. The collaboration Frascati/Rome–Wisconsin is also supported by the National Science Foundation through an international collaboration grant.

(a)Author to whom all correspondence should be addressed.

[1]For a recent review of the research in this area, see G. Margaritondo, Surf. Sci. (to be published).

[2]F. Capasso, Surf. Sci. **142**, 513 (1984), and references therein.

[3]P. J. Grunthaner, F. J. Grunthaner, M. H. Hecht, and N. M. Johnson, private communication.

[4]G. Margaritondo, C. Quaresima, F. Patella, F. Sette, C. Capasso, A. Savoia, and P. Perfetti, J. Vac. Sci. Technol. A **2**, 508 (1984).

[5]P. Perfetti, F. Patella, F. Sette, C. Quaresima, C. Capasso, A. Savoia, and G. Margaritondo, Phys. Rev. B **30**, 4533 (1984).

[6]A. D. Katnani, H. W. Sang, Jr., P. Chiaradia, and R. S. Bauer, J. Vac. Sci. Technol. B **3**, 608 (1982).

[7]S. P. Kowalczyk, E. A. Kraut, J. R. Waldrop, and R. W. Grant, J. Vac. Sci. Technol. **21**, 482 (1982), and references therein.

[8]A. D. Katnani, P. Chiaradia, Y. Cho, P. Mahowald, P. Pianetta, and R. S. Bauer, Phys. Rev. B **32**, 4071 (1985).

[9]D. W. Niles, G. Margaritondo, P. Perfetti, C. Quaresima, and M. Capozi, Appl. Phys. Lett. **47**, 1092 (1985); D. W. Niles, G. Margaritondo, P. Perfetti, C. Quaresima, and M. Capozi, J. Vac. Sci. Technol. (to be published).

[10]P. Perfetti, Surf. Sci. **168**, 507 (1986).

[11]C. Tejedor and F. Flores, J. Phys. C **11**, L19 (1978); F. Flores and C. Tejedor, J. Phys. C **12**, 731 (1979).

[12]J. Tersoff, Phys. Rev. B **30**, 4874 (1984).

[13]C. B. Duke and C. Mailhiot, Phys. Rev. B **33**, 1118 (1986).

[14]J. L. Freeouf and J. M. Woodall, Surf. Sci. (to be published).

[15]R. L. Anderson, Solid State Electron. **5**, 341 (1967).

[16]W. A. Harrison, J. Vac. Sci. Technol. **14**, 1016 (1977).

[17]F. Patella, F. Evangelisti, P. Fiorini, P. Perfetti, C. Quaresima, M. K. Kelly, R. A. Riedel, and G. Margaritondo, in *Proceedings of the International Conference on Optical Effects in Amorphous Semiconductors—1984*, edited by P. C. Taylor and S. G. Bishop, AIP Conference Proceedings No. 120 (American Institute of Physics, New York, 1984).

[18]W. R. Frensley and H. Kroemer, Phys. Rev. B **16**, 2642 (1977).

[19]R. T. Sanderson, *Inorganic Chemistry* (Reinhold, New York, 1967).

[20]R. T. Sanderson, *Chemical Bonds and Bond Energy* (Academic, New York, 1971).

[21]J. C. Carver, R. C. Gray, and D. M. Hercules, J. Am. Chem. Soc. **96**, 6851 (1984).

[22]K. L. Ngai and C. T. White, J. Appl. Phys. **52**, 320 (1981).

[23]K. Hübner, J. Phys. C **17**, 6553 (1984).

[24]R. Radzouk and B. E. Deal, J. Electrochem. Soc. **126**, 1573 (1979); H. Köster, Jr., I. N. Iassievich, and K. Hübner, Phys. Status Solidi B **115**, 409 (1983); A. Herms, J. R. Morante, J. Samitier, A. Cornet, P. Cartujo, and E. Lora-Tamayo, Surf. Sci. **168**, 665 (1986); G. Pananakakis and G. Kamarinos, Surf. Sci. **168**, 167 (1986).

[25]F. Capasso, A. Y. Cho, K. Mohammed, and P. W. Foy, Appl. Phys. Lett. **46**, 664 (1985); F. Capasso, K. Mohammed, and A. Y. Cho, J. Vac. Sci. Technol. B **3**, 1245 (1985).

SUBJECT INDEX

A

AlGaAs/GaAs 315
$Al_{1-x}Ga_xAs/GaAs$ 5, 12, 21, 70, 167, 196
$Al_xGa_{1-x}As/GaAs/Al_xGa_{1-x}As$ 173
Anderson model 116
Au/II-VI 178
Au/III-V 178

B

Band lineup 6, 15, 19, 25, 116, 214, 224, 268, 304, 311
  control of - 3, 23, 24, 25, 315, 322
  measurement of - 6
  pressure dependence of - 22

C

Capacitance-voltage characteristic (C-V) 6
Carrier mobility 52
CdS/Ge 322
CdS/Si 7, 322
CdSe/HgTe 200
CdTe/ZnTe 200
Charge neutrality 15, 214, 230, 249
Chemical vapor deposition 57
Common-anion rule 10, 12, 17, 177, 204
$CuAg_{1-x}InSe_2/Ge$ 13, 204
CuBr/GaAs 287
$CuIn_xGa_{1-x}Se_2/Ge$ 13, 204
Current-voltage characteristic (I-V) 6, 57

D

Device 17, 23, 26, 70, 99, 116, 315
Deep impurity levels 18, 284
Deep level model 17
Dielectric electronegativity approach 17
Dielectric midgap energy (DME) 249
Dielectric screening 220, 249
Double barrier 99

E

Effective mass 9, 175
  - filtering 105, 150

- polar/nonpolar 137
- relaxation 230
- states 177, 180, 230, 262
Internal photoemission 9

L

Landau level 312
Liquid-phase epitaxy (LPE) 5.

M

Materials
  graded gap - 70
  multilayer sawtooth - 83
Mean interstitial potential 13
Metal induced gap states (MIGS) 14, 18, 23, 25, 214, 218
Metal organic chemical vapor deposition (MOCVD) 5
Metallic intralayers 24, 304, 322, 325
Microstructures 49
Midgap energy 15, 21, 22, 214
Minibands 56, 99
Molecular beam epitaxy (MBE) 5, 23, 49, 70, 157, 196
MOSFET 177

N

Negative differential resistivity 6, 99

O

Optical absorption 9, 59, 175

P

Photocollection measurements 316
Photoconductivity 9
Photocurrent 167
Photoemission 153, 157, 204, 291
  internal - 167
  syncrotron radiation - 193, 204, 291, 322, 326
Photoluminescence 10, 12, 59, 196
Photothreshold 284

Q

Quantum dipole 218, 222
Quantum Hall effect 62

Quantum well 5, 49, 56, 99, 150, 196

S

Schottky model 4, 11, 15, 21, 177, 214, 218, 224, 291, 304
Screening effects 16
Si/II-VI 291, 305
Si/III-V 291, 305
Si/GaAs 258
Si/Ge 257
Si/SiO$_2$ 24, 326
Syncrotron radiation sources 8, 9
Solid state photomultiplier 6, 71
Spectroscopy
  deep-level transient - 7, 150
  Raman - 61
  photocurrent - 59
  photoemission - 163, 287
Surface 50
Superlattice 5, 51, 56, 99, 150, 268, 311
  strained-layer - 28

T

Techniques (Experimental)
  growth - 49
  magneto-optical - 26, 311
  optical - 7, 9
  photoemission 7, 16
Techniques (Theoretical)
  density-functional - 268
  first principles - 200, 268
  local density - 17, 268
  non local pseudopotential - 16, 268
  pseudo potential - 13, 208
  self-consistent - 16, 262
  tight binding - 14, 16, 200, 208, 224
Ternary/quaternary semiconductors 12, 204
Transmission electron microscopy (TEM) 5, 50
Tunneling (resonant) 5, 6, 56, 99

U

Unified defect model 24

W

Work function
  effective - 11, 177, 180
  metal - 4, 21, 22, 307

Z

AUTHOR INDEX

A

Abstreiter G. 9, 167
Altarelli M. 311
Anderson R.L. 4, 35

B

Bauer R.S. 9, 157

C

Caldas M.J. 284
Capasso F. 5, 6, 23, 70, 99, 315
Cardona M. 16, 249
Capozi M. 304, 322
Chang L.L. 311
Ching W.Y. 17
Cho A.Y. 99, 315
Christensen N.E. 16, 249
Ciszek T.F. 204
Classen L.M. 311
Cohen M.L. 16, 262
Colavita E. 304
Coluzza C. 325

D

Deb S.K. 204
Denley D. 153
Dingle R. 9, 173
Duke C.B. 11

E

Esaki L. 5, 56, 311

F

Fazzio A. 284
Flores F. 15, 21, 230
Freeman A.J. 17, 280
Freeouf J.L. 11, 177, 180
Frensley W.R. 13
Fortunato C. 325

G

Gossard A.C. 49, 196
Grant R.W. 9, 163, 287
Grunthaner P.J. 24
Gubanov G. 4

H

Harrison W. 13, 14, 15, 18, 20, 208, 224
Heiblum M. 21
Heine V. 15
Heinrich H. 17
Henry C.H. 173

K

Katnani A.D. 19, 291
Kilday D.G. 204
Kleinman D.A. 196
Kraut E.A. 163
Kroemer H. 4, 7, 13, 116

L

Lang D.V. 150
Langer J.M. 17

M

Maan J.C. 311
Mailhiot C. 11
Margaritondo G. 10, 11, 19, 193, 204, 222, 291, 304, 322, 325
Martin R.M. 16, 268
Massidda S. 17, 280
McMenamin J.C. 9, 157
Miller R.C. 12, 196
Mills K.A. 153
Min B.I. 17, 280
Mohammed K. 96, 315
Mönch W. 21, 307
Munteanu O. 196

N

Niles D.W. 11, 193, 304, 322
Nussbaum A. 17

P

Panish M.B. 150
Perfetti P. 9, 10, 153, 304, 322, 325
Pickett W. 16, 262
Prechtel U. 167

Q

Quaresima C. 304, 322, 325

R

Ruan Y.C. 17

S

Schlapp W. 167
Schokley W. 4
Sergent A.M. 150
Shirley D.A. 153

T

Tejedor C. 15, 230
Temkin H. 150
Tersoff J. 15, 21, 23, 214, 218, 224

V

Van de Walle C.G. 16, 268
Van Vachten J.A. 17

W

Waldrop J.R. 163, 287
Wei S.-H. 12, 200, 204
Weimann G. 167
Wiegman W. 173
Wyder P. 311
Woodall J.M. 11, 177, 180

Z

Zunger A. 12, 17, 200, 204, 284

PERSPECTIVES IN CONDENSED MATTER PHYSICS

Executive Editor: L. Miglio

Dipartimento di Fisica dell'Università di Milano
Via Celoria, 16 I-20133 MILANO
Fax + 39/2/2366583; **Telex** 334687 INFNMI; **Tel.** + 39/2/2392.408

Published Volumes

G. Margaritondo, *Electronic Structure of Semiconductor Heterojunctions*

Forthcoming Volumes

G. Jacucci, *Simulation Approach to Solids*
G.J. Iafrate, *Quantum Transport and New Semiconductor Devices*
G.M. Kalvius and W. Zinn, *Magnetic Materials*

finito di stampare nel mese
di ottobre 1988
dalla Nuova Timec s.r.l.
Albairate (MI)

Editoriale Jaca Book spa
Via Aurelio Saffi 19, 20123 Milano

spedizione in abbonamento
postale TR editoriale
aut. D/162247/PI/3
direzione PT Milano